Lecture Notes in Computer Science 13898

The series Lecture Notes in Computer Science (LNCS), including its subseries Lecture Notes in Artificial Intelligence (LNAI) and Lecture Notes in Bioinformatics (LNBI), has established itself as a medium for the publication of new developments in computer science and information technology research, teaching, and education.

LNCS enjoys close cooperation with the computer science R & D community, the series counts many renowned academics among its volume editors and paper authors, and collaborates with prestigious societies. Its mission is to serve this international community by providing an invaluable service, mainly focused on the publication of conference and workshop proceedings and postproceedings. LNCS commenced publication in 1973.

Marios Mavronicolas
Editor

Algorithms and Complexity

13th International Conference, CIAC 2023
Larnaca, Cyprus, June 13–16, 2023
Proceedings

Editor
Marios Mavronicolas
University of Cyprus
Nicosia, Cyprus

ISSN 0302-9743 ISSN 1611-3349 (electronic)
Lecture Notes in Computer Science
ISBN 978-3-031-30447-7 ISBN 978-3-031-30448-4 (eBook)
https://doi.org/10.1007/978-3-031-30448-4

This Springer imprint is published by the registered company Springer Nature Switzerland AG
The registered company address is: Gewerbestrasse 11, 6330 Cham, Switzerland

Preface

This volume contains the papers presented at the 13th International Symposium on Algorithms and Complexity (CIAC 2023), which was held on June 13–16, 2023, in Larnaca, Cyprus. This event follows the tradition of the biannual CIAC symposium.

The purpose of CIAC is to bring together researchers from computer science and mathematics to present and discuss original research on algorithms and complexity. It is intended to cover all important areas of the field, including but not limited to algorithm design and analysis, sequential, parallel and distributed algorithms, data structures, computational and structural complexity, lower bounds and limitations of algorithms, randomized and approximation algorithms, parameterized algorithms and parameterized complexity classes, smoothed analysis of algorithms, alternatives to the worst-case analysis of algorithms (e.g., algorithms with predictions), on-line computation and competitive analysis, streaming algorithms, quantum algorithms and complexity, algorithms in algebra, geometry, number theory and combinatorics, computational geometry, algorithmic game theory and mechanism design, algorithmic economics (including auctions and contests), computational learning theory, computational biology and bioinformatics, algorithmic issues in communication networks, algorithms for discrete optimization (including convex optimization) and algorithm engineering.

49 submissions to CIAC 2023 were received. Each submission was reviewed by at least three Program Committee members. The Program Committee decided to accept 25 papers. The program of CIAC 2023 featured three invited talks by three outstanding researchers in the general field of algorithms and complexity: Friedhelm Meyer auf der Heide, Pino Persiano and Paul Spirakis. We are very grateful to Friedhelm, Pino and Paul for joining us in Larnaca and for their invited talks.

The Program Committee of CIAC 2023 awarded a Best Paper Award jointly and equally to the excellent papers "On One-Sided Testing Affine Subspaces," by Nader Bshouty, and "Approximating Power Node-Deletion Problems," by Toshihiro Fujito, Kento Mukae and Junya Tsuzuki. These papers received highest scores in the reviewing process. We would like to warmly congratulate their authors for receiving the award. We are extremely thankful to Springer for its generous financial support that made this award possible.

Our sincere thanks go to all authors who submitted their research work to CIAC 2023. We would like to thank all Program Committee members, and also the external reviewers who assisted them, for their timely commitment and their thorough reviewing job. We are indebted to Easy Conferences Ltd., and particularly to Petros Stratis and Christos Therapontos, for their meticulous work in organizing CIAC 2023. The designers and the administrators of the Easy Chair conference system, which assisted tremendously the Program Committee in its work, deserve special thanks. We are honored that CIAC 2023 is embraced under the auspices of the European Association for Theoretical Computer Science (EATCS).

June 2023

Marios Mavronicolas

Organization

Program Committee

Kristoffer Arnsfelt Hansen	Aarhus University, Denmark
Hagit Attiya	Technion, Israel
Vittorio Bilo	University of Salento, Italy
Costas Busch	Augusta University, USA
Amin Coja-Oghlan	TU Dortmund, Germany
Gianlorenzo D'Angelo	Gran Sasso Science Institute, Italy
Ioannis Emiris	University of Athens & ATHENA Research Center, Greece
Antonio Fernandez Anta	IMDEA Networks, Spain
Nicola Galesi	Sapienza University of Rome, Italy
Alexandros Hollender	EPFL, Switzerland
Christos Kaklamanis	University of Patras & CTI, Greece
Naoyuki Kamiyama	Kyushu University, Japan
Marios Mavronicolas	University of Cyprus, Cyprus
Hendrik Molter	Ben-Gurion University of the Negev, Israel
Tobias Momke	University of Augsburg, Germany
Leonidas Palios	University of Ioannina, Greece
Vicky Papadopoulou	European University Cyprus, Cyprus
Giulio Ermanno Pibiri	Ca' Foscari, University of Venice, Italy
Thomas Sauerwald	University of Cambridge, UK
Maria Serna	Polytechnic University of Catalonia, Spain
Blerina Sinaimeri	LUISS Guido Carli University, Italy
Shay Solomon	Tel Aviv University, Israel
Paul Spirakis	University of Liverpool & CTI, UK
Daniel Stefankovic	University of Rochester, USA
Adrian Vladu	IRIF, CNRS & University of Paris, France
Peilin Zhong	Google Research, USA

Additional Reviewers

Balzotti, Lorenzo
Becker, Ruben
Bonacina, Ilario
Calzavara, Stefano
Carnevale, Daniele
Checa Nualart, Carles
Delfaraz, Esmaeil
Deligkas, Argyrios

Disser, Yann
Eiben, Eduard
Gianinazzi, Lukas
Goldreich, Oded
Gribling, Sander
Hahn-Klimroth, Maximilian
Huynh, Tony
Ioannakis, George
Karanikolas, Nikos
Karczmarz, Adam
Khadiev, Kamil
Klobas, Nina
Konstantopoulos, Charalampos
Korpas, Georgios
Krieg, Lena
Lauria, Massimo
Levin, Asaf
Li, Huan
Los, Dimitrios
Majumder, Atrayee
Mantzaflaris, Angelos
Marino, Andrea
Maystre, Gilbert
Melissourgos, Themistoklis
Mkrtchyan, Vahan
Muñoz, Martín
Otachi, Yota
Out, Charlotte
Psarros, Ioannis
Robinson, Peter
Rolvien, Maurice
Roshany, Aida
Rossi, Mirko
Ruderer, Michael
Scheftelowitsch, Olga
Silveira, Rodrigo
Skretas, George
Speckmann, Bettina
Talebanfard, Navid
Tong, Qianqian
Trehan, Amitabh
Wang, Chen
Wang, Qisheng
Zhang, Tianyi

Contents

Selected Combinatorial Problems Through the Prism of Random Intersection Graphs Models

Paul G. Spirakis[1,2](✉), Sotiris Nikoletseas[2,3], and Christoforos Raptopoulos[3]

[1] Department of Computer Science, University of Liverpool, Liverpool, UK
p.spirakis@liverpool.ac.uk
[2] Computer Technology Institute, Patras, Greece
nikole@ceid.upatras.gr
[3] Computer Engineering and Informatics Department, University of Patras, Patras, Greece
raptopox@ceid.upatras.gr

1 Introduction and Motivation

We discuss a simple, yet general family of models, namely *Random Intersection Graphs (RIGs)*, initially introduced by Karoński et al. [4] and Singer-Cohen [10]. In such models there is a universe $\mathcal{M}$ of *labels* and each one of n vertices selects a random subset of $\mathcal{M}$. Two vertices are connected if and only if their corresponding subsets of labels intersect. A formal definition is given below:

Definition 1 (Random Intersection Graph - $\mathcal{G}_{n,m,p}$ [4,10]**).** *Consider a universe $\mathcal{M} = \{1, 2, \ldots, m\}$ of labels and a set of n vertices V. Assign independently to each vertex $v \in V$ a subset S_v of $\mathcal{M}$, choosing each element $i \in \mathcal{M}$ independently with probability p and draw an edge between two vertices $v \neq u$ if and only if $S_v \cap S_u \neq \emptyset$. The resulting graph is an instance $G_{n,m,p}$ of the random intersection graphs model.*

In this model we also denote by L_i the set of vertices that have chosen label $i \in M$. Given a random instance $G_{n,m,p}$ of the random intersection graphs model, we will refer to $\{L_i, i \in \mathcal{M}\}$ as its *label representation*, and the corresponding matrix $\mathbf{R}$ with columns the incidence vectors of label sets assigned to vertices is called the *representation matrix*. Furthermore, we refer to the bipartite graph with vertex set $V \cup \mathcal{M}$ and edge set $\{(v, i) : i \in S_v\} = \{(v, i) : v \in L_i\}$ as the *bipartite random graph $B_{n,m,p}$ associated to $G_{n,m,p}$*. Notice that the associated bipartite graph is uniquely defined by the label representation.

It follows from the definition of the model that the (unconditioned) probability that a specific edge exists is $1 - (1 - p^2)^m$. Therefore, if mp^2 goes to infinity with n, then this probability goes to 1. We can thus restrict the range of the parameters to the "interesting" range of values $mp^2 = O(1)$ (i.e. the range of values for which the unconditioned probability that an edge exists does not go

M. Mavronicolas (Ed.): CIAC 2023, LNCS 13898, pp. 1–4, 2023.
https://doi.org/10.1007/978-3-031-30448-4_28

to 1). Furthermore, as is usual in the literature, we will assume that the number of labels is some power of the number of vertices, i.e. $m = n^\alpha$, for some $\alpha > 0$.

It is worth mentioning that the edges in $G_{n,m,p}$ are not independent. For example, there is a strictly positive dependence between the existence of two edges that share an endpoint (i.e. $\Pr(\exists\{u,v\}|\exists\{u,w\}) > Pr(\exists\{u,v\}))$. This dependence is stronger the smaller the number of labels $\mathcal{M}$ includes, while it seems to fade away as the number of labels increases. In fact, by using a coupling technique, the authors in [3] proved the equivalence (measured in terms of total variation distance) of uniform random intersection graphs and Erdős-Rényi random graphs, when $m = n^\alpha, \alpha > 6$. This bound on the number of labels was improved in [5], where it was proved that the total variation distance between the two models tends to 0 when $m = n^\alpha, \alpha > 4$. Furthermore, [9] proved the equivalence of sharp threshold functions among the two models for $\alpha \geq 3$. Similarity of the two models has been proved even for smaller values of α (e.g. for any $\alpha > 1$) in the form of various translation results (see e.g. Theorem 1 in [8]), suggesting that some algorithmic ideas developed for Erdős-Rényi random graphs also work for random intersection graphs (and also weighted random intersection graphs). These results suggest that random intersection graphs are quite general and that known techniques for random graphs can be used in the analysis of random intersection graphs with a large number of labels.

Motivation. Random intersection graphs may model several real-life applications quite accurately. In fact, there are practical situations where each communication agent (e.g. a wireless node) gets access only to some ports (statistically) out of a possible set of communication ports. When another agent also selects a communication port, then a communication link is implicitly established and this gives rise to communication graphs that look like random intersection graphs. RIG modeling is useful in the efficient blind selection of few encryption keys for secure communications over radio channels ([1]), as well as in k-Secret sharing between swarm mobile devices (see [2]). Furthermore, random intersection graphs are relevant to and capture quite nicely social networking. Indeed, a social network is a structure made of nodes tied by one or more specific types of interdependency, such as values, visions, financial exchange, friends, conflicts, web links etc. Other applications may include oblivious resource sharing in a distributed setting, interactions of mobile agents traversing the web, social networking etc. Even epidemiological phenomena (like spread of disease between individuals with common characteristics in a population) tend to be more accurately captured by this "proximity-sensitive" family of random graphs.

From an average case analysis algorithmic perspective, the number of labels m may be viewed as a parameter controlling the clique cover size of input graphs. It is worth noting that some combinatorial problems that are considered to be hard when the input is drawn from the Erdős-Rényi random graphs model are easily solved when the input is drawn from the random intersection graphs model and the representation matrix $\mathbf{R}$ is explicitly provided as part of the input (rather than just giving the constructed graph instance as input). One such example is the problem of finding a maximum clique in $G_{n,m,p}$, in the dense case $m = n^\alpha, \alpha < 1$. Furthermore, there are combinatorial optimization problems that can be naturally described as graph theoretical problems in generalizations of the aforementioned model of random intersection graphs. In this talk, we discuss some structural and algorithmic results regarding random intersection graphs and we present an interesting connection between the problem of discrepancy in random set systems and the problem of MAX CUT in weighted random intersection graphs.

2 Maximum Cliques in Random Intersection Graphs

We discuss the Single Label Clique Theorem (SLCT) from [6], which states that when the number of labels is less than the number of vertices, any large enough clique in a random instance of $\mathcal{G}_{n,m,p}$ is formed by a single label. This statement may seem obvious when p is small, but it is hard to imagine that it still holds for *all* "interesting" values for p. Indeed, when $p = o\left(\sqrt{\frac{1}{nm}}\right)$, it can be proved that $G_{n,m,p}$ almost surely has no cycle of size $k \geq 3$ whose edges are formed by k distinct labels (alternatively, the intersection graph produced by reversing the roles of labels and vertices is a tree). On the other hand, for larger p a random instance of $\mathcal{G}_{n,m,p}$ is far from perfect[1] and thus the proof of the SLCT is based on a careful contradiction argument regarding the non-existence of large multi-label cliques.

3 Maximum Cut and Discrepancy in Random Set Systems

A natural weighted version of the random intersection graphs model was introduced in [7], where to each edge $\{u, v\}$ we assign weight equal to the number of common labels chosen by u and v, namely $|S_u \cap S_v|$. In particular, the weight matrix of a random instance of the weighted random intersection graphs model $\overline{\mathcal{G}}_{n,m,p}$ is equal to $\mathbf{R}^T\mathbf{R}$, where the columns of $\mathbf{R}$ are the incidence vectors of label sets assigned to vertices; we denote the corresponding random instance by $G(V, E, \mathbf{R}^T\mathbf{R})$.

We initially present some results from [7] regarding the concentration of the weight of a maximum cut of $G(V, E, \mathbf{R}^T\mathbf{R})$ around its expected value, and then show that, when the number of labels is much smaller than the number of vertices (in particular, $m = n^\alpha, \alpha < 1$), a random partition of the vertices achieves asymptotically optimal cut weight with high probability. Furthermore, in the case $n = m$ and constant average degree (i.e. $p = \frac{\Theta(1)}{n}$), we show that with high probability, a majority type randomized algorithm outputs a cut with weight that is larger than the weight of a random cut by a multiplicative constant strictly larger than 1.

[1] A *perfect graph* is a graph in which the chromatic number of every induced subgraph equals the size of the largest clique of that subgraph. Consequently, the clique number of a perfect graph is equal to its chromatic number.

Finally, we present a connection between the computational problem of finding a (weighted) maximum cut in $G(V, E, \mathbf{R}^T\mathbf{R})$ and the problem of finding a 2-coloring that achieves minimum discrepancy for a set system Σ with incidence matrix $\mathbf{R}$ (i.e. minimum imbalance over all sets in Σ). This connection was exploited in [7] by proposing a (weak) bipartization algorithm for the case $m = n, p = \frac{\Theta(1)}{n}$ that, when it terminates, its output can be used to find a 2-coloring with minimum discrepancy in a set system with incidence matrix $\mathbf{R}$. In fact, with high probability, the latter 2-coloring corresponds to a bipartition with maximum cut-weight in $G(V, E, \mathbf{R}^T\mathbf{R})$.

References

1. Dolev, S., Gilbert, S., Guerraoui, R., Newport, C.C.: Secure communication over radio channels. In: Proceedings of the ACM Symposium on Principles of Distributed Computing (PODC), pp. 105–114 (2008)
2. Dolev, S., Lahiani, L., Yung, M.: Secret swarm unitreactive k-secret sharing. In: Proceedings of the 8th International Conference on Cryptology in India (INDOCRYPT), pp. 123–137 (2007)
3. Fill, J., Sheinerman, E., Singer-Cohen, K.: Random intersection graphs when $m = \omega(n)$: an equivalence theorem relating the evolution of the $G(n, m, p)$ and $G(n, p)$ models. Random Struct. Algorithms **16**(2), 156–176 (2000)
4. Karoński, M., Scheinerman, E.R., Singer-Cohen, K.B.: On random intersection graphs: the subgraph problem. Comb. Probab. Comput. **8**, 131–159 (1999)
5. Kim, J.H., Lee, S.J., Na, J.: On the total variation distance between the binomial random graph and the random intersection graph. Random Struct. Algorithms **52**(4), 662–679 (2018)
6. Nikoletseas, S.E., Raptopoulos, C.L., Spirakis, P.G.: Maximum cliques in graphs with small intersection number and random intersection graphs. Comput. Sci. Rev. **39**, 100353 (2021)
7. Nikoletseas, S., Raptopoulos, C., Spirakis, P.: Max cut in weighted random intersection graphs and discrepancy of sparse random set systems. In: Proceedings of the 32nd International Symposium on Algorithms and Computation (ISAAC), pp. 1–16 (2021)
8. Raptopoulos, C., Spirakis, P.: Simple and efficient greedy algorithms for bamilton cycles in random intersection graphs. In: Proceedings of the 16th International Symposium on Algorithms and Computation (ISAAC), pp. 493–504 (2005)
9. Rybarczyk, K.: Equivalence of a random intersection graph and $G(n, p)$. Random Struct. Algorithms **38**(1–2), 205–234 (2011)
10. Singer-Cohen, K.B.: Random intersection graphs, Ph.D. thesis, John Hopkins University (1995)

Unifying Gathering Protocols for Swarms of Mobile Robots

Jannik Castenow, Jonas Harbig, and Friedhelm Meyer auf der Heide(✉)

Heinz Nixdorf Institute and Computer Science Department, Paderborn University, Paderborn, Germany
{jannik.castenow,jonas.harbig,fmadh}@upb.de
https://www.hni.uni-paderborn.de/en/alg/

Abstract. In this paper, we survey recent results regarding the Gathering problem in the research area of distributed computing by mobile robots. We assume a simple, standard model of robots: they are point-shaped, live in a d- dimensional Euclidean space, are disoriented (do not agree on common directions or orientations), and have limited visibility (can only observe other robots up to a constant distance). The goal of Gathering is to gather all robots at a single, not predefined point. Our focus lies on unifying and extending existing work on gathering in the above model. For this, we derive core properties of protocols that guarantee Gathering and prove runtime bounds that improve upon previous work. This paper surveys results presented in [2,3,11] in which such core properties are derived in two different time models: a discrete, round-based time model and a continuous time model.

Keywords: mobile robots · gathering · continuous · discrete · runtime

1 Introduction

Model and Scenario. We consider a scenario where a swarm of n robots is distributed in a Euclidean space $\mathbb{R}^d$ of dimension $d \geq 1$. Each robot on its own has only very limited capabilities. Most crucially, the robots have a *limited visibility*, i.e., they can observe other robots only up to a constant distance. Throughout this work, we normalize this distance to 1. Moreover, each robot has its own local coordinate system, and the robots are *disoriented*, i.e., they do not agree on any direction and orientation of their local coordinate systems. The disorientation is even *variable*, meaning that the local orientation of the coordinate system of a fixed robot might change from time to time. Furthermore, the robots are *oblivious* (no persistent memory) and *silent* (no communication capability). We consider two models for *time*: time can either be discrete (the robots operate in rounds) or continuous (the movement of each robot is defined for each real point in time by a velocity vector). When time is discrete, we consider the robots to be

This work was partially supported by the German Research Foundation (DFG) under the project number ME 872/14-1.

M. Mavronicolas (Ed.): CIAC 2023, LNCS 13898, pp. 5–16, 2023.
https://doi.org/10.1007/978-3-031-30448-4_1

fully synchronized, i.e., all robots simultaneously start a new round (observe the environment, compute a target point, and move there). More details about the time models will be given in Sect. 2. All these robot features are in the literature denoted as the $\mathcal{OBLOT}$ model [8].

Typical tasks for such robot swarms are *formation problems*, where the robots have to move such that their positions form a predefined pattern. The patterns POINT and UNIFORM CIRCLE have evolved into the most important benchmark patterns in the literature. This stems from their high symmetry: it is well known that the patterns POINT and UNIFORM circle are the only patterns that *might* be formed from any initial configuration of the robots as it is only possible to form patterns that have the same or a higher symmetry as the initial configuration (see [13] for a precise definition of symmetries in this context).

The GATHERING Problem. In this work, we focus on the pattern POINT, which is in the literature commonly denoted as the GATHERING problem. The GATHERING problem is trivial if the robots have unlimited visibility (robots can observe *all* other robots) and operate fully synchronized. In case the robots operate in rounds, the robots can, for example, all move to the center of the smallest enclosing circle of all robots' positions and reach that position within one round [4]. If the robots are less synchronized and disoriented, the GATHERING problem is generally impossible to solve, even if the robots have unlimited visibility [12].

Considering robots with limited visibility living in the Euclidean plane that operate in fully synchronized rounds, the GO-TO-THE-CENTER protocol (GTC protocol), in which robots move toward the center of the smallest enclosing circle of all visible robots solves the GATHERING problem [1]. Later on, a time bound of $\mathcal{O}\left(n + \Delta^2\right)$ rounds has been proven, where Δ denotes the Euclidean diameter of the initial swarm [6]. The protocol has been generalized to the three-dimensional case with the same time bound in [2]. In this work, we focus on the class of λ-contracting protocols, which allows solving GATHERING within $\Theta(\Delta^2)$ rounds [3]. Notably, also the GTC protocol is λ-contracting.

Also, in the continuous time model, GATHERING can be solved by disoriented robots with limited visibility. In [9], e.g., the MOVE-ON-BISECTOR (MOB) protocol has been introduced in which robots that are located on the convex hull of all visible neighbors move with a speed of 1 along the angle bisector of vectors pointing to their neighbors of the convex hull. The MOB protocol gathers in time $\mathcal{O}(n)$ [5]. Moreover, the protocol gathers in time $\mathcal{O}\left(OPT \cdot \log n\right)$, where OPT denotes the optimal time a protocol with unlimited visibility needs. Hence, the price of locality, i.e., the additional time caused by the limited visibility, is at most a factor of $\log n$ in the continuous time model. In Sect. 3, we describe a class of continuous protocols, the *contracting protocols*. They perform gathering in time $\mathcal{O}\left(n \cdot Delta\right)$. Among others, MOB and a continuous version of GO-TO-THE-CENTER are contracting.

Scope and Outline. We survey recent general results about the GATHERING problem of disoriented robots with limited visibility in the $\mathcal{OBLOT}$ model, both for the round based and the continuous time model. For both time models, we introduce classes of protocols that solve the GATHERING problem efficiently. The classes encapsulate the core properties of efficient GATHERING protocols.

Most importantly, the classes of protocols can be applied to robots that live in a Euclidean space of arbitrary dimension d. We start in Sect. 2 with a general notion of robot formation protocols for the discrete and continuous time model. The notion serves as a basis to define the concrete classes of protocols afterward. Section 3 introduces continuous contracting protocols. For $d = 2$ and $d = 3$, the protocols have been introduced an analyzed in [2,11]. The case $d > 3$ has not been published yet but can be obtained by generalizing the proof technique derived in [2]. The discrete class of protocols denoted as *λ-contracting protocols* is the topic of Sect. 4. The class of protocols and their analysis have been recently published in [3]. We conclude the paper with open problems in Sect. 5.

2 Robot Formation Protocols

We consider a swarm of n robots $r_0, \ldots, r_{n-1}$. The robots start their protocol at time $t = 0$. When time is discrete, we consider time steps $t \in \mathbb{N}_0$. Otherwise, time is continuous, and we consider all points in time $t \in \mathbb{R}_{\geq 0}$. The position of a robot r_i at time t is denoted by $p_i(t)$ in a global coordinate system (not known to the robots). We collectively call the set of all robots' positions at time t the *configuration.* The *initial configuration*, i.e., the configuration at time $t = 0$, is assumed to be connected. More formally, at time $t = 0$, the initial Unit Ball Graph of the robots' positions is connected. In the Unit Ball Graph at time t, the vertices represent the robots, and two robots are connected via an edge if their distance is at most 1. Next, we describe continuous and discrete robot formation protocols, i.e., a formal way to describe the robots' behavior in two different time models: the continuous time model and the discrete, round-based, LCM model.

Continuous Protocols. At every point in time, each robot computes a target point and a speed between 0 and 1 and moves with the given speed towards this point. More formally, we obtain the following notion of continuous protocols.

Definition 1. *A* continuous robot formation protocol $\mathcal{P}$ *defines for each robot* r_i *and each* $t \in \mathbb{R}_{\geq 0}$ *a* target point $\text{target}_i^{\mathcal{P}}(t)$ *and a* speed $s_i^{\mathcal{P}}(t) \in [0, 1]$ *describing the robots current movement direction and speed. Together, both define the* velocity vector $v_i^{\mathcal{P}}(t) = \frac{s_i^{\mathcal{P}}(t)}{\|\text{target}_i^{\mathcal{P}}(t) - p_i(t)\|_2} \left(\text{target}_i^{\mathcal{P}}(t) - p_i(t)\right)$.

Often, we describe only the velocity vectors, however, typical protocols are designed such that each robot computes a target point and moves with a certain speed toward it. The function $p_i : \mathbb{R}_{\geq 0} \rightarrow \mathbb{R}^d$ is the *trajectory* of r_i. The trajectories are continuous but not necessarily differentiable. Occasionally, robots can change their direction and speed non-continuously (e.g., when a new robot becomes visible), leading to a non-differentiable trajectory. Typical protocols have right-differentiable trajectories. This way, we can interpret the velocity vectors as the (right) derivative of the trajectories, i.e., $v_i^{\mathcal{P}} : \mathbb{R}_{\geq 0} \rightarrow \mathbb{R}^d = \dot{p}_i$.

Discrete Protocols. When time is discrete, robots operate in LCM cycles consisting of the operations Look, Compute, and Move. During Look, each robot takes a snapshot of all visible robots in its local coordinate system; it computes a target point based on the snapshot during COMPUTE and finally moves there within the MOVE operation. We assume a *rigid* movement, i.e., the robots always reach their target points. The LCM cycles are also denoted as *rounds*. The timing of the cycles is fully synchronous ($\mathcal{F}$SYNC), i.e., each robot is active in every round, and all robots observe their environment and move simultaneously. Analogously to continuous robot formation protocols, we define *discrete* robot formation protocols, a convenient way to describe the properties of protocols.

Definition 2. *A* discrete robot formation protocol $\mathcal{P}$ *defines for each robot* r_i *and each* $t \in \mathbb{N}$ *a* target point $\text{target}_i^{\mathcal{P}}(t)$ *such that* $p_i(t+1) = \text{target}_i^{\mathcal{P}}(t)$.

3 Continuous Time Gathering

This section presents the class of *contracting protocols* for the continuous time model. The class itself is introduced in Sect. 3.1. Afterward, we present time bounds for contracting protocols in Sect. 3.2. We conclude the section with an exemplary contracting protocol in Sect. 3.3.

3.1 Contracting Protocols

The main idea of GATHERING protocols that consider robots with limited visibility is always to let robots move toward the inside of the swarm. Thereby, the global convex hull of all robots' positions gets smaller and smaller until the robots finally gather at a single point. In the continuous time model, we observe that such a simple and elegant criterion is sufficient to define efficient protocols to solve GATHERING. More precisely, we define the class of *contracting protocols* based on [10,11]. Intuitively, a robot formation protocol is contracting if robots located on the vertices of the global convex hull enclosing all robots' positions move with a speed of 1 inside the closed global convex hull. See Fig. 1 for a visualization.

We denote the global convex hull of all robots' positions at time t by $CH(t)$. Then, the following Definition 3 characterizes contracting protocols precisely.

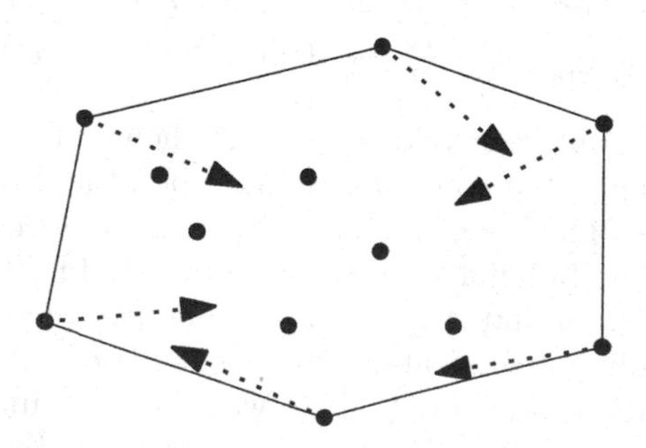

Fig. 1. A visualization of a (globally) contracting protocol. Robots that are located on the vertices of the global convex hull (marked by solid lines) move with a speed of 1 inside the closed convex hull (the velocity vectors are marked by arrows).

Definition 3. *A continuous robot formation protocol is called (globally)* contracting *if for every point in time t and every robot r_i whose position is a vertex of $CH(t)$, it holds that* $\text{target}_i^{\mathcal{P}}(t) \in CH(t)$ *and* $s_i^{\mathcal{P}}(t) = 1$.

Observe that Definition 3 refers to the global convex hull of all robots' positions. However, due to the limited visibility, robots compute their target points based on their (local) neighborhood. Nevertheless, typical target points such as the center of the smallest enclosing circle of a robot's neighborhood [1] lie inside the global convex hull and thus fulfill – at first sight – the criterion of being contracting. Still, the movements of the robots might disconnect the swarm. Consider, for instance, the case that a protocol divides the configuration into two connected components. The robots of the connected components might gather at two different points. Afterward, the robots are not aware that robots occupy a second position and remain in their current location. As a result, the protocol is not contracting since the robots do not move anymore, although they have not yet gathered at a single point. With this observation, we see that maintaining connectivity is an essential property of local protocols and define the following class of *locally contracting* protocols. The definition uses $CH_i(t)$ to describe the convex hull of all robots' positions visible by r_i at time t.

Definition 4. *A continuous robot formation protocol is called* locally contracting *if, for every point in time t,*

1. *the configuration is connected,*
2. *for every robot r_i whose position is a vertex of $CH_i(t)$,* $\text{target}_i^{\mathcal{P}}(t) \in CH_i(t)$ *and* $s_i^{\mathcal{P}}(t) = 1$.

Informally speaking, a locally contracting protocol must maintain the swarm's connectivity, and robots located on vertices of their *local* closed convex hull move with a speed of 1 inside. Since local convex hulls are completely contained in the global convex hull, every locally contracting protocol is also (globally) contracting.

Lemma 1. *Every locally contracting protocol is globally contracting.*

3.2 Results

For $d = 2$, contracting protocols have been analyzed in [11]. The authors prove that every contracting protocol gathers the robots in time $\mathcal{O}(n \cdot \Delta)$, where Δ represents the Euclidean diameter, i.e., the maximum distance of two robots, of the initial configuration. The analysis observes that the length of the boundary of the convex hull of all robots' positions is monotonically decreasing at a rate of $\Omega(1/n)$. Moreover, there is the GO-TO-THE-LEFT protocol that needs a time of $\Omega(n \cdot \Delta)$ to gather the robots when the robots are initially located on the vertices of a regular polygon with side length 1 (Theorem 1). The protocol assumes that the robots have a common understanding of left and right, and every robot moves with a speed of 1 towards its left neighbor. Thereby, every robot traces out a logarithmic spiral, and the robots finally gather at the center of the regular polygon. For an illustration, see Fig. 2.

Theorem 1. *There is a contracting protocol that gathers a connected swarm of n disoriented robots located in $\mathbb{R}^2$ with a viewing range of 1 in time $\Omega(n \cdot \Delta)$, where Δ denotes the Euclidean diameter of the initial configuration.*

Contracting protocols have been analyzed for $d = 3$ in [2]. The analysis idea is to project the original three-dimensional configuration onto a two-dimensional projection plane and to analyze the behavior of the robots in the projection. Note that the plane has to be chosen carefully; otherwise, some robots might not move in the projection (in case their velocity vectors are orthogonal to the plane). Hence, the analysis aims to find a plane where all robots move sufficiently fast. The authors of [2] prove an upper time bound of $\mathcal{O}\left(n^{3/2} \cdot \Delta\right)$. The lower bound of $\Omega\left(n \cdot \Delta\right)$ from the two-dimensional case still holds.

A rigorous generalization of the three-dimensional analysis, i.e., not only projecting once but $d - 1$ times, allows us to derive a time bound for arbitrary d. Every projection has to be chosen carefully and causes the velocity vectors of most robots to decrease. Hence, the time-bound (obtained by this analysis technique) depends on the dimension. Precisely, the time bound is stated as in the following theorem.

Theorem 2. *Every contracting protocol gathers a connected swarm of n disoriented robots located in $\mathbb{R}^d$ with a viewing range of 1 in time $\mathcal{O}(n^{\log(d)} \cdot \Delta)$, where Δ denotes the Euclidean diameter of the initial configuration.*

Notably, for $d = 2$, there is a contracting protocol known that has a significantly faster runtime. The MOVE-ON-BISECTOR (MOB) protocol gathers the robots in time $\mathcal{O}(n)$. In the following, we introduce a different contracting protocol.

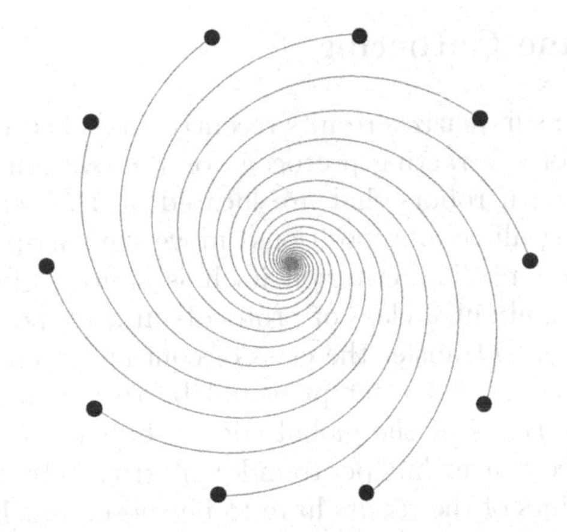

Fig. 2. A visualization of the GO-TO-THE-LEFT protocol. Initially, the robots are located on a regular 10-gon (the black dots). Then, all robots move with a speed of 1 towards their left neighbor. As a result, the robots move along logarithmic spirals toward the center of the polygon.

3.3 An Exemplary Contracting Protocol

Next, a continuous version of the GO-TO-THE-CENTER (GTC) protocol [1] will be considered a concrete example of a contracting protocol. The two-dimensional version of this protocol was adapted for the continuous-time model by [11]. In the discrete-time version of the GTC protocol, connectivity is maintained with the help of *limit circles*, i.e., robots sometimes do not move towards the center of the smallest enclosing circle of their neighborhood but stop earlier. Compared to the discrete-time version, no additional measures have to be taken to preserve connectivity in the continuous-time model, as it can be shown that this happens naturally. The protocol is summarized in Algorithm 1.

Algorithm 1. Continuous d-GTC

1: $\mathcal{R}_i(t) :=$ {positions of robots visible from r_i, including r_i at time t}
2: $\mathcal{S}_i(t) :=$ smallest enclosing hypersphere of $\mathcal{R}_i(t)$
3: $c_i(t) :=$ center of $\mathcal{S}_i(t)$
4: Move towards $c_i(t)$ with speed 1, or stay on $c_i(t)$ if r_i is already positioned on it.

With the same arguments as in [11], it can be seen that d-GTC maintains the connectivity of the swarm. As the center of the smallest enclosing hypersphere is a convex combination of the points it encloses [7], it follows that d-GTC is locally contracting.

Theorem 3. *The continuous d-GTC protocol is locally contracting and therefore gathers a connected swarm of n disoriented robots located in $\mathbb{R}^d$ with a viewing range of 1 in time $\mathcal{O}\left(n^{\log(d)} \cdot \Delta\right)$.*

4 Discrete Time Gathering

The following section summarizes results recently published in [3]. In Sect. 3, we have seen the class of contracting protocols for the continuous time model. A protocol is contracting if robots that are located on the vertices of the global convex hull (enclosing all robots' positions) move with a speed of 1 inside the global closed convex hull. The criterion itself is simple, elegant, and easy to verify. When thinking about a class of protocols in a round-based time model, the first idea would be to transfer the class of continuous contracting protocols to the discrete setting. A first attempt would be to demand that robots that are located on the vertices of the global convex hull move a certain distance inside the global closed convex hull per round. Unfortunately, in the discrete-time model, the target points of the robots have to be chosen much more carefully as otherwise the required time to solve GATHERING gets very high, or the robots do not gather at all. Consider for instance the GO-TO-THE-LEFT protocol (Sect. 3.2 and Fig. 2). The robots are located on the vertices of a regular polygon with a side length of 1. In the continuous variant, the robots move with a speed of 1 towards their left neighbor. Thereby, the robots move along logarithmic spirals towards the center of the regular polygon (Fig. 2). A discretization might be to demand that robots move to the positions of their left neighbor. This, however, does not work as the robots would move directly to the position of their left neighbor such that the resulting configuration is still the same regular polygon in the next round; the only difference is that each robot has moved one position to the left.

Moreover, the robots must also move a certain minimal distance to obtain reasonable time bounds. We consider again that the robots are located on the vertices of a regular polygon with side length 1. Given this configuration, e.g., the GTC protocol moves robots always to the midpoint between their two visible neighbors. Intuitively, also moving (by a constant factor) smaller distances lead to GATHERING in asymptotically the same time (cf. Fig. 3). However, the required time increases as soon as the distance is significantly smaller.

With these two observations, we have emphasized that the target points must be chosen carefully in the sense that the target points must lie *sufficiently far* inside the local convex hulls of the individual robots. In the following section, we formally define what sufficiently far means. More precisely, we define the class of λ-contracting protocols – a class of protocols that solves GATHERING efficiently.

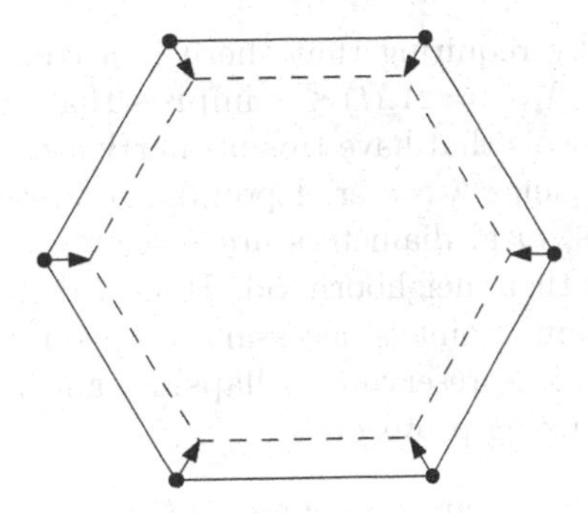

Fig. 3. Started on the vertices of a regular polygon, the robots have to move a certain minimal distance inside to guarantee a reasonable runtime.

4.1 λ-Contracting Protocols

Similar to locally contracting (continuous) protocols, we demand that any discrete protocol to solve GATHERING must be *connectivity preserving*, i.e., it always maintains the connectivity of the swarm. Next, we precisely define the meaning of *sufficiently far inside of local convex hulls*, leading to a characterization of valid target points.

Definition 5. *Let Q be a convex polytope with diameter diam and λ a constant with $\lambda \in (0, 1]$. A point $p \in Q$ is called to be* λ-centered *if it is the midpoint of a line segment completely contained in Q and has a length of $\lambda \cdot diam$.*

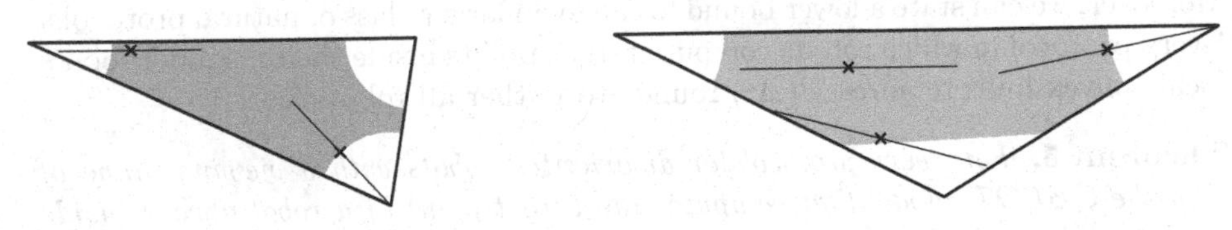

Fig. 4. A visualization of λ-centered points for the values of $\lambda = 4/7$ (left) and $\lambda = 4/11$ (right).

Two examples of λ-centered points are depicted in Fig. 4. Observe that moving to λ-centered points while maintaining connectivity does not necessarily enforce a final gathering of the protocols. Consider, for instance, two robots. A protocol that demands the two robots to move halfway towards the midpoint between themselves would compute 1/4-centered target points. However, the robots would only *converge* towards the same position. The robots must be guaranteed to eventually compute the same target point to obtain a final gathering. In the following, we denote by $N_i(t)$ the set of all visible robots (including itself) of r_i at time t. Moreover, $\Delta_i(t)$ is the Euclidean diameter of the robots in $N_i(t)$. We demand that two robots with the same view eventually compute the same target point by requiring that there is a constant $c < 1$, such that $N_i(t) = N_j(t)$ and that $\Delta_i(t) = \Delta_j(t) \leq c$ implies that the robots compute the same target point. Protocols that have this property are called *collapsing.* Being collapsing is reasonable since λ-centered points are always inside local convex hulls. Hence, the robots' local diameters are monotonically decreasing in case no further robot enters their neighborhood. Hence, demanding a threshold to enforce moving to the same point is necessary to ensure a final gathering. The combination of connectivity preserving, collapsing, and λ-centered points leads to the notion of λ-contracting protocols.

Definition 6. *A connectivity preserving and collapsing discrete robot formation protocol $\mathcal{P}$ is called* λ-contracting *if* $\text{target}_i^{\mathcal{P}}(t)$ *is a λ-centered point of $CH_i(t)$ for every robot r_i and every $t \in \mathbb{N}_0$.*

4.2 Results

The following theorem summarizes the main result about λ-contracting protocols. Notably, the obtained time bound is (in contrast to continuous contracting protocols) independent of the number of robots n and the dimension of the Euclidean space d.

Theorem 4. *Every λ-contracting protocol gathers a connected swarm of n disoriented robots located in $\mathbb{R}^d$ with a viewing range of 1 in the $\mathcal{OBLOT}$ model in $\mathcal{O}(\Delta^2)$ rounds.*

Theorem 4 raises the question of a lower bound for λ-contracting protocols: is the time bound of $\mathcal{O}(\Delta^2)$ rounds tight? The answer turns out to be positive. Moreover, we can state a lower bound for an even larger class of natural protocols. Every protocol in which robots compute target points inside their neighborhood's local convex hulls requires $\Omega(\Delta^2)$ rounds to gather all robots.

Theorem 5. *For every protocol for disoriented robots with a viewing range of 1 in the $\mathcal{OBLOT}$ model that computes the target point of a robot always inside of the local convex hull of all visible robots, there exists an initial configuration where the protocol requires $\Omega(\Delta^2)$ rounds to gather all robots.*

4.3 An Exemplary λ-Contracting Protocol

In Sect. 3.3, we have introduced the continuous variant of the GtC protocol as an example of continuous contracting protocols. Now, we introduce the discrete variant (for $d = 2$ originally introduced in [1] and for $d = 3$ in [2]). The core idea is the same as in the continuous variant: robots move towards the center of the smallest hypersphere enclosing their neighborhood. However, such a movement might cause a robot to get disconnected from the swarm. To overcome this, robots only move that far towards their target point such that the distance toward their neighbor always is at most one. The behavior is formally described in Sect. 4.3.

Algorithm 2. d-GTC (view of robot r_i)

1: $C_i(t) :=$ smallest enclosing hypersphere of $N_i(t)$
2: $c_i(t) :=$center of $C_i(t)$
3: $\forall r_j \in N_i(t) : m_j :=$ midpoint between r_i and r_j
4: D_j : hypersphere with radius $\frac{1}{2}$ centered at m_j
5: seg := line segment $\overline{p_i(t), c_i(t)}$
6: $A := \bigcap_{r_j \in N_i(t)} D_j \cap \text{seg}$
7: $x :=$ point in A closest to $c_i(t)$
8: $\text{target}_i^{\text{GTC}}(t) := x$

That the d-GTC protocol is connectivity preserving follows directly from its definition. With rigorous analysis, it can be proven that the resulting target point is $\sqrt{2}/16$-contracting, and we obtain the following theorem.

Theorem 6. *The d-GTC protocol is $\sqrt{2}/16$-contracting and therefore gathers a connected swarm of n disoriented robots located in $\mathbb{R}^d$ with a viewing range of 1 in $\Theta\left(\Delta^2\right)$ rounds.*

5 Outlook

The class of (continuous) contracting protocols gathers in time $\mathcal{O}\left(n^{\log(d)} \cdot \Delta\right)$, while a lower bound of $\Omega(n \cdot \Delta)$ is known. Further research could investigate the gap between the upper and the lower time-bound. We believe the real answer is a time bound of $\Theta\left(n \cdot \Delta\right)$ for every dimension d. Intuitively, the dependence on the dimension d is an artifact from the analysis ($d - 2$ projections), and the important parameters in the runtime are solely the number of robots and the diameter Δ. Furthermore, the two-dimensional contracting protocol MoB gathers in time $\mathcal{O}(n)$ which is in many cases severely faster than the bound of $\mathcal{O}\left(n \cdot \Delta\right)$ [5]. Consequently, a further research question is: what are the properties of continuous robot formation protocols that gather in time $\mathcal{O}(n)$? Is even a bound of $\mathcal{O}(\Delta)$ achievable? Our research about λ-contracting protocols hints that a dependence only on the diameter might be achievable since the class of λ-contracting protocols solves GATHERING in $\Theta\left(\Delta^2\right)$ rounds. Also, open questions remain regarding λ-contracting protocols and GATHERING of disoriented robots with limited visibility in general. The following major open question remains unanswered: is it possible to solve GATHERING of disoriented robots with limited visibility in the $\mathcal{OBLOT}$ model in $\mathcal{O}\left(\Delta\right)$ rounds? We could get closer to the answer: If there is such a protocol, it must often compute target points outside of the convex hulls of robots' neighborhoods.

References

1. Ando, H., Oasa, Y., Suzuki, I., Yamashita, M.: Distributed memoryless point convergence algorithm for mobile robots with limited visibility. IEEE Trans. Robot. Autom. **15**(5), 818–828 (1999)

2. Braun, M., Castenow, J., Meyer auf der Heide, F.: Local gathering of mobile robots in three dimensions. In: Richa, A.W., Scheideler, C. (eds.) SIROCCO 2020. LNCS, vol. 12156, pp. 63–79. Springer, Cham (2020). https://doi.org/10.1007/978-3-030-54921-3_4
3. Castenow, J., Harbig, J., Jung, D., Kling, P., Knollmann, T., Meyer auf der Heide, F.: A unifying approach to efficient (near)-gathering of disoriented robots with limited visibility. In: OPODIS. LIPIcs, vol. 253, pp. 15:1–15:25. Schloss Dagstuhl - Leibniz-Zentrum für Informatik (2022)
4. Cohen, R., Peleg, D.: Convergence properties of the gravitational algorithm in asynchronous robot systems. SIAM J. Comput. **34**(6), 1516–1528 (2005)
5. Degener, B., Kempkes, B., Kling, P., Meyer auf der Heide, F.: Linear and competitive strategies for continuous robot formation problems. ACM Trans. Parallel Comput. **2**(1), 2:1–2:18 (2015)
6. Degener, B., Kempkes, B., Langner, T., Meyer auf der Heide, F., Pietrzyk, P., Wattenhofer, R.: A tight runtime bound for synchronous gathering of autonomous robots with limited visibility. In: SPAA, pp. 139–148. ACM (2011)
7. Elzinga, D.J., Hearn, D.W.: The minimum covering sphere problem. Manage. Sci. **19**(1), 96–104 (1972)
8. Flocchini, P., Prencipe, G., Santoro, N.: Moving and computing models: Robots. In: Flocchini, P., Prencipe, G., Santoro, N. (eds.) Distributed Computing by Mobile Entities. Lecture Notes in Computer Science, vol. 11340, pp. 3–14. Springer, Cham (2019). https://doi.org/10.1007/978-3-030-11072-7_1
9. Gordon, N., Wagner, I.A., Bruckstein, A.M.: Gathering multiple robotic a(ge)nts with limited sensing capabilities. In: Dorigo, M., Birattari, M., Blum, C., Gambardella, L.M., Mondada, F., Stützle, T. (eds.) ANTS 2004. LNCS, vol. 3172, pp. 142–153. Springer, Heidelberg (2004). https://doi.org/10.1007/978-3-540-28646-2_13
10. Kling, P., Meyer auf der Heide, F.: Continuous protocols for swarm robotics. In: Flocchini, P., Prencipe, G., Santoro, N. (eds.) Distributed Computing by Mobile Entities. Lecture Notes in Computer Science, vol. 11340, pp. 317–334. Springer, Cham (2019). https://doi.org/10.1007/978-3-030-11072-7_13
11. Li, S., Markarian, C., Meyer auf der Heide, F., Podlipyan, P.: A continuous strategy for collisionless gathering. Theor. Comput. Sci. **852**, 41–60 (2021)
12. Prencipe, G.: Impossibility of gathering by a set of autonomous mobile robots. Theor. Comput. Sci. **384**(2–3), 222–231 (2007)
13. Suzuki, I., Yamashita, M.: Distributed anonymous mobile robots: formation of geometric patterns. SIAM J. Comput. **28**(4), 1347–1363 (1999)

The Complexity of Secure RAMs

Giuseppe Persiano[1,2](✉)

[1] Università di Salerno, Fisciano, Italy
[2] Google, Mountain View, US
giuper@gmail.com

Abstract. In this invited lecture, I survey the recent results on the complexity of Oblivious RAMs and of related cryptographic data structures and highlight the proof techniques employed.

In recent years, there has been significant work in studying data structures that provide privacy for the operations that are executed. These primitives aim to guarantee that observable access patterns to physical memory do not reveal substantial information about the queries and updates executed on the data structure.

The concept of an Oblivious RAMs (ORAMs) has been introduced by Goldreich and Ostrovsky [6]. An ORAM can be viewed an a secure implementation of the simplest data structure: an *array* (or a RAM) whose entries can be read and over-written. The typical setting is that of a client that has limited memory and outsources the storage of the array to a remote server and accesses the data stored in the array over a network. Clearly, to protect the confidentiality of the data, each entry can be encrypted before the upload and decrypted once it is downloaded. Still, the server sees the access pattern and from this deduce the type of algorithm that is being executed which in turn can reveal the interest of the client. An ORAM is a protocol between the client and server that hides the access pattern. The *obliviousness* guarantee of an Oblivious RAM requires that no adversary that picks two *challenge* sequences of operations of the same length and observes the access pattern incurred by the execution of one of the sequences still cannot determine which of the two sequences gave rise to the access pattern observed.

In recent years, ORAMs have been studied extensively to try and determine the optimal overhood (see [3,6,7,9,11] and references therein) that was reduced from $O(\log^3 n)$ to $O(\log n)$, for a RAM with n entries. Indeed, for b-bit entries on a server with memory cell (word) size of $\omega = \Theta(b)$ bits, the best known construction obtains logarithmic overhead $O(b/\omega \cdot \log n)$ [1] and requires only constant client memory.

Is this the best we can do?

The first logarithmic lower bounds were proven by Goldreich and Ostrovsky [6] of the form $\Omega((b/\omega) \cdot (\log n/\log c))$ where the client has storage of c bits. Boyle and Naor [2] pointed out that these lower bounds assumed the so-called *balls-and-bins* model with a non-encoding assumption on the underlying blocks.

M. Mavronicolas (Ed.): CIAC 2023, LNCS 13898, pp. 17–20, 2023.
https://doi.org/10.1007/978-3-031-30448-4_2

Larsen and Nielsen [10] were the first to prove lower bounds for the general case; i.e., without making any encoding assumption. They proved that a RAM of n entries each of b bits implemented by a server with a memory consisting of ω-bit words and a client with c bits of local memory requires $\Omega((b/\omega) \cdot \log(nb/c))$. This bound becomes increasingly weak as ω grows and Komargodski and Lin [8] proved a lower bound of $\Omega(\log(nb/c)/\log(\omega/b))$ for the case $\omega > b$.

In the hope of obtaining faster RAM that would still offer an adequate level of security, researchers have looked at weaker but still meaningful notions of security. In this talk we will overview three attempts and show that indeed any meaningful notion of security for RAMs seems to be as hard as Obliviousness.

Differentially Private RAMs. In various practical applications, including the field of privacy-preserving data analysis, the notion of *Differential Privacy* [5] is considered to offer an adequate level of protection. Differentially Private RAMs (DPRAMs) aim to provide privacy for individual operations, but may reveal information about a sequence consisting of many operations. In more detail, if an adversary receives two candidate equal-length operational sequences that differ in one operation and the access pattern incurred by the execution of one of the two sequences, the adversary should not be able to guess the identity of the executed sequence with too high probability. Unfortunately, DPRAMs incur in the same overhead as ORAM. Specifically, the $\Omega(b/\omega \cdot \log nb/c)$ lower bound for DPRAMs by Persiano and Yeo [15] showed that this is impossible when $b = \Omega(\omega)$ and, recently, this has been extended to $\Omega(\log(nb/c)/\log(\omega/b))$ which is significant for the case $\omega > b$ by [16].

Leaky RAMs. A second approach allows the RAM to leak some partial information about the sequence of operations. Currently, all known leaky RAMs with constant overhead reveal if two operations are performed on the same key or not. We denote this as *global key−equality pattern*. The result of [12] gives strong evidence that the leakage of the global key-equality pattern is inherent for any leaky RAM construction with $O(1)$ efficiency. In particular, they consider the slightly smaller leakage of *decoupled key-equality pattern* where leakage of key-equality between update and query operations is decoupled and the adversary only learns whether two operations of the *same type* are performed on the same key or not. They show that any leaky RAM with at most decoupled key-equality pattern leakage incurs $\Omega(b/w \cdot \log n)$ overhead.

Snapshot Adversaries. In some applications the server executing the access is not trusted but it could be temporarily compromised by an external adversary. Very recently, Du, Genkin and Grubbs [4] presented an ORAM construction with $O(\log \ell)$ overhead protecting against a *snapshot* adversary that observes the transcript of ℓ consecutive operations from a single breach. For small values of ℓ, this outperforms standard ORAMs. However, if one allows to have 3 breaches, it has been recently proved [14] that we go back to $\Omega(b/w \cdot \log(nb/c))$ overhead.

Open Problem. The following question is thus still open. Is there a *meaningful* notion of security for which RAMs require a sub-logarithmic, or maybe even constant, overhead?

Also, it would be interesting to look at different data structures. The research reported in [16] has a general framework to prove lower bounds for more sophisticated data structures.

Acknowledgments. Most of the work discussed in this invited lecture is co-authored with Sarvar Patel and Kevin Yeo.

References

1. Asharov, G., Komargodski, I., Lin, W.-K., Nayak, K., Peserico, E., Shi, E.: OptORAMa: optimal oblivious RAM. In: Canteaut, A., Ishai, Y. (eds.) EUROCRYPT 2020, Part II. LNCS, vol. 12106, pp. 403–432. Springer, Cham (2020). https://doi.org/10.1007/978-3-030-45724-2_14
2. Boyle, E., Naor, M.: Is there an oblivious RAM lower bound?. In: Sudan, M., (ed.) ITCS 2016, pp. 357–368. ACM (2016)
3. Devadas, S., van Dijk, M., Fletcher, C.W., Ren, L., Shi, E., Wichs, D.: Onion ORAM: a constant bandwidth blowup oblivious RAM. In: Kushilevitz, E., Malkin, T. (eds.) TCC 2016, Part II. LNCS, vol. 9563, pp. 145–174. Springer, Heidelberg (2016). https://doi.org/10.1007/978-3-662-49099-0_6
4. Du, Y., Genkin, D., Grubbs, P.: Snapshot-oblivious RAMs: sub-logarithmic efficiency for short transcripts. In: Dodis, Y., Shrimpton, T. (eds.) CRYPTO 2022. Lecture Notes in Computer Science, vol. 13510, pp. 152–181. Springer, Cham (2022). https://doi.org/10.1007/978-3-031-15985-5_6
5. Dwork, C., McSherry, F., Nissim, K., Smith, A.: Calibrating noise to sensitivity in private data analysis. In: Halevi, S., Rabin, T. (eds.) TCC 2006. LNCS, vol. 3876, pp. 265–284. Springer, Heidelberg (2006). https://doi.org/10.1007/11681878_14
6. Goldreich, O., Ostrovsky, R.: Software protection and simulation on oblivious RAMs. J. ACM (JACM) **43**(3), 431–473 (1996)
7. Goodrich, M.T., Mitzenmacher, M., Ohrimenko, O., Tamassia, R.: Privacy-preserving group data access via stateless oblivious RAM simulation. In: Rabani, Y. (ed.), 23rd SODA, pp. 157–167. ACM-SIAM (2012)
8. Komargodski, I., Lin, W.-K.: A logarithmic lower bound for oblivious RAM (for all parameters). In: Malkin, T., Peikert, C. (eds.) CRYPTO 2021, Part IV. LNCS, vol. 12828, pp. 579–609. Springer, Cham (2021). https://doi.org/10.1007/978-3-030-84259-8_20
9. Kushilevitz, E., Lu, S., Ostrovsky, R.: On the (in)security of hash-based oblivious RAM and a new balancing scheme. In: Rabani, Y. (ed.), 23rd SODA, pp. 143–156. ACM-SIAM (2012)
10. Larsen, K.G., Nielsen, J.B.: Yes, there is an oblivious RAM lower bound! In: Shacham, H., Boldyreva, A. (eds.) CRYPTO 2018, Part II. LNCS, vol. 10992, pp. 523–542. Springer, Cham (2018). https://doi.org/10.1007/978-3-319-96881-0_18
11. Patel, S., Persiano, G., Raykova, M., Yeo, K.: PanORAMa: oblivious RAM with logarithmic overhead. In: Thorup, M. (ed.) 59th FOCS, pp. 871–882. IEEE Computer Society Press (2018)
12. Patel, S., Persiano, G., Yeo, K.: Lower bounds for encrypted multi-maps and searchable encryption in the leakage cell probe model. In: Micciancio, D., Ristenpart, T. (eds.) CRYPTO 2020, Part I. LNCS, vol. 12170, pp. 433–463. Springer, Cham (2020). https://doi.org/10.1007/978-3-030-56784-2_15

13. Persiano, G., Yeo, K.: Limits of preprocessing for single-server PIR. In: SODA 2022, SIAM, pp. 2522–2548 (2022)
14. Persiano, G., Yeo, K.: Limits of breach-resistant and snapshot-oblivious RAMs. Unpublished manuscript (2023)
15. Persiano, G., Yeo, K.: Lower bounds for differentially private RAMs. In: Ishai, Y., Rijmen, V. (eds.) EUROCRYPT 2019, Part I. LNCS, vol. 11476, pp. 404–434. Springer, Cham (2019). https://doi.org/10.1007/978-3-030-17653-2_14
16. Persiano, G., Yeo, K.: Lower bound framework for differentially private and oblivious data structures. EUROCRYPT (2023)

The Power of the Binary Value Principle

Yaroslav Alekseev[1,2(✉)] and Edward A. Hirsch[1]

[1] Technion—Israel Institute of Technology, Haifa, Israel
tolstreg@gmail.com

[2] Steklov Institute of Mathematics, St. Petersburg, Russia

Abstract. The (extended) Binary Value Principle (eBVP, the equation $\sum_{i=1}^{n} x_i 2^{i-1} = -k$ for $k > 0$ and in the presence of $x_i^2 = x_i$) has received a lot of attention recently, several lower bounds have been proved for it [1,2,10]. It has been shown [2] that the probabilistically verifiable Ideal Proof System (IPS) [8] together with eBVP polynomially simulates a similar semialgebraic proof system. In this paper we consider Polynomial Calculus with the algebraic version of Tseitin's extension rule (Ext-PC). Contrary to IPS, this is a Cook–Reckhow proof system. We show that in this context eBVP still allows to simulate similar semialgebraic systems. We also prove that it allows to simulate the Square Root Rule [6] in a sharp contrast to the result of [1] that shows an exponential lower bound on the size of Ext-PC derivations of BVP from its square. On the other hand, we demonstrate that eBVP probably does not help in proving exponential lower bounds for Boolean formulas: we show that an Ext-PC (even with the Square Root Rule) derivation of any unsatisfiable Boolean formula in CNF from eBVP must be of exponential size.

Keywords: Proof complexity · Polynomial Calculus · Extension Rule

1 Introduction

Tseitin's extension rule [12] is a powerful concept that turns even very weak propositional proof systems into strong ones: it allows to introduce new variables for arbitrary formulas (it is enough to do this for the disjunction and the negation). In particular, it turns Resolution (a rather weak system for which superpolynomial lower bounds are known since [12]) into the powerful Extended Frege system [5] (a strong system for which we do not even know good enough candidates for superpolynomial lower bounds).

Surprisingly, in the context of algebraic proof systems an exponential lower bound for a system that uses Tseitin's rule was proved recently [1]. This system, Extended Polynomial Calculus (or Ext-PC), combines the algebraic version of the extension rule (so that we can introduce new variables for polynomials) with the Polynomial Calculus (PC) [4] system. While it has more power because it allows to talk about polynomials over any algebraically closed field (or, in the

The full version is available at https://arxiv.org/abs/2210.17429.

M. Mavronicolas (Ed.): CIAC 2023, LNCS 13898, pp. 21–36, 2023.
https://doi.org/10.1007/978-3-031-30448-4_3

Boolean setting, even just over a ring, such as $\mathbb{Z}$), the exponential lower bound has been proved for a system of polynomial equations that does not correspond to any Boolean formula (in particular, a formula in conjunctive normal form, as in Resolution). This system, called "the Binary Value Principle", is the equation $\sum_{i=1}^{n} x_i 2^{i-1} + 1 = 0$ along with the "Boolean axioms" $x_i^2 - x_i = 0$ for every variable x_i. It has also been used for proving other exponential lower bounds [2,10] and (as the Extended Binary Value Principle, eBVP) for demonstrating a polynomial simulation of polynomial inequalities by polynomial equations [2] for generalized proof systems that require polynomial identity testing for the verification (the algebraic system is the Ideal Proof System, IPS, of [8]). Note that polynomial inequalities are considered to be much more powerful than polynomial equations: for example, no exponential size lower bound is known even for the simplest proof system LS (motivated by the optimization procedure by László Lovász and Alexander Schrijver, see [7,11]).

Our Results. In this paper, we consider three questions about Ext-PC and eBVP, and prove three results:

1. How powerful is Ext-PC? We prove (Theorem 2) that together with eBVP it polynomially simulates a similar system that uses inequalities (namely, Ext-$\mathsf{LS}_{+,*}$ that is LS with extension variables, squares, and multiplication). This brings the result of [2] down to conventional proof systems from proof systems that use polynomial identity testing for proof verification. It is interesting how far we can weaken the proof systems to keep such simulation (formulas? bounded-depth formulas? sums of monomials?).
2. Grigoriev and Hirsch [6] introduced the square root rule that allows to conclude $f = 0$ from $f^2 = 0$. It would be needed for the implicational completeness of PC in the non-Boolean case. It is not needed at all in the Boolean context, however, it could shorten the proofs. It is impossible to simulate it polynomially in Ext-PC ([1] proves an exponential bit-size-of-coefficients lower bound on derivations of $\sum x_i 2^{i-1} + 1 = 0$ from $(\sum x_i 2^{i-1} + 1)^2 = 0$) and PC ([9] proves a linear degree lower bound on derivations of $\sum x_i + 1 = 0$ from $(\sum x_i + 1)^2 = 0$). We prove (Theorem 3) that eBVP allows one to simulate polynomially the square root rule in the case of Ext-PC.
3. Is it possible to use lower bounds for eBVP for proving lower bounds for formulas in conjunctive normal form? One could imagine deriving the translation of an unsatisfiable formula in conjunctive normal form (using the extension variables) from eBVP and concluding a lower bound for a formula in CNF. We prove an exponential lower bound (Corollary 2) on the size of derivations of such formulas from eBVP, showing an obstacle to this approach.

Our Methods. The divisibility method suggested in [1,2] allows to prove lower bounds on the size of algebraic proofs by analysing the scalars appearing in them. The simplest application of this method substitutes the input variables by the binary representations of all possible integers, and shows that the constant in the final contradiction in the proof over the integers divides all of them (if the system allows it). In this paper we further develop this method: we prove lower bounds for the derivation of a translation of an unsatisfiable formula in CNF (and not just a contradiction), so there is no single constant at the end. We show an exponential lower bound over the integers by counting the primes that divide the multiplicative constants in the derivation of every clause and Boolean equation. The lower bound for rationals follows using the translation of [1].

In order to show polynomial simulations we use the general approach suggested in [2]: to use bit arithmetic for proving facts about (semi)algebraic proofs. However, IPS [8] considered in that paper uses polynomial identity testing for proof verification, and thus allows to switch between the circuit representations of polynomials at no cost. Our setting is different: we need to simulate everything using the extension rule. Therefore, in order to simulate inequalities we derive gradually the facts that the values produced by bit arithmetic equal the values of polynomials in the original proof, and that these values are nonnegative. We also need to define the circuit representation, in particular for the extension variables, to reason about $\mathsf{Ext\text{-}LS}_{+,*}$ proofs. A somewhat similar approach works for the simulation of the square root rule; however, we need to derive that all individual bits of the zero are zeroes, and then take the square root.

2 Preliminaries

In this paper we work with polynomials over integers or rationals. We define *the size of a polynomial* roughly as the total length of the bit representation of its coefficients. Formally, let f be an arbitrary integer or rational polynomial in variables $\{x_1, \dots, x_n\}$.

- If $f \in \mathbb{Z}[x_1, \dots, x_n]$ then $Size(f) = \sum(\lceil \log |a_i| \rceil + 1)$, where a_i are the coefficients of f.
- If $f \in \mathbb{Q}[x_1, \dots, x_n]$ then $Size(f) = \sum(\lceil \log |a_i| \rceil + \lceil \log |b_i| \rceil + 1)$, where $a_i \in \mathbb{Z}$, $b_i \in \mathbb{N}$ and $\frac{a_i}{b_i}$ are the coefficients of f.

We also use *algebraic circuits*. Formally, an algebraic circuit is a dag whose vertices (gates) compute binary operations (addition and multiplication), thus gates have in-degree two; the inputs (or variables) and constants (nodes computing integers or rationals) are nodes of in-degree zero. Every gate of an algebraic circuit computes a polynomial in the input variables in a natural way; we sometimes identify a gate with the circuit consisting of all the nodes on which the gate depends (thus this gate is the output gate of such circuit).

The size of the circuit is the number of its gates plus the sum of the bit sizes of all constants. We will also be interested in the *syntactic length* of an algebraic circuit, defined for circuits over $\mathbb{Z}$: it is roughly a trivial upper bound on the number of bits of an integer computed by the circuit. The definition essentially follows [2], augmenting it with the multiplication.

Definition 1 (syntactic length of algebraic circuit). *Consider the gates of an algebraic circuit $G_1, \ldots, G_k$ in topological order. We define the syntactic length inductively:*

- *If G_i is an integer constant, then the syntactic length of G_i is $\lceil \log(|G_i|) \rceil$.*
- *If $G_i = G_j + G_k$, the syntactic length of G_j is t, and the syntactic length of G_k is s, then the syntactic length of G_i equals $\max(s, t) + 1$.*
- *If $G_i = G_j \cdot G_k$, the syntactic length of G_j is t, and the syntactic length of G_k is s, then the syntactic length of G_i equals $s + t + 3$.*

Note 1. 1. In the latter case the actual number of bits would be $s + t$; we state $s+t+3$ because this is how it is computed in our implementation of the integer multiplication in Sect. 4—however, it does not change much asymptotically, the resulting length changes at most polynomially.

2. Note that the circuit size cannot exceed its syntactic length.

2.1 Algebraic Proof Systems

In what follows, R denotes $\mathbb{Q}$ or $\mathbb{Z}$.

Definition 2 (Polynomial Calculus [4]). *Let $\Gamma = \{p_1, ..., p_m\} \subset R[x_1, ..., x_n]$ be a set of polynomials in variables $\{x_1, \ldots, x_n\}$ over R such that the system of equations $p_1 = 0, \ldots, p_m = 0$ has no solution. A Polynomial Calculus (PC_R) refutation of Γ is a sequence of polynomials $r_1, \ldots, r_s$ where $r_s = const \neq 0$ and for every l in $\{1, \ldots, s\}$, either $r_l \in \Gamma$ or r_l is obtained through one of the following derivation rules for $j, k < l$:*

- *$r_l = \alpha r_j + \beta r_k$, where $\alpha, \beta \in R$,*
- *$r_l = x_i r_k$.*

The size of the refutation is $\sum_{l=1}^{s} Size(r_l)$, and its degree is $\max_l deg(r_l)$.

Note 2. 1. In this paper we consider $\mathbb{Q}$ or $\mathbb{Z}$ as R in PC_R above or $\mathsf{Ext\text{-}PC}_R$ below. For both of these rings, we consider the Boolean case, where axioms $x_i^2 - x_i = 0$ are present for every variable x_i, and for this case our proof systems are complete.

2. Note that in the case $R = \mathbb{Q}$ one can assume $r_s = 1$, while in the case $R = \mathbb{Z}$ an arbitrary nonzero constant is needed to maintain the completeness.

Tseitin's extension rule allows to introduce new variables for arbitrary formulas. We use an algebraic version of this rule that allows to denote any polynomial by a new variable [1].

Definition 3 (Extended Polynomial Calculus, Ext-PC). *Let $\Gamma = \{p_1, \ldots, p_m\} \subset R[x_1, \ldots, x_n]$ be a set of polynomials in variables $\{x_1, \ldots, x_n\}$ over R such that the system of equations $p_1 = 0, \ldots, p_m = 0$ has no solution. An $\mathsf{Ext\text{-}PC}_R$ refutation of Γ is a Polynomial Calculus refutation of a set $\Gamma' = \{p_1, \ldots, p_m, y_1 - q_1(x_1, \ldots, x_n), y_2 - q_2(x_1, \ldots, x_n, y_1), \ldots, y_m - q_m(x_1, \ldots, x_n, y_1, \ldots, y_{m-1})\}$ where $q_i \in R[\bar{x}, y_1, \ldots, y_{i-1}]$ are arbitrary polynomials.*

We omit R from the notation of PC_R or $\mathsf{Ext\text{-}PC}_R$ when it is clear from the context. The size of the Ext-PC *refutation is equal to the size of the Polynomial Calculus refutation of Γ'.*

The square root rule [6] allows to conclude that $f = 0$ from $f^2 = 0$. We can consider it in the context of both PC and Ext-PC.

Definition 4 (PC$^{\surd}$, Ext-PC$^{\surd}$). *The proofs in* PC$^{\surd}$, Ext-PC$^{\surd}$ *follow Definitions 2, 3 but allow one more derivation rule in terms of Definition 2:*

- *derive r_l, if $r_l^2 = r_k$*

(derive a polynomial if its square has been already derived).

Note 3. As both $\mathbb{Q}$ and $\mathbb{Z}$ are domains, if $r^2 = 0$ for some $r \in R[\bar{x}]$, then $r = 0$.

The extended Binary Value Principle (eBVP) says that that the (nonnegative) integer value of a binary vector cannot be negative. To use this fact in the proof, we need to specify that such a polynomial can be replaced by 1 (in particular, if eBVP is present without a multiplier, it produces the contradiction $1 = 0$).

Definition 5 (Ext-PC+eBVP). Ext-PC+eBVP *uses exactly the same derivation rules as* Ext-PC *and one more rule:*

- *derive $r_l = g$ if for some polynomials $g, f_1, \ldots, f_t$ and integer constant $M > 0$ we have derived the polynomial $r_k = g \cdot (M + f_1 + 2f_2 + \ldots + 2^{t-1}f_t)$ along with polynomials $r_{k_1} = f_1^2 - f_1, \ldots, r_{k_t} = f_t^2 - f_t$, where $k, k_1, \ldots, k_t < l$.*

Note 4. We can define Ext-PC$^{\surd}$ + eBVP the same way.

2.2 A Semialgebraic Proof System

We will consider the following proof system that can be viewed as a generalization of the LS proof system [11] by the algebraic extension rule. Note that we could move the introduction of new variables to the beginning of the proof as we did in the definition of Ext-PC, however, it does not matter.

Definition 6 (Ext-LS$_{+,*}$). *Let $\Gamma = \{p_1, \ldots, p_k\} \subset R[x_1, \ldots, x_n]$ be a set of polynomials in variables $\{x_1, \ldots, x_n\}$ over R such that the system of equations $p_1 \geq 0, \ldots, p_k \geq 0$ has no solution. An* Ext-LS$_{+,*}$ *refutation of Γ is a sequence of polynomial inequalities $r_1 \geq 0, \ldots, r_m \geq 0$ where $r_m = -M$ ($M > 0$ is an integer constant) and each inequality r_l is obtained through one of the following inference rules:*

- *$r_l = p_j$ for some i, or $r_l = x_i$, or $r_l = 1 - x_i$, or $r_l = x_i^2 - x_i$, or $r_l = x_i - x_i^2$.*
- *$r_l = r_i \cdot r_j$ or $r_l = r_i + r_j$ for $i, j < l$. (Note that we can infer 1 as $x_i + (1 - x_i)$, thus we can multiply by any positive constant.)*

- *If variable y did not occur in polynomials $r_1, \ldots, r_{l-1}$, then we can derive a pair of polynomials $r_l = y - f$, $r_{l+1} = f - y$, where f is one of the basic operations (addition, multiplication, identity) applied to variables not including y, and constants.*
- $r_k = z^2$ *for any variable z, including the input variables x_i or the newly introduced variables.*

Note that the newly introduced variables are not necessarily Boolean. The size of the refutation is $\sum_{l=1}^{m} Size(r_l)$. The degree of the refutation is $\max_l deg(r_l)$.

Note 5. 1. Once again, in the case $R = \mathbb{Q}$ we could assume $M = 1$, while we need an arbitrary positive constant for $R = \mathbb{Z}$ in order to maintain completeness.

2. Note that while the definition of $\mathsf{Ext\text{-}LS}_{+,*}$ is written in a slightly different manner compared to Ext-PC, it is not difficult to see that $\mathsf{Ext\text{-}LS}_{+,*}$ polynomially simulates Ext-PC (in particular, conversion of equations to inequalities and of ideal inference to cone inference can be done similarly to [2, Sect. 4.1.1 of the Technical Report version]).

3 Circuit and Equational Representations

We will represent the polynomials of the $\mathsf{Ext\text{-}LS}_{+,*}$ derivation as circuits in the input variables. To do this, we define circuit representations of axioms and extension variables.

Definition 7 (Circuit representation: axioms). *For an axiom $f \geq 0$, we consider a natural circuit representation of the polynomial $f \in \mathbb{Z}[\overline{x}]$: $Z_{f,1} = x_1, \ldots, Z_{f,n} = x_n$, $Z_{f,n+1} = h_{f,1}(Z_{f,r_1}, Z_{f,t_1}), \ldots, Z_{f,n+s} = h_{f,s}(Z_{f,r_s}, Z_{f,t_s})$, where $r_i, t_i < n + i$ and $h_{f,i}$ is one of the basic operations (addition, multiplication) or a constant. We denote the resulting circuit by Z_f (where $Z_{f,n+s}$ is the output gate).*

Definition 8 (Circuit representation: extension variables). *Given a sequence of extension variables $y_1, \ldots, y_k$ introduced in some derivation by axioms $y_j = g_j(\overline{x}, y_1, \ldots, y_{j-1})$ (where $1 < j \leq k$), we can define their values by algebraic circuits computed in a natural way (the axioms are substituted into each other): define the sequence of circuits $Y_1(\overline{x}), \ldots, Y_k(\overline{x})$ by*

- $Y_1(\overline{x}) = g_1(\overline{x})$,
- *for each* $1 < j \leq k$, $Y_j(\overline{x}) = g_j(\overline{x}, Y_1(\overline{x}), \ldots, Y_{j-1}(\overline{x}))$.

We call Y_i the circuit representation of the extension variable y_i.

With the circuit representation of the extension variables and axioms, we can define the circuit representation of an $\mathsf{Ext\text{-}LS}_{+,*}$ proof.

Definition 9 (Circuit representation: $\mathsf{Ext\text{-}LS}_{+,*}$ refutation). *Given an $\mathsf{Ext\text{-}LS}_{+,*}$ refutation $p_1 \geq 0, \ldots, -M = p_m \geq 0$ of a system in variables x_i, we construct the circuit representation $P_1, \ldots, P_m$ of its polynomials inductively:*

- *If p_l is an axiom, P_l is the circuit representation of this axiom.*
- *If $p_l = x_i$, or $p_l = 1 - x_i$, then P_l is the simple circuit computing p_l.*
- *If $p_l = z^2$ for a variable z, then $P_l = Q \cdot Q$, where Q is the circuit representation of z (note that typically, z is an extension variable).*
- *If p_l is obtained using a binary operation $\circ$ (addition or multiplication) from p_i and p_j, we put $P_l = P_i \circ P_j$.*
- *If p_l introduces a new variable, or it is the Boolean axiom $x_i^2 - x_i$ (or $x_i - x_i^2$), we put $P_l = 0$.*

Note that the axioms and the extension variables appear in P_i's as subcircuits, and that the inputs of P_i's correspond to the original variables of the system.

Definition 10 (Equational representation). *Any algebraic circuit can be represented by equations (one equation per gate). More precisely, if we have gates $G_1, \ldots, G_m$ in topological order, then we can consider variables $\gamma_1, \ldots, \gamma_m$ with the corresponding set of polynomial equations:*

- *If $G_i = x_i$ or $1 - x_i$ for some input variable, then corresponding polynomial equation for the γ_i would be $\gamma_i = x_j$ or $\gamma_i = 1 - x_i$.*
- *If $G_i = G_k \circ G_\ell$, then the corresponding polynomial equation for the γ_i would be $\gamma_i = \gamma_k \circ \gamma_\ell$.*

We refer to this set of equations as the equational representation.

The following lemma is used in the simulation of Ext-LS$_{+,*}$. The proof is a simple induction, see full version.

Lemma 1. *Accordingly to Definitions 9 and 10, consider the circuit and equational representations of an* Ext-LS$_{+,*}$ *proof $p_1 \geq 0, \ldots, p_t \geq 0$. Consider any gate P_i corresponding to the equational representation with the output variable π_i. Then there is a polynomial-size (in the size of the original proof)* Ext-PC *derivation of $\pi_i = p_i$ using only the Boolean axioms and the definitions of extension variables of the* Ext-LS$_{+,*}$ *proof. The new extension variables needed in the* Ext-PC *derivation are those appearing in the equational representation.*

To simulate the square root derivation rule we need to consider a circuit representation of an arbitrary polynomial in extension variables, since a derivation in Ext-PC$^{\surd}$, unlike derivations in Ext-LS$_{+,*}$, does not correspond to an algebraic circuit (algebraic circuits do not use square root gates).

Definition 11 (Circuit representation: polynomials). *Consider a polynomial $g \in \mathbb{Z}[\overline{x}, \overline{y}]$, where $\overline{x}$ are the input variables and $\overline{y}$ are the variables introduced by the extension rule. Definition 8 defines the circuit representation $Y_1, \ldots, Y_m$ for the variables $y_1, \ldots, y_m$. Then we can consider any natural circuit $G'_1, \ldots, G'_t$ computing the polynomial g given variables $x_1, \ldots, x_n$, variables $y_1, \ldots, y_m$, and the constants. Substituting the subcircuits $Y_1, \ldots, Y_m$ in place of the inputs $y_1, \ldots, y_m$ of G'_i's, we get the* circuit representation *$G_1, \ldots, G_l$ of g.*

The syntactic length of the polynomial g is defined as the syntactic length of the circuit $G_1, \ldots, G_l$.

The following Lemma is proved similarly to Lemma 1, see the full version.

Lemma 2. *Consider any polynomial g in the extension variables $y_1, \ldots, y_k$ and the original Boolean variables $x_1, \ldots, x_n$, and consider the circuit representation $G_1, \ldots, G_l$ of this polynomial g from Definition 11.*

Then, if we consider an equational representation $\pi_1, \ldots, \pi_l$ of the circuit $G_1, \ldots, G_l$, then there is a polynomial-size (in the size of g) Ext-PC *derivation of the equation $g = \pi_l$.*

4 Explicit BIT Definition and Basic Lemmas

In our Ext-PC simulations in Sect. 5, we argue about individual bits of the values of the polynomials appearing in the Ext-$LS_{+,*}$ proof. In this section we construct the circuits corresponding to these bits and prove auxilary statements about our constructions. We basically follow [2] (Theorem 6.1 in the Technical Report version), however, there are important differences:

1. In the case of Ext-PC proofs, the circuits are used in the meta-language only. In the actual derivation, the bits are represented by extension variables defined through other extension variables, etc. (essentially computing the circuit value).
2. Contrary to [2], we cannot magically switch between different representations of polynomials, every step of the derivation has to be done syntactically.

The integers are represented in two's complement form (see the definition of VAL below). We use the following notation:

$\mathrm{BIT}_i(F)$: if $F(\overline{x})$ is a circuit in the variables $\overline{x}$, then $\mathrm{BIT}_i(F)$ is a new variable defined through other extension variables (and $\overline{x}$) that computes the i-th bit of the integer computed by F as a function of the input variables $\overline{x}$, where the variables $\overline{x}$ range over 0–1 values. The integer is represented in the two's complement form, that is, its highest bit is the sign bit.
$\mathrm{SIGN}(F)$ is used to denote this sign bit.
$\overline{\mathrm{BIT}}(F)$: a collection of new variables that compute the bit vector of F. Note that $\overline{\mathrm{BIT}}(F)$ also includes $\mathrm{SIGN}(F)$.
$\mathrm{VAL}(\overline{z})$: the evaluation polynomial that converts bit encoding of an integer $\overline{z}$ in two's complement representation to its integer value. Given $z_0, \ldots, z_{k-1}$,

$$\mathrm{VAL}(\overline{z}) = \sum_{i=0}^{k-2} 2^i \cdot z_i - 2^{k-1} \cdot z_{k-1}.$$

We construct the representation of $\mathrm{BIT}_i(F)$ by induction on the size of F.

Proof Strategy for the Simulation. Our plan for the simulation of Ext-$LS_{+,*}$ in Sect. 5.1 is as follows:

- Given an $\mathsf{Ext\text{-}LS}_{+,*}$ refutation $p_1(\overline{x},\overline{y}) \geq 0, \ldots, p_m(\overline{x},\overline{y}) \geq 0$, where $p_m = -M$ and $M \in \mathbb{N}$, we will consider the circuit representation $P_1, \ldots, P_m$ of polynomials $p_1, \ldots, p_m$ in order to speak about $\overline{\mathrm{BIT}}(P_i)$, and will introduce more extension variables according to the corresponding equational representation of P_i's.
- We will show by induction that we can derive the following statements in Ext-PC:
  1. $\mathrm{VAL}(\overline{\mathrm{BIT}}(P_i)) = p_i$.
  2. $\mathrm{SIGN}(P_i) = 0$.

  Then given the fact that $\mathrm{VAL}(\overline{\mathrm{BIT}}(P_m)) = p_m = -M$, where $M \in \mathbb{N}$, and $\mathrm{SIGN}(P_i) = 0$, we can apply eBVP to derive a contradiction in Ext-PC.

Before we accomplish this, we need to define BIT (using the definitions for basic arithmetic operation) and prove several useful lemmas about what can we derive in Ext-PC (basic facts about the values, the signs, etc.). These will be also useful for the simulation of the square root rule in Sect. 5.2.

Basic Arithmetic Operations. Circuit constructions of the basic operations that we will need for the BIT definition essentially follows the scheme of [2]. It consists of school-type carry-save addition $\overline{\mathrm{ADD}}$ producing a $(k+2)$-bit integer from two $(k+1)$-bit integers (including the sign bit), the absolute value $\overline{\mathrm{ABS}}$ producing a $(k+2)$-bit integer from a $(k+1)$-bit integer, and school-type multiplication "in shifts", $\overline{\mathrm{PROD}}$, producing a $(k+r+2)$-bit integer from a $(k+1)$-bit and an $(r+1)$-bit integers. See Section A.2 for details.

Definition of BIT. Following [2] we define the bit representation of the values of polynomials computed by algebraic circuits. In doing this, we construct another circuit. We identify its nodes with new variables that will appear in our Ext-PC + eBVP proof, and the defining equation for these variables are exactly the operations computed by the gates of the new circuit. Note that the inputs of this circuit are the same as the inputs of the original circuit.

Definition 12 (BIT).
Let $G_1 = f_1(\overline{x}), G_2 = f_2(\overline{x}, G_1), \ldots, G_m = f_m(\overline{x}, G_1, \ldots, G_{m-1})$ be a topological order of the gates of an algebraic circuit over variables $\overline{x}$.

*For each G_r we define $\mathrm{BIT}_i(G_r)$ to be a **new extension variable** with the corresponding polynomial equation so that $\mathrm{BIT}_i(G_r)$ computes the i-th bit of G_r:*

Case 1: *$G_r = x_j$ for an input x_j. Then, $\mathrm{BIT}_0(G_r) := x_j$, $\mathrm{BIT}_1(G_r) := 0$ (in this case there are just two bits).*
Case 2: *$G_r = \alpha$, for $\alpha \in \mathbb{Z}$. Then, $\mathrm{BIT}_i(G_r)$ is defined to be the i-th bit of α in two's complement notation.*
Case 3: *$G_r = G_k + G_l$. Then $\overline{\mathrm{BIT}}(G_r) := \overline{\mathrm{ADD}}(\overline{\mathrm{BIT}}(G_k), \overline{\mathrm{BIT}}(G_l))$, and $\mathrm{BIT}_i(y_r)$ is defined to be the i-th bit of $\overline{\mathrm{BIT}}(y_r)$.*
Case 4: *$G_r = G_k \cdot G_l$. Then $\overline{\mathrm{BIT}}(G_r) := \overline{\mathrm{PROD}}(\overline{\mathrm{BIT}}(G_k), \overline{\mathrm{BIT}}(G_l))$, and $\mathrm{BIT}_i(G_r)$ is defined to be the i-th bit of $\overline{\mathrm{BIT}}(G_r)$.*

Recall that in Cases 3, 4 the shorter number is padded to match the length of the longer number by copying the sign bit before applying $\overline{\mathrm{ADD}}$ or $\overline{\mathrm{PROD}}$.

The Binary Value Lemma. Similarly to [2], one can show that Ext-PC has short derivations of the fact that the BIT(G) circuit that we constructed computes the same binary value as the original circuit G. Moreover, it can be compactly proved in Ext-PC for the equational representation of BIT(G).

Lemma 3 (binary value lemma). *Let $y_1 = f_1(\overline{x}), y_2 = f_2(\overline{x}, y_1), \ldots, y_m = f_m(\overline{x}, y_1, \ldots, y_{m-1})$ be the equational representation of the algebraic circuit $G_1(\overline{x}) = f_1(\overline{x}), \ldots, G_m = f_m(\overline{x}, G_1(\overline{x}), \ldots, G_{m-1}(\overline{x}))$ over the variables $\overline{x} = \{x_1, \ldots, x_n\}$, and let t be the syntactic length of $G_1, \ldots, G_m$. Then, there is an* Ext-PC *proof (using only the Boolean axioms and the equations of the BIT encoding) of $y_i = \mathrm{VAL}(\overline{\mathrm{BIT}}(G_i))$ of size* poly(t) *for each $1 \leq i \leq m$.*

Useful Lemmas About the BIT Value. In this section we provide technical lemmas about individual bits in the bit representation.

Lemma 4. *For any vector of variables $r_0, \ldots, r_{k-1}, r_k$, there is a* poly($k$)*-size* Ext-PC + eBVP *derivation of $r_0 = \ldots = r_k = 0$ from $r_0^2 - r_0 = 0, \ldots, r_k^2 - r_k = 0$ and $r_0 + 2r_1 + \ldots + 2^{k-1}r_{k-1} - 2^k r_k = 0$.*

Proof. Multiply the last equation by r_k and replace r_k^2 by r_k. We get $(r_0 + 2r_1 + \ldots + 2^{k-1}r_{k-1} - 2^k)r_k = 0$, which has (the negation of) an instance of eBVP in the parentheses (for $r_i' = 1 - r_i$). It remains to apply the eBVP rule to prove that $r_k = 0$. After that we get

$$r_0 + 2r_1 + \ldots + 2^{k-1}r_{k-1} = 0.$$

Again, multiply this by r_{k-1} and replace r_{k-1}^2 by r_{k-1}. We get $(r_0 + 2r_1 + \ldots + 2^{k-2}r_{k-2} + 2^{k-1})r_{k-1} = 0$ with an instance of eBVP inside. After applying the eBVP rule we get that $r_{k-1} = 0$. We can continue in the same way for $r_{k-2}, \ldots, r_0$ getting $r_0 = \ldots = r_k = 0$. □

Lemma 5 (monotonicity of addition and multiplication). *For any two bit vectors $r_0, \ldots, r_{k-1}, r_k$ and $r_0', \ldots, r_{k-1}', r_k'$, there is a* poly($k$)*-size* Ext-PC *derivation of $\mathrm{SIGN}(\overline{\mathrm{PROD}}(\overline{r}, \overline{r}')) = 0$ and $\mathrm{SIGN}(\overline{\mathrm{ADD}}(\overline{r}, \overline{r}')) = 0$ from*

$$\begin{aligned} r_0^2 - r_0 = 0, \ldots, r_{k-1}^2 - r_{k-1} = 0, r_k^2 - r_k = 0; \qquad r_k = 0; \\ {r_0'}^2 - r_0' = 0, \ldots, {r_{k-1}'}^2 - r_{k-1}' = 0, {r_k'}^2 - r_k' = 0; \qquad r_k' = 0. \end{aligned}$$

Proof. See [2] (Lemma 6.7 in the Technical Report version), as the derivation presented in that paper is literally in Ext-PC. □

Lemma 6. *1. For any vector of variables $r_0, \ldots, r_{k-1}, r_k$, there is a* poly($k$)*-size* Ext-PC *derivation of $\mathrm{SIGN}(\overline{\mathrm{PROD}}(\overline{r}, \overline{r})) = 0$ from $r_0^2 - r_0 = 0, \ldots, r_{k-1}^2 - r_{k-1} = 0, r_k^2 - r_k = 0$.*

2. If additionally $\overline{\mathrm{PROD}}(\overline{r}, \overline{r}) = \overline{0}$ is given, there is a poly(k)*-size* Ext-PC *derivation of $r_0 = 0, \ldots, r_k = 0$.*

Proof (Sketch). The equation $\mathrm{SIGN}(\overline{\mathrm{PROD}}(\overline{r},\overline{r})) = 0$ is derived straightforwardly from the definition of $\overline{\mathrm{PROD}}$.

For the second part, we consider the definition of $\overline{\mathrm{PROD}}$ and show that the bits of $\overline{\mathrm{BIT}}(A)$ are equal to 0 one by one, starting from the least significant bit, full proof is given in the full version. □

5 Polynomial Simulations

1. Ext-PC$_{\mathbb{Z}}$ + eBVP *polynomially simulates* Ext-LS$_{+,*,\mathbb{Z}}$. To show it, we gradually apply Lemma 3 to the circuit representation of the Ext-LS$_{+,*,\mathbb{Z}}$ derivation.

Theorem 1 (derivation theorem). *Suppose we have a system of polynomial equations $f_1 = 0, \dots, f_k = 0$, and that there is an* Ext-LS$_{+,*,\mathbb{Z}}$ *refutation $p_1 \geq 0, \dots, p_m \geq 0$ of the corresponding system $f_1 \geq 0, f_1 \leq 0, \dots, f_k \geq 0, f_k \leq 0$.*

Consider its circuit representation according to Sect. 3. Denote the syntactic length of the circuit $P_1, \dots, P_m$ as t. Then, in terms of Sect. 3 there are poly(t)*-size* Ext-PC$_{\mathbb{Z}}$ + eBVP *derivations of the facts*

1. $p_1 = \mathrm{VAL}(\overline{\mathrm{BIT}}(P_1)), \dots, p_m = \mathrm{VAL}(\overline{\mathrm{BIT}}(P_m))$.
2. *Each* **sign bit** *in $\overline{\mathrm{BIT}}(P_i)$ equals 0 (written in the form of polynomial equation $s_i = 0$ where s_i is a variable, corresponding to the sign bit of $\overline{\mathrm{BIT}}(P_i)$).*

The axioms used in these derivations are the boolean axioms, the axioms defining extension variables, and (for the second statement) the input axioms.

Proof (Sketch). The first part of the theorem is an application of Lemmas 3, 1.

For the second part, we prove by induction that each **sign bit** in $\overline{\mathrm{BIT}}(P_i)$ equals 0. We apply Lemma 6 to prove that the sign bit of a square equals 0, and Lemma 4 proves that the sign bit of the encoding of any axiom is 0. This proves the induction base, and Lemma 5 allows to derive the nonnegativity of all next $\overline{\mathrm{BIT}}(P_i)$'s by the monotonicity of addition and multiplication. See the full version for the full proof. □

Definition 13 (Syntactic size of a refutation). *The syntactic size of an* Ext-LS$_{+,*,\mathbb{Z}}$ *refutation is the syntactic size of a corresponding circuit representation from Sect. 3.*

Theorem 2. *Consider a system of polynomial equations $f_1 = 0, \dots, f_k = 0$. Suppose there is an* Ext-LS$_{+,*,\mathbb{Z}}$ *refutation for the system $f_1 \geq 0, f_1 \leq 0, \dots, f_k \geq 0, f_k \leq 0$ of syntactic size S. Then there is an* Ext-PC$_{\mathbb{Z}}$ + eBVP *refutation for the system $f_1 = 0, \dots, f_k = 0$ of size at most* poly(S).

Proof (Sketch). Consider an Ext-LS$_{+,*}$ refutation $p_1 \geq 0, \dots, -M = p_k \geq 0$ of the system $f_1 \geq 0, f_1 \leq 0, \dots, f_k \geq 0, f_k \leq 0$. From Theorem 1(1) we know that there is a polynomial-size derivation of the equation $-M = p_k = \mathrm{VAL}(\overline{\mathrm{BIT}}(P_k))$. From Theorem 1(2) we know that the sign bit of $\overline{\mathrm{BIT}}(P_k)$ equals 0. These two facts together give us an instance of eBVP. See the full version for the details. □

2. Ext-PC$_{\mathbb{Z}}$ + eBVP *polynomially simulates* Ext-PC$_{\mathbb{Z}}^{\surd}$ + eBVP. Our strategy for the simulation of the square root rule is as follows.

- Suppose we want to derive $g = 0$ from $g^2 = 0$, for some polynomial g.
- We consider the bit representation $\overline{\mathrm{BIT}}(G^2)$ of g^2.
- Lemma 3 together with Lemma 2 provide a polynomial-size derivation of $\mathrm{VAL}(\overline{\mathrm{BIT}}(G^2)) = g^2$, thus we have $\mathrm{VAL}(\overline{\mathrm{BIT}}(G^2)) = 0$.
- From this, Lemma 4 provides a polynomial-size proof of $\overline{\mathrm{BIT}}(G^2) = \overline{0}$. Here we make use of eBVP.
- Now Lemma 6 provides a polynomial-size proof of $\overline{\mathrm{BIT}}(G) = \overline{0}$.
- From this we can derive that $g = \mathrm{VAL}(\overline{\mathrm{BIT}}(G)) = 0$.

In the full version we apply this strategy and hence get the following theorem.

Theorem 3. *Consider a system of polynomial equations* $f_1 = 0, \ldots, f_k = 0$*. If there is an* Ext-PC$_{\mathbb{Z}}^{\surd}$+eBVP *refutation for this system, then there is an* Ext-PC$_{\mathbb{Z}}$+ eBVP *refutation for the system* $f_1 = 0, \ldots, f_k = 0$ *that is at most polynomially longer wrt syntactic size than the initial refutation.*

6 eBVP Cannot be Used to Prove CNF Lower Bounds

Exponential lower bounds on the size of refutations of eBVP have been demonstrated for several proof systems including Ext-PC$^{\surd}$ [1]. However, they have a caveat: eBVP is not a translation of a Boolean formula in CNF. Is it still possible to use these bounds to prove an exponential lower bound for a formula in CNF? For example, one could provide a polynomial-size Ext-PC$^{\surd}$ derivation of a translation of an unsatisfiable Boolean formula in CNF from eBVP: together with the lower bound for eBVP, this would prove a bound for a formula in CNF. One could even introduce extension variables in order to describe such a formula.

In this section we show that this is not possible: any Ext-PC$^{\surd}$ derivation of an unsatisfiable CNF from eBVP$_n$ should have exponential size in n. We start with proving a lower bound over the integers. Then we extend this result to the rationals. The proof can be viewed as a generalization of the lower bound in [1]; however, the lower bound is proved not for the derivation of $M = 0$, but for the derivation of an arbitrary unsatisfiable CNF, possibly in the extension variables.

Formally, the derivation of an unsatisfiable formula in CNF from eBVP$_n$ in Ext-PC$_{\mathbb{Z}}^{\surd}$ starts with the equation $\sum_{i=1}^{n} x_i 2^{i-1} + M = 0$, the Boolean equations $x_i^2 - x_i = 0$, and the definitions of extension variables as axioms, and applies the rules of Polynomial Calculus and the square root rule until it reaches polynomial equations $C_1 \cdot p_1 = 0, \ldots, C_m \cdot p_m = 0$ and $C_1' \cdot (y_1^2 - y_1) = 0, \ldots, C_v' \cdot (y_v^2 - y_v) = 0$, where each C_i or C_j' is a nonzero integer constant and each p_i is the translation of a Boolean clause of the unsatisfiable CNF of the form

$$p_i = y_{j_1} \cdots y_{j_k} \cdot \neg y_{\ell_1} \cdots \neg y_{\ell_r}$$

for some extension variables y_t's and the "dual" extension variables $\neg y_t = 1 - y_t$. Since we work over the integers, we cannot assume that all C_i's and C_j''s equal

1 (we cannot divide), though if we derive polynomials multiplied by nonzero constants, it may still help in proving a lower bound for a CNF.

We start with formally defining how a substitution into the input variables changes polynomials that use extension variables:

Definition 14. *Suppose that that extension variables $y_1, \dots, y_m$ in an $\text{Ext-PC}_{\mathbb{Z}}^{\surd}$ derivation are introduced as $y_1 = q_1(x_1, \dots, x_n)$, $y_2 = q_2(x_1, \dots, x_n, y_1)$, $\dots$, $y_m = q_m(x_1, \dots, x_n, y_1, \dots, y_{m-1})$. Then for any variable y_i and any vector of bit values $\{b_1, \dots, b_n\} \in \{0,1\}^n$, we define substitution $y_i|_{x_1=b_1,\dots,x_n=b_n}$:*

- $y_1|_{x_1=b_1,\dots,x_n=b_n} := q_1(b_1, \dots, b_n)$.
- *For $i > 1$ we define* $y_i|_{x_1=b_1,\dots,x_n=b_n} := q_i(b_1, \dots, b_n, y_1|_{x_1=b_1,\dots,x_n=b_n}, \dots, y_{i-1}|_{x_1=b_1,\dots,x_n=b_n})$.

For any polynomial $f(x_1, \dots, x_n, y_1, \dots, y_m) \in \mathbb{Z}[\overline{x}, \overline{y}]$ we define $f|_{x_1=b_1,\dots,x_n=b_n}$ in the following way:

$$f|_{x_1=b_1,\dots,x_n=b_n} = f(b_1, \dots, b_n, y_1|_{x_1=b_1,\dots,x_n=b_n}, \dots, y_m|_{x_1=b_1,\dots,x_n=b_n}).$$

Before proving our bound, we observe a property of Boolean substitutions:

Lemma 7. *Consider an instance of* eBVP*: $M+x_1+2x_2+\dots+2^{n-1}x_n$. Consider any prime number $p < 2^n$ and the binary representation $b_1, \dots, b_k$ of any number $0 \le t < 2^n$ such that $t \equiv -M \pmod p$. Given an $\text{Ext-PC}_{\mathbb{Z}}^{\surd}$ derivation of the polynomial equation $f = 0$ from $M + x_1 + 2x_2 + \dots + 2^{n-1}x_n$ and the Boolean axioms $x_i^2 - x_i = 0$, the number $f|_{x_1=b_1,\dots,x_n=b_n}$ is divisible by p.*

Proof. After the substitution, the axioms (including definitions of new variables) are divisible by p. The statement follows by easy induction, see the full version.

Corollary 1. *Consider an instance of* eBVP *of the form $M + x_1 + 2x_2 + \dots + 2^{n-1}x_n$, any prime number $p < 2^n$ and the binary representation $b_1, \dots, b_k$ of any number $0 \le t < 2^n$ such that $t \equiv -M \pmod p$. Suppose we introduced extension variable y_i for which we have an $\text{Ext-PC}_{\mathbb{Z}}^{\surd}$ derivation of the polynomial equation $C' \cdot (y_i^2 - y_i) = 0$ from $M+x_1+2x_2+\dots+2^{n-1}x_n$. Then, either the number C' is divisible by p, or $y_i|_{x_1=b_1,\dots,x_n=b_n} \equiv 1 \pmod p$, or $y_i|_{x_1=b_1,\dots,x_n=b_n} \equiv 0 \pmod p$.*

Now we are ready to prove an exponential lower bound over the integers:

Theorem 4. *Given an $\text{Ext-PC}_{\mathbb{Z}}^{\surd}$ derivation of an unsatisfiable CNF from $M + x_1 + \dots + 2^{n-1}x_n = 0$ and the Boolean axioms, at least one of the following three conditions holds:*

- *The number of clauses in this CNF is at least $2^{n/3}$.*
- *We have derived a polynomial equation $C' \cdot (y_j^2 - y_j) = 0$ and the constant C' is divisible by at least $\Omega(2^{n/3})$ different prime numbers.*
- *There is a clause $C \cdot y_{j_1} \cdots y_{j_k} \cdot \neg y_{\ell_1} \cdots \neg y_{\ell_r}$ such that the constant C is divisible by at least $\Omega(2^{n/3})$ different prime numbers.*

Proof. Let $\mathcal{Y}$ be the set of variables occurring in our CNF.

Consider the set $\mathcal{P}$ of all prime numbers from $\{1, 2, \ldots, 2^n - 1\}$. Now consider any prime number $p \in \mathcal{P}$. As in Lemma 7, we can take an arbitrary $t \in \mathbb{Z}$, $0 \leq t < 2^n$, such that $t \equiv -M \pmod{p}$. Consider the binary representation $b_1, \ldots, b_n$ of this integer t. Corollary 1 says that for every $y_i \in \mathcal{Y}$ we have derived that $C'_i \cdot (y_i^2 - y_i) = 0$ and either C'_i is divisible by p, or $y_i|_{x_1=b_1,\ldots,x_n=b_n} \equiv 1 \pmod{p}$, or $y_i|_{x_1=b_1,\ldots,x_n=b_n} \equiv 0 \pmod{p}$. We fix now this particular equation for y_i in what follows.

Now suppose that for every $y_i \subset \mathcal{Y}$, the constant C'_i from equation $C'_i \cdot (y_i^2 - y_i) = 0$ is not divisible by p. Then we know that every number $y_i|_{x_1=b_1,\ldots,x_n=b_n}$ is 0/1 modulo p. Thus every number $\neg y_i|_{x_1=b_1,\ldots,x_n=b_n}$ is also 0/1 modulo p and

$$y_i|_{x_1=b_1,\ldots,x_n=b_n} \equiv 1 - \neg y_i|_{x_1=b_1,\ldots,x_n=b_n} \pmod{p}.$$

Then, since our CNF is *unsatisfiable*, we know that there is a clause $C \cdot y_{j_1} \cdots y_{j_k} \cdot \neg y_{\ell_1} \cdots \neg y_{\ell_r}$, such that

$$(y_{j_1} \cdots y_{j_k} \cdot \neg y_{\ell_1} \cdots \neg y_{\ell_r})|_{x_1=b_1,\ldots,x_n=b_n} \equiv 1 \pmod{p}.$$

On the other hand, from Lemma 7 we know that

$$C \cdot (y_{j_1} \cdots y_{j_k} \cdot \neg y_{\ell_1} \cdots \neg y_{\ell_r})|_{x_1=b_1,\ldots,x_n=b_n} \equiv 0 \pmod{p}.$$

Therefore, C is divisible by p.

Summarizing everything, we get that for every prime $p \in \mathcal{P}$ either we have derived a Boolean equation $C' \cdot (y^2 - y)$ where C' is divisible by p, or there is a clause $C \cdot y_{j_1} \cdots y_{j_k} \cdot \neg y_{\ell_1} \cdots \neg y_{\ell_r}$ where the constant C is divisible by p.

If the number of clauses in our CNF is at least $2^{n/3}$, the first condition of the theorem holds. Suppose we have derived an unsatisfiable CNF with less then $2^{n/3}$ clauses. Then we have less than $2^{n/3}$ different variables in our CNF since it is unsatisfiable, and we have derived less than $2^{n/3}$ equations of the form $C'_i \cdot (y_i^2 - y_i)$ and less than $2^{n/3}$ clauses of the form $C \cdot (y_{j_1} \cdots y_{j_k} \cdot \neg y_{\ell_1} \cdots \neg y_{\ell_r})$.

We showed that for any prime $p \in \mathcal{P}$ there is either an equation $C'_i \cdot (y_i^2 - y_i)$ such that C'_i is divisible by p or a clause $C \cdot (y_{j_1} \cdots y_{j_k} \cdot \neg y_{\ell_1} \cdots \neg y_{\ell_r})$ such that C is divisible by p. So, since the total number of those equations is less then $2^{n/3+1}$, there is a constant C (maybe $C = C'_i$) from one of those equations that is divisible by at least $\frac{|\mathcal{P}|}{2^{n/3+1}}$ prime numbers.

We know that the size of the set $\mathcal{P}$ is at least $C'' \cdot 2^n/n$ by the Prime Number Theorem for some constant C''. Thus the constant C should be divisible by at least $C'' \cdot \frac{2^n}{2^{n/3+1} \cdot n}$ prime numbers, which is sufficient to satisfy the second or the third condition of the theorem. □

Corollary 2. *Any* Ext-PC$_{\mathbb{Z}}^{\surd}$ *derivation of an unsatisfiable CNF in n variables from* eBVP$_n$ *requires size* $\Omega(2^{n/3})$.

Proof. If the number of clauses in this CNF is less than $2^{n/3}$, then by Theorem 4 there is a constant C in our derivation divisible by at least $\Omega\left(2^{n/3}\right)$ different prime numbers, hence C has $\Omega(2^{n/3})$ bits. □

A similar lower bound over the rationals is proved by translating derivations over rationals into derivations over integers (see the full version).

7 Further Research

A long-standing open question in semialgebraic proof complexity is to prove a superpolynomial lower bound for a rather weak proof system (called LS after Lovász and Schrijver), even for its most basic version [11] (degree two, no squares axioms ($f^2 \geq 0$), no extension variables, multiplication by x, $1 - x$ and nonnegative constants only). Recently lower bounds on very strong proof systems have been proved for systems of polynomial equations (based on eBVP) that do *not* come from Boolean formulas. Does this generalization help to prove superpolynomial lower bounds for polynomial inequalities, for example, for LS?

We have shown a polynomial simulation of $\text{Ext-LS}_{+,*}$ proofs in Ext-PC augmented by the eBVP rule, which was already known for stronger systems IPS vs CPS [2]. How can we weaken the basic system so that the statement remains true? For example, following [3] we can simulate binary arithmetic in logarithmic depth (by formulas), which, unfortunately, gives only $\log^2 n$ depth proofs. Is it possible to do better?

Acknowledgements. We are grateful to Ilario Bonacina and Dima Grigoriev for fruitful discussions, and to Yuval Filmus for his detailed comments on an earlier draft of this paper.

This research has been partially supported by the European Union's Horizon 2020 research and innovation programme under grant agreement No 802020-ERC-HARMONIC. The first author is supported by Lady Davys Fellowship.

References

1. Alekseev, Y.: A lower bound for polynomial calculus with extension rule. In: CCC-2021. Leibniz International Proceedings in Informatics (LIPIcs), vol. 200, pp. 21:1–21:18. Leibniz-Zentrum für Informatik, Dagstuhl, Germany (2021)
2. Alekseev, Y., Grigoriev, D., Hirsch, E.A., Tzameret, I.: Semi-algebraic proofs, IPS lower bounds and the τ-conjecture: Can a natural number be negative? In: STOC-2020. pp. 54–67 (2020). Technical details can be found in ECCC TR19-142
3. Buss, S.R.: Polynomial size proofs of the propositional pigeonhole principle. J. Symb. Log. **52**(4), 916–927 (1987)
4. Clegg, M., Edmonds, J., Impagliazzo, R.: Using the Groebner basis algorithm to find proofs of unsatisfiability. In: Proceedings of the 28th Annual ACM Symposium on the Theory of Computing (Philadelphia, PA, 1996), pp. 174–183. ACM (1996)
5. Cook, S.A., Reckhow, R.A.: The relative efficiency of propositional proof systems. J. Symb. Log. **44**(1), 36–50 (1979)
6. Grigoriev, D., Hirsch, E.A.: Algebraic proof systems over formulas. Theoret. Comput. Sci. **303**(1), 83–102 (2003)
7. Grigoriev, D., Hirsch, E.A., Pasechnik, D.V.: Complexity of semialgebraic proofs. Mosc. Math. J. **2**(4), 647–679, 805 (2002)
8. Grochow, J.A., Pitassi, T.: Circuit complexity, proof complexity, and polynomial identity testing: the ideal proof system. J. ACM **65**(6), 37:1–37:59 (2018)
9. Part, F., Thapen, N., Tzameret, I.: First-order reasoning and efficient semi-algebraic proofs. In: 36th Annual ACM/IEEE Symposium on Logic in Computer Science, LICS 2021, Rome, Italy, 29 June–2 July 2021, pp. 1–13. IEEE (2021)

10. Part, Fedor, Tzameret, Iddo: Resolution with counting: Dag-like lower bounds and different moduli. Comput. Complex. **30**(1), 1–71 (2021). https://doi.org/10.1007/s00037-020-00202-x
11. Pudlák, P.: On the complexity of the propositional calculus. In: Sets and Proofs (Leeds 1997). London Mathematical Society Lecture Note Series, vol. 258, pp. 197–218. Cambridge University Press, Cambridge (1999)
12. Tseitin, G.S.: On the complexity of derivation in the propositional calculus. Zapiski nauchnykh seminarov LOMI **8**, 234–259 (1968). English translation of this volume: Consultants Bureau, N.Y., 1970, pp. 115–125

Independent Set Under a Change Constraint from an Initial Solution

Yuichi Asahiro[1], Hiroshi Eto[2], Kana Korenaga[2], Guohui Lin[3], Eiji Miyano[2(✉)], and Reo Nonoue[2]

[1] Kyushu Sangyo University, Fukuoka, Japan
asahiro@is.kyusan-u.ac.jp
[2] Kyushu Institute of Technology, Iizuka, Japan
{eto,miyano}@ai.kyutech.ac.jp,
{korenaga.kana518,nonoue.reo265}@mail.kyutech.jp
[3] University of Alberta, Edmonton, Canada
guohui@ualberta.ca

Abstract. In this paper, we study a type of incremental optimization variant of the Maximum Independent Set problem (MaxIS), called Bounded-Deletion Maximum Independent Set problem (BD-MaxIS): Given an unweighted graph $G = (V, E)$, an initial feasible solution (i.e., an independent set) $S^0 \subseteq V$, and a non-negative integer k, the objective of BD-MaxIS is to find an independent set $S \subseteq V$ such that $|S^0 \setminus S| \leq k$ and $|S|$ is maximized. The original MaxIS is generally NP-hard, but, it can be solved in polynomial time for perfect graphs (and therefore, comparability, co-comparability, bipartite, chordal, and interval graphs). In this paper, we show that BD-MaxIS is NP-hard even if the input is restricted to bipartite graphs, and hence to comparability graphs. On the other hand, fortunately, BD-MaxIS on co-comparability, interval, convex bipartite, and chordal graphs can be solved in polynomial time. Finally, we study the computational complexity on very similar variants of the Minimum Vertex Cover and the Maximum Clique problems for graph subclasses.

1 Introduction

Background. Motivated by the practice-oriented research on the railroad blocking problem, the following general framework of *incremental optimization* problems with initial solutions was introduced [20]: Let P be an optimization problem with a starting feasible solution S^0, and let $\mathcal{F}$ be the set of all feasible solutions for P. For a new feasible solution $S \in \mathcal{F}$, the increment from S^0 to S is the amount of change given by a function $f(S, S^0) : \mathcal{F} \times \mathcal{F} \to \mathbb{R}$, which we refer to as the increment function. Suppose that k is a given bound on the total amount of change permitted. We call S an incremental solution if it satisfies the inequality $f(S, S^0) \leq k$. The goal is to find an incremental solution S^* that results in the maximum improvement in the objective function value.

M. Mavronicolas (Ed.): CIAC 2023, LNCS 13898, pp. 37–51, 2023.
https://doi.org/10.1007/978-3-031-30448-4_4

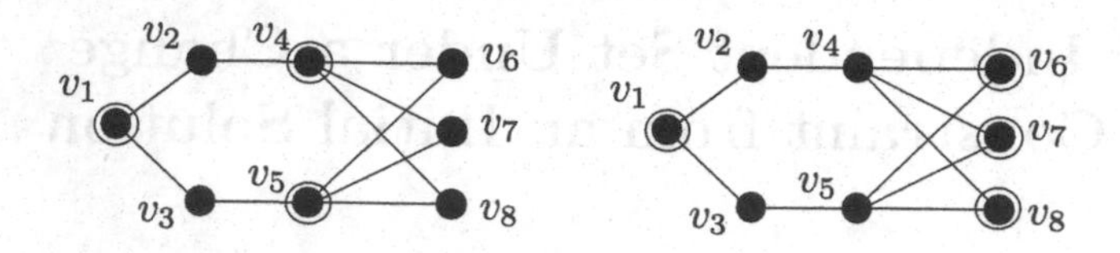

Fig. 1. Given a graph G, an initial solution $\{v_1, v_4, v_5\}$, and $k = 2$ as input (left), an optimal solution is $\{v_1, v_6, v_7, v_8\}$.

In this paper we study a type of incremental optimization of the MAXIMUM INDEPENDENT SET problem (MaxIS for short). The original MaxIS is one of the most important and most investigated combinatorial optimization problems in theoretical computer science. The input of MaxIS is an unweighted graph $G = (V, E)$, where V and E are the sets of vertices and edges in G, respectively. An *independent set* of G is a subset $S \subseteq V$ of vertices such that for every pair $u, v \in S$, the edge $\{u, v\}$ is not in E. The goal of MaxIS is to find an independent set of maximum cardinality. The problem MaxIS is a well-studied algorithmic problem, and actually it is one of the Karp's 21 fundamental NP-hard problems [14]. Furthermore, it is well known that MaxIS remains NP-hard even for substantial restricted graph classes such as cubic planar graphs [6], triangle-free graphs [19], and graphs with large girth [17]. Fortunately, however, it is also known that the problem can be solved in polynomial time if the input graph is restricted to, for example, graphs with constant treewidth [5] (and therefore, outerplanar, series-parallel, cactus graphs, and so on), perfect graphs [12] (and therefore, chordal [7], comparability [10], co-comparability, bipartite graphs, and so on), circular-arc graphs [8], and many other graph classes.

Our Problem and Contributions. Throughout this paper, we let S^0 and S denote an initial solution (i.e., an initial independent set) and a solution obtained by our algorithm. We define the increment function as $f(S, S^0) = |S^0 \setminus S|$, the number of vertices in S^0 but not in S, which is the number of deleted vertices from the initial solution S^0. The obtained solution S must satisfy the inequality $|S^0 \setminus S| \leq k$. That is, the number of vertices deleted from the initial solution S^0 is bounded by the given bound k. The function f can be seen as a "change-constraint" function. Now, we can define our problem as follows:

> BOUNDED-DELETION MAXIMUM INDEPENDENT SET (BD-MaxIS)
> **Input:** An unweighted graph $G = (V, E)$, an initial feasible solution (i.e., an independent set) $S^0 \subseteq V$, and a non-negative integer k.
> **Goal:** The goal is to find an independent set $S \subseteq V$ such that $|S^0 \setminus S| \leq k$ and $|S|$ is maximized.

See Fig. 1 for an example. If a graph G of eight vertices, an initial solution $\{v_1, v_4, v_5\}$, and $k = 2$ are given as input, then $\{v_1, v_6, v_7, v_8\}$ is an optimal solution, which is obtained by deleting two vertices $\{v_4, v_5\}$ and adding three vertices $\{v_6, v_7, v_8\}$. If $k = 1$, then the initial solution $\{v_1, v_4, v_5\}$ is optimal

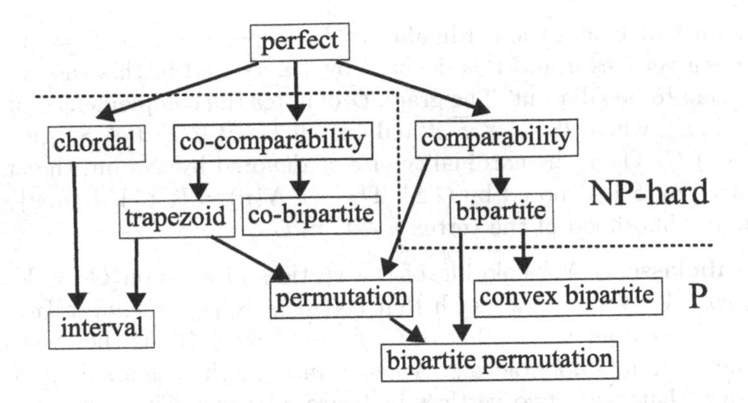

Fig. 2. Computational complexity of BD-MaxIS on graph classes. For example, "perfect → comparability" means that the perfect graph class is a superclass of the comparability graph class.

since one vertex-deletion does not make it possible to insert two or more new independent vertices.

One sees that BD-MaxIS is generally NP-hard since if $k \geq |S^0|$, then we can completely change the solution, and thus BD-MaxIS includes the classical MaxIS as a special case (or simply, MaxIS is the case where S^0 is empty and $k = 0$). Hence, our work focuses on the computational complexity of BD-MaxIS on polynomial-time solvable graph classes such as perfect, comparability, co-comparability, bipartite, chordal graphs, and so on.

Our main results are summarized in the following list and Fig. 2:

(1) BD-MaxIS is NP-hard even if the input is restricted to bipartite graphs. Since every bipartite graph is comparability and perfect, BD-MaxIS on comparability graphs, or perfect graphs is also NP-hard.
(2) BD-MaxIS can be solved in $O(k|V|^2)$ time for co-comparability graphs. If the input graph is an interval graph, then there is an $O(k|V| + |E|)$-time algorithm for BD-MaxIS.
(3) BD-MaxIS can be solved in $O(k|E|)$ time for convex bipartite graphs.
(4) BD-MaxIS can be solved in $O(k^2(|V| + |E|)^2)$ time for chordal graphs.

Other well-known graph classes including trapezoid, co-bipartite, permutation, and bipartite permutation are also polynomial-time solvable from the results (2), (3), and (4).

2 Preliminaries

Notation. Let $G = (V, E)$ be a simple (without multiple edge or self-loop edge), unweighted, and undirected graph, where V and E are sets of vertices and edges, respectively. We sometimes denote by $V(G)$ and $E(G)$ the vertex and the edge sets of G, respectively. Unless otherwise described, n and m denote the cardinality of V and the cardinality of E, respectively, for $G = (V, E)$. An edge between vertices u and v is denoted by $\{u, v\}$, and in this case vertices u and v are said to be adjacent. The graph $\overline{G}$ denotes the complement graph of G, i.e., $\overline{G} = (V, \overline{E})$, where $\{u, v\} \in \overline{E}$ if and only if $\{u, v\} \notin E$. Let $S \subseteq V$ be a set of vertices of G. Then, the cardinality of S is denoted by $|S|$ and the subgraph of G induced by S is denoted by $G[S]$. The set $N(u) = \{v \in V \mid \{u, v\} \in E\}$ is called the neighborhood of the vertex $u \in V$ in G.

Graph subclasses. A k-coloring of the vertices of a graph $G = (V, E)$ is a mapping $col : V \to \{1, \ldots, k\}$ such that $col(u) \neq col(v)$ whenever $\{u, v\}$ is an edge in G. The chromatic number of G, denoted by $\chi(G)$, is the least number k such that G admits a k-coloring. A *clique* in a graph G is a subset $S \subseteq V$ of vertices such that every two vertices in S are adjacent. The clique number of G, denoted by $\omega(G)$, is the number of vertices in a maximum clique of G. An independent set in a graph is a set of vertices no two of which are adjacent. The independence number of G, denoted by $\alpha(G)$, is the size of a largest independent set in G.

A graph G is called *perfect* if $\chi(H) = \omega(H)$ for every induced subgraph H of G. A graph is called *chordal* if every cycle of length at least four contains a chord, which is an edge that is not part of the cycle but connects two vertices of the cycle. A graph G is called *bipartite* if its chromatic number is at most two. Consider a bipartite graph G with the vertex set $V \cup W$ and its 2-coloring col, where V and W are the disjoint sets of vertices such that $col(V) = 1$ and $col(W) = 2$. The bipartite graph G is *convex* if the vertices in W can be ordered in such a way that, for each $v \in V$, the neighborhood $N(v)$ of v are consecutive in W. The ordering of the vertices in W is said to be *convex*, and G is said to be *convex with respect to* W. A graph G is called *co-bipartite* if its complement graph $\overline{G}$ is bipartite. A graph is called *comparability* if there exists a partial order $<_\sigma$ on its vertices such that two vertices u and v are adjacent in the graph if and only if $u <_\sigma v$ or $v <_\sigma u$. A graph G is called *co-comparability* if its complement graph $\overline{G}$ is a comparability graph. A graph is called *permutation* if it can be represented by a permutation $\pi : \{1, \ldots, n\} \to \{1, \ldots, n\}$ in such a way that two vertices $i < j$ are adjacent if and only if $\pi(i) > \pi(j)$. A graph is called *bipartite permutation* if it is both bipartite and permutation.

3 NP-Hardness of BD-MaxIS on Bipartite Graphs

Given an unweighted graph G, the goal of the MAXIMUM CLIQUE problem (MaxClique) is to find a clique $Q \subseteq V$ of maximum cardinality [14]. Let q-Clique be the decision version of MaxClique, i.e., given a graph G and an integer q, q-Clique is to determine if there is a clique of size q in G:

Theorem 1. *BD-MaxIS is NP-hard even if the input is restricted to bipartite graphs.*

Proof. We show that the NP-complete problem q-Clique is polynomial-time reducible to BD-MaxIS on bipartite graphs. Suppose that the input of q-Clique is $G^0 = (V^0, E^0)$, where $V^0 = \{v_1^0, \dots, v_n^0\}$ of n vertices and $E^0 = \{e_1^0, \dots, e_m^0\}$ of m edges. Then, we construct the following bipartite graph $G = (V_v \cup V_e, E)$ of BD-MaxIS by subdividing every edge in E^0 to two edges:

$$\begin{aligned}
V_v &= \{v_1, v_2, \dots, v_n\}, \\
V_e &= \{e_1, e_2, \dots, e_m\}, \text{ and} \\
E &= \{\{v_i, e_s\}, \{v_j, e_s\} \mid e_s^0 = \{v_i^0, v_j^0\} \in E^0\}.
\end{aligned}$$

That is, the constructed graph G is so-called an *incidence graph* of G^0, and thus G must be bipartite. Then, we set an initial solution $S^0 = V_v$ and an integer $k = q$. This completes the reduction. One sees that each edge in E connects a vertex in V_v with a vertex in V_e. Therefore, $S^0 = V_v$ must be a (feasible) independent set. The reduction can be clearly executed in polynomial time.

For the above construction of G, we show that G contains an independent set S such that $|S^0 \backslash S| \leq k$ and $|S| \geq |V| - k + k(k-1)/2$ if and only if G^0 contains a clique Q^0 such that $|Q^0| \geq q$.

(1) Suppose that G^0 contains a clique Q^0 of size q, and $Q^0 = \{v_1^0, \dots, v_q^0\}$, without loss of generality. Then, let $R = \{v_1, \dots, v_q\}$ be the subset of the corresponding q vertices in the initial independent set V_v. Since there must be an edge between every pair of v_i^0 and v_j^0 in Q^0 of G^0, we can find a set, say, A, of $q(q-1)/2$ isolated vertices in V_e by deleting all the vertices in R corresponding to Q^0. Let $S = (S^0 \backslash R) \cup A$. One can see that (i) $S^0 \backslash S = R$, and thus $|S^0 \backslash S| = q = k$, and (ii) $S \backslash S^0 = A$ and $|S \backslash S^0| = q(q-1)/2 = k(k-1)/2$. Namely, $|S| = |V| - k + k(k-1)/2$.

(2) Suppose that the size of a maximum clique in G^0 is at most $q-1$. Let $R = \{v_1, \dots, v_q\}$ be an arbitrary subset of q vertices in the initial independent set V_v. Then, we consider the corresponding set $R^0 = \{v_1^0, \dots, v_q^0\}$ of q vertices in G^0 of q-Clique and the subgraph $G[R^0]$ induced by R^0 in G^0. Since the size of the maximum clique in G is at most $q-1$, $G[R^0]$ contains at most $q(q-1)/2-1$ edges. It follows that we can only obtain the new independent set of at most $q(q-1)/2-1 = k(k-1)/2-1$ vertices by deleting any subset of $q = k$ vertices from V_v, i.e., the size of any independent set is at most $|V| - k + k(k-1)/2 - 1$. This completes the proof. □

Since comparability graphs and perfect graphs are superclasses of bipartite graphs [11], we obtain the following corollary:

Corollary 1 *BD-MaxIS is NP-hard even if the input is restricted to comparability graphs, or perfect graphs.*

4 Polynomial-Time Solvable Graph Subclasses of BD-MaxIS

4.1 Co-comparability Graphs

In this section, for BD-MaxIS on co-comparability graphs, we design a polynomial-time algorithm, while BD-MaxIS on perfect graphs is NP-hard as

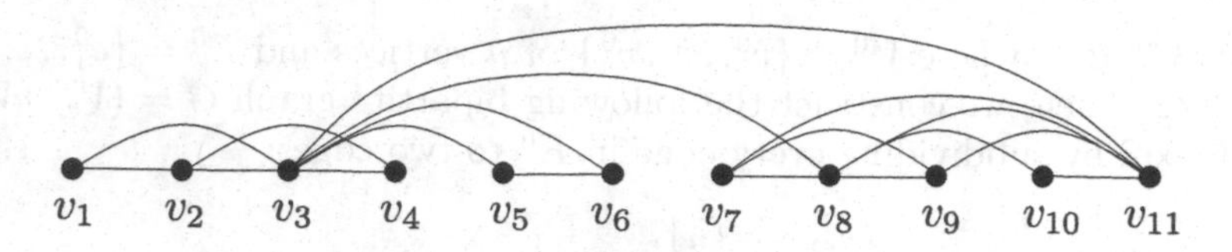

Fig. 3. Umbrella-free vertex ordering. For example, consider three vertices v_3, v_7, and v_8. Since there is an edge between v_3 and v_8, there is an edge $\{v_7, v_8\}$.

shown in the previous section. Before the detailed description of our algorithm ALG_CoC, we give the *vertex ordering characterization* of co-comparability graphs.

Vertex ordering characterization. A *vertex ordering* of $G = (V, E)$ is a bijection $\sigma : V \leftrightarrow \{1, 2, \ldots, n\}$, i.e., for $v \in V$, $\sigma(v)$ denotes the unique position of v in σ, $\sigma(u) \neq \sigma(v)$ for $u \neq v$. For two vertices u and v, we write that $u <_\sigma v$ if and only if $\sigma(u) < \sigma(v)$. For two vertices $u, v \in V$, we say that u is left (resp., right) to v in σ if $u <_\sigma v$ (resp., $v <_\sigma u$). A *vertex ordering characterization* is an ordering on the vertices of a graph that satisfies certain properties. If every $G \in \mathcal{G}$ has a total ordering of its vertices that satisfies some property, then we say that the graph class $\mathcal{G}$ has a vertex ordering characterization on the property, which is often used to design polynomial-time algorithms. The co-comparability graph has the following vertex ordering characterization:

Proposition 1 ([15]). *A graph $G = (V, E)$ is a* co-comparability *graph if and only if there exists a vertex ordering σ of its vertices such that for every triple of vertices u, v, and w such that if $u <_\sigma v <_\sigma w$ and $\{u, w\} \in E$, then $\{u, v\} \in E$ or $\{v, w\} \in E$ (or both).*

The vertex ordering σ that satisfies the above proposition is called an *umbrella-free ordering* since σ does not contain an *umbrella*, which is a triple of vertices $u <_\sigma u <_\sigma w$ with $\{u, w\} \in E$ but $\{u, v\}, \{v, w\} \notin E$. For example, see Fig. 3. McConnell and Spinrad presented an algorithm to compute such a vertex ordering in $O(n + m)$ time [16].

Algorithm. Our algorithm ALG_CoC for BD-MaxIS on co-comparability graphs is based on a dynamic programming along the vertex ordering of co-comparability graphs. Given a co-comparability graph $G = (V, E)$, we first compute an umbrella-free vertex ordering σ of V in $O(n + m)$ time. Suppose that the ordering σ is $v_1 <_\sigma v_2 <_\sigma \cdots <_\sigma v_n$. In order to make the description of our algorithm easier, we add an isolated dummy vertex v_0 so that $v_0 <_\sigma v_1$ into the leftmost position (i.e., the 0th position). Let $V_{i..j} = \{v_i, v_{i+1}, \ldots, v_j\}$ be the set of the $j - 1 + 1$ consecutive vertices, v_i through v_j. Also, let $N_L(v_i) = N(v_i) \cap V_{0..(i-1)} = \{v_j \in V_{0..(i-1)} \mid \{v_i, v_j\} \in E\}$ is called the left neighborhood of v_i. Let δ_i be the subscript of the leftmost vertex in $N_L(v_i)$. If $N_L(v_i) = \emptyset$, then $\delta_i = i$. Let $\overline{N}_L(v_i) = \{v_j \in V_{\delta_i..(i-1)} \mid \{v_j, v_i\} \notin E\}$. See Fig. 3 again. For example, $N_L(v_{11}) = \{v_3, v_7, v_8, v_9, v_{10}\}$, $\overline{N}_L(v_{11}) = \{v_4, v_5, v_6\}$, and $\delta_{11} = 3$.

Now, consider the following two values *pick* and j: The former value $pick \in \{0,1\}$ indicates whether v_i is picked into a (partial) solution S or not. The latter value $j \in \{0,1,\ldots,k\}$ indicates the number of deleted vertices from the initial solution S^0 in order to construct S. For the ith vertex v_i, we define $IS(i, pick, j)$ to be the value of a maximum independent set in the induced subgraph $G[V_{1..i}]$ satisfying the following: (i) If $pick = 1$, then a partial solution S for $G[V_{1..i}]$ includes the ith vertex v_i; otherwise, S does not include v_i. (ii) The number $|(V_{1..i} \cap S^0)\setminus S|$ of deleted vertices so far is exactly j.

Let $\#S(i_1, i_2)$ be the number of vertices in $S^0 \cap V_{i_1..i_2}$, i.e., the number of vertices in $\{v_{i_1}, \ldots, v_{i_2}\}$ which are picked into the initial solution S^0. Initially we set $IS(0, pick, j) = 0$ for $pick = 0, 1$, and $j = 0, 1, \ldots k$. The recursive formula of our DP-based algorithm ALG_CoC is divided into the following two cases, (Case 1) v_i is not in the initial solution S^0, i.e., $v_i \notin S^0$, and (Case 2) v_i is in S^0, i.e., $v_i \in S^0$.

(Case 1) Suppose that $v_i \notin S^0$. The recursive formula is defined as follows:

$$IS(i, pick, j) = \begin{cases} \max\{IS(i-1,0,j), IS(i-1,1,j)\} \\ \qquad\qquad\qquad\qquad \text{if } pick = 0; \\ 1 + \max\Big\{ \max_{v_\ell \in \overline{N}_L(v_i)} \{IS(\ell, 1, j - \#S(\ell+1, i-1))\}, \\ \qquad\qquad IS(\delta_i, 0, j - \#S(\delta_i+1, i-1))\Big\} \\ \qquad\qquad\qquad\qquad \text{if } pick = 1 \text{ and } \delta_i \neq i; \\ 1 + \max\{IS(i-1,0,j), IS(i-1,1,j)\} \\ \qquad\qquad\qquad\qquad \text{if } pick = 1 \text{ and } \delta_i = i. \end{cases}$$

(1) Consider the case where v_i is not picked into the solution S. Then, the number $|(V_{1..i} \cap S^0)\setminus S|$ of deleted vertices at v_i is equal to the number $|(V_{1..(i-1)} \cap S^0)\setminus S|$ at v_{i-1}. Furthermore, one sees that the value of the maximum independent set in the induced subgraph $G[V_{1..i}]$ is equal to the value of a maximum independent set in the induced subgraph $G[V_{1..(i-1)}]$.

(2) Suppose that v_i is picked into the solution S. Then, the value of the maximum independent set in $G[V_{1..i}]$ increases by one. One sees that all the left neighborhood of v_i cannot be picked into S, but $v_\ell \in \overline{N}_L(v_i)$ can be possibly picked into S since v_ℓ is not adjacent to v_i. (i) If all the vertices $v_\ell \in \overline{N}_L(v_i)$ are not picked into S, then $IS(i, 1, j)$ (now $pick = 1$) is equal to the value of a maximum independent set in the induced subgraph $G[V_{1..\delta_i}]$ which is stored into $IS(\delta_i, 0, j - \#S(\delta_i, i-1))$ since all the vertices of $S \cap V_{\delta_i, i-1}$ must *not* be included in the solution S. (ii) For ease of exposition, take a look at five vertices v_3, v_4, v_5, v_6, and v_{11} in Fig. 3. If v_{11} is in S, then v_3 is not in S. Suppose that v_4 and v_6 in $\overline{N}_L(v_{11})$ is picked into S and v_5 is not in S. Since v_5 is not in S, $IS(5, 0, j)$ can be obtained from $\max\{IS(4,0,j), IS(4,1,j)\}$ if v_5 is in the initial solution S^0, and from $\max\{IS(4,0,j-1), IS(4,1,j-1)\}$ if v_5 is not in S^0. That is, if v_5 is not in S, then the current information of

v_5 can be obtained from the information of the left vertex v_4. Therefore, it is enough to verify the information of $v_\ell \in \overline{N}_L(v_i)$ only when v_ℓ is picked into S. This is the main reason why our DP-based algorithm works in polynomial time if the vertex ordering characterization is umbrella-free.

(3) Suppose that v_i is picked into the solution S, and $N_L(v_i) = \emptyset$. Then, $IS(i, pick, j)$ can be computed from the two values $IS(i-1, 0, j)$ and $IS(i-1, 1, j)$ of the left vertex v_{i-1}.

(Case 2) Suppose that $v_i \in S^0$. One sees that "v_i is not picked" means that v_i must be deleted from the initial solution S^0. The recursive formula is almost the same as the formula in (Case 1), but, the number of deleted vertices is different if v_i is not picked into the solution S:

$$IS(i, pick, j) = \begin{cases} \max\{IS(i-1, 0, j-1), IS(i-1, 1, j-1)\} \\ \qquad\qquad\qquad\qquad \text{if } pick = 0; \\ 1 + \max\Big\{\max_{v_\ell \in \overline{N}_L(v_i)} \{IS(\ell, 1, j - \#S(\ell+1, i-1))\}, \\ \qquad\qquad IS(\delta_i, 0, j - \#S(\delta_i + 1, i-1))\Big\} \\ \qquad\qquad\qquad\qquad \text{if } pick = 1 \text{ and } \delta_i \neq i; \\ 1 + \max\{IS(i-1, 0, j), IS(i-1, 1, j)\} \\ \qquad\qquad\qquad\qquad \text{if } pick = 1 \text{ and } \delta_i = i. \end{cases}$$

Our algorithm ALG_CoC computes the value of $IS(i, pick, j)$ and stores it into a three-dimensional table IS of size $(n+1) \times 2 \times (k+1) = O(kn)$. Then, finally, ALG_CoC returns $\max_{0 \le j \le k} \{IS(n, 0, j), IS(n, 1, j)\}$.

Theorem 2. *Given an n-vertex co-comparability graph G and a non-negative integer k, BD-MaxIS can be solved in $O(kn^2)$ time.*

Proof. Given the co-comparability graph G, we can obtain its umbrella-free ordering in $O(n^2)$ time by using the method proposed in [16]. Clearly, each table entry takes $O(n)$ time to compute. Since the table size is $O(kn)$, the running time of ALG_CoC is $O(kn^2)$. □

4.2 Interval Graphs

Since every interval graph is co-comparability, BD-MaxIS on interval graphs can be solved in $O(kn^2)$ time by ALG_CoC. Fortunately, however, we can provide a faster algorithm ALG_Int if the following vertex ordering characterization of interval graphs, known as an *interval ordering*, is given:

Proposition 2 ([18]). *A graph $G = (V, E)$ is an interval graph if and only if there exists an ordering σ of its vertices such that for every triple of vertices u, v, and w such that if $u <_\sigma v <_\sigma w$ and $\{u, w\} \in E$, then $\{u, v\} \in E$.*

Theorem 3. *Suppose that we are given the interval ordering of an n-vertex interval graph G and a non-negative integer k as input. Then, BD-MaxIS can be solved in $O(kn)$ time. (The proof will appear in the full version of this paper).*

Since the interval ordering of interval graphs with n vertices and m edges can be obtained in $O(n+m)$ [3], we obtain the following corollary:

Corollary 2. *Given an interval graph with n vertices and m edges, and a non-negative integer k, BD-MaxIS can be solved in* $O(kn+m)$ *time.*

4.3 Convex Bipartite Graphs

As shown in Sect. 3, BD-MaxIS on bipartite graphs is NP-hard. One of the famous subclasses of bipartite graphs is the convex bipartite graph class. In this section we show that BD-MaxIS on convex bipartite graphs can be solved in polynomial-time. Here, we give our notation and additional terminology.

Let $G = (V, W, E)$ be a convex bipartite graph with respect to W. Suppose that V and W have n_1 and n_2 vertices, $V = \{v_1, v_2, \ldots, v_{n_1}\}$ and $W = \{w_1, w_2, \ldots, w_{n_2}\}$, where the convex vertex ordering σ is $w_1 <_\sigma w_2 <_\sigma \cdots <_\sigma w_{n_2}$. The vertex ordering of vertices in V is given later. See Fig. 4. For example, the neighborhood $N(v_3) = \{w_3, w_4, w_5, w_6, w_7\}$ of v_3 contains five consecutive vertices. Let w_i^ℓ and w_i^r be the leftmost and the rightmost vertices in $N(v_i)$ of v_i, respectively. Assume that n_1 vertices in $V = \{v_1, \ldots, v_{n_1}\}$ are sorted such that $w_1^r <_\sigma \cdots <_\sigma w_{n_1}^r$ holds by the vertex ordering σ, with ties broken arbitrarily. The ordering can be computed in $O(n_1 \log n_1)$. For the convex bipartite graph in Fig. 4, $w_1^r = w_2^r = w_5$, $w_3^r = w_7$, $w_4^r = w_9$, and $w_5^r = w_{10}$. As for the ith vertex v_i, $e_v^\ell(i) = \{v_i, w_i^\ell\}$ and $e_v^r(i) = \{v_i, w_i^r\}$ are called the leftmost and the rightmost edges of v_i, respectively. The other edges are called middle edges of v_i. If v_i is incident to one edge only, then the edge is also regarded as the rightmost edge. Now we define a mapping $right_v : E \to \{0,1\}$ such that if an edge e is the rightmost edge, then $right_v(e) = 1$; otherwise, $right_v(e) = 0$. Similarly, let v_i^ℓ and v_i^r be the leftmost and the rightmost vertices in $N(w_i)$ of w_i, respectively. As for the ith vertex w_i, $e_w^\ell(i) = \{v_i^\ell, w_i\}$ and $e_w^r(i) = \{v_i^r, w_i\}$ are called the leftmost and the rightmost edges of w_i, respectively. The other edges are called middle edges of w_i. If w_i is incident to one edge only, then the edge is regarded as the leftmost edge. Again, we define a mapping $left_w : E \to \{0,1\}$ such that if an edge e is the leftmost edge, then $left_w(e) = 1$; otherwise, $left_w(e) = 0$. Take a look at w_5 in Fig. 4. One sees that the neighborhood $N(w_5)$ of w_5 is v_1, v_2, v_3, and v_4. Then, for example, $right_v(\{v_1, w_5\}) = 1$ and $right_v(\{v_2, w_5\}) = 1$, but, $right_v(\{v_3, w_5\}) = 0$. Also, $left_w(\{v_1, w_5\}) = 1$.

Algorithm. Our algorithm ALG_CB for BD-MaxIS on convex bipartite graphs with respect to W follows the convex ordering of W, roughly from the leftmost edge to the rightmost edge. More precisely, ALG_CB uses the *edge* ordering σ_e such that $\{w_1, v_{1_1}\} <_{\sigma_e} \{w_1, v_{1_2}\} <_{\sigma_e} \cdots <_{\sigma_e} \{w_1, v_{1_{|N(w_1)|}}\} <_{\sigma_e} \{w_2, v_{2_1}\} <_{\sigma_e} \cdots <_{\sigma_e} \{w_2, v_{2_{|N(w_2)|}}\} <_{\sigma_e} \cdots <_{\sigma_e} \{w_{n_2}, v_{n_{2|N(w_{n_2})|}}\}$, where $N(w_i) = \{v_{i_1}, \ldots, v_{i_{|N(w_i)|}}\}$ and $v_{i_1} <_\sigma \cdots <_\sigma v_{i_{|N(w_i)|}}$ for $1 \le i \le n_2$. That is, the leftmost $|N(w_1)|$ edges of the edge ordering are incident to w_1, the next $|N(w_2)|$ edges are incident to w_2, and so on.

Let $[pick_i, pick_{i_q}] \in \{[0,0], [0,1], [1,0]\}$ be a status of the edge $\{w_i, v_{i_q}\}$ such that if $pick_i = 1$ (resp., $pick_i = 0$), then w_i is picked (resp, not picked) into

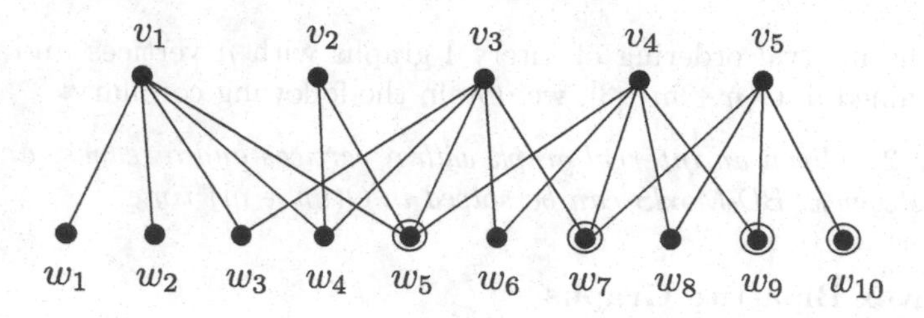

Fig. 4. Convex bipartite

the solution S and if $pick_{i_q} = 1$ (resp., $pick_{i_q} = 0$), then v_{i_q} is picked (resp, not picked) into the solution S for $1 \leq i \leq n_2$ and $1 \leq q \leq |N(w_i)|$. For the ith vertex w_i, we define $IS([i, i_q], [pick_i, pick_{i_q}], j)$ to be the value of a maximum independent set in the induced subgraph $G[\{w_1, \ldots, w_i\} \cup N(w_1) \cup \ldots \cup N(w_{i-1}) \cup \{v_{i_1}, \ldots, v_{i_q}\}]$ satisfying that the number of deleted vertices is exactly j, where $[pick_i, pick_{i_q}]$ is $[0,0]$, $[0,1]$, or $[1,0]$.

In order to make the description of our algorithm easier, we add two dummy vertices v_0 and w_0 into the leftmost positions in V and W, respectively. Initially we set $IS([0,0],[0,0],j) = IS([0,0],[0,1],j) = IS([0,0],[1,0],j) = 0$ for $j = 0, 1, \ldots, k$.

Consider a vertex v_i and its neighbor vertices in $N(v_i)$. If v_i is in S, then any vertex in $N(v_i)$ cannot be picked into S. Conversely, if at least one vertex in $N(v_i)$ is in S, then v_i cannot be picked into S. See Fig. 4 again. Consider six vertices $W_{1..6} = \{w_1, w_2, w_3, w_4, w_5, w_6\}$ and their four neighbor vertices $V_{1..4} = \{v_1, v_2, v_3, v_4\}$, and also 13 edges between $W_{1..6}$ and $V_{1..4}$. If every vertex in $W_{1..6}$ is fixed to be picked or not into S, then the status $v_1 \in S$ or $v_1 \notin S$, and $v_2 \in S$ or $v_2 \notin S$ can be fixed since $\{v_1, w_5\}$ and $\{v_2, w_5\}$ are the rightmost edges and $w_5 <_\sigma w_6$. On the other hand, for example, the status $v_3 \in S$ or $v_3 \notin S$ cannot be fixed since it depends on whether w_7 is picked or not into S. Therefore, roughly speaking, as for vertices in W, (i) if w_i is picked into S, then the size of S increases by one; on the other hand, as for vertices in V, (ii) if any neighbor vertex in $N(v_i) \setminus \{w_i^r\}$ are not picked into S, then the size of S is incremented when $w_i^r \in S$ is determined.

The recursive formula of our DP-based algorithm `ALG_CB` is divided into the following three cases, (Case 1) $w_i, v_{i_q} \notin S^0$, (Case 2) $w_i \notin S^0$ but $v_{i_q} \in S^0$, and (Case 3) $w_i \in S^0$ but $v_{i_q} \notin S^0$. Furthermore, each of the three cases (Case 1), (Case 2), and (Case 3) has four sub-cases $(right_v(\{w_i, v_{i_q}\}), left_w(\{w_i, v_{i_q}\})) = (0,0)$, $(0,1)$, $(1,0)$, and $(1,1)$. Note that if an edge $\{i, i_q\} \notin E$, then we set $IS([i, i_q], [pick_i, pick_{i_q}], j) = 0$ in the right-hand side of the recursive formula. Here we show only (Case 1) since (Case 2) and (Case 3) are very similar to (Case 1); (Case 2) and (Case 3) will appear in the full version of this paper.

(Case 1) $w_i, v_{i_q} \notin S^0$.

(i) Suppose that $right_v(\{w_i, v_{i_q}\}) = 0$ and $left_w(\{w_i, v_{i_q}\}) = 0$.

$$IS([i,i_q],[pick_i,pick_{i_q}],j)$$
$$=\begin{cases}\max\{IS([i,i_{q-1}],[0,0],j),IS([i,i_{q-1}],[0,1],j),\\ \qquad IS([i-1,i_q],[0,0],j),IS([i-1,i_q],[1,0],j)\} & \text{if } pick_i=0 \text{ and } pick_{i_q}=0\\ \max\{IS([i,i_{q-1}],[1,0],j),IS([i-1,i_q],[0,0],j),\\ \qquad IS([i-1,i_q],[1,0],j)\} & \text{if } pick_i=1 \text{ and } pick_{i_q}=0\\ \max\{IS([i,i_{q-1}],[0,0],j),IS([i,i_{q-1}],[0,1],j),\\ \qquad IS([i-1,i_q],[0,1],j)\} & \text{if } pick_i=0 \text{ and } pick_{i_q}=1\end{cases}$$

(ii) Suppose that $right_v(\{w_i,v_{i_q}\})=1$ and $left_w(\{w_i,v_{i_q}\})=0$.

$$IS([i,i_q],[pick_i,pick_{i_q}],j)$$
$$=\begin{cases}\max\{IS([i,i_{q-1}],[0,0],j),IS([i,i_{q-1}],[0,1],j),\\ \qquad IS([i-1,i_q],[0,0],j),IS([i-1,i_q],[1,0],j)\} & \text{if } pick_i=0 \text{ and } pick_{i_q}=0\\ \max\{IS([i,i_{q-1}],[1,0],j),IS([i-1,i_q],[0,0],j),\\ \qquad IS([i-1,i_q],[1,0],j)\} & \text{if } pick_i=1 \text{ and } pick_{i_q}=0\\ 1+\max\{IS([i,i_{q-1}],[0,0],j),IS([i,i_{q-1}],[0,1],j),\\ \qquad IS([i-1,i_q],[0,1],j)\} & \text{if } pick_i=0 \text{ and } pick_{i_q}=1\end{cases}$$

(iii) Suppose that $right_v(\{w_i,v_{i_q}\})=0$ and $left_w(\{w_i,v_{i_q}\})=1$.

$$IS([i,i_q],[pick_i,pick_{i_q}],j)$$
$$=\begin{cases}\max\{IS([i-1,i_q],[0,0],j),IS([i-1,i_q],[1,0],j)\} & \text{if } pick_i=0 \text{ and } pick_{i_q}=0\\ 1+\max\{IS([i-1,i_q],[0,0],j),IS([i-1,i_q],[1,0],j)\} & \text{if } pick_i=1 \text{ and } pick_{i_q}=0\\ IS([i-1,i_q],[0,1],j) & \text{if } pick_i=0 \text{ and } pick_{i_q}=1\end{cases}$$

(iv) Suppose that $right_v(\{w_i, v_{i_q}\}) = 1$ and $left_w(\{w_i, v_{i_q}\}) = 1$.

$$IS([i, i_q], [pick_i, pick_{i_q}], j) = \begin{cases} \max\{IS([i-1, i_q], [0,0], j), IS([i-1, i_q], [1,0], j)\} \\ \qquad\qquad \text{if } pick_i = 0 \text{ and } pick_{i_q} = 0 \\ 1 + \max\{IS([i-1, i_q], [0,0], j), IS([i-1, i_q], [1,0], j)\} \\ \qquad\qquad \text{if } pick_i = 1 \text{ and } pick_{i_q} = 0 \\ 1 + IS([i-1, i_q], [0,1], j) \\ \qquad\qquad \text{if } pick_i = 0 \text{ and } pick_{i_q} = 1 \end{cases}$$

Theorem 4. *Given an m-edge convex bipartite graph G and a non-negative integer k, BD-MaxIS can be solved in $O(km)$ time.*

Proof. Our algorithm `ALG_CB` computes the value of $IS([i, i_q], [pick_i, pick_{i_q}], j)$ and stores it into a two-dimensional table IS of size $(m+1) \times 3 \times (k+1) = O(km)$. Then, finally, `ALG_CB` returns

$$\max_{0 \le j \le k} \Big\{ IS([n_2, n_1], [0,0], j), IS([n_2, n_1], [0,1], j), IS([n_2, n_1], [1,0], j) \Big\}.$$

Since each table entry takes $O(1)$ time to compute, the running time of `ALG_CB` is $O(km)$. □

4.4 Chordal Graphs

The class of chordal graphs is one of the important subclasses of perfect graphs. Indeed, chordal graphs have attracted interest in graph theory since several combinatorial optimization problems that are intractable turn to be tractable on chordal graphs. In this section we provide a polynomial-time algorithm for BD-MaxIS on chordal graphs, which is again based on a dynamic programming for the *clique tree* representation of chordal graphs.

Clique Tree. Let $\mathcal{Q}_G$ be the set of all maximal cliques in a graph G, and let $\mathcal{Q}_v \subseteq \mathcal{Q}_G$ be the set of all maximal cliques that contain a vertex $v \in V(G)$. It is known [4,9] that G is chordal if and only if there exists a tree $T = (\mathcal{Q}_G, E(T))$ such that each node[1] of T corresponds to a maximal clique in $\mathcal{Q}_G$ and T has the *induced subtree property*, i.e., the subtree $T[\mathcal{Q}_v]$ induced by $\mathcal{Q}_v$ is connected for every vertex $v \in V(G)$. Such a tree is called a *clique tree* of G, and it can be constructed in linear time [1]. Given a chordal graph $G = (V, E)$, we construct a clique tree T and then a *rooted* clique tree $T(Q_r)$ of G by selecting an arbitrary node in T as a root Q_r. For the rooted clique tree $T(Q_r)$ of G and a node Q_i in $T(Q_r)$, $T(Q_i)$ represents the subtree rooted at Q_i. See Fig. 5, where the left is a chordal graph G of 11 vertices, and the right is its rooted clique tree $T(Q_r)$.

[1] We will refer to a node in a tree in order to distinguish it from a vertex in a graph.

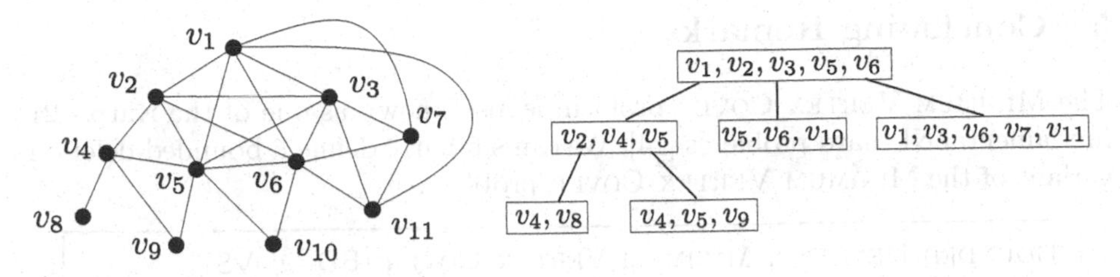

Fig. 5. (Left) Chordal graph G, and (Right) its rooted clique tree T_r with a root $G[\{v_1, v_2, v_3, v_4, v_5\}]$

Let $V(Q_i)$ and $V(T(Q_i))$ be the set of vertices in the node Q_i and the union of vertices in all nodes of the subtree $T(Q_i)$ rooted at Q_i, respectively.

In this paper, we consider a *weak clique tree* representation of a chordal graph [13]. Each node of the original clique tree must be a maximal clique, but each node of the weak clique tree is just a clique. It is known [13] that every chordal graph $G = (V, E)$ has a weak clique tree T such that T is a *binary* tree with $O(n)$ nodes and the sum of all cardinalities of its nodes is $O(n + m)$. Furthermore, every weak clique tree of a chordal graph is still a *tree decomposition*, i.e., satisfies the induced subtree property. Therefore, the dynamic programming using the weak clique tree works well.

Algorithm. Given a chordal graph $G = (V, E)$, we first compute a rooted weak clique tree $T(Q_r)$ of G in $O(n+m)$ time. For the sake of notational convenience, let $T = T(Q_r)$, $T_i = T(Q_i)$, and let $V_T = \{Q_1, Q_2, \ldots, Q_{|V_T|}\}$ be the set of nodes in T. Suppose that Q_{i_ℓ} and Q_{i_r} in V_T respectively are the left and the right children of Q_i, if exist. Recall that $V(T_i)$ $(= V(T(Q_i)))$ is the union of vertices in all nodes in the subtree T_i rooted at Q_i, and $V(T) = V(G)$.

Let S_{T_i} be an independent set in the subtree induced by $V(T_i)$, i.e., S_{T_r} is an independent set S of G. For a node Q_i in T, $S_i \subseteq V(Q_i)$, and $j_i \in \{0, \ldots, k\}$, we define $IS(i, S_i, j_i)$ to be the maximum size of the independent set S_{T_i} in T_i satisfying that $S_{T_i} \cap V(Q_i) = S_i$ and the number $|S^0 \setminus S_{T_i}|$ of deleted vertices from T_i is exactly j_i. A high level description of the recursive formula used by our algorithm `ALG_Cho` is very similar to the formula given in [2], although we have to count the number of deleted vertices $|S^0 \setminus S_{T_i}|$: The algorithm `ALG_Cho` computes the values of $IS(i, S_i, j_i)$ for all nodes Q_i in T. This can be done in a typical bottom-up manner in the weak clique tree. Then, finally, `ALG_Cho` returns a maximum independent set satisfying the deletion constraint at the root node Q_r. Each table value $IS(i, S_i, j_i)$ is computed after the table values of the two children are obtained. Note that each node Q_i in T is a clique, and thus we can pick at most one vertex from Q_i. It follows that for every node Q_i, the number of possible choices as S_i is at most $|V(Q_i)| + 1$ including $S_i = \emptyset$. Further details are omitted here, but, we can obtain the following theorem:

Theorem 5. *Given an n-vertex chordal graph G and a non-negative integer k, BD-MaxIS can be solved in $O(k^2(n + m)^2)$ time.*

5 Concluding Remarks

The Minimum Vertex Cover problem is also known as one of the Karp's 21 fundamental NP-hard problems [14]. We can similarly define a bounded-deletion variant of the Minimum Vertex Cover problem:

> Bounded-Deletion Minimum Vertex Cover (BD-MinVS)
> **Input:** An unweighted graph $G = (V, E)$, an initial feasible solution (i.e., a vertex cover) $S^0 \subseteq V$, and a non-negative integer k.
> **Goal:** The goal is to find a vertex cover $S \subseteq V$ such that $|S^0 \backslash S| \leq k$ and $|S|$ is minimized.

Although details are omitted here, we can show the following results by a similar polynomial-time reduction in the proof of Theorem 1:

Corollary 3. *BD-MinVS is NP-hard even if the input is restricted to bipartite graphs, comparability graphs, or perfect graphs.*

Similarly, we can consider a deletion-bounded variant of the classical and famous NP-hard MaxClique [14]:

> Bounded-Deletion Maximum Clique (BD-MaxClique)
> **Input:** An unweighted graph $G = (V, E)$, an initial feasible solution (i.e., a clique) $S^0 \subseteq V$, and a non-negative integer k.
> **Goal:** The goal is to find a clique set $S \subseteq V$ such that $|S^0 \backslash S| \leq k$ and $|S|$ is maximized.

Every independent set S in the complement graph $\overline{G}$ of a graph G forms a clique induced by S in G. Therefore, we can show the following:

Corollary 4. *BD-MaxClique is NP-hard even if the input is restricted to co-bipartite graphs, co-comparability graphs, or perfect graphs.*

Future work is to show the tractability/intractability of BD-MaxClique on graph classes, such as chordal graphs, interval graphs, and permutation graphs.

Acknowledgments. This work is partially supported by NSERC Canada, and JSPS KAKENHI Grant Numbers JP17K00024, JP21K11755 and JP22K11915.

References

1. Blair, J.R.S., Peyton, B.: An introduction to chordal graphs and clique trees. In: George, A., Gilbert, J.R., Liu, J.W.H. (eds.) Graph Theory and Sparse Matrix Computation. The IMA Volumes in Mathematics and its Applications, vol. 56, pp. 1–29. Springer, New York (1993). https://doi.org/10.1007/978-1-4613-8369-7_1
2. Bodlaender, H.L.: A tourist guide through treewidth. Acta Cybern. **11**, 1–21 (1993)

3. Booth, K.S., Lueker, G.S.: Testing for the consecutive ones property, interval graphs and graph planarity using PQ-tree algorithm. J. Comput. Syst. Sci. **13**, 335–379 (1976)
4. Buneman, P.: A characterization of rigid circuit graphs. Discret. Math. **9**, 205–212 (1974)
5. Cygan, M., et al.: Parameterized Algorithms. Springer, Cham (2015). https://doi.org/10.1007/978-3-319-21275-3
6. Garey, M.R., Johnson, D.S., Stockmeyer, L.: Some simplified NP-complete graph problems. Theor. Comput. Sci. **1**, 237–267 (1976)
7. Gavril, F.: Algorithms for minimum coloring, maximum clique, minimum covering by cliques, and maximum independent set of chordal graph. SIAM J. Comput. **1**, 180–187 (1972)
8. Gavril, F.: Algorithms on circular-arc graphs. Networks **4**, 357–369 (1974)
9. Gavril, F.: The intersection graphs of subtrees in trees are exactly the chordal graphs. J. Comb. Theor. Ser. B **16**, 47–56 (1974)
10. Golumbic, M.C.: The complexity of comparability graph recognition and coloring. Computing **18**, 199–208 (1977). https://doi.org/10.1007/BF02253207
11. Golumbic, M.C.: Algorithmic Graph Theory and Perfect Graphs. Annals of Discrete Mathematics, vol. 57. North-Holland Publishing Co., Amsterdam (2004)
12. Grötschel, M., Lovász, L., Schrijver, A.: The ellipsoid method and its consequences in combinatorial optimization. Combinatorica **1**(2), 169–197 (1981). https://doi.org/10.1007/BF02579273
13. Kammer, F.: Treelike and chordal graphs: algorithms and generalizations. University of Augsburg (2012)
14. Karp, R.M.: Reducibility among combinatorial problems. In: Miller, R.E., Thatcher, J.W., Bohlinger, J.D. (eds.) Complexity of Computer Computations. The IBM Research Symposia Series, pp. 85–103. Springer, Boston (1972). https://doi.org/10.1007/978-1-4684-2001-2_9
15. Kratsch, D., Stewart, L.: Domination on cocomparability graphs. SIAM J. Discrete Math. **6**(3), 400–417 (1993)
16. McConnell, R.M., Spinrad, J.P.: Modular decomposition and transitive orientation. Discrete Math. **201**(1), 189–241 (1999)
17. Murphy, O.J.: Computing independent sets in graphs with large girth. Discrete Appl. Math. **35**, 167–170 (1992)
18. Olariu, S.: An optimal greedy heuristic to color interval graphs. Inf. Process. Lett. **37**(1), 21–25 (1991)
19. Poljak, S.: A note on stable sets and coloring of graphs. Comment. Math. Univ. Carolin. **15**, 307–309 (1974)
20. Seref, O., Ahuja, R.K., Orlin, J.B.: Incremental network optimization: theory and algorithms. Oper. Res. **57**(3), 586–594 (2009)

Asynchronous Fully-Decentralized SGD in the Cluster-Based Model

Hagit Attiya(✉) and Noa Schiller

Department of Computer Science, Technion, Haifa, Israel
{hagit,noa.schiller}@cs.technion.ac.il

Abstract. This paper presents fault-tolerant asynchronous *Stochastic Gradient Decent* (*SGD*) algorithms. SGD is widely used for approximating the minimum of a cost function Q, a core part of optimization and learning algorithms. Our algorithms are designed for the *cluster-based* model, which combines message-passing and shared-memory communication layers. Processes may fail by *crashing*, and the algorithm inside each cluster is *wait-free*, using only reads and writes.

For a *strongly convex* Q, our algorithm *can withstand partitions of the system.* It provides convergence rate that is the maximal distributed acceleration over the optimal convergence rate of *sequential* SGD.

For arbitrary smooth functions, the convergence rate has an additional term that depends on the maximal difference between the parameters at the same iteration. (This holds under standard assumptions on Q). In this case, the algorithm obtains the same convergence rate as sequential SGD, up to a logarithmic factor. This is achieved by using, at each iteration, a *multidimensional approximate agreement* algorithm, tailored for the cluster-based model.

The general algorithm communicates with nonfaulty processes belonging to clusters that include a majority of all processes. We prove that this condition is necessary when optimizing some non-convex functions.

Keywords: Cluster-based model · Distributed learning · Asynchronous computing · Multi-dimensional approximate agreement · Stochastic gradient descent

1 Introduction

An *optimization* problem attempts to minimize the value of a *cost function* $Q : \mathbb{R}^d \to \mathbb{R}$, that is, find $x^* \in \arg\min_{x \in \mathbb{R}^d} Q(x)$. Among their many uses, optimization problems play a key role in machine and deep learning [19], often using *stochastic gradient descent* (SGD). SGD [29] repeatedly applies the update rule $\mathbf{x}_{t+1} = \mathbf{x}_t - \eta_t G(\mathbf{x}_t, z_t)$, in each iteration t. The arguments for this rule are the *learning parameter* $\mathbf{x}_t$, the *learning rate* η_t, and a random sample z_t from *data*

The full version [6] contains all omitted lemmas and proofs, as well as additional material. This work was supported by Pazy grant 226/20 and the Israel Science Foundation grants 380/18 and 22/1425.

M. Mavronicolas (Ed.): CIAC 2023, LNCS 13898, pp. 52–66, 2023.
https://doi.org/10.1007/978-3-031-30448-4_5

distribution $\mathcal{D}$; $G(\mathbf{x}_t, z_t)$ computes the *stochastic gradient* of $\mathbf{x}_t$ and z_t, which is an *unbiased estimator* of the true gradient $\nabla Q(\mathbf{x}_t)$. Intuitively, the gradient points to the direction of the steepest slope at that point, and its opposite direction gives the biggest (local) decrease in function value. When the function Q is strongly convex, $\mathbf{x}_t$ will converge to the unique minimum of Q [9]; otherwise, it will converge to a point with zero gradient [16].

In learning applications, SGD is applied to a function Q of high dimension d, using many stochastic gradients [11]. The convergence of the basic SGD algorithm can be improved by *mini-batch SGD*, which computes b stochastic gradients using b samples, drawn uniformly at random from $\mathcal{D}$. The average of these b gradients has variance that is a factor of b smaller than σ^2, the variance of a single stochastic gradient, implying a linear reduction in the number of iterations for convergence. Since gradients are computed independently, (mini-batch) SGD is a prime target for large-scale distributed and parallel computing. Mini-batch SGD provides a baseline for measuring performance in the distributed setting.

In an iteration of a typical distributed SGD algorithm, a *worker* performs some local computation and then, all computed values are aggregated to collectively compute parameters for the next iteration [8]. This can be done in a *centralized* manner, where a *parameter server* aggregates all the computation done by the workers (e.g., [2–4,14,22,28]), or in a *decentralized* manner, where each worker holds a copy of the parameters (e.g., [13,20,21,23,24]). A straightforward, synchronized implementation of (mini-batch) SGD requires locks or barriers to ensure that workers proceed in lock-step, thereby harming the performance.

This paper considers completely decentralized and asynchronous SGD in a *cluster-based model* [27] that combines both shared memory and message passing (Fig. 1): processes are partitioned to disjoint clusters, each sharing a memory space, accessed with reads and writes; additionally, all processes can communicate by message passing. Processes may fail by *crashing*, that is, stopping to take steps. This model is interesting from a practical point-of-view, as it captures several system architectures, for example *high-performance computing* systems [7].

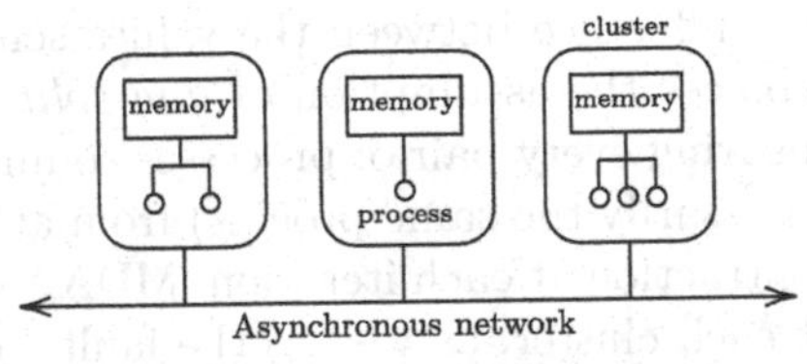

Fig. 1. Cluster-based model

Our first main contribution shows that when the function Q is strongly convex, a *simple asynchronous* algorithm that collects n_b messages in each iteration, matches the convergence rate of a sequential mini-batch SGD algorithm (for strongly-convex functions) with batch size n_b [1,9]. Our analysis of this algorithm is relatively simple, and leverages the strong convexity of Q to prove convergence, despite the fact that each process holds a local copy of the learning parameters. Specifically, we prove (Theorem 1) that if Q is a smooth strongly-convex function, then the convergence rate of the algorithm after T iterations with parameter n_b is $O(1/n_bT)$. Progress is ensured as long as n_b is not larger than the number of non-failed processes.

The algorithm for strongly-convex functions does not rely on communication within clusters and it applies to general message-passing systems. Even more importantly, because n_b is arbitrary and does not depend on the number of possible failures f, the algorithm works even if the system partitions. Roughly, since Q is strongly convex, it has a single minimum, and thus, all processes will independently converge to this minimum.

Our second main contribution (Theorem 4) shows that in general, when Q is not strongly convex, any SGD algorithm requires that the set of non-failed processes *represents* a majority of the processes, when a process represents all the processes in its cluster. In the special case where communication is only through message passing, this reduces to requiring $n > 2f$.

Our third main contribution is a general SGD algorithm, under the same assumption that the set of non-failed processes represents a majority of the processes. It has a weaker convergence guarantee, relative to the baseline, with an additional term Δ, depending on the difference between the learning parameters of the different processes during the algorithm execution. We show that if Q is a smooth function, then the convergence rate of the algorithm after T iterations with parameter $n_b \leq n - f$ is $\mathcal{O}\big(1/\sqrt{n_b T} + T\Delta\big)$. The first term matches the convergence rate achieved in standard analysis for non-convex objectives [16].

Unlike the strongly convex case, where the difference between the learning parameters is intrinsically bounded, here we bound Δ using *multidimensional approximate agreement* (*MDAA*) [25]. In MDAA, processes start with inputs in $\mathbb{R}^d$, and the outputs of nonfaulty processes should be "close together" and in the *convex hull* of their inputs. We use a shared-memory adaptation of [15] to bound the difference between the values sent from the same cluster. (Shared memory replaces the assumption of *non-split* communication patterns used in [15]). By ensuring every pair of processes communicate with a representative process (not necessarily the same process) from at least one common cluster, we ensure good contraction at each iteration. MDAA encapsulates the use of the shared memory at each cluster, as well as the fault-tolerance required from the algorithm. This algorithm is interesting by itself, beyond its application in distributed learning.

Each MDAA iteration contains several *communication rounds*, where each process sends a message and receives responses representing $n - f$ processes. We prove that our general SGD algorithm can match the convergence rate of the sequential algorithm, *up to a logarithmic blowup* in the number of communication rounds. Specifically (Theorem 2), for a smooth function Q, the convergence rate of the algorithm after R communication rounds is $\widetilde{\mathcal{O}}\big(1/\sqrt{n_b R}\big)$.

In our algorithms, each process serves as both a computation and a communication thread, executing the computation and sending its result to the other processes. Within the cluster, the algorithm is wait-free and uses only reads and writes to the cluster's shared memory; no locks or barriers are used. Since processes operate in an independent manner, the algorithms achieve a speedup in the total number of processes and not the number of clusters. Our algorithms also improve the resilience relative to a pure message passing model, since a process can *represent* its cluster [5].

Related Work. We only discuss the most relevant papers from the vast literature on distributed SGD algorithms; a more extensive survey appears in [6].

Elastic consistency [26] is a framework that assumes that the difference between the parameter used to compute the stochastic gradient by a process and the actual global parameter is bounded. SGD converges under this assumption, for both convex and non-convex objective functions. They prove that several popular frameworks obey the above assumption. One example is in shared-memory [4,28], where processes access the same learning parameter stored in memory and update it, one coordinate at a time, using fetch&add. Another example is with message-passing [2,22], where the parameter server may receive stale gradients, i.e., gradients computed using old parameters from previous iterations. Both cases assume *bounded asynchrony*, with a maximum delay τ on the staleness of gradients.

Our algorithms are completely asynchronous, and do not assume any bound on the difference between the iterations different processes are in at any point in the execution. Like many classical distributed algorithms that proceed in asynchronous rounds, our algorithms ignore *stale* messages from earlier rounds. The only SGD algorithms (to our knowledge) that proceed in a similar manner [13,14] handle malicious failures, where processes may behave arbitrarily bad, and do not bound the convergence rate as a function of the number of processes. While the algorithms in [13,14] could also tolerate crash failures, they are not optimized for this case and do not achieve a speedup in the number of workers.

Some decentralized algorithms, e.g., [23,24], assume the communication is dictated by a graph. We consider systems with a full communication graph, where all processes may send messages to each other.

Approximate agreement was originally defined over the reals [12]. *Multidimensional approximate agreement* was defined for asynchronous systems with malicious failures [25], requiring that the outputs of nonfaulty processes' are close together and in the *convex-hull* of their inputs. They prove that the optimal fault-tolerance for this problem is $n > f(d+2)$, for inputs in $\mathbb{R}^d$. The lower bound can be circumvented by using *averaging agreement* [13], which allows to prove the convergence of the *average* of the outputs of nonfaulty nodes.

2 Preliminaries

2.1 Model of Computation

There are n *processes*, $1, \dots, n$, which are partitioned into $m \leq n$ disjoint *clusters*, $P_1, \dots, P_m$. Formally, $P_i \subseteq \{1, \dots, n\}$, $P_i \cap P_j = \emptyset$ for every $1 \leq i < j \leq m$, and $\bigcup_{i=1}^{m} P_i = \{1, \dots, n\}$. Given a process i, $\mathsf{cluster}(i)$ is its cluster, i.e., $\mathsf{cluster}(i) = P_j$ such that $i \in P_j$. Processes may *crash* and stop taking steps. A process is *nonfaulty* if it never stops taking steps. For simplicity, a process keeps taking (empty) steps even after it completes the algorithm. A *shared memory* is associated with each cluster, and it is accessed with read and write operations, only by the processes in the cluster. In addition, each process can send messages to each other process, using an asynchronous, reliable communication link. This means that messages between nonfaulty processes are eventually delivered, but there is no upper bound on the delivery time. Figure 1 shows $n = 6$ processes organized in $m = 3$ clusters: $P_1 = \{1, 2\}, P_2 = \{3\}$ and $P_3 = \{4, 5, 6\}$. The maximal number of processes that can crash is denoted f, $1 \le f \le n$.

A process i *represents* all the processes in $\mathsf{cluster}(i)$. Let f^* be the maximal integer such that any set $P \subseteq [n]$ of size $n - f^*$ represents a majority of the processes, i.e., $|\cup_{i \in P} \mathsf{cluster}(i)| \ge \lfloor n/2 \rfloor + 1$. Intuitively, this is the exact number of failures a system can withstand without *partitioning*, i.e., the situation where two disjoint sets of processes run without communication. In the special case of a pure message-passing system (with singleton clusters) requiring $f \le f^*$ amounts to $f < n/2$. Note that $f^* \ge \lfloor (n-1)/2 \rfloor$, and we have:

Lemma 1 ([5]). *If $f \le f^*$, then any two sets $P, Q \subseteq [n]$, each representing $n - f$ processes, must include a process from the same cluster.*

We assume a *weak adversary* that does not observe the local state of the processes, and is oblivious to their local coin flips when scheduling the processes.

2.2 Stochastic Gradient Descent

The Euclidean norm of a vector $\mathbf{x} = (x_1, ..., x_d) \in \mathbb{R}^d$ is $\|\mathbf{x}\|_2 \triangleq \sqrt{\sum_{i=1}^d |x_i|^2}$; we use the standard notation, $\|\mathbf{x}\|_2^2 \triangleq (\|\mathbf{x}\|_2)^2 = \sum_{i=1}^d |x_i|^2$. The *variance* of a random vector $\mathbf{x}$ is $\mathbb{V}[\mathbf{x}] \triangleq \mathbb{E}\left[\|\mathbf{x} - \mathbb{E}[\mathbf{x}]\|_2^2\right]$.

Each process can access the same data distribution $\mathcal{D}$, and *loss function* $\ell(\mathbf{x}, z)$, which takes a *learning parameter* $\mathbf{x} \in \mathbb{R}^d$ and a data point $z \in \mathcal{D}$. Given a learning parameter $\mathbf{x} \in \mathbb{R}^d$, the *cost function* Q is: $Q(\mathbf{x}) \triangleq \mathbb{E}_{z \sim \mathcal{D}}[\ell(\mathbf{x}, z)]$. A distributed *Stochastic Gradient Descent (SGD)* algorithm collectively minimizes the cost function Q, i.e., it finds $\mathbf{x}^* \in \arg\min_{\mathbf{x} \in \mathbb{R}^d} Q(\mathbf{x})$.

The cost function is *differentiable* and *smooth*, i.e., for a constant $L \in \mathbb{R}^+$, $\forall \mathbf{x}, \mathbf{y} \in \mathbb{R}^d$, $\|\nabla Q(\mathbf{x}) - \nabla Q(\mathbf{y})\|_2 \le L\|\mathbf{x} - \mathbf{y}\|_2$, where $\nabla Q(\mathbf{x}) \in \mathbb{R}^d$ is the *gradient* of Q at $\mathbf{x}$. The gradient at $\mathbf{x} \in \mathbb{R}^d$ can be estimated by the *stochastic gradient* $G(\mathbf{x}, z) = \nabla \ell(\mathbf{x}, z) \in \mathbb{R}^d$, calculated at a data point z that is drawn uniformly at random from $\mathcal{D}$. The stochastic gradient is an *unbiased estimator* of the true gradient $\mathbb{E}_{z \sim \mathcal{D}}[G(\mathbf{x}, z)] = \nabla Q(\mathbf{x})$. In addition, the estimations have *bounded variance*, i.e., there is a non-negative constant $\sigma \in \mathbb{R}$ such that $\mathbb{V}_{z \sim \mathcal{D}}[G(\mathbf{x}, z)] = \mathbb{E}_{z \sim \mathcal{D}}\left[\|G(\mathbf{x}, z) - \nabla Q(\mathbf{x})\|_2^2\right] \le \sigma^2$. These are standard assumptions in SGD analysis [9,10,16].

At the end of the algorithm, each nonfaulty process i outputs an estimate of the learning parameter, $\mathbf{x}^i \in \mathbb{R}^d$. We require the algorithm to *externally converge with expected error* $\epsilon > 0$ (simply called *to converge* in the optimization literature). The convergence requirement expresses the quality of the solution relative to a minimal one. It varies according to the assumptions on the cost function, whether it is strongly-convex (Sect. 3) or not (Sect. 4).

The algorithm *internally converge with expected error* $\delta > 0$, if for every pair of nonfaulty processes i and j:

$$\mathbb{E}\left[\left\|\mathbf{x}^i - \mathbf{x}^j\right\|_2^2\right] \leq \delta \tag{1}$$

3 Strongly-Convex Cost Functions

We start by presenting the algorithm when the function Q is strongly convex. Formally, Q is *μ-strongly convex*, for $\mu > 0$, if for every $\mathbf{x}, \mathbf{y} \in \mathbb{R}^d$, $Q(\mathbf{y}) \geq Q(\mathbf{x}) + \langle \nabla Q(\mathbf{x}), \mathbf{y} - \mathbf{x}\rangle + \frac{\mu}{2}\|\mathbf{y} - \mathbf{x}\|_2^2$. A strongly convex cost function has a single minimum, denoted $\mathbf{x}^*$. In this case, an algorithm externally converges if for every nonfaulty process i, $\mathbb{E}\left[\left\|\mathbf{x}^i - \mathbf{x}^*\right\|_2^2\right] \leq \epsilon$. Following standard analysis [9], the convergence rate of vanilla SGD in this case is $\mathcal{O}(1/T)$ after T iterations. That is, $\mathbb{E}\left[\|\mathbf{x}_T - \mathbf{x}^*\|_2^2\right] \leq C/T$ for a constant C depending on μ, L, σ and $\|\mathbf{x}_1 - \mathbf{x}^*\|_2^2$. This implies that $\mathbb{E}\left[\|\mathbf{x}_T - \mathbf{x}^*\|_2^2\right] \leq \epsilon$ after $T = \mathcal{O}(\epsilon^{-1})$ iterations. Using a minibatch of size b gives a linear speedup in convergence rate to $\mathcal{O}(1/bT)$. This will serve as the baseline for the external convergence rate.

Algorithm 1 works in iterations, corresponding to those of sequential SGD. A process starts an iteration t with a local learning parameter, and computes a new one for the next iteration. First, the process computes a stochastic gradient using its current learning parameter, performs a local SGD step, and sends the updated learning parameter to all the other processes. After receiving learning parameters from iteration t from n_b processes, the process averages all the parameters. In the last iteration, each process outputs the learning parameter it has computed. The algorithm ignores *stale* gradients from previous iterations, i.e., gradients computed using the learning parameter of a previous round.

Algorithm 1. Cluster-based SGD, for strongly-convex function: code for process i

Global input: initial point $\mathbf{x}_1$
1: $\mathbf{x}_1^i \leftarrow \mathbf{x}_1$
2: **for** $t = 1 \ldots T$ **do**
3: draw uniformly at random $z \in \mathcal{D}$
4: $\mathbf{g}_t^i \leftarrow G(\mathbf{x}_t^i, z)$
5: $\mathbf{y}_t^i \leftarrow \mathbf{x}_t^i - \eta_t \mathbf{g}_t^i$
6: broadcast $\langle t, \mathbf{y}_t^i \rangle$ to all processes
7: wait to receive n_b messages of the form $\langle t, - \rangle$
8: $\mathbf{x}_{t+1}^i \leftarrow$ avg(received learning parameters)
9: output $\mathbf{x}_{T+1}^i$

The value of n_b can be arbitrary; we only assume that $n_b \leq n - f$, to ensure progress. As we show, the larger n_b is, the better the convergence of the algorithm is. The learning parameter for the first iteration of all processes is $\mathbf{x}_1 \in \mathbb{R}^d$. The *learning rate* for iteration t is η_t, and for each t, it is the same for all processes; the learning rate is decreasing, i.e., $\eta_t = \mathcal{O}(1/t)$ for every iteration t.

Let V_t, $t \geq 1$, be the set of processes that compute learning parameters for iteration $t+1$ in Line 8. We can prove that the average of b stochastic gradients, each computed using different learning parameters, has variance of at most σ^2/b. Since Line 8 in Algorithm 1 averages the received parameters and the weak adversary cannot determine the learning parameters that arrive at each process, this implies that for any iteration $t \geq 1$ and set of processes $S \subseteq V_t$, $\mathbb{V}\left[\frac{1}{|S|}\sum_{i\in S}\mathbf{g}_t^i\right] \leq \frac{\sigma^2}{|S|}$.

The next theorem shows that for strongly-convex functions our algorithm achieves the same convergence rate (both internal and external) as the sequential baseline. Intuitively, as there is a single minimum, all processes will converge to this point independently. Hence, despite different processes holding different learning parameters, the expected gradients will point in the same direction. This allows us to obtain terms similar to the ones in the classical strongly-convex analysis [9], and achieve the same convergence rate. For the full proof and additional details, see [6].

Theorem 1. *Let Q be an L-smooth and μ-strongly convex function with a single minimum $\mathbf{x}^*$, then for decreasing learning rate $\eta_t = \frac{\beta}{\gamma+t} \leq \frac{1}{L}$ for some constants $\beta > \frac{1}{\mu}$ and $\gamma > 0$,*

$$\max_{i,jinV_T} \mathbb{E}\left[\left\|\mathbf{x}_{T+1}^i - \mathbf{x}_{T+1}^j\right\|_2^2\right] \leq \frac{4\eta^2\sigma^2}{(\eta\mu - 1)(\gamma + T + 1)n_b} \quad \text{(Internal convergence)}$$

$$\max_{iinV_T} \mathbb{E}\left[\left\|\mathbf{x}_{T+1}^i - \mathbf{x}^*\right\|_2^2\right] \leq \frac{(\gamma+1)\|\mathbf{x}_1 - \mathbf{x}^*\|_2^2}{(\gamma + T + 1)n_b} + \frac{\eta^2\sigma^2}{(\eta\mu - 1)(\gamma + T + 1)n_b} \quad \text{(External convergence)}$$

Neglecting dependencies on $\|\mathbf{x}_1 - \mathbf{x}^*\|_2^2$, μ, L and σ, this means that Algorithm 1 converges externally in $\mathcal{O}(\epsilon^{-1}/n_b)$ iterations, and internally in $\mathcal{O}(\delta^{-1}/n_b)$ iterations; that is, the rates are the same. Since each iteration takes a single communication round, we get the same upper bound on the number of rounds.

4 Non-Convex Cost Functions

In the general case, where the function Q is non-convex, the algorithm has to converge to a point with zero gradient; that is, for every nonfaulty process i:

$$\mathbb{E}\left[\left\|\nabla Q(\mathbf{x}^i)\right\|_2^2\right] \leq \epsilon \tag{2}$$

We assume that Q is lower bounded by Q^*, i.e., for every $\mathbf{x} \in \mathbb{R}^d$, $Q(\mathbf{x}) \geq Q^*$ [9].

When the function Q is not strongly convex, processes that obtain disjoint estimations at an iteration may compute *diverging* learning parameters. For this reason, we need to ensure that processes communicate with intersecting sets of *clusters*. To further expedite the contraction rate, and reduce the distance between the learning parameters, we end each iteration with *multidimensional approximate agreement* (*MDAA*). The input to MDAA is the local learning parameter, and its output serves as the learning parameter for the next iteration.

Formally, in *multidimensional approximate agreement* [25], each process i starts with input $\mathbf{x}_i \in \mathbb{R}^d$ and outputs a value $\mathbf{y}_i \in \mathbb{R}^d$, such that:

Convexity: The outputs are in the *convex hull* of the inputs, that is, they are a *convex combination* of the outputs.

q-Contraction: The outputs are contracted by a factor of q relative to the inputs, that is, for every pair of nonfaulty processes i, j, $\|\mathbf{y}_i - \mathbf{y}_j\|_2^2 \leq q \operatorname{diam}(\mathbf{x}_1, \ldots \mathbf{x}_n)$, where the squared Euclidean *diameter* of a set $A \subseteq \mathbb{R}^d$ is $\operatorname{diam}(A) \triangleq \max_{\mathbf{x},\mathbf{y} \in A} \|\mathbf{x} - \mathbf{y}\|_2^2$.

Standard approximate agreement [12,25] requires ϵ-*agreement*, that is, the distance between outputs is at most ϵ. We only require contraction *relative to the diameter of the inputs*, rather than a predefined maximal distance.

Algorithm 2. Cluster-based SGD: code for process i

Global input: initial point $\mathbf{x}_1$ and random iteration τ

1: $\mathbf{x}_1^i \leftarrow \mathbf{x}_1$
2: **for** $t = 1 \ldots T$ **do**
3: draw uniformly at random $z \in \mathcal{D}$
4: broadcast $\langle t, G(\mathbf{x}_t^i, z) \rangle$ to all processes
5: wait to receive n_b messages of the form $\langle t, - \rangle$
6: $\mathbf{g}_t^i \leftarrow$ avg(received stochastic gradients)
7: $\mathbf{y}_t^i \leftarrow \mathbf{x}_t^i - \eta_t \mathbf{g}_t^i$
8: $\mathbf{x}_{t+1}^i \leftarrow \mathsf{MDAA}_t(\mathbf{y}_t^i, q)$
9: output $\mathbf{x}_\tau^i$

Algorithm 2 deals with non-convex functions. One difference from Algorithm 1 is in Line 8, calling MDAA with contraction parameter q. Another difference is that processes send the stochastic gradients they computed, average the received gradients to a *mini-batch* stochastic gradient and then use it to perform a local SGD step. Every process outputs the learning parameter of the *same* iteration $\tau \in [T]$, which is drawn uniformly at random; this is a typical practice in SGD algorithms for non-convex objective functions [9,16]. The convergence rate of (vanilla) minibatch SGD with batch size b is $\mathcal{O}\left(1/\sqrt{bT}\right)$, i.e., $\mathbb{E}\left[\|\nabla Q(\mathbf{x}_\tau)\|_2^2\right] \leq C/\sqrt{bT}$, for constant C which depends on L, σ and $(Q(\mathbf{x}_1) - Q^*)$. This implies that $\mathbb{E}\left[\|\nabla Q(\mathbf{x}_\tau)\|_2^2\right] \leq \epsilon$ after $T = \mathcal{O}(\epsilon^{-2}/\sqrt{b})$ iterations. As in the strongly-convex case, this will serve as our baseline. The algorithm is similar to those in [13], but they use aggregation rules which are resilient against malicious failures, and *averaging agreement*, while we rely on the convexity property ensured by MDAA.

The parameter $n_b \in [n]$ determines how many messages a process waits for in every iteration; to ensure progress, we require that $n_b \leq n - f$, as in the strongly-convex case. Furthermore, we assume that $f \leq f^*$ to guarantee the convergence of the MDAA algorithm. Section 5 presents a cluster-based MDAA algorithm, with $\mathcal{O}(\log q^{-1})$ communication rounds. Let V_t, $t \geq 1$, be the set of processes that compute learning parameters for iteration $t+1$ (Line 8). Similarly to the strongly-convex case, we have the next lemma (see [6]):

Lemma 2. *For every iteration $t \geq 1$ and process $i \in V_t$, $\mathbb{V}[\mathbf{g}_t^i] \leq \sigma^2/n_b$.*

The proof for external convergence uses the *effective gradient* [14], which is defined as the difference between two consecutive iterations parameters. We bound the difference between the effective gradient and the true gradient in each iteration, depending on the diameter of the learning parameters in the same iteration. This allows to prove convergence, as the effective change between two consecutive iterations is close enough to the true gradient. The next lemma shows that external convergence depends on the maximal distance between the learning parameters at the different processes, and uses a fixed learning rate of $\eta = \sqrt{n_b/T}$. Note that the first two terms are the classical error rates in the non-convex case [9]. (All proofs for this section appear in [6]).

Lemma 3. *Let Q be an L-smooth cost function. Then for $T \geq 16L^2 n_b$, constant learning rate $\eta = \frac{\sqrt{n_b}}{\sqrt{T}}$ and any process $i \in V_T$,*

$$\mathbb{E}\left[\left\|\nabla Q(\mathbf{x}_\tau^i)\right\|_2^2\right] \leq \frac{4(Q(\mathbf{x}_1) - Q^*)}{\sqrt{n_b T}} + \frac{4\sigma^2 L}{\sqrt{n_b T}} + \left(\frac{8T}{n_b} + 8L^2\right) \max_{t\in[T]} \max_{i,j\in V_{t-1}} \mathbb{E}\left[\left\|\mathbf{x}_t^i - \mathbf{x}_t^j\right\|_2^2\right]$$

Lemma 3 shows that external convergence depends on internal convergence. (Recall from Sect. 3, that in the special case where the function is strongly convex, both external and internal convergence are achieved naturally.) The proof of this lemma only uses the *convexity* property of MDAA and gives motivation for adding the *contraction* property to ensure this term will be sufficiently small.

Internal Convergence. The next lemma bounds the diameter of the learning parameters of iteration $t+1$ relative to the diameter of the previous iteration t. It is proved by first bounding the diameter after each process performs a local SGD step, and then using the contraction property of MDAA.

Lemma 4. *For every iteration $t \geq 1$,*

$$\max_{i,j\in V_t} \mathbb{E}\left[\left\|\mathbf{x}_{t+1}^i - \mathbf{x}_{t+1}^j\right\|_2^2\right] \leq q(2 + 2L^2\eta_t^2) \max_{i,j\in V_{t-1}} \mathbb{E}\left[\left\|\mathbf{x}_t^i - \mathbf{x}_t^j\right\|_2^2\right] + q\frac{4\sigma^2\eta_t^2}{n_b}$$

When $q(2 + 2L^2\eta_t^2) < 1$ we get contraction relative to the previous iteration, with an additive term. Assuming that $\eta_t \leq \frac{1}{L}$, yields that $2 + 2L^2\eta_t^2 \leq 4$. Hence, for any $q < \frac{1}{4}$ this term is smaller than 1. Since the additive term also depends on q, we can use it to control its magnitude. The next lemma bounds the distance between the learning parameters at each iteration, using a constant learning rate η and contraction parameter $q \approx \eta$.

Lemma 5. *Consider Algorithm 2 with constant learning rate $\eta \leq \min\{\frac{1}{2}, \frac{1}{L}\}$ and $q = \frac{\eta}{4}$. Then for every iteration $t \geq 1$,*

$$\max_{i,j\in V_t} \mathbb{E}\left[\left\|\mathbf{x}_{t+1}^i - \mathbf{x}_{t+1}^i\right\|_2^2\right] \leq \frac{2\sigma^2\eta^3}{n_b}$$

Lemma 5 implies that setting $\eta = \mathcal{O}\left(\sqrt[3]{n_b\delta/\sigma^2}\right)$ yields internal convergence. Using the same learning rate as in Lemma 3, $\eta = \sqrt{n_b/T}$, we get that $\max_{i,j\in V_{\tau-1}} \mathbb{E}\left[\left\|\mathbf{x}_\tau^i - \mathbf{x}_\tau^j\right\|_2^2\right] = \mathcal{O}\left(n_b^{1/2}T^{-3/2}\right)$. This yields internal convergence in $\mathcal{O}\left(n_b^{1/3}\delta^{-2/3}\right)$ iterations.

External Convergence. Finally, by bounding the distance of the learning parameters at each iteration, we prove that Algorithm 2, using an MDAA algorithm with $\mathcal{O}(\log T)$ communication rounds, has convergence rate that matches the sequential SGD algorithm, up to a logarithmic factor in the number of rounds. In a nutshell, the proof of the theorem uses the bound on the distance between the learning parameters from Lemma 5 in Lemma 3 (See [6]).

Theorem 2. *Let Q be an L-smooth cost function. Consider Algorithm 2 with $T \geq \max\{16L^2 n_b, 4n_b\}$, constant learning rate $\eta = \frac{\sqrt{n_b}}{\sqrt{T}}$ and $q = \frac{\eta}{4}$. Then, after $R = \mathcal{O}(T \log T)$ communication rounds for some constant C and every process $i \in V_T$*

$$\mathbb{E}\left[\left\|\nabla Q(\mathbf{x}_\tau^i)\right\|_2^2\right] \leq \frac{C(Q(\mathbf{x}_1) - Q^*)\log R}{\sqrt{n_b R}} + \frac{5C\sigma^2 L \log R}{\sqrt{n_b R}} + \frac{4C\sigma^2 \log R}{\sqrt{n_b R}}$$

Theorem 2 shows that Algorithm 2 converge externally in $\widetilde{\mathcal{O}}(\epsilon^{-2}/n_b)$ communication rounds, and by Lemma 3, internally in $\widetilde{\mathcal{O}}({n_b}^{1/3}\delta^{-2/3})$ communication rounds. In both cases, dependencies on $(Q(\mathbf{x}_1) - Q^*)$, L and σ are neglected. In the non-convex case, internal convergence is faster than external convergence.

5 Cluster-Based MDAA

Algorithm 3. Cluster-based MDAA: code for process i in cluster c

```
MDAA(x, q):
1: x_1^i ← x
2: for r = 1 ... R = ⌈log_{23/24} q⌉ do
3:     y_r^i ← SMMDAA_r(x_r^i, 1/6)
         ▷ Shared-memory algorithm inside cluster c
4:     broadcast ⟨r, y_r^i⟩ to all processes
5:     wait to receive messages of the form ⟨r, −⟩
6:     representing n − f processes
7:     Rcv ← set of received values
8:     x_{r+1}^i ← MidExtremes(Rcv)
9: output x_{R+1}^i
```

Algorithm 3 solves MDAA in the cluster-based model. The algorithm leverages inter-cluster communication to increase the number of failures that can be tolerated, and only requires read and write operations. The algorithm works in rounds, each starting with *shared-memory* MDAA (SMMDAA) within each cluster (Line 3). This algorithm satisfies the two properties defined above (convexity and contraction), among the inputs of each cluster separately. The processes of each cluster communicate in this algorithm using only the common shared memory. (See [6] for full details and proof). This algorithm allows processes to wait to *receive messages representing a majority of the processes at each round, rather than waiting for a majority of the processes,* which is the usual practice in crash-tolerant message-passing algorithms. This guarantees that every pair of processes receive a value from a common cluster. Any two values sent from the same cluster have smaller diameter, compared to the diameter of all the process values, since they are the output of the inter-cluster SMMDAA algorithm. After collecting sufficiently many messages, the process computes the next round value using the *MidExtremes* rule [15], which returns the average of the two values that realize the maximum Euclidean distance among all received vectors. Formally, for $X \subseteq \mathbb{R}^d$

$$\mathsf{MidExtremes}(X) = (\mathbf{a}+\mathbf{b})/2, \text{ where } (\mathbf{a},\mathbf{b}) = \underset{(\mathbf{a},\mathbf{b})\in X^2}{\arg\max} \|\mathbf{a}-\mathbf{b}\|_2.$$

Note that at each round the value is a convex combination of the previous rounds' values, hence, this algorithm satisfies the convexity property. In each round of our MDAA algorithm, the diameter of the values is reduced by a fixed *contraction rate* relative to the diameter of the values in the previous round. Let $\mathbf{x}_r$ be the set of round-r outputs, the *round-r contraction rate* is an upper bound $\rho_r \geq \mathrm{diam}(\mathbf{x}_{r+1})/\mathrm{diam}(\mathbf{x}_r)$. We consider algorithms with a uniform upper bound $\rho < 1$ on the contraction rate of all rounds. Contraction rate ρ ensures q-contraction within $\lceil \log_\rho q \rceil \approx \log_2(1/q)$ rounds (assuming $q, \rho < 1$).

We assume that $f \leq f^*$, which by Lemma 1 implies that two sets of processes, each representing $n-f$ processes, must include a process from the same cluster (not necessarily the same process). This is optimal, since even *one-dimensional* approximate agreement cannot be solved when $f > f^*$ [5].

We explain our cluster-based MDAA algorithm in the one-dimensional case; the full proof [6] covers the multidimensional case. Intuitively, rather than requiring that collected sets intersect, it suffices to assume that they contain "close enough" values. When $d = 1$, MidExtremes returns the *MidPoint*, i.e., for a set $X \subseteq \mathbb{R}$, $\mathsf{MP}(X) = (\min(X) + \max(X))/2$. We also have that $\mathrm{diam}(X) = \max(X) - \min(X)$ for $X \subseteq \mathbb{R}$. Let $A, B \subseteq U \subseteq \mathbb{R}$, A and B stand for values collected in the same round by two different processes (Line 7), and U is all the possible round values. For any pair of values $a \in A$ and $b \in B$, we have $\mathsf{MP}(A) \leq (a + \max(U))/2$ and $\mathsf{MP}(B) \geq (\min(U) + b)/2$, which implies:

$$\mathsf{MP}(A) - \mathsf{MP}(B) \leq \frac{1}{2}\underbrace{(\max(U) - \min(U))}_{\mathrm{diam}(U)} + \frac{1}{2}(a-b) \tag{3}$$

When A and B intersect, the second term in (3) zeros out, which gives contraction rate of $1/2$ when any two sets of collected values intersect. (In the general d-dimensional case, MidExtremes has contraction rate of $7/8$ under this assumption [15]). The assumption that any two collected sets of size $n-f$ must intersect is not guaranteed when only assuming $f \leq f^*$. Instead, we can use an SMMDAA algorithm with contraction parameter of $1/2$. Assuming there are values $a \in A$ and $b \in B$ such that $a - b \leq \mathrm{diam}(U)/2$, by (3), $|\mathsf{MP}(A) - \mathsf{MP}(B)| \leq 3/4\,\mathrm{diam}(U)$, leading to a slightly worse contraction rate of $3/4$.

The full proof of the algorithm relates the SMMDAA contraction parameter q_{sm} and the global contraction rate ρ of the outer algorithm in the general d-dimensional case, which is $\rho \leq 11/12 + q_{sm}/4$. Using $q_{sm} = 1/6$, we get an MDAA algorithm with contraction rate $23/24$.

Theorem 3. *Algorithm 3 satisfies q-contraction within $\lceil \log_{23/24} q \rceil$ communication rounds.*

6 Impossibility of Asynchronous SGD with System Partitions

A non-convex function Q may have several *stationary points*, i.e., $\mathbf{x} \in \mathbb{R}^d$ such that $\nabla Q(\mathbf{x}) = 0$. A stationary point can be either global or local minimum or maximum, or a *saddle point*, which is not a local extremum of the function. Although SGD converges to a stationary point (see (2)), it may happen that starting from the same initial point, SGD converges to different stationary points in different random executions. If these stationary points are γ apart, for a large enough γ, and the probability of reaching them is large enough, then we can prove that no distributed SGD implementation, satisfying both internal and external convergence, can tolerate *partitioning*. (Recall that in our model, the system is *partitioned* if $f > f^*$, and the correct processes may represent a minority of the processes.) Intuitively, this is because each partition can converge individually to points that are γ apart from each other, violating internal convergence.

Definition 1, below, formalizes the phenomenon where the SGD algorithm converge γ apart in two different random executions with probability at least p.

Let $\mathcal{A}^{seq}(Q, T, \mathbf{x}_0)$ be a sequential SGD algorithm optimizing the function Q for T iterations starting from $\mathbf{x}_0$. Let $\mathbf{x}_{seq}$ be a random variable, corresponding to the output of the sequential algorithm. For $\beta \in \mathbb{R}$ and point $\mathbf{x} \in \mathbb{R}^d$, let $E_{seq}(\beta, \mathbf{x})$ be the event that $\|\mathbf{x}_{seq} - \mathbf{x}\|_2 \leq \beta$. For a set of points $S \subseteq \mathbb{R}^d$, $E_{seq}(\beta, S)$ denotes the event that for some $\mathbf{x} \in S$, $\|\mathbf{x}_{seq} - \mathbf{x}\|_2 \leq \beta$, i.e., $E_{seq}(\beta, S) = \bigcup_{\mathbf{x} \in S} E_{seq}(\beta, \mathbf{x})$. Note that for a stationary point $\mathbf{x}$, using smoothness $\|\nabla Q(\mathbf{x}_{seq})\|_2^2 \leq L^2 \|\mathbf{x}_{seq} - \mathbf{x}\|_2^2$. This implies that if the algorithm converges near a stationary point, then this point has small squared gradient. We denote the event that SGD converges to a point in a set of stationary points S using the event $E_{seq}(\beta, S)$, for some small β.

Definition 1. *Function Q is* (γ,p)-split *for $\gamma > 0$ and $p \in (0, 1)$, if there are two sets of points $S_1, S_2 \subseteq \mathbb{R}^d$ where $\min_{\mathbf{x}_1 \in S_1, \mathbf{x}_2 \in S_2} \|\mathbf{x}_1 - \mathbf{x}_2\|_2 \geq d$ and for some $\alpha, \beta > 0$ such that $(d - \alpha - \beta) \geq \gamma$, $\mathbb{P}[E_{seq}(\alpha, S_1)]\,\mathbb{P}[E_{seq}(\beta, S_2)] \geq p$.*

We also need a slightly more formal model of distributed computation. A *configuration* C consists of the local state of each process, pending messages that were not received yet and the shared memory state of each cluster. An *event* is either a delivery of some message by process i or some operation on its cluster shared memory. A *step* consists of some local computation, possibly a *coin flip*, and a single event. By applying an event preformed by process i to configuration C, we obtain a new configuration with a new local state for process i, possibly removing or adding messages from the pending messages buffer and the updated shared memory state of i's cluster.

Given a configuration C, for every process there is a fixed probability for every step it can take from C. An *execution tree* $\mathcal{T}$ is a directed weighted tree where each node is a configuration and the edges are all the possible steps that can be taken from this configuration. The weight on each edge is exactly the probability for the step to be taken from the parent configuration. The root of the execution tree is an *initial configuration.* Any path in the execution tree, beginning from the root, defines a legal *execution.* The probability over the execution tree for an execution to occur is the product of weights along the path that defines the execution. Given an infinite execution tree $\mathcal{T}$ and an execution α in $\mathcal{T}$, a process i *crashes* in α if it stops taking steps from some point in α. $P(\alpha)$ is the set of all non-crashed processes in α.

Let $\mathcal{A}$ be an algorithm in the cluster-based model and let $\mathcal{T}$ be an execution tree of algorithm $\mathcal{A}$. Let $\mathbf{x}^i(\mathcal{T}) \in \mathbb{R}^d$ be a random variable, corresponding to the output of process i from algorithm $\mathcal{A}$ over execution tree $\mathcal{T}$. If i crashed then this value is $\perp$. For a set of processes P, let $E(\beta, \mathbf{x}, \mathcal{T}, P)$ be the event that for some process $i \in P$, such that $\mathbf{x}^i(\mathcal{T}) \neq \perp$, $\left\|\mathbf{x}^i(\mathcal{T}) - \mathbf{x}\right\|_2 \leq \beta$. For a set of points S, let $E(\beta, S, \mathcal{T}, P) = \bigcup_{\mathbf{x} \in S} E(\beta, \mathbf{x}, \mathcal{T}, P)$.

To ensure the distributed algorithm implements an SGD algorithm, we require it to preserve the same convergence distribution as the sequential algorithm, as explained next.

We say that $\mathcal{A}$ *preserves the convergence distribution over an execution tree* $\mathcal{T}$, if for any $\beta > 0$, set of points S and set of processes P, $\mathbb{P}[E(\beta, S, \mathcal{T}, P)] \geq \mathbb{P}[E_{seq}(\beta, S)]$. We say that $\mathcal{A}$ *distributively implements* $\mathcal{A}^{seq}(Q, T, \mathbf{x}_0)$ *over an execution tree* $\mathcal{T}$ if it converge externally (2) and internally (1) over the outputs $\left\{\mathbf{x}^i(\mathcal{T})\right\}_{i=1}^n$, and it preserves the convergence distribution over $\mathcal{T}$.

The proof of the next theorem adapts the impossibility proofs of [5,18] to probabilistic algorithms by using *probabilistic indistinguishability* [17] (see [6]). The result is proved for a *weak* adversary, which cannot observe the local coin flips, shared memory states and the messages sent during the execution.

Theorem 4. *If a function* Q *is* (γ, p)*-split with* $p > \frac{\delta}{2\gamma^2}$ *and* $\gamma > \sqrt{\delta/2}$*, then no algorithm* $\mathcal{A}$ *distributively implements* $\mathcal{A}^{seq}(Q, T, \mathbf{x}_0)$ *over all execution trees* $\mathcal{T}$ *that contain an execution* α *such that* $|\cup_{i \in P(\alpha)} \mathsf{cluster}(i)| \leq \lfloor n/2 \rfloor$.

7 Summary

We present crash-tolerant asynchronous SGD algorithms for the cluster-based model. For strongly convex functions, our algorithm obtains maximal speedup of the convergence rate over the sequential algorithm, and tolerates any number of failures. For other functions, we employ multidimensional approximate agreement to bring parameters close together in each iteration. This algorithm requires that the set of non-failed processes represents a majority of the processes. We prove that this condition is necessary for optimizing certain functions.

Our results assume processes fail only by crashing, which is an adequate model for several computing systems. Moreover, concentrating on crash failures allows to obtain good bounds on the convergence rate, as well as optimal bounds on the ratio of faulty processes. This also leads to simpler and more modular proofs. We believe that the cluster-based model with crash failures offers a blueprint for optimization algorithms for *high-performance computing* systems.

These architectures include many multi-processor computers, each running multiple threads that share a memory space, which are connected by a network.

Future work could explore the use of specific properties of objective functions, beyond strong convexity, to improve the algorithms. Another direction is to have more cooperative inter-cluster computation, e.g., using fetch&add.

References

1. Agarwal, A., Bartlett, P.L., Ravikumar, P., Wainwright, M.J.: Information-theoretic lower bounds on the oracle complexity of stochastic convex optimization. IEEE Trans. Inf. Theor. **58**(5), 3235–3249 (2012)
2. Agarwal, A., Duchi, J.C.: Distributed delayed stochastic optimization. In: Advances in Neural Information Processing Systems (2011)
3. Alistarh, D., Allen-Zhu, Z., Li, J.: Byzantine stochastic gradient descent. In: Advances in Neural Information Processing Systems (2018)
4. Alistarh, D., De Sa, C., Konstantinov, N.: The convergence of stochastic gradient descent in asynchronous shared memory. In: Proceedings of the 37th Symposium on Principles of Distributed Computing, pp. 169–178 (2018)
5. Attiya, H., Kumari, S., Schiller, N.: Optimal resilience in systems that mix shared memory and message passing. In: 24th International Conference on Principles of Distributed Systems, pp. 16:1–16:16 (2020)
6. Attiya, H., Schiller, N.: Asynchronous fully-decentralized SGD in the cluster-based model (2022). https://arxiv.org/abs/2202.10862
7. Barney, B.: Introduction to parallel computing tutorial. https://hpc.llnl.gov/documentation/tutorials/introduction-parallel-computing-tutorial
8. Ben-Nun, T., Hoefler, T.: Demystifying parallel and distributed deep learning: an in-depth concurrency analysis. ACM Comput. Surv. **52**(4), 1–43 (2019)
9. Bottou, L., Curtis, F.E., Nocedal, J.: Optimization methods for large-scale machine learning. SIAM Rev. **60**(2), 223–311 (2018)
10. Bubeck, S.: Convex optimization: algorithms and complexity. Found. Trends Mach. Learn. **8**(3–4), 231–357 (2015)
11. Dean, J., et al.: Large scale distributed deep networks. In: Advances in Neural Information Processing Systems (2012)
12. Dolev, D., Lynch, N.A., Pinter, S.S., Stark, E.W., Weihl, W.E.: Reaching approximate agreement in the presence of faults. J. ACM **33**(3), 499–516 (1986)
13. El-Mhamdi, E.M., Farhadkhani, S., Guerraoui, R., Guirguis, A., Hoang, L.N., Rouault, S.: Collaborative learning in the jungle (decentralized, Byzantine, heterogeneous, asynchronous and nonconvex learning). In: Advances in Neural Information Processing Systems (2021)
14. El-Mhamdi, E., Guerraoui, R., Guirguis, A., Hoang, L.N., Rouault, S.: Genuinely distributed Byzantine machine learning. In: Proceedings of the 39th Symposium on Principles of Distributed Computing, pp. 355–364 (2020)
15. Függer, M., Nowak, T.: Fast multidimensional asymptotic and approximate consensus. In: 32nd International Symposium on Distributed Computing (DISC 2018), pp. 27:1–27:16 (2018)
16. Ghadimi, S., Lan, G.: Stochastic first- and zeroth-order methods for nonconvex stochastic programming. SIAM J. Optim. **23**(4), 2341–2368 (2013)
17. Goren, G., Moses, Y., Spiegelman, A.: Brief announcement: probabilistic indistinguishability and the quality of validity in Byzantine agreement. In: 35th International Symposium on Distributed Computing (DISC 2021), pp. 57:1–57:4 (2021)

18. Hadzilacos, V., Hu, X., Toueg, S.: Optimal register construction in M&M systems. In: 23rd International Conference on Principles of Distributed Systems (OPODIS 2019), pp. 28:1–28:16 (2020)
19. LeCun, Y., Bengio, Y., Hinton, G.: Deep learning. Nature **521**(7553), 436–444 (2015)
20. Li, S., Ben-Nun, T., Girolamo, S.D., Alistarh, D., Hoefler, T.: Taming unbalanced training workloads in deep learning with partial collective operations. In: Proceedings of the 25th ACM SIGPLAN Symposium on Principles and Practice of Parallel Programming, pp. 45–61 (2020)
21. Li, S., et al.: Breaking (global) barriers in parallel stochastic optimization with wait-avoiding group averaging. IEEE Trans. Parallel Distrib. Syst. **32**(7), 1725–1739 (2021)
22. Lian, X., Huang, Y., Li, Y., Liu, J.: Asynchronous parallel stochastic gradient for nonconvex optimization. In: Advances in Neural Information Processing Systems (2015)
23. Lian, X., Zhang, C., Zhang, H., Hsieh, C.J., Zhang, W., Liu, J.: Can decentralized algorithms outperform centralized algorithms? A case study for decentralized parallel stochastic gradient descent. In: Advances in Neural Information Processing Systems (2017)
24. Lian, X., Zhang, W., Zhang, C., Liu, J.: Asynchronous decentralized parallel stochastic gradient descent. In: Proceedings of the 35th International Conference on Machine Learning, pp. 3043–3052 (2018)
25. Mendes, H., Herlihy, M., Vaidya, N., Garg, V.K.: Multidimensional agreement in Byzantine systems. Distrib. Comput. **28**(6), 423–441 (2014). https://doi.org/10.1007/s00446-014-0240-5
26. Nadiradze, G., Markov, I., Chatterjee, B., Kungurtsev, V., Alistarh, D.: Elastic consistency: a practical consistency model for distributed stochastic gradient descent. In: Proceedings of the AAAI Conference on Artificial Intelligence, vol. 35, no. 10, pp. 9037–9045 (2021)
27. Raynal, M., Cao, J.: One for all and all for one: Scalable consensus in a hybrid communication model. In: 2019 IEEE 39th International Conference on Distributed Computing Systems (ICDCS), pp. 464–471 (2019)
28. Recht, B., Re, C., Wright, S., Niu, F.: Hogwild!: a lock-free approach to parallelizing stochastic gradient descent. In: Advances in Neural Information Processing Systems, pp. 693–701 (2011)
29. Robbins, H., Monro, S.: A stochastic approximation method. Ann. Math. Stat. **22**(3), 400–407 (1951)

Non-crossing Shortest Paths Lengths in Planar Graphs in Linear Time

Lorenzo Balzotti(✉) and Paolo G. Franciosa

Dipartimento di Scienze Statistiche, Sapienza Università di Roma, p.le Aldo Moro 5, 00185 Rome, Italy
{lorenzo.balzotti,paolo.franciosa}@uniroma1.it

Abstract. Given a plane graph it is known how to compute the union of non-crossing shortest paths. These algorithms do not allow neither to list each single shortest path nor to compute length of shortest paths. Given the union of non-crossing shortest paths, we introduce the concept of *shortcuts* that allows us to establish whether a path is a shortest path by checking local properties on faces of the graph. By using shortcuts we can compute the length of each shortest path, given their union, in total linear time, and we can list each shortest path p in $O(\max\{\ell, \ell \log\log \frac{k}{\ell}\})$, where ℓ is the number of edges in p and k the number of shortest paths.

Keywords: shortest paths · undirected planar graphs · non-crossing paths

1 Introduction

The problem of finding non-crossing shortest paths in a plane graph (i.e., a planar graph with a fixed planar embedding) has its primary applications in VLSI layout [6,14,15], where two paths are *non-crossing* if they do not cross each other in the chosen embedding. Thanks to Reif [19], it also appears as a basic step in the computation of maximum flow in a planar network and related problems [1,5]. The *non-crossing shortest path* (NCSP) problem can be formalized as follows: given an undirected plane graph G with non-negative edge lengths and k terminal pairs that lie on a specified face boundary, w.l.o.g, the external face boundary, find k non-crossing shortest paths in G, each connecting a terminal pair. It is assumed that terminal pairs appear in the infinite face so that non-crossing paths exist; this property can be easily verified in linear time.

We deal with a problem strictly linked to the NCSP problem. Our input is an undirected plane graph U composed by the union of k non-crossing shortest paths in a plane graph G whose extremal vertices lie on the same face of G. Thus U arises from the union of paths $p_1, \ldots, p_k$, where, for each $i \in [k]$, p_i is a shortest path from x_i to y_i in G, x_i and y_i are on the infinite face f^∞ of G, p_i and p_j are non-crossing for every $i, j \in [k]$. We stress that we know U but we ignore the p_i's; indeed, algorithms solving the NCSP problem in [3,4,20] compute the union of the non-crossing shortest paths without listing every single path.

M. Mavronicolas (Ed.): CIAC 2023, LNCS 13898, pp. 67–81, 2023.
https://doi.org/10.1007/978-3-031-30448-4_6

We show how to compute the lengths of the p_i's shortest paths in linear time, i.e., in time proportional to the number of edges/vertices in U. In this way we prove that if there exists an algorithm solving the NCSP problem, then we can compute the lengths of shortest non-crossing paths in the same time complexity. We also explain an efficient way to list each path.

Our problem was already discussed in the geometrical case [16,18] under Euclidean distances, but not in a weighted plane graph which has a more general structure.

State of the Art. Takahashi *et al.* [20] proposed an algorithm able to compute k non-crossing shortest paths that requires $O(n \log n)$ time, where n is the size of G. In the same article it is also analyzed the case where the terminal pairs lie on two different face boundaries, and this case is reduced to the previous one within the same computational complexity. The complexity of their solution can be reduced to $O(n \log k)$ by plugging in the linear time algorithm by Henzinger *et al.* [11] for computing a shortest path tree in a planar graph. In the unweighted case, by using the result of Eisenstat and Klein [8], Balzotti and Franciosa showed that k non-crossing shortest paths can be found in $O(n)$ time [4].

Our problem is a special case of computing distances between vertices lying on the same face of a plane undirected graph. This problem can be solved in $O(n \log n)$ by Klein's algorithm [13], that has been recently improved [7] to $O(n \log |f|)$, where f is the face of G where terminal pairs lie.

The algorithm proposed in [20] first computes the union of the k shortest paths, which is claimed to be a forest. The second step relies on the data structure due to Gabow and Tarjan [10] for efficiently solving least common ancestor (LCA) queries in a forest, in order to obtain distances between the terminal pairs in $O(n)$ time.

Actually, the union of the k shortest paths may in general contain cycles. An instance is shown in Fig. 1, in which the unique set of three shortest paths forms a cycle, hence the distances between terminal pairs cannot always be computed by solving LCA queries in a forest. This limitation was noted first in [9,18].

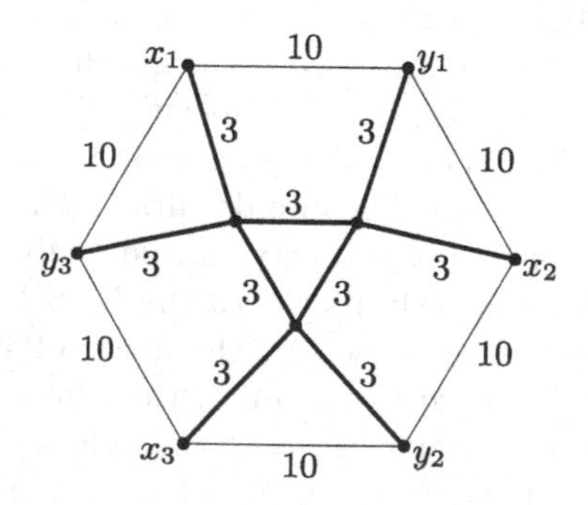

Fig. 1. In this example the union of shortest paths from x_i to y_i, for $i = 1, 2, 3$, contains a cycle (the union is highlighted with bold edges).

Ericksonn and Nayyeri [9] generalized the work in [20] to the case in which the k terminal pairs lie on h face boundaries. They proved that k non-crossing paths, if they exists, can be found in $2^{O(h^2)} n \log k$ time. The authors also stated that the union of non-crossing shortest paths can always be covered with two (possibly non edge-disjoint) forests so that each path is contained in at least one forest. They do not describe how to obtain such a decomposition. This is in contrast with the example in Fig. 2, where we report the union of 15 non-crossing shortest paths that cannot be covered with two forests so that each path is contained in at least one forest (it can be easily proved by enumeration). Recently, Balzotti [2] proved that the union of non-crossing shortest paths can always be covered with four (possibly not edge-disjoint) forests so that each path is contained in at least one forest, and he also showed that four forests are necessary for some instances. We stress that the theoretical result in [2] does not provide a linear time algorithm for computing the covering forests, thus it does not allow to compute path lengths in linear time exploiting lower common ancestors computation [10] as claimed in [20].

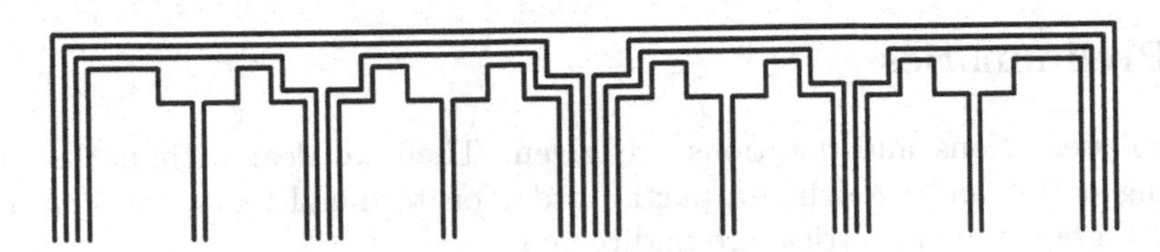

Fig. 2. Union of 15 non-crossing shortest paths that cannot be decomposed into two forests so that each path is contained in at least one forest (parallel adjacent segments represent overlapping paths).

Our Contribution. In this article we exploit the novel concept of *shortcuts*, that are portions of the boundary of a face that allow us to modify a path without increasing (and possibly decreasing) its length. We show that it is possible to establish whether a path is a shortest path by looking at the presence of shortcuts. Hence while being a shortest path is a global property, we can verify it locally by checking a single face at a time for the presence of shortcuts adjacent to the path, ignoring the rest of the graph. Notice that this is only possible when the input graph is the union of non-crossing shortest paths, not for general plane graphs. Without this property, finding one distance is as difficult as finding a shortest path on U; where we recall that U is the graph arising from the union of all non-crossing shortest paths.

Shortcuts allow us to compute the lengths of non-crossing shortest paths in total linear time. Thus we extend the result in [20] also in the case in which U contains cycles. Our novel simple technique does not require the result by Gabow and Tarjan [10]. Moreover, we also propose an algorithm for listing the sequence of edges in a shortest path p joining a terminal pair in $O(\max\{\ell, \ell \log\log(\frac{k}{\ell})\})$, where ℓ is the number of edges in p.

In this way we prove that if there exists an algorithm able to compute the union of non-crossing shortest paths whose extremal vertices lie on the same face of a plane undirected graph, then we can compute the lengths of these paths in the same time complexity.

The algorithm we propose can be easily implemented and it does not require sophisticated data structures. We follow the same approach of Polishchuck and Mitchell [18], that was inspired by Papadopoulou's work [16]. They solve the problem of finding k non-crossing shortest paths in a polygon with n vertices, where distances are defined according to the Euclidean metric.

Structure. The article is organized as follows: in Sect. 2 we give preliminary definitions and notations that will be used in the whole article. In Sect. 3 we deal with shortcuts and in Sect. 4 we use shortcuts in algorithm `ImplicitPaths` for describing an implicit representation of non-crossing shortest paths. This representation is used to compute distances between terminal pairs in linear time, and, in Sect. 5, it is also used to list the non-crossing shortest paths. Finally, in Sect. 6 conclusions are given and we mention some open problems.

2 Preliminaries

General definitions and notations are given. Then we deal with paths, non-crossing paths, and we define a partial order on terminal pairs, the *genealogy tree*. All graphs in this article are undirected.

2.1 Definitions and Notations

Let $G = (V(G), E(G))$ be a plane graph, i.e., a planar graph with a fixed planar embedding. We denote by F_G the set of faces and by f_G^∞ its infinite face. When no confusions arise we use the term face to denote both the border cycle and the finite region bounded by the border cycle, and the infinite face is simply denoted by f^∞.

We use standard union and intersection operators on graphs.

Definition 1. *Given two graphs $G = (V(G), E(G))$ and $H = (V(H), E(H))$, we define the following operations and relations:*

- $G \cup H = (V(G) \cup V(H), E(G) \cup E(H))$;
- $G \cap H = (V(G) \cap V(H), E(G) \cap E(H))$;
- $H \subseteq G \Longleftrightarrow V(H) \subseteq V(G) \wedge E(H) \subseteq E(G)$;
- $G \setminus H = (V(G), E(G) \setminus E(H))$.

Given a graph $G = (V(G), E(G))$, an edge e and a vertex v we write, for short, $e \in G$ in place of $e \in E(G)$ and $v \in G$ in place of $v \in V(G)$. An *ab path* is a path whose extremal vertices are a and b.

We use round brackets to denote ordered sets. For example, $\{a, b, c\} = \{c, a, b\}$ and $(a, b, c) \neq (c, a, b)$. Moreover, for every $r \in \mathbb{N}$ we denote by $[r]$ the set $\{1, \ldots, r\}$.

Let $\omega : E(G) \to \mathbb{R}^+$ be a weight function on edges. The weight function is extended to a subgraph H of G so that $\omega(H) = \sum_{e \in E(H)} \omega(e)$.

We assume that the input of our problem is a plane undirected graph $U = \bigcup_{i \in [k]} p_i$, where p_i is a shortest $x_i y_i$ path in a plane graph G, and the terminal pairs $\{(x_i, y_i)\}_{i \in [k]}$ lie on the infinite face f^∞ of G. We stress that we work with a fixed embedding of U. W.l.o.g. we assume that U is connected, otherwise it suffices to work on each connected component.

For a (possibly not simple) cycle C, we define the *region bounded by* C the maximal subgraph of U whose infinite face has C as boundary. If R is a subgraph of U, then we denote by ∂R the infinite face of R. Moreover, we define $\mathring{R} = R \setminus \partial R$.

Let γ_i be the path in f^∞ that goes clockwise from x_i to y_i, for $i \in [k]$. We assume also that pairs $\{(x_i, y_i)\}_{i \in [k]}$ are *well-formed*, i.e., for all $j, \ell \in [k]$ either $\gamma_j \subseteq \gamma_\ell$ or $\gamma_j \supseteq \gamma_\ell$ or γ_j and γ_ℓ share no edges. We note that if terminal pairs are well-formed, then there exists a set of pairwise non-crossing shortest $x_i y_i$ paths. The reverse is not true if some paths are subpaths of the infinite face of G; this case is not interesting in the applications and has never been studied in literature, where the terminal pairs are always assumed to be well-formed. The well-formed property can be easily verified in linear time, since it corresponds to checking that a string of parentheses is balanced, and it can be done by a sequential scan of the string. We also assume that the terminal pairs are distinct, i.e., there does not exist any pair $i, j \in [k]$ such that $\{x_i, y_i\} = \{x_j, y_j\}$.

Given $i \in [k]$, we denote by *i-path* an $x_i y_i$ path. It is always useful to see each i-path as oriented from x_i to y_i, for $i \in [k]$, even if the path is undirected. For an i-path p, we define Left_p as the left portion of U with respect to p, i.e., the finite region bounded by the cycle formed by p and γ_i; similarly, we define Right_p as the right portion of U with respect to p, i.e., the finite region bounded by the cycle formed by p and $f^\infty \setminus \gamma_i$.

For an i-path p and a j-path q, we say that *q is to the right of p* if $q \subseteq \mathrm{Right}_p$, similarly, we say that *q is to the left of p* if $q \subseteq \mathrm{Left}_p$. Given $R \subseteq U$ and an i-path $p \subseteq R$, for some $i \in [k]$, we say that p is the *leftmost i-path in R* if p is to the left of q for each i-path $q \subseteq R$. Similarly, we say that p is the *rightmost i-path in R* if p is to the right of q for each i-path $q \subseteq R$.

2.2 Paths and Non-crossing Paths

Given an ab path p and a bc path q, we define $p \circ q$ as the (possibly not simple) path obtained by the concatenation of p and q. Given a simple path p and two vertices u, v of p, we denote by $p[u, v]$ the subpath of p with extremal vertices u and v.

Now we introduce the operator $\rtimes$, explained in Fig. 3, that allows us to replace a subpath in a path.

Definition 2. *Let p be a simple ab path, let $u, v \in V(p)$ such that a, u, v, b appear in this order in p and let q be a uv path. We denote by $p \rtimes q$ the (possibly not simple) path $p[a, u] \circ q \circ p[v, b]$.*

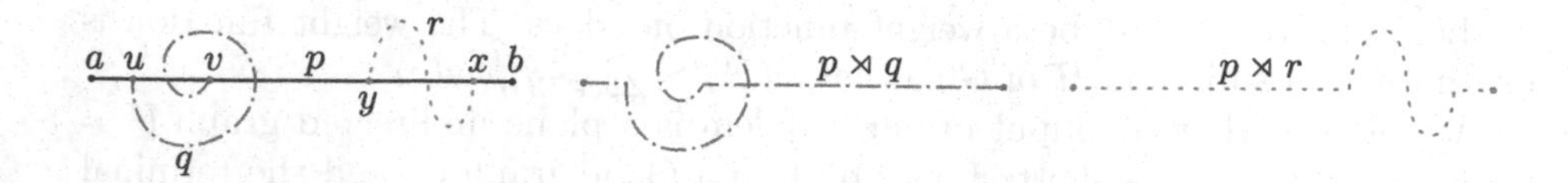

Fig. 3. Illustrating operator $\rtimes$.

We say that two paths in a plane graph G are *non-crossing* if the curves they describe in the graph embedding do not cross each other; a combinatorial definition of non-crossing paths can be based on the *Heffter-Edmonds-Ringel rotation principle* [17]. We stress that this property depends on the embedding of the graph. Non-crossing paths may share vertices and/or edges. We also define a class of paths that will be used later.

Definition 3. *Two paths p and q are* single-touch *if $p \cap q$ is a (possibly empty) path.*

Examples of non-crossing paths and single-touch paths are given in Fig. 4.

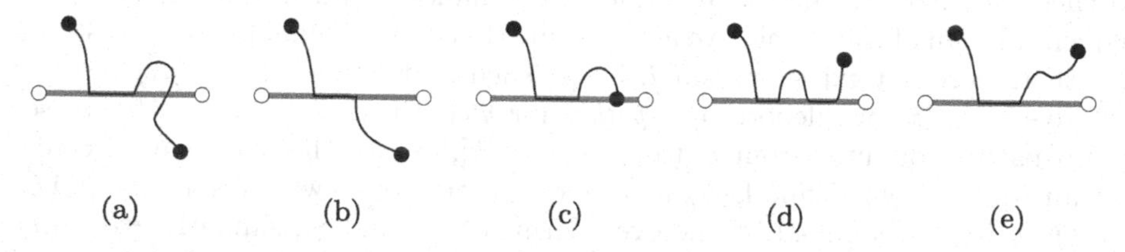

Fig. 4. Paths in (a) and (b) are crossing, while paths in (c), (d), (e) are non-crossing. Moreover, paths in (a), (c) and (d) are not single-touch, while paths in (b) and (e) are single-touch.

Our algorithm builds a set of single-touch paths even if the shortest p_i's paths in G composing the input graph $U = \bigcup_{i \in [k]} p_i$ are not pairwise single-touch. This may happen if there are more shortest paths in G joining the same pair of vertices. Uniqueness of shortest paths can be easily ensured by introducing small perturbations in the weight function of G. We wish to point out that the technique we describe in this article does not rely on perturbation, but we break ties by choosing rightmost or leftmost paths. This implies that our results can also be used in the unweighted case, as done in [4]. Note that the single-touch property does not depend on the embedding, and if the terminal-pairs are well-formed, then it implies the non-crossing property. This is explained in the following remark, and, for this reason, we can say that the solution found by our algorithm holds for any feasible planar embedding of the graph.

Remark 1. If $\{\pi_i\}_{i \in [k]}$ is a set of simple single-touch paths, where π_i is an i-path, for $i \in [k]$, then $\{\pi_i\}_{i \in [k]}$ is a set of pairwise non-crossing paths for all the embeddings of U such that the terminal pairs $\{(x_i, y_i)\}_{i \in [k]}$ are well-formed.

2.3 Genealogy Tree

Given a well-formed set of pairs $\{(x_i, y_i)\}_{i\in[k]}$, we define here a partial ordering as in [20] that represents the inclusion relation between γ_i's. This relation intuitively corresponds to an *adjacency* relation between non-crossing shortest paths joining each pair.

Choose an arbitrary i^* such that there are neither x_j nor y_j, with $j \neq i^*$, walking on f^∞ from x_{i^*} to y_{i^*} (either clockwise or counterclockwise), and let e^* be an arbitrary edge on that walk. For each $j \in [k]$, we can assume that $e^* \notin \gamma_j$, indeed if it is not true, then it suffices to switch x_j and y_j. We say that $i \preceq j$ if $\gamma_i \subseteq \gamma_j$. We define the *genealogy tree* T_g of a set of well formed terminal pairs as the transitive reduction of poset $([k], \preceq)$.

Figure 5 shows an example of well-formed terminal pairs, and the corresponding genealogy tree for $i^* = 1$. From now on, in all figures we draw f^∞ by a solid light grey line.

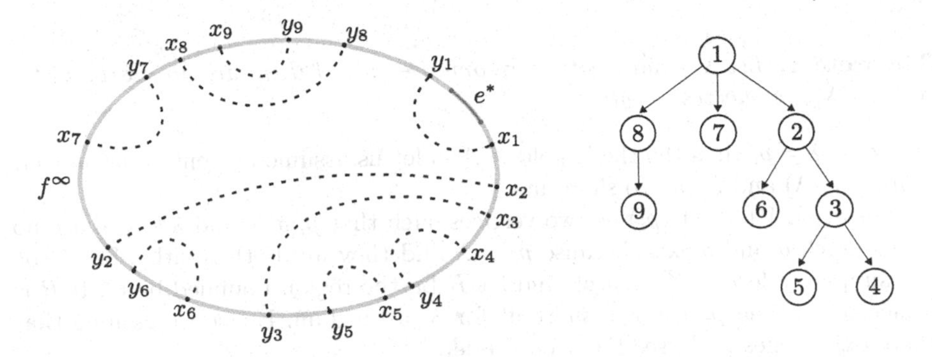

Fig. 5. On the left a set of well-formed terminal pairs. If we choose $i^* = 1$, then we obtain the genealogy tree on the right.

3 Shortcuts

Now we introduce *shortcuts*, that are the main tool of algorithm `ImplicitPaths` introduced in Sect. 4, and the most important theoretical novelty of this article.

Roughly speaking, a shortcut appears if there exists a face f adjacent to a path p so that we can modify p going around f without increasing its length. We show that we can decide whether a path is a shortest path by looking at the existence of shortcuts: in this way, we can check a global property of a path p—i.e., being a shortest path—by checking a local property—i.e., the presence of shortcuts in faces adjacent to p. This result is not true for general plane graphs, but it only holds when the input graph is the union of shortest paths joining well-formed terminal pairs on the same face. Shortcuts are the main tool of algorithm `ImplicitPaths` described in Sect. 4, and the most important theoretical novelty of this article.

Now we can formally define shortcuts, which are clarified in Fig. 6. The main application of shortcuts is stated in Theorem 1.

Definition 4. *Given a path p and a face f containing two vertices $u, v \in p$, we say that a uv subpath q of ∂f not contained in p is a* shortcut *for p if $\omega(p \rtimes q) \leq \omega(p)$.*

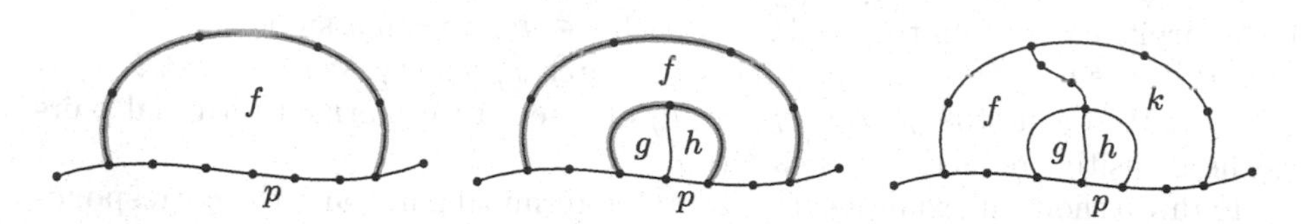

Fig. 6. All edges have unit weight. On the left, highlighted in orange, there is a shortcut for p contained in ∂f. In the middle there are two shortcuts for p both contained in ∂f. On the right there are no shortcuts for p. (Color figure online)

Theorem 1. *Let λ be an i-path, for some $i \in [k]$. If there are no shortcuts for λ, then λ is a shortest i-path.*

Proof. If $\lambda = p_i$ then the thesis holds. Thus let us assume by contradiction that $\omega(p_i) < \omega(\lambda)$ and λ has no shortcuts.

Let $a, b \in V(\lambda) \cap V(p_i)$ be two vertices such that $p_i[a, b]$ and $\lambda[b, a]$ share no edges (such a and b exist because $p_i \neq \lambda$ and they are both i-path). Let C be the simple cycle $p_i[a, b] \circ \lambda[b, a]$, and let R be the region bounded by C. If R is a face of U, then $p_i[a, b]$ is a shortcut for λ, absurdum. Hence we assume that there exist edges in $\mathring{R}$, see Fig. 7 on the left.

Either $R \subsetneq \text{Left}_{p_i}$ or $R \subsetneq \text{Right}_{p_i}$. W.l.o.g., we assume that $R \subsetneq \text{Left}_{p_i}$. Being $U = \bigcup_{j \in [k]} p_j$, for every edge $e \in \mathring{R}$ there exists at least one path $q \in P$ such that $e \in q$, where $P = \bigcup_{i \in [k]} \{p_i\}$. Moreover, the extremal vertices of q are in γ_i because paths in P are non-crossing and $R \subsetneq \text{Left}_{p_i}$.

Now we show by construction that there exist a path $p \in P$ and a face f such that $f \subsetneq R$, ∂f intersects λ on vertices and $\partial f \setminus \lambda \subsetneq p$; thus $\partial f \setminus \lambda$ is a shortcut for λ because p is a shortest path.

For all $q \in P$ such that $q \subsetneq \text{Left}_{p_i}$ we assume that $\text{Left}_q \subsetneq \text{Left}_{p_i}$ (if it is not true, then it suffices to switch the extremal vertices of q).

For each $q \subsetneq \text{Left}_{p_i}$, let $F_q = \{f \in F \mid \partial f \subsetneq R \cap \text{Left}_q\}$, where F is the set of the faces of U. To complete the proof, we have to find a path p such that $|F_p| = 1$, indeed, the unique face f in F_p satisfies $\partial f \setminus \lambda \subsetneq p$, and thus $\partial f \setminus \lambda$ is a shortcut for λ.

Now, let $e_1 \in \mathring{R}$ and let $q_1 \in P$ be such that $e_1 \in q_1$. Being $e_1 \in \mathring{R}$, then $|F_{q_1}| < |F_{p_i}|$ and $|F_{q_1}| > 0$ because $e_1 \in q_1$, see Fig. 7 on the right. If $|F_{q_1}| = 1$, the proof is completed, otherwise we choose $e_2 \in \mathring{R} \cap \mathring{\text{Left}}_{q_1}$ and $q_2 \in P$ such that $e_2 \in q_2$. It holds that $|F_{q_2}| < |F_{q_1}|$ and $|F_{q_2}| > 0$ because $e_2 \in q_2$. By repeating this reasoning, and being $U = \bigcup_{j \in [k]} p_j$, we find a path p such that $|F_p| = 1$. $\square$

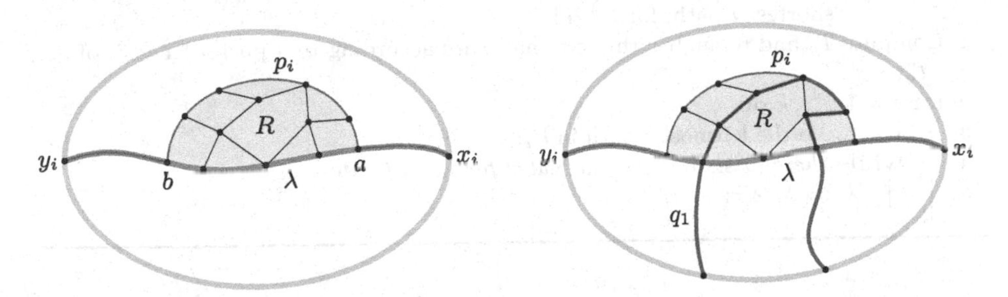

Fig. 7. Paths and regions used in Theorem 1's proof. Path λ is in green, p_i in red, q_1 in blue and region R is highlighted in grey. It holds that $|F_{q_1}| = 4$. (Color figure online)

Given a path p, we say that a path q is a *right shortcut for* p if q is a shortcut for p and $q \subseteq \text{Right}_p$. The following corollary can be proved by the same approach of Theorem 1 and is more useful for our purposes.

Corollary 1. *Let λ be an i-path, for some $i \in [k]$. If there are no right shortcuts for λ, then there does not exist any path $\lambda' \subseteq Right_\lambda$ satisfying $\omega(\lambda') \leq \omega(\lambda)$.*

4 Computing Lengths in Linear Time

In Theorem 2 we show that the distances between terminal pairs can be computed in $O(|E(U)|)$ time by knowing U. This is the main result of this article. To achieve it, we introduce algorithm `ImplicitPaths`, that gives us an implicit representation of non-crossing shortest paths used in the proof of Theorem 2. The implicit representation is described in Remark 3.

The main idea behind algorithm `ImplicitPaths` is the following. We build a set of shortest i-paths $\{\lambda_i\}_{i \in [k]}$, by finding λ_i at iteration i, where the terminal pairs are numbered according to a postorder visit of T_g. In particular, at iteration i we find the rightmost shortest i path in $U_i = \bigcap_{j \in [i-1]} \text{Right}_{\lambda_j}$ in the following way: first we set λ_i as the leftmost i-path in U_i, then we update λ_i by moving right through right shortcuts (the order in which shortcuts are chosen is not relevant). When λ_i has no more right shortcuts, then it is the rightmost shortest i-path in U_i by Corollary 1. For the first iteration, we define U_1 as U.

Lemma 1. *Let $\{\lambda_i\}_{i \in [k]}$ be the set of paths given by algorithm `ImplicitPaths`. Then*

1.(1) λ_i is the rightmost shortest i-path in U_i, for $i \in [k]$;
1.(2) $\{\lambda_i\}_{i \in [k]}$ is a set of single-touch paths.

Algorithm `ImplicitPaths`:

Input: an undirected plane graph U composed by the union of k non-crossing shortest paths in a plane graph G each one joining a terminal pair on the infinite face of G

Output: an implicit representation of a set of paths $\{\lambda_1, \ldots, \lambda_k\}$, where λ_i is a shortest i-path, for $i \in [k]$

1 Compute T_g and renumber the terminal pairs according to a postorder visit of T_g;
2 **for** $i = 1, \ldots, k$ **do**
3   Let λ_i be the leftmost i-path in U_i;
4   **while** *there exists a right shortcut τ for λ_i in U_i* **do**
5         $\lambda_i := \lambda_i \rtimes \tau$;

Proof. We proceed by induction to prove the first statement. Trivially λ_1 is the rightmost shortest 1-path in $U_1 = U$ because of Corollary 1. Let us assume that λ_j is the rightmost shortest j-path in U_j, for $j \in [i-1]$, we have to prove that λ_i is the rightmost shortest i-path in U_i.

In Line 3, we initialize λ_i as the leftmost i-path in U_i. By induction and the postorder visit, at this step, there does not exist in U_i any i-path p to the left of λ_i shorter than λ_i. Otherwise λ_i would cross a path λ_j, for some $j < i$, implying that λ_j is not a shortest j-path. We conclude by the while cycle in Line 4 and Corollary 1.

Statement 1.(2) follows from 1.(1); indeed, if λ_i and λ_j are not single-touch, for some $i, j \in [k]$, then 1.(1) is denied either for λ_i or for λ_j. □

Given $i \in [k]$ we define C_i as the set of children of i in the genealogy tree, moreover, we say that λ_j is a *child of* λ_i if $j \in C_i$.

Before stating our main result, we introduce a trivial consequence of non-crossing property and Jordan's Curve Theorem [12], indeed, every i-path π satisfies that $\pi \circ \gamma_i$ is a closed curve.

Remark 2. Let $\{\pi_i\}_{i\in[k]}$ be a set of non-crossing i-paths. Let $i, j, \ell \in [k]$. If π_i, π_j and π_ℓ share a common edge, then at least two among $\{i, j, \ell\}$ are a couple of ancestor/descendant in the genealogy tree.

Theorem 2. *Given an undirected plane graph U composed by the union of k non-crossing shortest paths in a plane graph G each one joining a terminal pair on the infinite face of G, we can compute the length of each shortest path in $O(|E(U)|)$ total time.*

Proof. We show that during the execution of algorithm `ImplicitPaths` we can also compute the length of λ_i, for all $i \in [k]$, in linear total time. If we also prove that algorithm `ImplicitPaths` can be executed in linear time, then the thesis follows from Lemma 1.

We define, for all $i \in [k]$, $\lambda_{i,0}$ as λ_i after Line 3, i.e., the leftmost i-path in $U_i = \bigcap_{j\in[i-1]} \text{Right}_{\lambda_j}$. We have to show that all the $\lambda_{i,0}$'s and all the shortcuts required in Line 4 can be computed in total linear time.

Let's start dealing with the shortcuts. We do a linear preprocessing that visits clockwise every face. Let f be a face of U: after this preprocessing, the lengths of the clockwise path and of the counterclockwise path in ∂f joining any given pair of vertices in f can both be computed in constant time. Now, if the intersection between ∂f and λ_i, for some $i \in [k]$, is contained in λ_j, for some $j \in C_i$, then we know that there are no right shortcuts in f for λ_i, otherwise they would be right shortcuts for λ_j. Thus we ask for a right shortcut in f for λ_i if and only if λ_i visits at least one edge in ∂f that is not contained in its children or $\lambda_i \cap \partial f$ is contained in an least two children of λ_i (consequently at least one more edge of ∂f is visited). We can check this last case in constant time by verifying whether the vertex v joining two children belongs to exactly one face in U_i or not; this check can be made in constant time because of the embedding by looking whether the degree of v in U_i is 2 or not. If v belongs to exactly one face f, then, by above, we check for a shortcut in f.

In this way, during the execution of algorithm `ImplicitPaths` we ask for a shortcut in f at most $O(|E(f)|)$ times thank also to Remark 2. This implies that finding all the shortcuts requires total linear time.

Now we prove that all the $\lambda_{i,0}$'s can be computed in total linear time. We stress that all the λ_i's and all the $\lambda_{i,0}$'s are represented as list, in this way we can join two paths in constant time. We recall that $\lambda_{i,0}$ is the leftmost i-path in $U_i = \bigcap_{j\in[i-1]} \text{Right}_{\lambda_j}$, C_i is the set of children of i, and γ_i is the clockwise path on the infinite face of G from x_i to y_i. Let $Y_i = \bigcup_{j\in C_i} \lambda_j$ and let f_i be the infinite face of $Y_i \cup \gamma_i$. We observe that, by its definition, $\lambda_{i,0}$ is the counterclockwise i-path on f_i. Clearly, if all the λ_j's, for $j \in C_i$, are vertex disjoint, then λ_j is contained in f_i, for all $j \in C_i$. If the λ_j's are not vertex disjoint, then some edges of Y_i are not in f_i, and by construction, they are not in f_ℓ, for all $i \preceq \ell$. Thus we can see the sequence of the f_i's as an updating graph for which if an edge is deleted at iteration i, then it does not appear again in f_ℓ for all $i \preceq \ell$. Hence, thanks to Remark 2, every edge appears at most two times in this construction. Consequently, all the $\lambda_{i,0}$'s can be computed in total linear time because also to their list representation.

We have proved that algorithm `ImplicitPaths` requires linear time. We use the same argument to compute paths' lengths. Let $i \in [k]$ and $j \in C_i$. At iteration i we know $\omega(\lambda_j)$, and we compute $\omega(\lambda_{i,0} \cap \lambda_j)$ by subtracting from $\omega(\lambda_j)$ the length of edges of λ_j that are not in $\lambda_{i,0}$. In this way we can compute the lengths of $\lambda_{i,0}$ for all $i \in [k]$ in total linear time because every edge is considered at most two times thanks to Remark 2. Being the shortcuts computable in linear time, the thesis follows. □

By following Theorem 2's proof we obtain the following implicit representation of the λ_i's.

Remark 3. Paths λ_i's computed by algorithm `ImplicitPaths` are implicitly represented as follows: $\lambda_i = q_1 \circ \lambda_{j_1}[a_1, b_1] \circ q_2 \circ \lambda_{j_2}[a_2, b_2] \circ \ldots \circ q_r \circ \lambda_{j_r}[a_r, b_r] \circ q_{r+1}$

where $\{j_1, j_2, \ldots, j_r\} \subseteq C_i$, $q_1, q_2, \ldots, q_{r+1}$ are path in U such that $E(q_\ell \cap \lambda_z) = \emptyset$ for all $\ell \in [r+1]$ and $z \in C_i$. Note that the q_ℓ's can be empty.

Now we explain the implicit representation of the λ_i's. If i is a leaf of the genealogy tree, then λ_i is given explicitly. Otherwise we give explicitly edges that do not belong to the children of λ_i, that is the q_i's paths, and we give the intersection path between λ_i and one of its child by specifying the extremal vertices of this intersection. This representation requires linear space thanks to Remark 2.

5 Listing Paths

We study the problem of listing the edges in λ_i, for some $i \in [k]$, after the execution of algorithm `ImplicitPaths`. We want to underline the importance of the single-touch property. In Fig. 8, in (a) four shortest paths are drawn (the graph is unit-weighted), we observe that the single-touch property is clearly not satisfied. A single-touch version of the previous four paths is drawn in (b); it can be obtained by algorithm `ImplicitPaths`. It is clear that the problem of listing the edges in a path in this second case is easier. We stress that in the general case the union of a set of single-touch paths can form cycles, see Fig. 1 for an example.

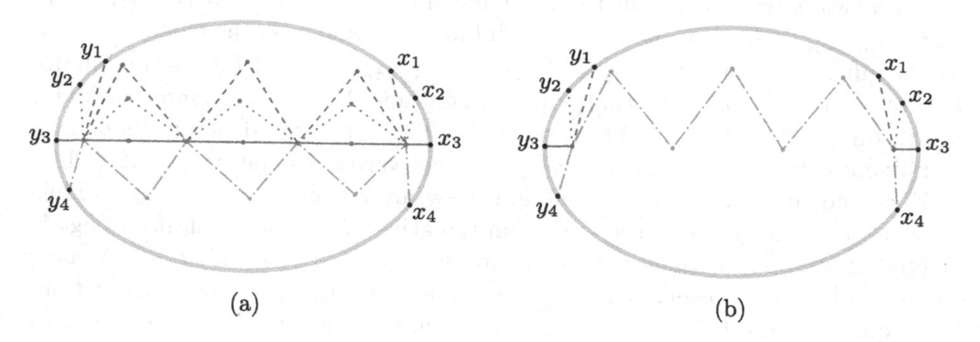

Fig. 8. (a) the union of shortest i-paths, for $i \in [4]$, in unit-weighted graph, each different path has different style, (b) the union of $\{\lambda_i\}_{i\in[4]}$, the output paths of algorithm `ImplicitPaths`.

Theorem 3. *After $O(n)$ time preprocessing, each shortest path λ_i, for $i \in [k]$, can be listed in $O(\max\{\ell_i, \ell_i \log\log(\frac{k}{\ell_i})\})$ time, where ℓ_i is the number of edges of λ_i.*

Proof. For any $i \in [k]$, we denote by $\overrightarrow{\lambda_i}$ the oriented version of λ_i from x_i to y_i. During the execution of algorithm `ImplicitPaths`, we introduce a function `Mark` that marks an arc d with i if and only if the d is used for the first time in the execution of algorithm ImplicitPaths at iteration i. It means that Mark$(d) = i$ if and only if d belongs to $\overrightarrow{\lambda_i}$ and d does not belong to $\overrightarrow{\lambda_j}$, for all $j \preceq i$ and $j \neq i$. This function can be executed within the same time bound of algorithm ImplicitPaths. Now we explain how to find arcs in $\overrightarrow{\lambda_i}$.

Let us assume that $(d_1, \ldots, d_{\ell_i})$ is the ordered sequence of arcs in $\overrightarrow{\lambda_i}$. Let $v = \text{head}[d_{j-1}]$, and let us assume that $deg(v) = r$ in the graph $\bigcup_{j\in[k]} \lambda_j$. We claim that if we know d_{j-1}, then we find d_j in $O(\log\log r)$ time. First we order the outgoing arcs in v in clockwise order starting in d_{j-1}, thus let $\text{Out}_v = (g_1, \ldots, g_r)$ be this ordered set (this order is given by the embedding of the input plane graph). We observe that all arcs in Out_v that are in Left_{λ_i} are in $\overrightarrow{\lambda_w}$ for some $w \le i$, thus Mark$(d) \le i$ for all $d \in \text{Out}_v \cap \text{Left}_{\lambda_i}$. Similarly, all arcs in Out_v that are Right_{λ_i} are in $\overrightarrow{\lambda_z}$ for some $z \ge i$, thus Mark$(d) \ge i$ for all $d \in \text{Out}_v \cap \text{Right}_{\lambda_i}$. Using this observation, we have to find the unique $l \in [r]$ such that Mark$(g_l) \le i$ and Mark$(g_{l+1}) > i$. This can be done in $O(\log\log r)$ by using a van Emde Boas tree [21].

Being the $\overrightarrow{\lambda_i}$'s pairwise single-touch, then $\sum_{v\in V(\lambda_i)} deg(v) \le 2k$, where the equality holds if and only if every $\overrightarrow{\lambda_j}$, for $j \neq i$, intersects on vertices $\overrightarrow{\lambda_i}$ exactly two times, that is the maximum allowed by the single-touch property.

Finally, if $2k \le \ell_i$, then we list $\overrightarrow{\lambda_i}$ in $O(\ell_i)$ because the searches of the correct arcs do not require more than $O(k)$ time, otherwise we note that

$$\sum_{\substack{j=1,\ldots,\ell \\ a_1+\ldots+a_\ell \le 2k}} \log\log a_j \le \ell \log\log\left(\frac{2k}{\ell}\right),$$

so the time complexity follows. □

6 Conclusions

In this article we extend the result of Takahashi *et al.* [20] by computing the lengths of non-crossing shortest paths in undirected plane graphs also in the general case when the union of shortest paths is not a forest. Moreover, we provide an algorithm for listing the sequence of edges of each path in $O(\max\{\ell, \ell\log\log(\frac{k}{\ell})\})$, where ℓ is the number of edges in the shortest path.

We also introduced shortcuts on non-crossing shortest paths in plane graphs. They are a useful tool of interest itself.

All results of this article can be easily applied in a geometric setting, where it is asked to search for paths in polygons instead of plane graphs. The same results can be extended to the case of terminal pairs lying on two distinct faces, by the same argument shown in [20].

We left open the problem of listing a shortest path in time proportional to its length and finding the union of non-crossing shortest paths joining k terminal pairs lying on the same face of a plane graph in $o(n\log k)$ time.

References

1. Ausiello, G., Franciosa, P.G., Lari, I., Ribichini, A.: Max flow vitality in general and st-planar graphs. Networks **74**, 70–78 (2019)
2. Balzotti, L.: Non-crossing shortest paths are covered with exactly four forests. CoRR (2022)
3. Balzotti, L., Franciosa, P.G.: Non-crossing shortest paths in undirected unweighted planar graphs in linear time. J. Graph Algorithms Appl. **26**, 589–606 (2022)
4. Balzotti, L., Franciosa, P.G.: Non-crossing shortest paths in undirected unweighted planar graphs in linear time. In: Kulikov, A.S., Raskhodnikova, S. (eds.) CSR 2022. LNCS, vol. 13296, pp. 77–95. Springer, Cham (2022). https://doi.org/10.1007/978-3-031-09574-0_6
5. Balzotti, L., Franciosa, P.G.: How vulnerable is an undirected planar graph with respect to max flow. In: Mavronicolas, M. (ed.) CIAC 2023. LNCS, vol. 13898, pp. 82–96. Springer, Cham (2023). https://doi.org/10.1007/978-3-031-30448-4_7
6. Bhatt, S.N., Leighton, F.T.: A framework for solving VLSI graph layout problems. J. Comput. Syst. Sci. **28**, 300–343 (1984)
7. Das, D., Kipouridis, E., Gutenberg, M.P., Wulff-Nilsen, C.: A simple algorithm for multiple-source shortest paths in planar digraphs. In: 5th Symposium on Simplicity in Algorithms, SOSA, Virtual Conference, pp. 1–11. SIAM (2022)
8. Eisenstat, D., Klein, P.N.: Linear-time algorithms for max flow and multiple-source shortest paths in unit-weight planar graphs. In: Symposium on Theory of Computing Conference, STOC 2013, pp. 735–744. ACM (2013)
9. Erickson, J., Nayyeri, A.: Shortest non-crossing walks in the plane. In: Proceedings of the Twenty-Second Annual ACM-SIAM Symposium on Discrete Algorithms, SODA 2011, pp. 297–208. SIAM (2011)
10. Gabow, H.N., Tarjan, R.E.: A linear-time algorithm for a special case of disjoint set union. J. Comput. Syst. Sci. **30**, 209–221 (1985)
11. Henzinger, M.R., Klein, P.N., Rao, S., Subramanian, S.: Faster shortest-path algorithms for planar graphs. J. Comput. Syst. Sci. **55**, 3–23 (1997)
12. Jordan, C.: Cours d'analyse de l'École polytechnique, vol. 1. Gauthier-Villars et fils (1893)
13. Klein, P.N.: Multiple-source shortest paths in planar graphs. In: Proceedings of the Sixteenth Annual ACM-SIAM Symposium on Discrete Algorithms, pp. 146–155. SIAM (2005)
14. Leighton, F.T.: Complexity Issues in VLSI: Optimal Layouts for the Shuffle-Exchange Graph and Other Networks. MIT Press (1983)
15. Leighton, F.T.: New lower bound techniques for VLSI. Math. Syst. Theory **17**, 47–70 (1984)
16. Papadopoulou, E.: k-Pairs non-crossing shortest paths in a simple polygon. Int. J. Comput. Geom. Appl. **9**, 533–552 (1999)
17. Pisanski, T., et al.: Topological graph theory. In: Handbook of Graph Theory, Discrete Mathematics and Its Applications, pp. 610–786. Chapman & Hall/Taylor & Francis (2003)
18. Polishchuk, V., Mitchell, J.S.B.: Thick non-crossing paths and minimum-cost flows in polygonal domains. In: Proceedings of the 23rd ACM Symposium on Computational Geometry, pp. 56–65. ACM (2007)
19. Reif, J.H.: Minimum s-t cut of a planar undirected network in $O(n \log^2(n))$ time. SIAM J. Comput. **12**, 71–81 (1983)

20. Takahashi, J., Suzuki, H., Nishizeki, T.: Shortest noncrossing paths in plane graphs. Algorithmica **16**, 339–357 (1996)
21. van Emde Boas, P.: Preserving order in a forest in less than logarithmic time and linear space. Inf. Process. Lett. **6**, 80–82 (1977)

How Vulnerable is an Undirected Planar Graph with Respect to Max Flow

Lorenzo Balzotti(✉) and Paolo G. Franciosa

Dipartimento di Scienze Statistiche, Sapienza Università di Roma, p.le Aldo Moro 5, 00185 Rome, Italy
{lorenzo.balzotti,paolo.franciosa}@uniroma1.it

Abstract. We study the problem of computing the vitality of edges and vertices with respect to the st-max flow in undirected planar graphs, where the vitality of an edge/vertex is the st-max flow decrease when the edge/vertex is removed from the graph. This allows us to establish the vulnerability of the graph with respect to the st-max flow.

We give efficient algorithms to compute an additive guaranteed approximation of the vitality of edges and vertices in planar undirected graphs. We show that in the general case high vitality values are well approximated in time close to the time currently required to compute st-max flow $O(n \log \log n)$. We also give improved, and sometimes optimal, results in the case of integer capacities. All our algorithms work in $O(n)$ space.

Keywords: planar graphs · undirected graphs · max flow · vitality · vulnerability

1 Introduction

Max flow problems have been intensively studied in the last 60 years, we refer to [1,2] for a comprehensive bibliography. Currently, the best known algorithms for general graphs [20,26] compute the max flow between two vertices in $O(mn)$ time, where m is the number of edges and n is the number of vertices.

Italiano *et al.* [19] presented an algorithm for max flow that solves the problem in $O(n \log \log n)$ time for undirected planar graphs. For directed st-planar graphs (i.e., graphs allowing a planar embedding with s and t on the same face) finding a max flow was reduced by Hassin [16] to the single source shortest path (SSSP) problem, that can be solved in $O(n)$ time by the algorithm in [17]. For the planar directed case, Borradaile and Klein [12] presented an $O(n \log n)$ time algorithm. In the special case of directed planar unweighted graphs, an $O(n)$ time algorithm was proposed by Eisenstat and Klein [13].

The effect of edges deletion on the max flow value has been studied since 1963, only a few years after the seminal paper by Ford and Fulkerson [14] in 1956. Wollmer [30] presented a method for determining the most vital edge (i.e., the edge whose deletion causes the largest decrease of the max flow value) in a railway

M. Mavronicolas (Ed.): CIAC 2023, LNCS 13898, pp. 82–96, 2023.
https://doi.org/10.1007/978-3-031-30448-4_7

network. A more general problem was studied in [28], where an enumerative approach is proposed for finding the k edges whose simultaneous removal causes the largest decrease in max flow. Wood [31] showed that this problem is NP-hard in the strong sense, while its approximability has been studied in [4,27].

In this paper we deal with the computation of *vitality* of edges and vertices with respect to the value of an st-max flow in an undirected planar graph G, denoted by MF_G or MF if no confusion arises, where s and t are two fixed vertices. The vitality of an edge e (resp., of a vertex v) measures the st-max flow decrease observed after the removal of edge e (resp., all edges incident on v) from the graph.

A reasonable measure of the overall vulnerability of a network can be the number of edges/vertices with high vitality. So, if all edges and vertices have small vitality, then the graph is robust. We stress that verifying the robustness/vulnerability of the graph by using previous algorithms requires to compute the exact vitality of all edges and/or vertices. We refer to [3,24,25] for surveys on several kind of robustness and vulnerability problems discussed by an algorithmic point of view.

A survey on vitality with respect to the max flow problems can be found in [5]. In the same paper, it is shown that for st-planar graphs (both directed or undirected) the vitality of all edges and all vertices can be found in optimal $O(n)$ time. Ausiello *et al.* [6] proposed a recursive algorithm that computes the vitality of all edges in an undirected unweighted planar graph in $O(n \log n)$ time.

Formally, the st-max flow vitality of a set $x \in (V(G) \setminus \{s,t\}) \cup E(G)$, denoted by $vit(x)$, is equal to $MF_G - MF_{G-x}$, where $G - x$ is the graph obtained from G by removing set x.

The vitality has not been studied directly in [19,23], but their dynamic algorithm leads to the following result.

Theorem 1 ([19,23]). *Let G be a planar graph with positive edge capacities. Then it is possible to compute the vitality of h single edges or the vitality of a set of h edges in* $O(\min\{\frac{hn}{\log n} + n \log\log n,\ hn^{2/3}\log^{8/3} n + n\log n\})$.

Our Contribution. We propose fast algorithms for computing an additive guaranteed approximation of the vitality of all edges and vertices whose capacity is less than an arbitrary threshold c. Later, we explain that these results can be used to obtain a useful approximation of vitality for general distribution of capacities and in the case of power-law distribution. We stress that in real world applications we are usually interested in finding edges and or vertices with high vitality, i.e., edges or vertices whose removal involves relevant decrease on the max flow value.

Our main results are summarized in the following two theorems. For a graph G, we denote by $E(G)$ and $V(G)$ its set of edges and vertices, respectively. Let $c : E(G) \rightarrow \mathbb{R}^+$ be the *edge capacity function*, we define the capacity $c(v)$ of a vertex v as the sum of the capacities of all edges incident on v. Moreover, for every $x \in E(G) \cup V(G)$ we denote by $vit(x)$ its vitality with respect to the st-max flow. We show that we can compute a value $vit^{\delta}(x)$ in $(vit(x) - \delta, vit(x)]$ for any $\delta > 0$.

Theorem 2. *Let G be a planar graph with positive edge capacities. Then for any $c, \delta > 0$, we can compute a value $vit^{\delta}(e) \in (vit(e) - \delta, vit(e)]$ for all $e \in E(G)$ satisfying $c(e) \leq c$, in $O(\frac{c}{\delta}n + n \log\log n)$ time.*

Theorem 3. *Let G be a planar graph with positive edge capacities. Then for any $c, \delta > 0$, we can compute a value $vit^{\delta}(v) \in (vit(v) - \delta, vit(v)]$ for all $v \in V(G)$ satisfying $c(v) \leq c$, in $O(\frac{c}{\delta}n + n \log n)$ time.*

All our algorithms work in $O(n)$ space. To explain the result stated in Theorem 2, we note that in the general case the capacities are not bounded by any function of n. Despite this in many cases we can assume c/δ constant, implying that the time complexity of Theorem 2 is equal to the best current time bound for computing the st-max flow. The following remark is crucial, where $c_{max} = \max_{e \in E(G)} c(e)$.

Remark 1. [Bounding capacities]. We can bound all edge capacities higher than MF to MF, obtaining a new bounded edge capacity function. This change has no impact on the st-max flow value or the vitality of any edge/vertex. Thus w.l.o.g., we can assume that $c_{max} \leq MF$.

By using Remark 1 we can explain why c/δ can be assumed constant, we study separately the case of general distribution of capacities and the case of power-law distribution.

- *General distribution* (after bounding capacities as in Remark 1). If we set $c = c_{max}$ and $\delta = c/k$, for some constant k, then we obtain the capacities with an additive error less than MF/k, because of Remark 1. In many applications this error is acceptable even for small values of k, e.g., $k = 10, 50, 100$. In this way we obtain small percentage error of vitality for edges with high vitality—edges whose vitality is comparable with MF—while edges with small vitality—edges whose vitality is smaller than MF/k—are badly approximated. We stress that we are usually interested in high capacity edges, and that with these choices the time complexity is $O(n \log\log n)$, that is the time currently required for the computation of the st-max flow.
- *Power-law distribution* (after bounding capacities as in Remark 1). The previous method cannot be applied to power-law distribution because most of the edges have capacity lower than MF/k, even for high value of k. Thus we have to separate edges with high capacity and edges with low capacity. Let $c = \frac{c_{max}}{\ell}$ for some constant ℓ and let $H_c = \{e \in E(G) \mid c(e) > c\}$. By power-law distribution, $|H_c|$ is small even for high values of ℓ, and thus we compute the exact vitality of edges in H_c by Theorem 1. For edges with capacity less than c, we set $\delta = c/k$, for some constant k. By Remark 1 we compute the vitality of these edges with an additive error less than $\frac{MF}{k\ell}$. Again, the overall time complexity is equal or close to the time currently required for the computation of the st-max flow.

The result in Theorem 2 is useful even in the case in which $c = MF$ and $vit^{\delta}(e) = 0$ for all $e \in E(G)$. This implies that all edges have vitality in $[0, \delta]$, where δ is the acceptable error in Theorem 2. Thus we certify that all edges in the network have low vitality, so the network is robust.

To apply the same arguments to vertex vitality we need some observations. If G's vertices have maximum degree d, then, after bounding capacities as in Remark 1, we have $\max_{v \in V(G)} c(v) \leq dMF$. Otherwise, we note that a real-world planar graph is expected to have few vertices with high degree (it is also implied by Euler's formula for planar graphs). The exact vitality of these vertices can be computed by Theorem 1 or by the following result.

Theorem 4. *Let G be a planar graph with positive edge capacities. Then for any $S \subseteq V(G)$, we can compute $vit(v)$ for all $v \in S$ in $O(|S|n + n \log\log n)$ time.*

If we denote by $E_S = \sum_{v \in S} deg(v)$, where $deg(v)$ is the degree of vertex v, then the result in Theorem 4 is more efficient than the result given in Theorem 1 if either $|S| < \log n$ and $E_S > |S| \log n$ or $|S| \geq \log n$ and $E_S > \frac{|S|n^{1/3}}{\log^{8/3}}$.

Small Integer Case. In the case of integer capacity values that do not exceed a small constant, or in the more general case in which capacity values are integers with bounded sum we also prove the following corollaries by using the results in [9,10,22].

Corollary 1. *Let G be a planar graph with integer edge capacity and let L be the sum of all the edge capacities. Then*

- *for any $H \subseteq E(G) \cup V(G)$, we can compute $vit(x)$ for all $x \in H$, in $O(|H|n + L)$ time,*
- *for any $c \in \mathbb{N}$, we can compute $vit(e)$ for all $e \in E(G)$ satisfying $c(e) \leq c$, in $O(cn + L)$ time.*

Corollary 2. *Let G be a planar graph with unit edge capacity. Let $n_{>d}$ be the number of vertices whose degree is greater than d. We can compute the vitality of all edges in $O(n)$ time and the vitality of all vertices in $O(\min\{n^{3/2}, n(n_{>d} + d + \log n)\})$ time.*

Corollary 3. *Let G be a planar graph with unit edge capacity where only a constant number of vertices have degree greater than a fixed constant d. Then we can compute the vitality of all vertices in $O(n)$ time.*

Our Approach. We adopt Itai and Shiloach's approach [18], that first computes a modified version D of a dual graph of G, then reduces the computation of the max flow to the computation of non-crossing shortest paths between pairs of vertices of the infinite face of D. We first study the effect on D of an edge or a vertex removal in G, showing that computing the vitality of an edge or a vertex can be reduced to computing some distances in D (see Proposition 2 and Proposition 3).

Then we determine required distances by solving SSSP instances. To decrease the cost we use a divide-and-conquer strategy: we slice D in regions delimited by some of the non-crossing shortest paths computed above. We choose non-crossing shortest paths with similar lengths, so that we compute an additive guaranteed approximation of each distance by looking into a single region instead of examining the whole graph D (see Lemma 2).

Finally we have all the machinery to compute an approximation of required distances of Proposition 2 and Proposition 3 and obtain edge and vertex vitalities.

Structure of the Paper. In Sect. 2 we report main results about how to compute max flow in planar graphs; we focus on the approach in [18] on which our algorithms are based. In Sect. 3 we show preliminary results that allow us to compute edge and vertex vitality. In Sect. 4 we explain our divide-and-conquer strategy. In Sect. 5 we state our main result about edge vitality and in Sect. 6 conclusions and open problems are given.

Vertex vitality and corollaries about small integer capacities are explained in the full version of the paper [8].

2 Max Flow in Planar Graphs

In this section we report some well-known results concerning max flow, focusing on planar graphs.

Given a connected undirected graph $G = (V(G), E(G))$ with n vertices, we denote an edge $e = \{i, j\} \in E(G)$ by the shorthand notation ij, and we define $\text{dist}_G(u, v)$ as the length of a shortest path in G joining vertices u and v. Moreover, for two sets of vertices $S, T \subseteq V(G)$, we define $\text{dist}_G(S, T) = \min_{u \in S, v \in T} \text{dist}_G(u, v)$. We write for short $v \in G$ and $e \in G$ in place of $v \in V(G)$ and $e \in E(G)$, respectively. We say that a path p is an ab path if its extremal vertices are a and b.

Let $s, t \in G$, $s \neq t$, be two fixed vertices. A *feasible flow* in G assigns to each edge $e = ij \in G$ two real values $x_{ij} \in [0, c(e)]$ and $x_{ji} \in [0, c(e)]$ such that: $\sum_{j:ij \in E(G)} x_{ij} = \sum_{j:ij \in E(G)} x_{ji}$, for each $i \in V(G) \setminus \{s, t\}$. The *flow from s to t* under a feasible flow assignment x is defined as $F(x) = \sum_{j:sj \in E(G)} x_{sj} - \sum_{j:sj \in E(G)} x_{js}$. The maximum flow from s to t, denoted by *MF*, is the maximum value of $F(x)$ over all feasible flow assignments x.

An *st-cut* is a partition of $V(G)$ into two subsets S and T such that $s \in S$ and $t \in T$. The capacity of an *st*-cut is the sum of the capacities of the edges $ij \in E(G)$ such that $|S \cap \{i, j\}| = 1$ and $|T \cap \{i, j\}| = 1$. The well known Min-Cut Max-Flow theorem [14] states that the maximum flow from s to t is equal to the capacity of a minimum *st*-cut for any weighted graph G.

We denote by $G - e$ the graph G after the removal of edge e. Similarly, we denote by $G - v$ the graph G after the removal of vertex v and all edges adjacent to v.

Definition 1. *The vitality $vit(e)$ (resp., $vit(v)$) of an edge e (resp., vertex v) with respect to the maximum flow from s to t, according to the general concept* *of vitality in [21], is defined as the difference between the maximum flow in G and the maximum flow in $G - e$ (resp., $G - v$).*

We deal with planar undirected graphs. A *plane graph* is a planar graph with a fixed embedding. The dual of a plane undirected graph G is an undirected planar multigraph G^* whose vertices correspond to faces of G and such that for each edge e in G there is an edge $e^* = \{u^*, v^*\}$ in G^*, where u^* and v^* are the vertices in G^* that correspond to faces f and g adjacent to e in G. Length $w(e^*)$ of e^* equals the capacity of e, moreover, for a subgraph H of G^* we define $w(H) = \sum_{e \in H} w(e)$.

We fix a planar embedding of the graph, and we work on the dual graph G^* defined by this embedding. A vertex v in G generates a face in G^* denoted by f_v^*. We choose in G^* a vertex v_s^* in f_s^* and a vertex v_t^* in f_t^*. A cycle in the dual graph G^* that separates vertex v_s^* from vertex v_t^* is called an *st-separating cycle*. Moreover, we choose a shortest path π in G^* from v_s^* to v_t^*.

Proposition 1 ([18,29]). *A (minimum) st-cut in G corresponds to a (shortest) cycle in G^* that separates vertex v_s^* from vertex v_t^*.*

2.1 Itai and Shiloach's Approach/decomposition

According to the approach by Itai and Shiloach in [18] used to find a min-cut by searching for minimum st-separating cycles, graph G^* is "cut" along the fixed shortest path π from v_s^* to v_t^*, obtaining graph D_G, in which each vertex v_i^* in π is split into two vertices x_i and y_i; when no confusion arises we omit the subscript G. In Fig. 1 there is a plane graph G in black continuous lines and in Fig. 2 on the right graph D. Now we explain the construction of the latter.

Let us assume that $\pi = \{v_1^*, v_2^*, \ldots, v_k^*\}$, with $v_1^* = v_s^*$ and $v_k^* = v_t^*$. For convenience, let π_x be the duplicate of π in D whose vertices are $\{x_1, \ldots, x_k\}$ and let π_y be the duplicate of π in D whose vertices are $\{y_1, \ldots, y_k\}$. For any $i \in [k]$, where $[k] = \{1, \ldots, k\}$, edges in G^* incident on each v_i^* from below π are moved to y_i and edges incident on v_i^* from above π are moved to x_i. Edges incident on v_s^* and v_t^* are considered above or below π on the basis of two dummy edges: the first joining v_s^* to a dummy vertex α inside face f_s^* and the second joining v_t^* to a dummy vertex β inside face f_t^*. In Fig. 1 there is a graph G in black continuous line, G^* in red dashed lines and shortest path π from v_1^* to v_k^*. In Fig. 2, on the left there are the graph G and G^* of Fig. 1 where path π is doubled.

For each $e^* \in \pi$, we denote by e_x^* the copy of e^* in π_x and e_y^* the copy of e^* in π_y. Note that each $v \in V(G) \setminus \{s,t\}$ generates a face f_v^D in D. There are not faces f_s^D and f_t^D because the dummy vertices α and β are inside faces f_s^* and f_t^*, respectively. Both faces f_s^* and f_t^* "correspond" in D to the leftmost x_1y_1 path and to the rightmost x_ky_k path, respectively. Since we are not interested in removing vertices s and t, then faces f_s^D and f_t^D are not needed in D. In Fig. 2, on the right there is graph D built on G in Fig. 1.

If $e^* \notin \pi$, then we denote the corresponding edge in D by e^D. Similarly, if $v_i^* \notin \pi$ (that is, $i > k$), then we denote the corresponding vertex in D by v_i^D.

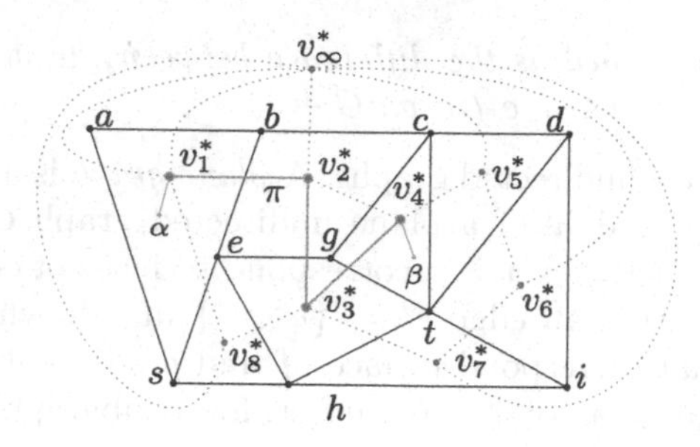

Fig. 1. Graph G in black continuous line, G^* in red dashed lines, shortest path π from v^*_s (v^*_1) to v^*_t (v^*_k) in green, α and β are dummy vertices. (Color figure online)

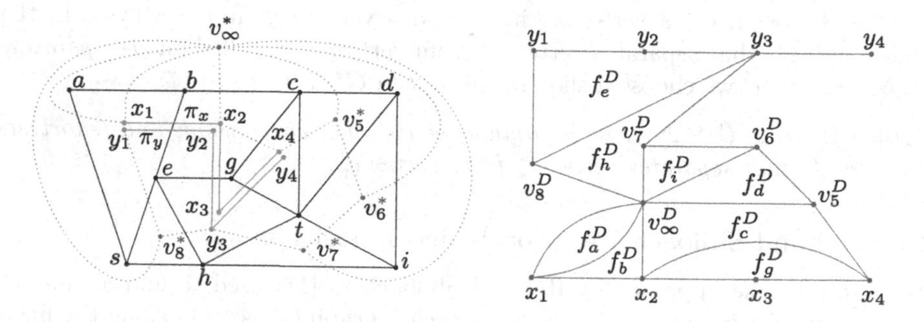

Fig. 2. On the left green path π is doubled into paths π_x and π_y, and edges incident on x_1, y_1, x_4, y_4 in G^* are moved according to the dummy vertices α and β in Fig. 1. On the right graph D. (Color figure online)

3 Preliminary Results

In this section we show preliminary but crucial results (Proposition 2 and Proposition 3) that allow us to compute edge and vertex vitality. In Subsect. 3.1 we show the effects in G^* and D of removing an edge or a vertex from G and in Subsect. 3.2 we state the two main propositions about edge and vertex vitality.

3.1 Effects on G^* and D of Deleting an Edge or a Vertex of G

We observe that removing an edge e from G corresponds to contracting endpoints of e^* into one vertex in G^*. With respect to D, if $e^* \notin \pi$, then the removal of e corresponds to the contraction into one vertex of endpoints of e^D. If $e^* \in \pi$, then both copies of e^* have to be contracted. In Fig. 3 we show the effects of removing edge eg from graph G in Fig. 1.

Let v be a vertex of $V(G)$. Removing v corresponds to contracting vertices of face f^*_v in G^* into a single vertex. In D, if f^*_v and π have no common vertices, then all vertices of f^D_v are contracted into one. Otherwise f^*_v intersects π on vertices $\bigcup_{i \in I}\{v^*_i\}$ for some non empty set $I \subseteq [k]$. Then all vertices of f^D_v are contracted into one vertex, all vertices of $\bigcup_{i \in I}\{x_i\}$ not belonging to f^D_v are contracted into another vertex and all vertices of $\bigcup_{i \in I}\{y_i\}$ not belonging to f_v^D are contracted into a third vertex. For convenience, we define $q^x_{f_v^D} = (\bigcup_{i \in I_v}\{x_i\}) \setminus V(f_v^D)$ and $q^y_{f_v^D} = (\bigcup_{i \in I_v}\{y_i\}) \setminus V(f_v^D)$. To better understand these definitions, see Fig. 4. In Fig. 5 it is shown what happens when we remove vertex g of graph G in Fig. 1.

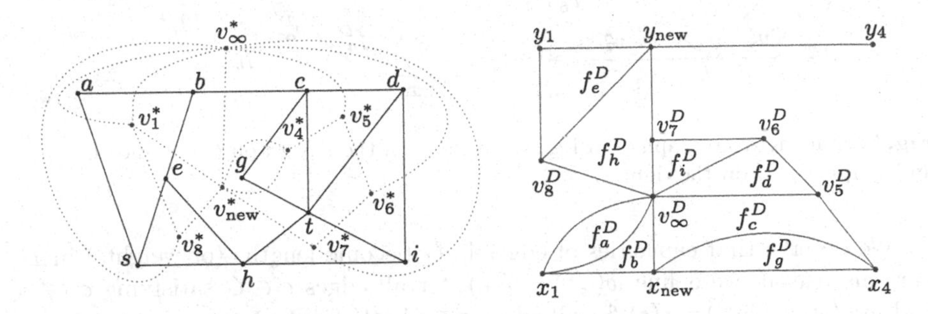

Fig. 3. Starting from graph G in Fig. 1, we show on the left graph $G-eg$ and $(G-eg)^*$, and graph D_{G-eg} on the right.

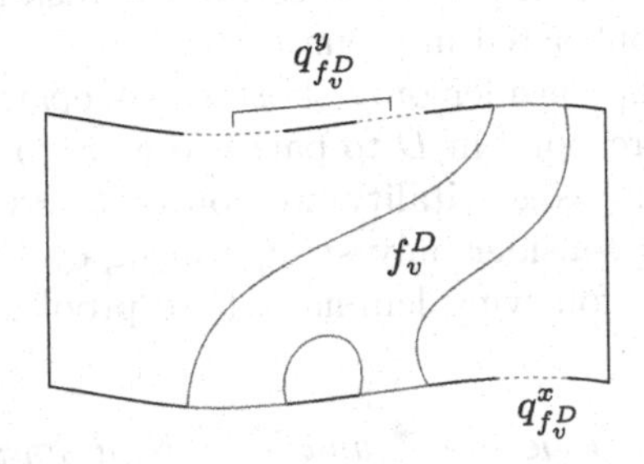

Fig. 4. A face f_v^D, for some $v \in V(G)$, and sets $q^x_{f_v^D}$ and $q^y_{f_v^D}$. Removing v from G corresponds in D to contracting vertices of f_v^D, $q^x_{f_v^D}$ and $q^y_{f_v^D}$ in three distinct vertices.

3.2 Vitality vs. Distances in D

The main results of this subsection are Proposition 2 and Proposition 3. The first proposition shows which distances in D are needed to obtain edge vitality and in the latter proposition we do the same for vertex vitality. In Subsect. 3.1 we have proved that removing an edge or a vertex from G corresponds to contracting in single vertices some sets of vertices of D. The main result of Proposition 2 and Proposition 3 is that we can consider these vertices individually.

Let e be an edge of G. The removal of e from G corresponds to the contraction of endpoints of e^* into one vertex in G^*. Thus if an st-separating cycle γ of G^* contains e^*, then the removal of e from G reduces the length of γ by $w(e^*)$. Thus e has strictly positive vitality if and only if there exists an st-separating cycle γ in G^* whose length is strictly less than $MF + w(e^*)$ and $e^* \in \gamma$. This is the main idea to compute the vitality of all edges. Now we have to translate it to D.

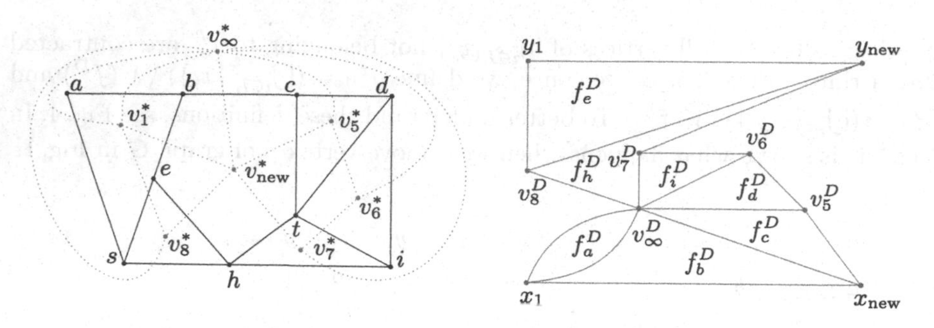

Fig. 5. Starting from graph G in Fig. 1, we show on the left graph $G-g$ and $(G-g)^*$, and graph D_{G-g} on the right.

We observe that capacities of edges in G become lengths (or weights) in D. For this reason, we define $w(e^D) = c(e)$, for all edges $e \in G$ satisfying $e^* \notin \pi$ and $w(e_x^D) = w(e_y^D) = c(e)$ for all edges $e \in G$ satisfying $e^* \in \pi$.

For $i \in [k]$, we define $d_i = \text{dist}_D(x_i, y_i)$. We observe that $MF = \min_{i\in[k]} d_i$. For a subset S of $V(D)$ and any $i \in [k]$ we define $d_i(S) = \min\{d_i, \text{dist}_D(x_i, S) + \text{dist}_D(y_i, S)\}$. We observe that $d_i(S)$ represents the distance in D from x_i to y_i if all vertices of S are contracted into one.

Itai and Shiloach [18] consider only shortest st-separating cycles that cross π exactly once, that correspond in D to paths from x_i to y_i, for some $i \in [k]$. In our approach, to compute edge vitality, we contract vertices of an edge of G^*. Despite this we can still consider only st-separating cycles that cross π exactly once as explained in the following lemma, whose proof is omitted due to page limit.

Lemma 1. *Let e^* be an edge of G^* and let γ be a simple st-separating cycle such that crosses π exactly once and $e^* \in \gamma$. After contracting vertices of e^* into one vertex, then γ becomes the union of two simple cycles and exactly one of them, called γ', is an st-separating cycle that crosses π exactly once. It holds that $vit(e) = MF - w(\gamma')$.*

For every $x \in V(G) \cup E(G)$ we define MF_x as the max flow in graph $G - x$. By definition, $vit(x) = MF - MF_x$ and, trivially, x has strictly positive vitality if and only if $MF_x < MF$.

Proposition 2. *For every edge e of G, if $e^* \notin \pi$, then $MF_e = \min_{i\in[k]}\{d_i(e^D)\}$. If $e^* \in \pi$, then $MF_e = \min_{i\in[k]} \{ \min\{d_i(e_x^D), d_i(e_y^D)\}\}$.*

Proof. Let e be an edge of G. If $vit(e) = 0$, then $MF_e = MF$ and the thesis trivially holds. Hence let us assume $vit(e) > 0$, then Lemma 1 there exists an st-separating cycle in G^* that crosses π exactly once satisfying $w(\gamma) < MF + w(e^*)$ and $e^* \in \gamma$. If $e^* \notin \pi$, then e corresponds in D to edge e^D, thus the thesis holds. If $e^* \in \pi$, then we note that every path in D containing both e_x^D and e_y^D corresponds in G^* to an st-separating cycle that passes through e^* twice, thus its length is equal or greater than $MF + 2c(e)$. Thus we consider only paths that contain e_x^D or e_y^D but not both. The thesis follows. □

Note that if f_v^* and π have some common vertices, then one among $q^x_{f_v^D}$ and $q^y_{f_v^D}$ could be empty. For this reason, we set $d_i(\emptyset) = +\infty$, for all $i \in [k]$. The proof of the following proposition is similar to Proposition 2's proof and it is reported in the full version of the paper.

Proposition 3. *For every vertex v of G, if f_v^* and π have no common vertices, then $MF_v = \min_{i\in[k]}\{d_i(f)\}$, where $f = f_v^D$, otherwise*

$$MF_v = \min \left\{ \begin{array}{l} \min_{i\in[k]}\{d_i(f)\}, \\ \min_{i\in[k]}\{d_i(q_f^x)\}, \\ \min_{i\in[k]}\{d_i(q_f^y)\}, \\ dist_D(f, q_f^x), \\ dist_D(f, q_f^y) \end{array} \right\}. \quad (1)$$

4 Slicing Graph D Preserving Approximated Distances

In this section we explain our divide-and-conquer strategy. We slice graph D along shortest x_iy_i's paths. If these paths have lengths that differ at most δ, then we have a δ additive approximation of distances required in Proposition 2 and Proposition 3 by looking into a single slice instead of the whole graph D. This result is stated in Lemma 2. These slices can share boundary vertices and edges, implying that their dimension might be $O(n^2)$. In Lemma 3 we compute an implicit representation of these slices in linear time.

From now on, we mainly work on graph D, thus we omit the superscript D unless we refer to G or G^*. To work in D we need a shortest x_iy_i path and its length, for all $i \in [k]$. In the following theorem we show time complexities for obtaining elements in D. We say that two paths are *single-touch* if their intersection is still a path.

Given two graphs $A = (V(A), E(A))$ and $B = (V(B), E(B))$ we define $A \cup B = (V(A) \cup V(B), E(A) \cup E(B))$ and $A \cap B = (V(A) \cap V(B), E(A) \cap E(B))$.

Theorem 5 ([11,15,19]). *If G is a positive edge-weighted planar graph,*

- *we can compute $U = \bigcup_{i\in[k]} p_i$ and $w(p_i)$ for all $i \in [k]$, where p_i is a shortest x_iy_i path in D and $\{p_i\}_{i\in[k]}$ is a set of pairwise non-crossing single-touch paths, in $O(n \log\log n)$ time—see [19] for computing U and [7] for computing $w(p_i)$'s,*
- *for every $I \subseteq [k]$, we can compute $\bigcup_{i\in I} p_i$ in $O(n)$ time—see [15] by noting that U is a forest and the paths can be found by using nearest common ancestor queries.*

From now on, for each $i \in [k]$ we fix a shortest x_iy_i path p_i, and we assume that $\{p_i\}_{i\in[k]}$ is a set of pairwise single-touch non-crossing shortest paths. Let $U = \bigcup_{i\in[k]} p_i$, see Fig. 6(a).

Given an ab path p and a bc path q, we define $p \circ q$ as the (possibly not simple) ac path obtained by the union of p and q. Each p_i's splits D into two parts as shown in the following definition and in Fig. 6(b).

Definition 2. *For every $i \in [k]$, we define Left$_i$ as the subgraph of D bounded by the cycle $\pi_y[y_1, y_i] \circ p_i \circ \pi_x[x_i, x_1] \circ l$, where l is the leftmost x_1y_1 path in D. Similarly, we define Right$_i$ as the subgraph of D bounded by the cycle $\pi_y[y_i, y_k] \circ r \circ \pi_x[x_k, x_i] \circ p_i$, where r is the rightmost x_ky_k path in D.*

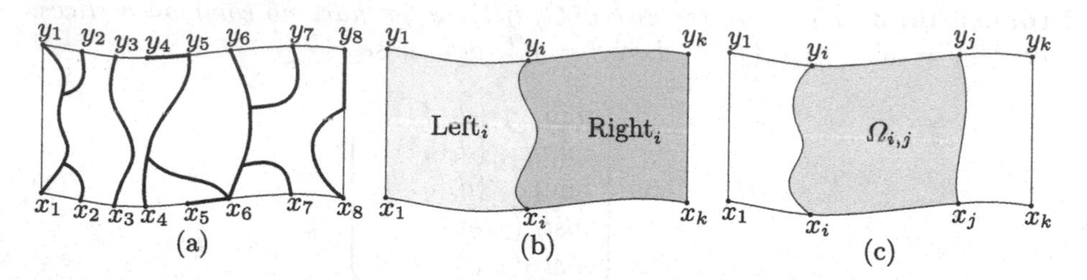

Fig. 6. In (a) the graph U in bold and in (b) subgraphs Left$_i$ and Right$_i$ are highlighted. In (b) subgraph $\Omega_{i,j}$, for some $i < j$.

Based on Definition 2, for every $i, j \in [k]$, with $i < j$, we define $\Omega_{i,j} =$ Right$_i \cap$ Left$_j$, see Fig. 6(c). We classify (x_i, y_i)'s pairs according to the difference between d_i and MF. Each class contains pairs for which this difference is about r times δ; where $\delta > 0$ is an arbitrarily fixed value.

For every $r \in \mathbb{N}$, we define $L_r = (\ell_1^r, \ldots, \ell_{z_r}^r)$ as the ordered list of indices in $[k]$ such that $d_j \in [MF + \delta r, MF + \delta(r + 1))$, for all $j \in L_r$, and $\ell_j^r < \ell_{j+1}^r$ for all $j \in [z_r - 1]$. It is possible that $L_r = \emptyset$ for some $r > 0$ (it holds that $L_0 \neq \emptyset$). If no confusion arises, we omit the superscript r; thus we write ℓ_i in place of ℓ_i^r.

The following lemma is the key of our slicing strategy. In particular, Lemma 2 can be applied for computing distances required in Proposition 2 and Proposition 3, since the vertex set of a face or an edge of D is always contained in a slice. An application is in Fig. 8.

Lemma 2. *Let $r > 0$ and let $L_r = (\ell_1, \ell_2, \ldots, \ell_z)$. Let S be a set of vertices of D with $S \subseteq \Omega_{\ell_i, \ell_{i+1}}$ for some $i \in [z - 1]$. Then*

$$\min_{\ell \in L_r} d_\ell(S) > \min\{d_{\ell_i}(S), d_{\ell_{i+1}}(S)\} - \delta.$$

Moreover, if $S \subseteq$ Left$_{\ell_1}$ (resp., $S \subseteq$ Right$_{\ell_z}$) then $\min_{\ell \in L_r} d_\ell(S) > d_{\ell_1}(S) - \delta$ (resp., $\min_{\ell \in L_r} d_\ell(S) > d_{\ell_z}(S) - \delta$).

Proof. We need the following crucial claim.

a) Let $i < j \in L_r$. Let L be a set of vertices in Left$_i$ and let R be set of vertices in Right$_j$. Then $d_i(L) < d_j(L) + \delta$ and $d_j(R) < d_i(R) + \delta$.

Proof of a): we prove that $d_i(L) < d_j(L) + \delta$. By symmetry, it also proves that $d_j(R) < d_i(R) + \delta$. Let us assume by contradiction that $d_i(L) \geq d_j(L) + \delta$.

Let α (resp., ϵ, μ, ν) be a path from x_i (resp., y_i, x_j, y_j) to z_α (resp., z_ϵ, z_μ, z_ν) whose length is $d(x_i, L)$ (resp. $d(y_i, L)$, $d(x_j, L)$, $d(y_j, L)$), see Fig. 7 on the left. Being $x_j, y_j \in \text{Right}_i$ and $L \subseteq \text{Left}_i$, then μ and ν cross p_i. Let v be the vertex that appears first in $p_i \cap \mu$ starting from x_j on μ and let u be the vertex that appears first in $p_i \cap \nu$ starting from y_j on ν. An example of these paths is in Fig. 7 on the left. Let $\zeta = p_i[y_i, u]$, $\theta = p_i[u, v]$, $\beta = p_i[x_i, v]$, $\kappa = \mu[x_j, v]$, $\iota = \nu[y_j, u]$, $\eta = \nu[u, z_\nu]$ and $\gamma = \mu[v, z_\mu]$, see Fig. 7 on the right.

Now $w(\beta)+w(\gamma) \geq w(\alpha)$, otherwise α would not be a shortest path from x_i to L. Similarly $w(\zeta)+w(\eta) \geq w(\epsilon)$. Moreover, being $w(\zeta)+w(\theta)+w(\beta) = d_i$, then $w(\theta) \leq d_i - w(\alpha)+w(\gamma)-w(\epsilon)+w(\eta)$. Being $d_i(L) \geq d_j(L)+\delta$, then $w(\alpha)+w(\epsilon) \geq w(\mu)+w(\nu)+\delta$, this implies $w(\alpha)+w(\epsilon) \geq w(\kappa)+w(\gamma)+w(\iota)+w(\eta)+\delta$.

It holds that $w(\theta)+w(\kappa)+w(\iota) \leq d_i - w(\alpha)+w(\gamma)-w(\epsilon)+w(\eta)+w(\alpha)+w(\epsilon)-w(\gamma)-w(\eta)-\delta = d_i - \delta < d_j$ because $i, j \in L_r$ imply $|d_i - d_j| < \delta$. Thus $\kappa \circ \theta \circ \iota$ is a path from x_j to y_j strictly shorter than d_j, absurdum. End Proof of a).

Being $S \subseteq \text{Right}_{\ell_j}$ for all $j < i$ and $S \subseteq \text{Left}_{\ell_{j'}}$ for all $j' > i+1$, then the first part of the thesis follows from a). The second part follows also from a) by observing that if $S \subseteq \text{Left}_{\ell_1}$, then $S \subseteq \text{Left}_{\ell_i}$ for all $i \in L_r$. □

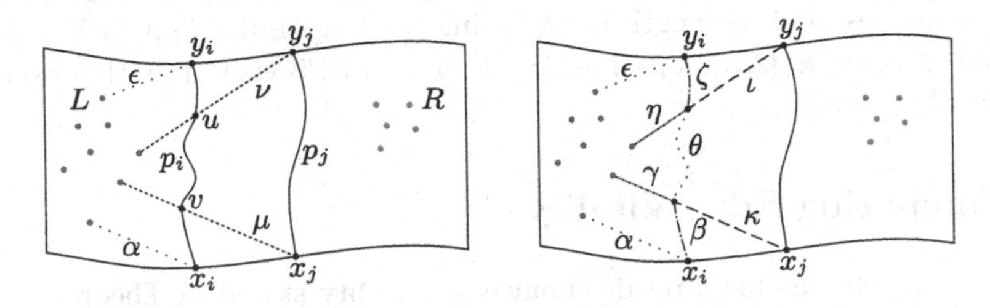

Fig. 7. Example of paths and subpaths used in the proof of a).

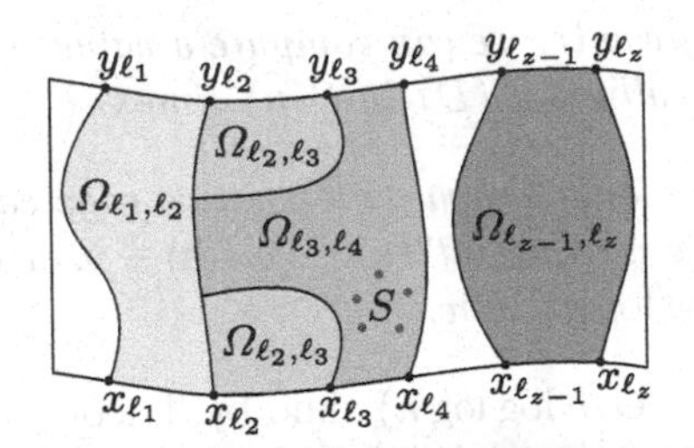

Fig. 8. By Lemma 2, it holds that $\min_{\ell \in L_r} d_\ell(S) \geq \min\{d_{\ell_3}(S), d_{\ell_4}(S)\} - \delta$.

To compute distances in D we have to solve some SSSP instances in some $\Omega_{i,j}$'s subsets. These subsets can share boundary edges, thus the sum of their edges might be $O(n^2)$. We note that, by the single-touch property, if an edge e belongs to $\Omega_{i,j}$ and $\Omega_{j,\ell}$ for some $i < j < \ell \in [k]$, then $e \in p_j$.

To overcome this problem we introduce subsets $\widetilde{\Omega}_{i,j}$ in the following way: for any $i < j \in [k]$, if $p_i \cap p_j$ is a non-empty path q, then we define $\widetilde{\Omega}_{i,j}$ as $\Omega_{i,j}$ in which we replace path q by an edge with the same length; note that the single-touch property implies that all vertices in q but its extremal have degree two. Otherwise, we define $\widetilde{\Omega}_{i,j} = \Omega_{i,j}$. Note that distances between vertices in $\widetilde{\Omega}_{i,j}$ are the same as in $\Omega_{i,j}$. It the following lemma we show how to compute some $\widetilde{\Omega}_{i,j}$'s in $O(n)$ time.

Lemma 3. *Let $A = (a_1, a_2, \ldots, a_z)$ be any increasing sequence of indices of $[k]$. It holds that $\sum_{i \in [z-1]} |E(\widetilde{\Omega}_{a_i,a_{i+1}})| = O(n)$. Moreover, given U, we can compute $\widetilde{\Omega}_{a_i,a_{i+1}}$, for all $i \in [z-1]$, in $O(n)$ total time.*

Proof. For convenience, we denote by Ω_i the set $\Omega_{a_i,a_{i+1}}$, for all $i \in [z-1]$. We note that if $e \in \Omega_i \cap \Omega_{i+1}$, then $e \in p_{i+1}$. Thus, if e belongs to more than two Ω_i's, then e belongs to exactly two $\widetilde{\Omega}$'s because it is contracted in all other Ω_i's by definition of the $\widetilde{\Omega}_i$'s. Thus $\sum_{i \in [z-1]} |E(\widetilde{\Omega}_i)| = O(n) + O(z) = O(n)$ because $z \leq k \leq n$.

To obtain all $\widetilde{\Omega}_i$'s, we compute $U_z = \bigcup_{a \in A} p_a$ in $O(n)$ time by Theorem 5. Then we preprocess all trees in U_z in $O(n)$ time by using Gabow and Tarjan's result [15] in order to obtain the intersection path $p_{a_i} \cap p_{a_{i+1}}$ via lowest common ancestor queries, and its length in $O(1)$ time with a similar approach. Finally, we build $\widetilde{\Omega}_i$ in $O(|E(\widetilde{\Omega}_i)|)$, for all $i \in [z-1]$, with a BFS visit of Ω_i that excludes vertices of $p_{a_i} \cap p_{a_{i+1}}$. □

5 Computing Edge Vitality

Now we can give our main result about edge vitality stated in Theorem 2. We need the following preliminary lemma that is an easy consequence of Lemma 2 and Lemma 3, it's proof is in the full version of the paper.

Lemma 4. *Let $r \in \mathbb{N}$, given U, we can compute a value $\alpha_r(e) \in [\min_{i \in L_r}\{d_i(e)\}, \min_{i \in L_r}\{d_i(e)\} + \delta)$ for all $e \in E(D)$ in $O(n)$ time.*

Theorem 2 *Let G be a planar graph with positive edge capacities. Then for any $c, \delta > 0$, we can compute a value $vit^\delta(e) \in (vit(e) - \delta, vit(e)]$ for all $e \in E(G)$* i *$c(e) \leq c$,* in *$O(\frac{c}{\delta}n + n \log\log n)$ time.*

Proof. We compute U in $O(n \log\log n)$ time by Theorem 5. If $d_i > MF + c(e)$, then $d_i(e^D) > MF$, so we are only interested in computing (approximate) values of $d_i(e^D)$ for all $i \in [k]$ satisfying $d_i < MF + c$. By Lemma 4, for each $r \in \{0, 1, \ldots, \lceil \frac{c}{\delta} \rceil\}$, we compute $\alpha_r(e^D) \in [\min_{i \in L_r} d_i(e^D), \min_{i \in L_r} d_i(e^D) + \delta)$, for all $e^D \in E(D)$, in $O(n)$ time. Then, for each $e^D \in E(D)$, we compute $\alpha(e^D) = \min_{r \in \{0,1,\ldots,\frac{c}{\delta}\}} \alpha_r(e^D)$; it holds that $\alpha(e^D) \in [\min_{i \in [k]}\{d_i(e^D)\}, \min_{i \in [k]}\{d_i(e^D)\} + \delta)$. Then, by Proposition 2, for each $e \in E(G)$ satisfying $c(e) \leq c$, we compute a value $vit^\delta(e) \in (vit(e) - \delta, vit(e)]$ in $O(1)$ time. □

6 Conclusions and Open Problems

We proposed algorithms for computing an additive guaranteed approximation of the vitality of all edges or vertices with bounded capacity with respect to the max flow from s to t in undirected planar graphs. These results are relevant for determining the vulnerability of real world networks, under various capacity distributions.

It is still open the problem of computing the exact vitality of all edges of an undirected planar graph within the same time bound as computing the max flow value, as is already known for the st-planar case.

References

1. Ahuja, R.K., Magnanti, T.L., Orlin, J.B.: Network Flows (1988)
2. Ahuja, R.K., Magnanti, T.L., Orlin, J.B., Reddy, M.: Applications of network optimization. In: Handbooks in Operations Research and Management Science, vol. 7, pp. 1–83 (1995)
3. Alderson, D.L., Brown, G.G., Carlyle, W.M., Cox, L.A., Jr.: Sometimes there is no "most-vital" arc: assessing and improving the operational resilience of systems. Mil. Oper. Res. **18**, 21–37 (2013)
4. Altner, D.S., Ergun, Ö., Uhan, N.A.: The maximum flow network interdiction problem: valid inequalities, integrality gaps, and approximability. Oper. Res. Lett. **38**, 33–38 (2010)
5. Ausiello, G., Franciosa, P.G., Lari, I., Ribichini, A.: Max flow vitality in general and st-planar graphs. Networks **74**, 70–78 (2019)
6. Ausiello, G., Franciosa, P.G., Lari, I., Ribichini, A.: Max-flow vitality in undirected unweighted planar graphs. CoRR, abs/2011.02375 (2020)
7. Balzotti, L., Franciosa, P.G.: Computing lengths of non-crossing shortest paths in planar graphs. CoRR, abs/2011.04047 (2020)
8. Balzotti, L., Franciosa, P.G.: Max flow vitality of edges and vertices in undirected planar graphs. CoRR, abs/2201.13099 (2022)
9. Balzotti, L., Franciosa, P.G.: Non-crossing shortest paths in undirected unweighted planar graphs in linear time. J. Graph Algorithms Appl. **26**, 589–606 (2022)
10. Balzotti, L., Franciosa, P.G.: Non-crossing shortest paths in undirected unweighted planar graphs in linear time. In: Kulikov, A.S., Raskhodnikova, S. (eds.) CSR 2022. LNCS, vol. 13296, pp. 77–95. Springer, Cham (2022). https://doi.org/10.1007/978-3-031-09574-0_6
11. Balzotti, L., Franciosa, P.G.: Non-crossing shortest paths lengths in planar graphs in linear time. In: Mavronicolas, M. (ed.) CIAC 2023. LNCS, vol. 13898, pp. 67–81. Springer, Cham (2023). https://doi.org/10.1007/978-3-031-30448-4_6
12. Borradaile, G., Klein, P.N.: An $O(n \log n)$ algorithm for maximum st-flow in a directed planar graph. J. ACM **56**, 9:1–9:30 (2009)
13. Eisenstat, D., Klein, P.N.: Linear-time algorithms for max flow and multiple-source shortest paths in unit-weight planar graphs. In: Symposium on Theory of Computing Conference, STOC 2013, pp. 735–744. ACM (2013)
14. Ford, L.R., Fulkerson, D.R.: Maximal flow through a network. Can. J. Math. **8**, 399–404 (1956)
15. Gabow, H.N., Tarjan, R.E.: A linear-time algorithm for a special case of disjoint set union. J. Comput. Syst. Sci. **30**, 209–221 (1985)

16. Hassin, R.: Maximum flow in (s, t) planar networks. Inf. Process. Lett. **13**, 107 (1981)
17. Henzinger, M.R., Klein, P.N., Rao, S., Subramanian, S.: Faster shortest-path algorithms for planar graphs. J. Comput. Syst. Sci. **55**, 3–23 (1997)
18. Itai, A., Shiloach, Y.: Maximum flow in planar networks. SIAM J. Comput. **8**, 135–150 (1979)
19. Italiano, G.F., Nussbaum, Y., Sankowski, P., Wulff-Nilsen, C.: Improved algorithms for min cut and max flow in undirected planar graphs. In: Proceedings of the 43rd ACM Symposium on Theory of Computing, pp. 313–322. ACM (2011)
20. King, V., Rao, S., Tarjan, R.E.: A faster deterministic maximum flow algorithm. J. Algorithms **17**, 447–474 (1994)
21. Koschützki, D., Lehmann, K.A., Peeters, L., Richter, S., Tenfelde-Podehl, D., Zlotowski, O.: Centrality indices. In: Brandes, U., Erlebach, T. (eds.) Network Analysis. LNCS, vol. 3418, pp. 16–61. Springer, Heidelberg (2005). https://doi.org/10.1007/978-3-540-31955-9_3
22. Kowalik, L., Kurowski, M.: Short path queries in planar graphs in constant time. In: Proceedings of the 35th Annual ACM Symposium on Theory of Computing, pp. 143–148. ACM (2003)
23. Łącki, J., Sankowski, P.: Min-cuts and shortest cycles in planar graphs in $O(n \log \log n)$ time. In: Demetrescu, C., Halldórsson, M.M. (eds.) ESA 2011. LNCS, vol. 6942, pp. 155–166. Springer, Heidelberg (2011). https://doi.org/10.1007/978-3-642-23719-5_14
24. Mattsson, L.-G., Jenelius, E.: Vulnerability and resilience of transport systems–a discussion of recent research. Transp. Res. Part A: Policy Pract. **81**, 16–34 (2015)
25. Murray, A.T.: An overview of network vulnerability modeling approaches. GeoJournal **78**, 209–221 (2013)
26. Orlin, J.B.: Max flows in $O(nm)$ time, or better. In: Symposium on Theory of Computing Conference, STOC 2013, pp. 765–774. ACM (2013)
27. Phillips, C.A.: The network inhibition problem. In: Proceedings of the Twenty-Fifth Annual ACM Symposium on Theory of Computing, pp. 776–785. ACM (1993)
28. Ratliff, H.D., Sicilia, G.T., Lubore, S.: Finding the n most vital links in flow networks. Manag. Sci. **21**, 531–539 (1975)
29. Reif, J.H.: Minimum s-t cut of a planar undirected network in $O(n \log^2(n))$ time. SIAM J. Comput. **12**, 71–81 (1983)
30. Wollmer, R.D.: Some Methods for Determining the Most Vital Link in a Railway Network. Rand Corporation (1963)
31. Wood, R.K.: Deterministic network interdiction. Math. Comput. Model. **17**, 1–18 (1993)

Maximum Flows in Parametric Graph Templates

Tal Ben-Nun[1], Lukas Gianinazzi[1(✉)], Torsten Hoefler[1], and Yishai Oltchik[2]

[1] ETH Zurich, Universitätstrasse 6, Zürich, Switzerland
{talbn,glukas,htor}@inf.ethz.ch
[2] NVIDIA, Tel-Aviv, Israel
yoltchik@nvidia.com

Abstract. Execution graphs of parallel loop programs exhibit a nested, repeating structure. We show how such graphs that are the result of nested repetition can be represented by succinct parametric structures. This *parametric graph template* representation allows us to reason about the execution graph of a parallel program at a cost that only depends on the program size. We develop structurally-parametric polynomial-time algorithm variants of maximum flows. When the graph models a parallel loop program, the maximum flow provides a bound on the data movement during an execution of the program. By reasoning about the structure of the repeating subgraphs, we avoid explicit construction of the instantiation (e.g., the execution graph), potentially saving an exponential amount of memory and computation. Hence, our approach enables graph-based dataflow analysis in previously intractable settings.

Keywords: Graph algorithms · Graph theory · Maximum flow

1 Introduction

Parallel program analysis approaches to optimize data movement and program transformation commonly rely on graph algorithms [8, 24, 25, 29]. These problems

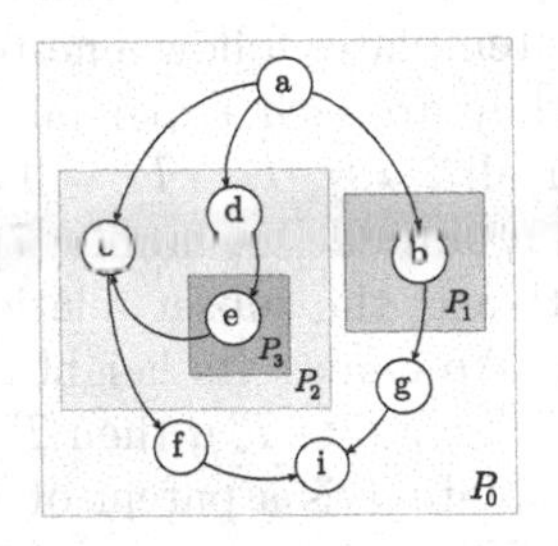

(a) Param. Graph Template $\mathcal{G}$.

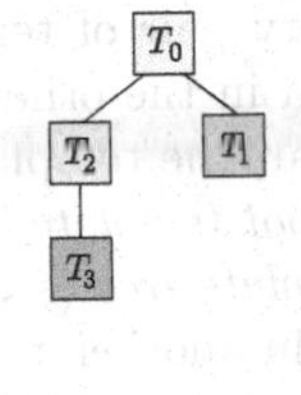

(b) Template tree of $\mathcal{G}$.

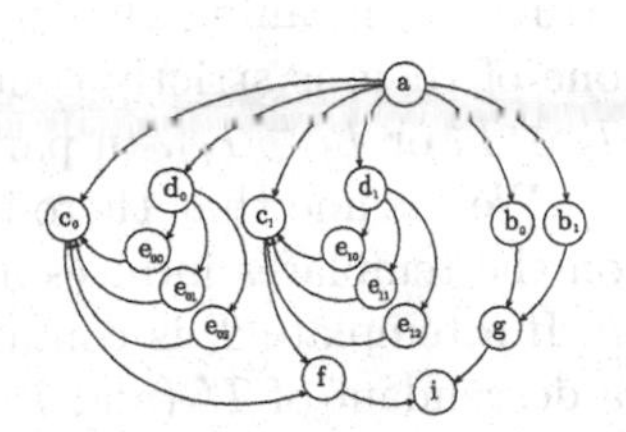

(c) Instantiation of $\mathcal{G}$.

Fig. 1. Illustration of the parametric graph template $\mathcal{G}$ with templates $T_0 = \{a, b, c, d, e, f, g, i\}$, $T_1 = \{b\}$, $T_2 = \{c, d, e\}$, $T_3 = \{e\}$; parameters $P_0 = 1, P_1 = 2, P_2 = 2, P_3 = 3$; and $h = 2$.

M. Mavronicolas (Ed.): CIAC 2023, LNCS 13898, pp. 97–111, 2023.
https://doi.org/10.1007/978-3-031-30448-4_8

concern an execution graph, where vertices model computation and edges model data movement. Maximum flows provide an algorithmic measure of the overall data movement. Such execution graphs contain highly repetitive substructures. Other application areas also face repeating graph structures, for example, computational biology [19] and network topology [6,23,30].

The naive approach is to directly work on the graphs and apply classic algorithms. However, this is prohibitively slow. For example, execution graphs can have billions of vertices or have a parametric size. Another approach is to design domain-specific representations and solutions [17,31]. Having a more general-purpose framework would allow sharing progress across domains.

We observe that many application-relevant graphs follow a model of *nested repetition*, where a small *template graph* is repeated a *parametric* number of times [5]. In this work, we propose a representation of such hierarchically repeating graphs, which we call *parametric graph templates*, and provide algorithms for extensions to the classical graph problem of maximum s-t flow.

The main challenge lies in avoiding the naive solution of materializing the full graph (which we call *instantiation*) and using a classic algorithm, which would negate any time savings. Instead, we carefully study the structural relationship between the template and the potentially exponentially larger instantiated graph. We discover and exploit symmetries in the instantiation process. This allows us to answer graph problems with a runtime that only depends on the size of the succinct representation, enabling asymptotic time and space savings compared to a naive approach that explicitly performs the nested repetition.

1.1 Parametric Graph Templates

Next, we introduce our model and give some examples. Our goal is to represent graphs with a hierarchically repeating structure, where the number of repetitions depends on some parameters. This will allows us to represent parallel loop programs and their executions. A parametric graph template with k parameters $\mathcal{G} = (G, \mathcal{T}, \mathcal{P})$ contains a (potentially weighted) and directed *template graph* $G = (V, E)$ with n vertices V, m edges E and edge weights $w : E \mapsto \mathbb{R}$, a list of *templates* $\mathcal{T} = T_0, T_2, \ldots, T_{k-1}$, each with $\emptyset \neq T_i \subseteq V$, and a list of positive integer *parameters* $\mathcal{P} = P_0, \ldots, P_{k-1}$ (see Fig. 1a). The templates follow a nested structure, meaning that for every pair of templates they are either disjoint or one of them is strictly contained in the other one (for all $i \neq j$, $T_i \cap T_j = \emptyset$ or $T_i \subset T_j$ or $T_j \subset T_i$). In particular, the templates form a laminar set family [7].

We assume that there is a *root template* $T_0 = V$. Hence, the subset relation on the templates induces a *template tree* (see Fig. 1b). We denote its height by h. If a template T is contained in another template T' (i.e., $T \subset T'$), then T is a descendant of T' (and T' is an ancestor of T). A template T is a parent of T' (and T' is a child of T) if T' is the direct descendant of T.

To create an *instantiation* of a parametric graph template $\mathcal{G}$, repeatedly rewrite it as follows (see Fig. 1c). As long as there is more than one template, pick a leaf template T_i. For each vertex v in T_i create P_i copies $v_1, \ldots, v_{P_i}$ called instances of v, replacing v in V. The set of vertices with the same subscript are called an *instance* of T_i. For each edge $e = (u, v)$ with both endpoints in T_i, create

P_i instances $e_1 = (u_1, v_1), \ldots, e_{P_i} = (u_{P_i}, v_{P_i})$, replacing e in E. For each edge $e = (u, v)$ with one endpoint u in T_i, create P_i instances $e_1 = (u_1, v), \ldots, e_{P_i} = (u_{P_i}, v)$, replacing e in E. Proceed symmetrically for each edge $e = (u, v)$ with one endpoint v in T_i. Then, remove the template T_i and its parameter P_i.

In Sect. 3 we will represent nested parallel loop programs as parametric graph templates. In such a representation, cuts and flows correspond to data movement in a parallel execution.

1.2 Related Work

Graph Grammars. [4,10,14,15,27] describe a (possibly infinite) language of graphs compactly with a set of construction rules. There is a wide variety of such ways of constructing a graph, differing in expressive power. A classic problem for graph grammars is to decide whether a graph can be constructed from a given grammar (parsing). In contrast to graph grammars, we are not primarily concerned with expressing an infinite set of graphs, but instead with a succinct representation of a graph and algorithmic aspects of solving graph problems efficiently on this succinct representation.

Hierarchical Graphs. [13] model graphs where edges expand to other, possibly hierarchical graphs. They are a variant of context-free hyperedge replacement grammars that incorporate a notion of hierarchy. The authors consider graph transformations (i.e., replacing subgraphs within other subgraphs). However, their method does not include parametric replication. This makes it unsuitable for modeling variably-sized execution graphs.

Nested Graphs. [28] allow "hypernodes" to represent other nested graphs. The authors focus on the case where a node represents a *fixed* nested graph. This precludes nested graphs from effectively representing graphs of a parametric size.

Edge-Weight Parametric Problems. Several graph problems have been generalized to the edge-weight parametric case, where edge weights are functions of one or several parameters μ_i. This includes maximum *s*-*t* flow/minimum *s*-*t* cut [3,18,20], (global) minimum cut [2,21] and shortest paths [16,22]. The solution is then a piecewise characterization of the solution space. Usually, only linear (or otherwise heavily restricted) dependency of the edge weights on the parameters have been solved. For *edge-weight parametric minimum s-t cuts*, the problem can be solved in polynomial time when each edge e has weight $\min(c(e), \mu)$ for constants $c(e)$ and a single parameter μ [3]. Granot, McCormick, Queyranne, and Tardella explore other tractability conditions [20].

1.3 Problem Statement

We approach parametric graph templates from an algorithmic perspective. The goal is to solve classical graph problems for *fixed parameters*, but in time that is strongly polynomial in the size of the parametric graph template. We focus on the classic problem of maximum *s*-*t* flow, which has an interpretation in terms of data movement for program-derived graphs and operations research [1]. For an execution graph, a maximum s-t flow corresponds to a upper bound on the data movement between vertices s and t when they are placed on different processors.

An s-t flow f assigns every edge e a nonnegative real flow $f(e) \leq w(e)$. The sum $\sum_{e=(u,v)} f(e) - \sum_{e=(v,w)} f(e)$ is the *net flow* of the vertex v. A flow has to have net flow 0 for all vertices except s and t. The value of the flow is the net flow of the source. A maximum flow is a flow of maximum value.

The maximum s-t flow problem has a natural generalization to parametric graph templates when s and t are vertices in the root template: Instantiate the graph and compute a maximum flow between the only instance of s and the only instance of t. There are multiple possibilities for how to interpret the case when s and t have multiple instances. One interpretation is as a multiple-source and multiple-target flow problem, where all instances of s are treated as sources and all instances of t as sinks. We call this a *maximum all-s-t flow.* Another interpretation considers the maximum flow between a fixed instance of s and a fixed instance of t. We call this a *maximum single-s-t flow.*

1.4 Our Results

We show how to efficiently represent a class of parallel loop programs as parametric graph templates and how properties of data movement in the parallel loop programs relates to cuts and flows in the parametric graph templates.

Then, we demonstrate that maximum s-t flow can be solved asymptotically faster than instantiating the parametric graph template. In particular, it is possible to obtain a runtime that is similar to the runtime on the template graph.

For maximum all-s-t flow, our algorithms match the runtime of a maximum s-t algorithm such as Orlin's $O(mn)$ time algorithm [26]. We solve this problem using a technique called *Edge Reweighting.* It observes that scaling the edge weights in the graph template solves the problem. For maximum single-s-t flow and minimum cuts, there is an overhead proportional to the height h of the template tree. In addition to Edge Reweighting, we use a technique called *Partial Instantiation.* We observe that a carefully chosen part of the instantiated graph can give sufficient information to extrapolate the result to the rest of the graph. How this part is chosen depends on the problem.

2 Preliminaries

We proceed to introduce definitions, notation, and assumptions that we use throughout this work.

Template a Vertex Belongs to. If a vertex v is in a template T_i and v is in no other template that is a descendant of T_i, then v *belongs to* T_i. We denote the unique template that v belongs to by $T(v)$.

Template an Edge Belongs to. If both endpoints of an edge belong to a template T_i, then this edge belongs to template T_i. We denote the number of vertices and edges that belong to a template T_i by n_i and m_i, respectively.

Cross-Template Edges. An edge (u, v) where u and v belong to different templates is *cross-template.*

No Jumping. We assume there are no edges that 'jump' layers in the template hierarchy. Specifically, if (u, v) is a cross-template edge, then $T(u)$ is a parent or child of $T(v)$. This rule ensures that a path in the graph corresponds to a walk in the template tree. It comes without loss of generality for cut and flow problems, as an edge that jumps layers can be split into multiple edges (all of weight ∞ except the edge connected to the vertex that belongs to the deeper template in the template tree). For graphs that model programs, this assumption corresponds to disallowing jumps to arbitrary program locations.

Boundary Vertices. Consider a vertex u and v where $T(v)$ is a parent of $T(u)$. If there is an edge from u to v or from v to u in the template graph, then v is a *boundary vertex* of $T(u)$.

Template Graph of a Template. The subgraph of the template graph G induced by a template T_i is called the *template graph of* T_i.

Instance Tree. We extend the nomenclature of templates to instances. The template hierarchy can be transferred onto the instances, where an instance I is a descendant of an instance I' if the template T that instantiated I is a descendant of the template T' that instantiated I'. Similarly, we extend the notions of ancestor, parent, and child to the instances, creating an *instance tree.* Two instances that have the same parent instance are *siblings.* If a vertex v is contained in an instance I, but it is not contained in any other descendant of I, the vertex v belongs to the instance I. If b_i is an instance of a boundary vertex b of a template T, then b_i is a boundary vertex of the instance that b_i belongs to. The instance of the root template is the *root instance.* For a vertex v in the instantiation, we write $T(v)$ for the template of the instance that v belongs to.

Isomorphism. Two parametric graph templates $\mathcal{G}_1$ and $\mathcal{G}_2$ are *isomorphic* if they instantiate isomorphic graphs. Two isomorphic $\mathcal{G}_1$ and $\mathcal{G}_2$ can have different parameters, templates, and their template graphs need not be isomorphic.

Cycles. Acyclic graphs are easier to handle for many algorithmic problems. In parametric graph templates, we consider two different notions of what constitutes a cycle. The simplest notion of cycles comes from considering cycles in the template graph. If it does not contain any cycles, then the instantiation does neither (and vice versa). A path $p_1, \dots p_k$ in the template graph that contains three vertices p_i, p_j, p_k with $i < j < k$ and $T(i) = T(k)$ but $T(i) \neq T(j)$ is a template-cycle. We say a parametric graph template is *template-acyclic* if it does not contain a template-cycle. This notion is incomparable to the notion of acyclic parametric graph templates. There are acyclic parametric graph templates that are not template-acyclic (consider a path whose nodes alternate between belonging to some template and its child). Note that a template-acyclic graph can have cycles (consider a cycle whose vertices belong to the same template).

3 Templates of Parallel Loop Programs

We show that a broad class of parallel programs can be modeled as parametric graph templates, such that the parametric graph template corresponds to the source code of the program and an instantiation of the parametric graph template corresponds to an execution of the program. This allows us to analyze properties of the execution of a program by considering a parametric graph template of a size comparable to the source of the program.

The parametric graph templates we consider can model nested loop programs, for example Projective Nested Loops [12] and Simple Overlap Access Programs [24]. The program receives a set of multi-dimensional input arrays $A_1, \ldots, A_k$. The goal is to output a multi-dimensional array B. The program can use several multi-dimensional temporary arrays $C_1, \ldots C_{k'}$. For any array D, its size in the i-th dimension is $\text{size}_i(D)$.

Roughly speaking, we allow any composition of elementary operations and parallel nested loops where the loop bounds only depend on the sizes of the input arrays. We allow parallel reduction to aggregate the results of a loop. We do not allow data-dependent control flow, but we allow the locations of memory accesses to be data-dependent. Examples of algorithms that can be represented this way include matrix multiplication, convolution, and cross-correlation.

We call the resulting parametric graph templates *parallel loop graph templates.* Next, we describe their syntax. Then, we describe a semantic for these parametric graph templates. Finally, we relate the data movement of the parallel loop programs with their templates' instantiations.

3.1 Syntax

The vertices of the template graph are annotated with types corresponding to their function in the program. Each template graph contains the *input memory vertices* $A_1, \ldots, A_k$, the *output memory vertex* B, and the *temporary memory vertices* $C_1, \ldots, C_{k'}$. The memory vertices can have arbitrary in-degree and out-degree and belong to the root template. Other vertices have out-degree 1, except if stated otherwise. The outgoing edge is called the *output edge.* To disambiguate the inputs to a vertex, the incoming edges are numbered consecutively. We refer to inputs in this *input order.* We consider the following control flow constructs. These are boundary vertices.

Parfor. A PARFOR (parallel for loop) vertex has no input edge. Its output edge leads to a vertex in a child template.

Reduce(Op), where OP is an associative and commutative operator. Has a single input edge from a vertex in a child template. The output edge leads to a non-memory vertex.

Copy. A COPY vertex v has arbitrary outdegree. For each of its output edges (v, u), the template $T(u)$ is not a parent of $T(v)$ and u is not a memory vertex.

We consider the following memory constructs, which are boundary vertices.

'→'. A (pass-through) → vertex has in-degree and out-degree 1. At most one of the two neighbors can be a memory vertex.

Read. A Read vertex has a first input edge from a memory vertex or a → vertex and one or more other input edges from a non-memory vertex. Its output edge leads to a non-memory vertex.

Write. A Write vertex has two or more input edges from non-memory vertices. Its output edge leads to a memory or → vertex.

We consider the following types of operator vertices. They cannot be connected to memory vertices and are not boundary vertices.

(Op), for OP $\in \{+, -, *, \div\}$, which has in-degree 2.

[c], for any representable constant c.

3.2 Semantics

A *well-formed* program has an acyclic template graph. For a well-formed program, a *serial execution* is any topological order of the instantiation of the parametric graph template. Each d-dimensional input array A_i initially contains some current value $A_i[j_1]....[j_d]$ at each position $(j_1, \dots, j_d)$, where the d-th dimension of A_i has size $\text{size}_d(A_i)$. All other arrays contains 0 at each of their positions. The arrays do not alias each other.

The semantics of a serial execution is given by applying the following rules to each vertex in the serial execution. Before evaluating the rules, contract all edges which have at least one → vertex neighbor (these exist to transfer values from inside the template hierarchy to the memory vertices in the root template).

Parfor. A PARFOR vertex with k children outputs one value of the permutation of $\{0, \dots, k-1\}$ to each child in an injective way.

Copy. Given input x, COPY outputs x to all its children.

Reduce(Op), for OP $\in \{+, *\}$. Given the inputs $x_1, \dots, x_k$, REDUCE(OP) outputs the result of applying OP repeatedly in an arbitrary order to the inputs.

Read. Given inputs $A_i, j_1, \dots, j_d$, if A_i has dimension d and for all $j_k \in \{j_1, \dots, j_d\}$ we have $0 \leq j_k < \text{size}_k(A_i)$, a READ vertex outputs the current value of $A_i[j_1]....[j_d]$. Otherwise, the result of the serial execution is undefined.

Write. Given inputs $x, j_1, \dots, j_d$, a WRITE vertex outputs x. This has the side effect of updating the current value of the array into which the output edge leads: Say it leads to A_i. Then, if A_i has dimension d and for all $j_k \in \{j_1, \dots, j_d\}$ we have $0 \leq j_k < \text{size}_k(A_i)$, the current value of $A_i[j_1]....[j_d]$ becomes x. Otherwise, the result of the serial execution is undefined.

(Op), for OP $\in \{+, -, *, \div\}$. Given x, y, outputs x OP y.

[c] outputs the constant c.

In a serial execution, we say that two reads or writes u, v are *totally ordered* if there is a path from u to v or from v to u in the instantiation. A *data race* occurs if there are two writes W_1, W_2 with the same right input (i.e., index) that connect to the same memory vertex and W_1 and W_2 are not totally ordered. The output of a well-formed program is *well-defined* if none of its serial executions has an undefined result or contains a data race.

3.3 Applications of Flows and Cuts

To model dataflow in the parallel loop programs, we set the weight of the PARFOR edges to 0 and the weight of the other edges to 1. Loop indices can be recomputed and thus do not cause data movement. The parallel loop graph template encodes all the data movement in its edges. However, it cannot resolve the aliasing of array locations. Hence, the weight of the edges going across a partition of the vertices provides an *upper*bound on the data movement:

Observation 1. *Consider a partition $(V_0, \ldots, V_p)$ of the vertices in an instantiation of a parallel loop graph template. The value total weight of the edges with endpoints in different partitions is an upper bound on the data movement incurred when the partitions are allocated to distinct processors.*

Note that in our formulation of parallel loop graph templates, vertices corresponding to arrays are placed on a single processor. Thus, to model the distribution of an array across multiple processors, a vertex must be created for each processor that holds its subarray (this subarray can be discontiguous though).

Since a maximum *s*-*t* flow equals the value of a minimum *s*-*t* cut, the maxflow provides a partition of the loop program with small data movement:

Observation 2. *If a parallel loop graph template has a maximum all-s-t flow of value x, then there is a partition of the parallel loop program which incurs at most x data movement and in which all instances of s are executed on a different processor as all instances of t.*

We can get a similar statement for maximum single-*s*-*t* flows.

4 Template Maximum Flows

Next, we turn to the first algorithmic question on parametric graph templates. Our goal here is to solve the maximum *s*-*t* flows problem on a parametric graph template *without explicitly instantiating it.* Instead, the goal is to get a runtime that is polynomial in the size of the graph template. Our algorithms use a series of observations on the structure of maximum flows in parametric graphs which allow us to produce transformed parametric graph templates, on which the answer can be efficiently computed.

We will approach the problem by considering the case where s and t are in the root template first. Then, we show how to reduce both the maximum all-*s*-*t* flow and the maximum single-*s*-*t* flow problem to an instance of this simpler problem. Throughout, we assume that all vertices are reachable from s and can reach t, as otherwise they cannot carry flow.

In the template-acyclic case, the maximum single-s-t flow is trivially zero *except* when s and t are in the same instance of the least common ancestor of $T(s)$ and $T(t)$ in the template tree. Therefore, in the acyclic case it makes sense to restrict our attention to this case where the flow is not trivially zero. In the case where there are template-cycles, it matters which instances of s and t are picked. These can be identified by numbering the instances they belong to.

4.1 Edge-Reweighting

An efficient way to solve a problem on parametric graph templates is to show how it relates to a problem on the template graph with *scaled weights*. The idea is that an edge that intersects template T_i can be used P_i times and can therefore be used to carry P_i times the amount of flow. We will see that this observation holds as long as s and t are in the root template or if we consider the maximum all-s-t flow problem. We call this approach *Edge-Reweighting.*

Algorithm: Edge-Reweighting. Transform the parametric graph template $\mathcal{G} = (G, \mathcal{T}, \mathcal{P})$ with edge weights w into a graph G' with edge weights w'. The *reweighted graph* G' has the same vertex and edge set as the template graph G, but the weights are scaled as follows: Multiply the weight of an edge in the template graph by the product of the parameters of the templates that contain at least one endpoint of the edge. That is, let $I(e)$ be the index set of all templates that contain at least one of the endpoints of e. Then, the weight of $w'(e)$ is $w(e) \prod_{i \in I(e)} P_i$. To implement this in linear time $O(m)$, precompute in a pre-order traversal of the template tree for each template the product of all the ancestors' parameters.

4.2 Source and Sink Belong to the Root Template

Our goal is to show that when the source s and sink t belong to the root template, then a maximum s-t flow in the reweighted graph equals the value of a maximum all-s-t flow. If s and t belong to the root template (which is instantiated once), then a maximum single-s-t flow equals a maximum all-s-t flow and we call it a maximum s-t flow for short.

The linear programming *dual* of a maximum s-t flow is a *minimum s-t cut* [11]. We will use strong duality [9] in our proof, which means that it suffices to identify an s-t flow and a minimum s-t cut of equal value to prove that they are optimal. We argue that Edge Reweighting preserves the value of the dual minimum all-s-t cut. Hence, it also preserves the maximum all-s-t flow value.

The following shows us how to construct an s-t cut C' in the transformed graph G' from an s-t cut C in $\mathcal{G}$ of the same value. Together with the other (easier) direction of the proof, this shows that the transformed graph G' has the same maximum s-t flow.

Lemma 1. *In a parametric graph template $\mathcal{G} = (G, \mathcal{T}, \mathcal{P})$, if s and t are in the root template, there is a minimum s-t cut of the instantiation of $\mathcal{G}$ where every instance of every vertex is on the same side of the cut.*

Lemma 2. *If a parametric graph template $\mathcal{G}$ has a minimum s-t cut of value μ and s and t are in the root template, then the graph G' constructed by edge reweighting has a minimum s-t cut of value μ.*

The proofs of Lemma 1 and Lemma 2 are omitted due to space constraints.

4.3 Instance Merging

We show how to merge all instances of a vertex v in a parametric graph template by transforming it into parametric graph template of almost the same size (the overhead is an additive $O(nh)$). We will use this technique to reduce the general case for maximum all-s-t flow to the case where s and t are in the root.

The idea is that merging all instances of a vertex s is akin to moving the vertex from the template $T(s)$ it belongs to into the root template (so that it belongs to the root template). The *no jumping rule* only allows edges to go from parent templates to children templates (or vice versa), we need to introduce *dummy edges* and *dummy vertices* along the way. The dummy edges have ∞ weight. An original edge (u, s) will be transformed into a path $u, d_1, \ldots, d_k, s$ for dummies $d_1, \ldots, d_k$ (symmetrically for an edge (s, u)).

Algorithm: Instance-Merging. Given a parametric graph template $\mathcal{G}$ and a vertex s, repeat the following until s is in the root template:

1. For any cross-template edge (u, s), introduce a dummy vertex d in the template $T(s)$ that s belongs to. Replace the edge (u, s) by two edges $e_1 = (u, d)$ and $e_2 = (d, s)$. The weight of the edge e_1 is the same as the weight of the edge e, but the weight of the edge e_2 is set to ∞. Proceed symmetrically for any cross-template edge (s, u).
2. Move the vertex s from the template $T(s)$ to the parent of the template $T(s)$ (i.e., remove s from the set $T(s)$).

4.4 Maximum All-s-t Flow

To solve maximum all s-t Flow, all we would need to do is use Instance Merging on s and then on t to ensure that they are both in the root template. Then, we could use the edge reweighing Lemma 2. This approach would cost $O(nm+n^2h)$ time. We can avoid this overhead by observing that edge reweighting works directly for maximum all-s-t Flow (even when s and t are not in the root template).

Lemma 3. *Edge reweighting of a parametric graph template $\mathcal{G}$ produces a reweighted graph G' where the value of the maximum s-t flow of G' equals the value of the maximum all-s-t flow of $\mathcal{G}$.*

Proof. Instance Merge s and then t in $\mathcal{G}$ to produce a parametric graph template $\mathcal{G}'$. By definition, all instances of s (and t respectively) must be on the same side of a minimum all-s-t cut, this parametric graph template $\mathcal{G}'$ has the same minimum all-s-t cut value as the original parametric graph template $\mathcal{G}$. Edge reweighting $\mathcal{G}'$ gives us a graph $\hat{G}$. From Lemma 2 we know that a minimum s-t cut of $\hat{G}$ corresponds to the minimum all-s-t cut of $\mathcal{G}'$ (which puts all instances of the same vertex on the same side of the cut).

An ∞-weight edge never crosses a minimum s-t cut and therefore such dummy edges (introduced by the instance merging) from $\hat{G}$ can be contracted, yielding a graph G'. This graph G' is the same graph that we get from edge reweighting the original parametric graph template $\mathcal{G}$.

Now, the results follows:

Theorem 1. *Computing a maximum all-s-t flow in parametric graph template takes* $O(mn)$ *time.*

4.5 Partial Instantiation

The technique of *partial instantiation* revolves around instantiating only part of the parametric graph template, depending on the problem at hand. The goal is to choose the partial instantiation such that the remaining problem is solvable by using the symmetry of the problem (e.g., using edge-reweighting). Partial instantiation can be seen as an example of the more general technique of *retemplating.* The intuition of retemplating is that in certain cases, it suffices to change the representation of the parametric graph template into another isomorphic parametric graph template to significantly simplify the problem at hand.

Next, show how to move a single vertex s from deep in the template tree to the root, without changing the instantiated graph. This solves the maximum s-t problems when s (or t) belongs to a template that is deep in the template tree (See Sect. 4.6). We call this technique *Upwards Partial Instantiation* from s. For simplicity, let us start with the special case of template-acyclic graphs.

In a template-acyclic parametric graph template, once a path goes from an instance of a template T_i to its parent, it never enters another instance of T_i again. This property implies that, when considering the reachable subgraph from a vertex that is an instance of s, we can simply "merge" $T(s)$ and all the templates that are ancestors of the template $T(s)$ in the template tree. Formally, this corresponds to deleting $T(s)$ and all the templates that are ancestors of $T(s)$ (except the root) from the parametric graph templates' list of templates.

If the parametric graph template has template-cycles, our goal remains to transform the parametric graph template into an equivalent graph where a particular instance of a vertex s is in the root template.

Algorithm: Partial Instantiation. Repeat the following until all templates from $T(s)$ to the root have parameter 1:

1. Consider the topmost template T that contains s and has parameter greater than 1. Let P_s be the number of instances of the template T.
2. Instantiate the template T twice. Create a new parametric graph template that has the two instances as templates, where the first template has parameter 1 and the second template has parameter $P_s - 1$. The vertices in the second template are relabeled (s is in the one with parameter 1).

Now, merge $T(s)$ and all the templates that are ancestors of $T(s)$, leaving s in the root template.

Because this process performs the same rewriting of the parametric graph template as instantiation, just in a different order and stopping early, this process creates an isomorphic parametric graph template. Every iteration adds at most n vertices and m edges and there are at most h iterations. We conclude that:

Observation 3. *Upwards Partial Instantiation from s produces an isomorphic parametric graph template with at most h additional templates and $O(nh)$ vertices and $O(mh)$ edges in the template graph.*

4.6 Maximum Single-s-t Flow

We give a partial instantiation and edge reweighting approach to maximum single-s-t flow. For there to be a flow through some instance, it must lie along an s-t path. Hence, we can use Upwards Partial Instantiation twice to ensure that s and t lie in the root template. Then, we use Edge Reweighting.

Algorithm: Single-s-t Flow. We solve maximum single-s-t flow as follows:

1. Perform upwards partial instantiation from s.
2. Perform upwards partial instantiation from t.
3. Construct an edge-reweighted graph G'.
4. Run a maximum s-t flow algorithm on the partially instantiated and reweighted graph G'.

Theorem 2. *Computing a maximum single-s-t flow in parametric graph template takes $O(mnh)$ time.*

5 Allowing Edges Between Sibling Templates

So far, we have disallowed any edges between instances of the same template. This limits the types of graphs which have a small template graph. For example, a path of length n requires a template graph with n nodes. We can extend the model by allowing an instance to have edges *to another instance of the same template*. These edges can, for example, more efficiently model sequential chains (paths), convolutional networks, and grids. We call these edges *sibling edges* (because they connect siblings). A *sibling edge* (u,v) of template T_i connects (in the template graph of T_i) a vertex u that belongs to template T_i to a vertex v that *also* belongs to T_i. Every sibling edge $e = (u, v)$ of template T_i is associated with a *bijective* (and computable) *sibling function* $f_e : \{0, \ldots, P_i{-}1\} \rightarrow \{0, \ldots, P_i{-}1\}$ which tells us that if the head of edge e is in instance j of the template T_i, then the tail of edge e is in instance $f(j)$ of the template T_i.

Note that in the model with sibling edges, a path of length n can be represented with two nodes instead of n nodes and a 1-dimensional cross-correlation of two n-dimensional signal can be represented with 2 nodes.

The structural Lemma 1 for edge reweighting still holds with sibling edges. Hence, the results on maximum all-s-t flow and maximum single-s-t flow hold analogously in the presence of sibling edges within the same bounds Theorems 1 and 2.

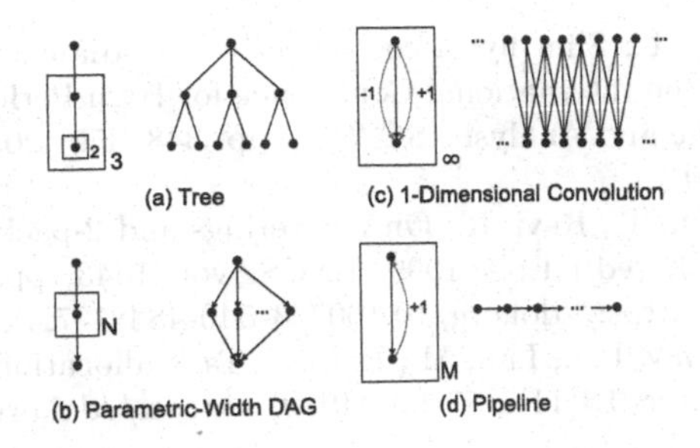

Fig. 2. Graph Templates and their expanded/instiantiated counterparts. An edge e labelled with $\pm\Delta$ indicates a sibling edge e with sibling function $f_e(x) = x + \Delta$

6 Conclusion

In this work, we explored the notion of structural parameterization in graphs. We show how graph templates correspond to the computation graphs of parallel programs. Our model leads to a $O(mn)$ time algorithm for a template version of maximum s-t flow (and hence minimum s-t cuts). These flows provide upper bounds on the data movement of partitions of certain parallel loop programs.

Other interesting problems would include partitions into multiple parts and subgraph isomorphism. Moreover, future work could explore lower bounds for parametric graph template algorithms.

Acknowledgements. This work received support from the PASC project DaCeMI and from the European Research Council under the European Union's Horizon 2020 programme (Project PSAP, No. 101002047), as well as funding from EuroHPC-JU under grant DEEP-SEA, No. 955606.

References

1. Ahuja, R.K., Magnanti, T.L., Orlin, J.B.: Network Flows: Theory, Algorithms, and Applications. Prentice-Hall Inc. (1993)
2. Aissi, H., Mahjoub, A.R., McCormick, S.T., Queyranne, M.: Strongly polynomial bounds for multiobjective and parametric global minimum cuts in graphs and hypergraphs. Math. Program. **154**(1-2), 3–28 (2015). https://doi.org/10.1007/s10107-015-0944-8
3. Aneja, Y.P., Chandrasekaran, R., Nair, K.: Parametric min-cuts analysis in a network. Discret. Appl. Math. **127**(3), 679–689 (2003). https://doi.org/10.1016/S0166-218X(02)00496-1
4. Bauderon, M., Courcelle, B.: Graph expressions and graph rewritings. Math. Syst. Theory **20**(2–3), 83–127 (1987). https://doi.org/10.1007/BF01692060
5. Ben-Nun, T., de Fine Licht, J., Ziogas, A.N., Schneider, T., Hoefler, T.: Stateful dataflow multigraphs: a data-centric model for performance portability on heterogeneous architectures. In: Proceedings of the International Conference for High Performance Computing, Networking, Storage and Analysis, SC 2019. Association for Computing Machinery, New York (2019). https://doi.org/10.1145/3295500.3356173

6. Besta, M., Hoefler, T.: Slim fly: a cost effective low-diameter network topology. In: Proceedings of the International Conference for High Performance Computing, Networking, Storage and Analysis, SC 2014, pp. 348–359 (2014). https://doi.org/10.1109/SC.2014.34
7. Cheriyan, J., Jordán, T., Ravi, R.: On 2-coverings and 2-packings of laminar families. In: Nešetřil, J. (ed.) ESA 1999. LNCS, vol. 1643, pp. 510–520. Springer, Heidelberg (1999). https://doi.org/10.1007/3-540-48481-7_44
8. Chu, W.W., Holloway, L.J., Lan, M., Efe, K.: Task allocation in distributed data processing. Computer **13**(11), 57–69 (1980). https://doi.org/10.1109/MC.1980.1653419
9. Chvátal, V.: Linear Programming. Series of Books in the Mathematical Sciences. W. H. Freeman (1983)
10. Courcelle, B.: An axiomatic definition of context-free rewriting and its application to NLC graph grammars. In: Cori, R., Wirsing, M. (eds.) STACS 1988. LNCS, vol. 294, pp. 237–247. Springer, Heidelberg (1988). https://doi.org/10.1007/BFb0035848
11. Dantzig, G.B., Fulkerson, D.R.: On the Max Flow Min Cut Theorem of Networks. RAND Corporation, Santa Monica (1955)
12. Dinh, G., Demmel, J.: Communication-optimal tilings for projective nested loops with arbitrary bounds. In: Scheideler, C., Spear, M. (eds.) 32nd ACM Symposium on Parallelism in Algorithms and Architectures, SPAA 2020, Virtual Event, USA, 15–17 July 2020, pp. 523–525. ACM (2020). https://doi.org/10.1145/3350755.3400275
13. Drewes, F., Hoffmann, B., Plump, D.: Hierarchical graph transformation. J. Comput. Syst. Sci. **64**(2), 249–283 (2002). https://doi.org/10.1006/jcss.2001.1790
14. Ehrig, H., Pfender, M., Schneider, H.J.: Graph-grammars: an algebraic approach. In: 14th Annual Symposium on Switching and Automata Theory, Iowa City, Iowa, USA, 15–17 October 1973, pp. 167–180 (1973). https://doi.org/10.1109/SWAT.1973.11
15. Engelfriet, J.: Context-free NCE graph grammars. In: Csirik, J., Demetrovics, J., Gécseg, F. (eds.) FCT 1989. LNCS, vol. 380, pp. 148–161. Springer, Heidelberg (1989). https://doi.org/10.1007/3-540-51498-8_15
16. Erickson, J.: Maximum flows and parametric shortest paths in planar graphs. In: Proceedings of the Twenty-First Annual ACM-SIAM Symposium on Discrete Algorithms, SODA 2010, Austin, Texas, USA, 17–19 January 2010, pp. 794–804 (2010). https://doi.org/10.1137/1.9781611973075.65
17. Feautrier, P.: Some efficient solutions to the affine scheduling problem. I. One-dimensional time. Int. J. Parallel Program. **21**(5), 313–347 (1992). https://doi.org/10.1007/BF01407835
18. Gallo, G., Grigoriadis, M.D., Tarjan, R.E.: A fast parametric maximum flow algorithm and applications. SIAM J. Comput. **18**(1), 30–55 (1989). https://doi.org/10.1137/0218003
19. Ginsburg, A., Ben-Nun, T., Asor, R., Shemesh, A., Ringel, I., Raviv, U.: Reciprocal grids: a hierarchical algorithm for computing solution X-ray scattering curves from supramolecular complexes at high resolution. J. Chem. Inf. Model. **56**(8), 1518–1527 (2016)
20. Granot, F., McCormick, S.T., Queyranne, M., Tardella, F.: Structural and algorithmic properties for parametric minimum cuts. Math. Program. **135**(1–2), 337–367 (2012). https://doi.org/10.1007/s10107-011-0463-1

21. Karger, D.R.: Enumerating parametric global minimum cuts by random interleaving. In: Proceedings of the 48th Annual ACM SIGACT Symposium on Theory of Computing, STOC 2016, Cambridge, MA, USA, 18–21 June 2016, pp. 542–555 (2016). https://doi.org/10.1145/2897518.2897578
22. Karp, R.M., Orlin, J.B.: Parametric shortest path algorithms with an application to cyclic staffing. Discret. Appl. Math. **3**(1), 37–45 (1981). https://doi.org/10.1016/0166-218X(81)90026-3
23. Kim, J., Dally, W.J., Scott, S., Abts, D.: Technology-driven, highly-scalable dragonfly topology. In: 2008 International Symposium on Computer Architecture, pp. 77–88 (2008). https://doi.org/10.1109/ISCA.2008.19
24. Kwasniewski, G., et al.: Pebbles, graphs, and a pinch of combinatorics: towards tight I/O lower bounds for statically analyzable programs. In: Agrawal, K., Azar, Y. (eds.) 33rd ACM Symposium on Parallelism in Algorithms and Architectures, Virtual Event, USA, 6–8 July 2021, SPAA 2021, pp. 328–339. ACM (2021). https://doi.org/10.1145/3409964.3461796
25. Lattner, C., Adve, V.: LLVM: a compilation framework for lifelong program analysis transformation. In: International Symposium on Code Generation and Optimization, CGO 2004, pp. 75–86 (2004). https://doi.org/10.1109/CGO.2004.1281665
26. Orlin, J.B.: Max flows in O(nm) time, or better. In: Symposium on Theory of Computing Conference, STOC 2013, Palo Alto, CA, USA, 1–4 June 2013, pp. 765–774 (2013). https://doi.org/10.1145/2488608.2488705
27. Pavlidis, T.: Linear and context-free graph grammars. J. ACM **19**(1), 11–22 (1972). https://doi.org/10.1145/321679.321682
28. Poulovassilis, A., Levene, M.: A nested-graph model for the representation and manipulation of complex objects. ACM Trans. Inf. Syst. **12**(1), 35–68 (1994). https://doi.org/10.1145/174608.174610
29. Shen, C., Tsai, W.: A graph matching approach to optimal task assignment in distributed computing systems using a minimax criterion. IEEE Trans. Comput. **34**(3), 197–203 (1985). https://doi.org/10.1109/TC.1985.1676563
30. Valadarsky, A., Shahaf, G., Dinitz, M., Schapira, M.: Xpander: towards optimal-performance datacenters. In: Proceedings of the 12th International on Conference on Emerging Networking EXperiments and Technologies, CoNEXT 2016, pp. 205–219. Association for Computing Machinery, New York (2016). https://doi.org/10.1145/2999572.2999580
31. Vasilache, N., et al.: Tensor comprehensions: framework-agnostic high-performance machine learning abstractions. CoRR abs/1802.04730 (2018)

Dynamic Coloring on Restricted Graph Classes

Sriram Bhyravarapu[1(✉)], Swati Kumari[2], and I. Vinod Reddy[2(✉)]

[1] The Institute of Mathematical Sciences, HBNI, Chennai, India
sriramb@imsc.res.in
[2] Department of Electrical Engineering and Computer Science, IIT Bhilai, Raipur, India
{swatik,vinod}@iitbhilai.ac.in

Abstract. A proper k-coloring of a graph is an assignment of colors from the set $\{1, 2, \ldots, k\}$ to the vertices of the graph such that no two adjacent vertices receive the same color. Given a graph $G = (V, E)$, the Dynamic Coloring problem asks to find a proper k-coloring of G such that for every vertex $v \in V(G)$ of degree at least two, there exists at least two distinct colors appearing in the neighborhood of v. The minimum integer k such that there is a dynamic coloring of G using k colors is called the dynamic chromatic number of G and is denoted by $\chi_d(G)$.

The problem is NP-complete in general, but solvable in polynomial time on several restricted families of graphs. In this paper, we study the problem on restricted classes of graphs. We show that the problem can be solved in polynomial time on chordal graphs and biconvex bipartite graphs. On the other hand, we show that it is NP-complete on star-convex bipartite graphs, comb-convex bipartite graphs and perfect elimination bipartite graphs. Next, we initiate the study on Dynamic Coloring from the parameterized complexity perspective. First, we show that the problem is fixed-parameter tractable when parameterized by neighborhood diversity or twin-cover. Then, we show that the problem is fixed-parameter tractable when parameterized by the combined parameters clique-width and the number of colors.

Keywords: proper coloring · fixed-parameter tractable · dynamic coloring · neighborhood diversity · twin-cover · bipartite graphs

1 Introduction

A *vertex coloring* (or proper coloring) of a graph G is an assignment of colors to the vertices of the graph such that no two adjacent vertices are assigned the same color. The minimum number of colors required for a proper coloring of G is called the chromatic number of G denoted by $\chi(G)$. Given a graph $G = (V, E)$ and an integer $k \in \mathbb{N}$, a proper k-coloring $f : V(G) \to [k]$ is called a *dynamic coloring*, if for every vertex $v \in V(G)$ of degree at least 2, there are at least two distinct colors appearing in the neighborhood of v, i.e., $|f(N(v))| \geq 2$, where the

M. Mavronicolas (Ed.): CIAC 2023, LNCS 13898, pp. 112–126, 2023.
https://doi.org/10.1007/978-3-031-30448-4_9

set $N(v)$ denotes the neighbors of v in G. The smallest integer k such that there is a dynamic coloring of G using k colors is called the *dynamic chromatic number* of G and is denoted by $\chi_d(G)$. Note that for any graph G, $\chi_d(G) \geq \chi(G)$.

Dynamic coloring was introduced by Montgomery [14] and it is NP-complete on general graphs [13]. The problem has been studied on various restricted graph classes. For example, polynomial time algorithms are obtained for trees [14] and graphs with bounded tree-width [13] while it is NP-hard even for planar bipartite graphs with maximum degree at most three and arbitrarily high girth [16]. Finding upper bounds of $\chi_d(G)$ for planar graphs have been studied in several papers. It was shown in [3] that $\chi_d(G) \leq 5$ if G is a planar graph. Later in [10] it was shown that if G is a connected planar graph with $G \neq C_5$, then $\chi_d(G) \leq 4$. Dynamic coloring of graphs has been studied extensively by several authors, see for instance [1,3,8,14,16].

In this paper, we study the decision version of the Dynamic Coloring problem, which is stated as follows.

> Dynamic Coloring
> **Input:** A graph $G = (V, E)$ and a positive integer k.
> **Question:** Does G have a dynamic coloring using at most k colors?

In the first part of this paper, we study the computational complexity of Dynamic Coloring on restricted families of graphs. We show that the problem can be solved in polynomial time on chordal graphs. As the problem is NP-complete on bipartite graphs [13], we study its complexity on sub-classes of bipartite graphs and close the gap between classes of graphs that are NP-complete and P time solvable. It is known that $\chi_d(G)$ is unbounded [8] when G is a bipartite graph. We show that $\chi_d(G) \leq 4$, when G is a biconvex bipartite graph and give a polynomial time algorithm by exploiting its structural properties. We also show that the problem is NP-complete on several sub-classes of bipartite graphs, a hierarchy of which is illustrated in Fig. 1.

In the second part of this paper, we study Dynamic Coloring from the viewpoint of parameterized complexity [4,5]. In parameterized complexity, each problem instance is associated with an integer, say k, called parameter. A parameterized problem is said to be fixed-parameter tractable (FPT) with respect to a parameter k if it can be solved in time $f(k)n^{O(1)}$, where n is the input size and f is a computable function only depending on the parameter k.

There may be many parameterizations for Dynamic Coloring. The most natural parameter to consider is the "solution size", which in this case is the number of colors. As the problem is NP-complete [13] even when the number of colors is three, we do not expect to have an FPT algorithm with solution size as the parameter. There are parameters which are selected based on the structure of the graph, called "structural parameterizations". For instance, vertex cover, tree-width, neighborhood diversity, etc. The hierarchy of a few structural graph parameters is illustrated in Fig. 2.

We study the parameterized complexity of Dynamic Coloring with respect to several structural graph parameters. Tree-width is one of the most used

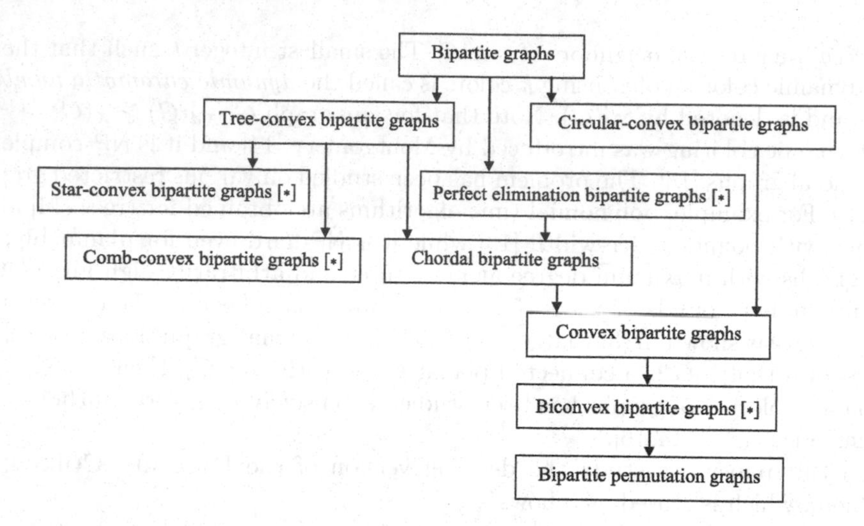

Fig. 1. Hierarchy showing the relationship between sub-classes of bipartite graphs. The graph classes considered in this paper are indicated by $*$.

structural parameters when dealing with NP-hard graph problems. Li et al. in [16] showed that DYNAMIC COLORING is FPT when parameterized by tree-width, following its formulation in monadic second order logic. One disadvantage with tree-width is that it is unbounded for dense graphs (e.g., cluster graphs). In this paper, we consider the parameters twin-cover and neighborhood diversity, which also include dense graphs. We show that DYNAMIC COLORING is fixed-parameter tractable when parameterized by twin-cover or neighborhood diversity.

Next, we consider the graph parameter clique-width, which is suitable when dealing with hard graph problems on dense graphs. Clique-width is a generalization of the parameters twin-cover and neighborhood diversity, in the sense that graphs of bounded neighborhood diversity or graphs of bounded twin-cover also have bounded clique-width. We show that DYNAMIC COLORING is FPT when parameterized by the combined parameters clique-width and the number of colors. Hence studying the parameterized complexity of DYNAMIC COLORING with respect to the above mentioned parameters reduces the gap between tractability and intractability. We summarize our contribution below.

1. In Sect. 3, we show that DYNAMIC COLORING can be solved in polynomial time on chordal graphs.
2. In Sect. 4, we show that DYNAMIC COLORING is polynomial time solvable on bipartite permutation graphs, a sub-class of biconvex graphs. We extend this algorithm to design a polynomial time algorithm for biconvex graphs, in Sect. 5.
3. In Sect. 6, we show NP-completeness results on star-convex bipartite graphs, comb-convex bipartite graphs and perfect elimination bipartite graphs

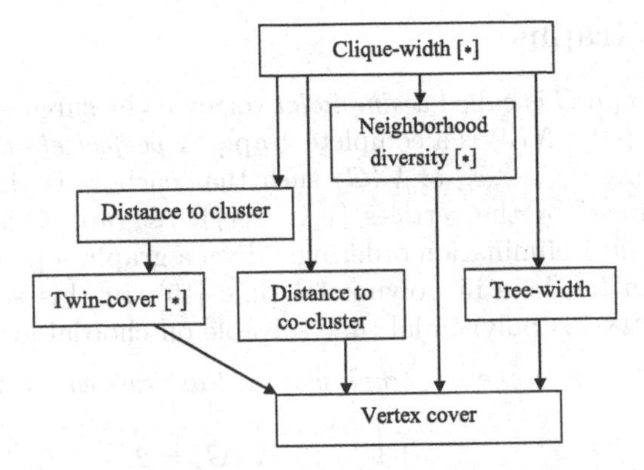

Fig. 2. Hasse diagram of a few structural graph parameters. An edge from a parameter k_1 to a parameter k_2 means that there is a function f such that for all graphs G, we have $k_1(G) \leq f(k_2(G))$. The parameters considered in this paper are indicated by $*$.

strengthening the NP-completeness result of DYNAMIC COLORING on bipartite graphs.

4. In Sects. 7 and 8, we show that DYNAMIC COLORING is FPT when parameterized by neighborhood diversity, twin-cover or the combined parameters clique-width and the number of colors.

2 Preliminaries

For $k \in \mathbb{N}$, we use $[k]$ to denote the set $\{1, 2, \ldots, k\}$. All graphs we consider in this paper are undirected, connected, finite and simple. For a graph $G = (V, E)$, we denote the vertex set and edge set of G by $V(G)$ and $E(G)$ respectively. We use n to denote the number of vertices and m to denote the number of edges of a graph. An edge between vertices x and y is denoted as xy for simplicity. For a subset $X \subseteq V(G)$, the graph $G[X]$ denotes the subgraph of G induced by the vertices of X.

For a vertex set $X \subseteq V(G)$, we denote by $G \setminus X$, the graph obtained from G by deleting all vertices of X and their incident edges. For a vertex $v \in V(G)$, by $N(v)$, we denote the set $\{u \in V(G) \mid uv \in E(G)\}$ and we use $N[v]$ to denote the set $N(v) \cup \{v\}$. The neighborhood of a vertex set $S \subseteq V(G)$ is $N(S) = (\cup_{v \in V(G)} N(v)) \setminus S$. A graph is *bipartite* if its vertex set can be partitioned into two disjoint sets such that no two vertices in the same set are adjacent. We say a vertex v of degree at least two to have satisfied *dynamic coloring property* if there exist at least two vertices in the neighborhood of v that are colored distinctly.

Due to space constraints, the proofs of the Theorems marked ($\star$) are presented in the full version of the paper.

3 Chordal Graphs

A vertex v of a graph G is called a *simplicial* vertex if the subgraph of G induced by the vertex set $\{v\} \cup N(v)$ is a complete graph. A *perfect elimination ordering* is a vertex ordering $v_1, \ldots, v_n$ of $V(G)$ such that each v_i is simplicial in $G[i]$, the subgraph induced by the vertices $\{v_1, \ldots, v_i\}$. A graph G is *chordal* if and only if it has a perfect elimination ordering. Given a graph, a perfect elimination ordering of G can be done in polynomial time [15]. In this section, we show DYNAMIC COLORING is polynomial time solvable on chordal graphs.

Theorem 1. *Let G be a chordal graph with at least two edges, then*

$$\chi_d(G) = \begin{cases} 3 & \text{if } \omega(G) = 2 \\ \omega(G) & \text{if } \omega(G) \geq 3 \end{cases}$$

where $\omega(G)$ is the size of a largest clique of G.

Proof. Let $G = (V, E)$ be a chordal graph. Hence G is C_k-free, for $k \geq 4$. If $\omega(G) = 2$, then G is C_k-free for $k \geq 3$. That is, G is a tree, hence from [14] we know $\chi_d(G) = 3$. We now deal with the case when $\omega(G) \geq 3$.

Let $v_1, \ldots, v_n$ be a perfect elimination ordering (PEO) of G. That is, each v_i is simplicial in $G[i]$. Let $f : V(G) \to [\omega(G)]$ be a coloring of G defined as follows: assign $f(v_1) = 1$ and $f(v_2) = 2$. For each v_i, where $i \geq 3$, let $D_i = N(v_i) \cap \{v_1, \ldots, v_{i-1}\}$ be the neighbors of v_i in $G[i]$ and T_i be the set of colors used to color the vertices of D_i. Then,

$$f(v_i) = \begin{cases} \min\{[\omega(G)] \setminus \{f(v_j), f(v_k)\}\} & \text{if } D_i = \{v_j\} \text{ and } D_j = \{v_k\}, \text{ where } k < j < i \\ \min\{[\omega(G)] \setminus T_i\} & \text{otherwise} \end{cases}$$

Clearly, f is a proper coloring as each v_i is greedily assigned a color (minimum) not appearing in T_i. Next, we show that f is dynamic coloring. For a vertex v_i, if $|D_i| \geq 2$ then $|T_i| \geq 2$ (i.e., $|f(N(v_i))| \geq 2$). If $|D_i| = 1$, and let $D_i = \{v_j\}$, then v_i has only one neighbor v_j in $G[i]$.

If v_i has no neighbor in the set $\{v_{i+1}, \ldots, v_n\}$ then the degree of v_i in G is one. Suppose, v_i has a neighbor in the set $\{v_{i+1}, \ldots, v_n\}$. Let v_p be the first neighbor (according to PEO) of v_i in the set $\{v_{i+1}, \ldots, v_n\}$. If $|D_p| = 1$, then by case 1, $f(v_p) \neq f(v_j)$, $f(v_p) \neq f(v_i)$ and $f(v_i) \neq f(v_j)$, hence v_i has at least two neighbors with distinct colors. Else if $|D_p| = 2$, then $D_p = \{v_j, v_i\}$, again by case 2, the three vertices v_j, v_i and v_p are colored with three distinct colors. Hence v_i has at two neighbors with distinct colors. Note that $|D_p|$ cannot be greater than 2 because of the choice of p.

It is easy to see that $\chi_d(G) \geq \omega(G)$ as we need at least $\omega(G)$ colors to properly color the largest clique of G. Since any vertex v_i has $\ell = |D_i|$ many neighbors in $G[i]$, at least one of the colors $1, 2, \cdots, \ell+1$ is not used in T_i. Hence our algorithm finds a coloring of G with at most $\max_i\{|T_i| + 1\}$ colors, which is at most $\omega(G)$.

Hence $\chi_d(G) = \omega(G)$. The time required for the above coloring procedure is $O(n^2)$. □

4 Bipartite Permutation Graphs

In this section, we show that DYNAMIC COLORING is polynomial time solvable on bipartite permutation graphs. We start the section with some basic definitions and notations that are needed to describe the algorithm.

Definition 1. (Chain Graph [6]). *A bipartite graph $G = (A \cup B)$ is called a chain graph if for every two vertices $u_1, u_2 \in A$ we have either $N(u_1) \subseteq N(u_2)$ or $N(u_2) \subseteq N(u_1)$.*

That is, there is an ordering of the vertices of A, say $u_1, u_2, \ldots u_{|A|}$, such that $N(u_i) \subseteq N(u_{i+1})$, $1 \leq i < |A|$. As a consequence, we can also find an ordering of the vertices of B, say $v_1, v_2, \ldots v_{|B|}$, such that $N(v_i) \subseteq N(v_{i+1})$, $1 \leq i < |B|$.

We can see that each part of a chain graph can be linearly ordered under the inclusion of their neighborhoods. We say a vertex ordering σ of A is increasing if $x <_\sigma y$ implies $N(x) \subseteq N(y)$, and decreasing if $x <_\sigma y$ implies $N(y) \subseteq N(x)$.

Definition 2. (Multi-chain Ordering [2,6]). *Given a connected graph $G = (V, E)$, we arbitrarily choose a vertex as $v_0 \in V(G)$ and construct distance layers $L_0, L_1, \ldots, L_p$ from v_0. The layer L_i, where $i \in [p]$, represents the set of vertices that are at a distance i from v_0 and p is the largest integer such that $L_p \neq \emptyset$.*

We say that these layers form a multi-chain ordering *of G if for every two consecutive layers L_i and L_{i+1}, where $i \in \{0, 1, \ldots, p-1\}$, we have that $G[L_i \cup L_{i+1}]$ forms a chain graph.*

We say a graph G *admits multi-chain ordering* if there exists a vertex $v_0 \in V(G)$ such that the distance layers form a multi-chain ordering. An illustration of a multi-chain ordering of a graph is given in Fig. 3. It is known [6] that all connected permutation graphs and interval graphs admit multi-chain ordering. We first observe the following on multi-chain ordering.

Observation 2. *If a graph G admits a multi-chain ordering with $p+1$ layers, then there exists a vertex v in L_i, $i \in [p]$, such that $N(v) \supseteq L_{i+1}$.*

Definition 3. (Bipartite Permutation Graph [2]). *A connected graph $G = (V, E)$ is bipartite permutation if and only if $V(G)$ can be partitioned into $q+1$ disjoint independent sets $L_0, L_1, \ldots, L_q$ (in this order) in such a way that*

1. *Any two vertices in non-consecutive sets are non adjacent.*
2. *Any two consecutive sets L_{i-1} and L_i induce a chain graph, denoted by G_i.*
3. *For each $i \in \{1, 2, \ldots, q-1\}$, there is an ordering of vertices of the set L_i such that it is non-increasing in G_i and non-decreasing in G_{i+1}. For the set L_0 (resp. L_q), there is a non-decreasing (resp. non increasing) ordering of vertices of L_0 (resp. L_q) in G_1 (resp. G_q).*

Observation 3. *If c is any dynamic coloring of a bipartite permutation graph that uses exactly three colors then at most two colors are used in any layer L_i, i.e., $|c(L_i)| \leq 2$.*

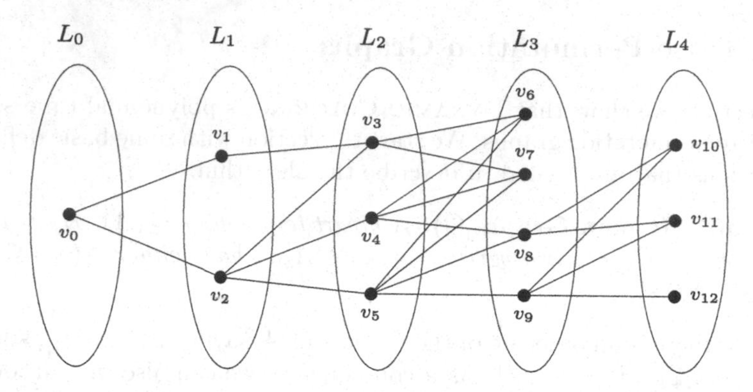

Fig. 3. An illustration of a multi-chain ordering of a graph. The ordering of the vertices in L_2 is $\sigma_2 = v_3, v_4, v_5$. In the chain graph $G[L_1 \cup L_2]$ we have that $N(v_3) \supseteq N(v_4) \supseteq N(v_5)$ and in $G[L_2 \cup L_3]$ we have that $N(v_3) \subseteq N(v_4) \subseteq N(v_5)$.

Proof. Suppose there exists a layer L_i such that $|c(L_i)| = 3$. From Observation 2, there is a vertex $v \in L_{i-1}$ such that $N(v) \supseteq L_i$. Since v is adjacent to all vertices in L_i and $|c(L_i)| = 3$, there exists a color assigned to a vertex in L_i that is same as $c(v)$. This is a contradiction to the fact that c is a proper coloring and thus a dynamic coloring. □

We now show a polynomial time algorithm to decide if the dynamic chromatic number of a bipartite permutation graph is three.

Lemma 1. *Given a bipartite permutation graph G, there is a polynomial algorithm to decide if $\chi_d(G) = 3$.*

Proof. Let $G = (V, E)$ be a bipartite permutation graph and $L_0, L_1, \ldots, L_p$ be the distance layers in a multi-chain ordering of G constructed from an arbitrarily chosen vertex $v_0 \in V(G)$, according to Definition 3. If $\chi_d(G) = 3$, then from Observation 3, we have that in any dynamic coloring $c : V(G) \to \{c_1, c_2, c_3\}$ of G, at most two colors are used for assigning colors in any layer L_i of G. This leaves us with the following cases: either $|c(L_i)| = 1$ or $|c(L_i)| = 2$. At each layer L_i, $0 \leq i \leq p$, we maintain all possible colorings of L_i in the following manner.

If L_i uses exactly one color, then we have three possible ways of coloring L_i. That is, we have three colorings where each coloring of L_i represents all its vertices being assigned the same color (one of the three colors).

If L_i uses exactly two colors, say c_1 and c_2, then we guess four vertices: the first and the last vertices in the ordering σ_i of L_i that are assigned the colors c_1 and c_2. Using this guess, we extend the coloring to the remaining vertices of L_i as follows. Let x_1^i and y_1^i (resp. x_2^i and y_2^i) be the first and the last vertices in σ_i which are colored with c_1 (resp. c_2). Let $w \in L_i \setminus \{x_1^i, y_1^i, x_2^i, y_2^i\}$. From the above description, we have that either $x_1^i < w < y_1^i$ or $x_2^i < w < y_2^i$. If $x_1^i < w < y_1^i$ then we color w with c_1 else we color it with c_2.

Let $c : V(G) \to \{c_1, c_2, c_3\}$ be a dynamic coloring of G. Let $c(L_i) = \{c_1, c_2\}$ and $\{x_1^i, y_1^i, x_2^i, y_2^i\}$ be the first and the last vertices in σ_i that are assigned the colors c_1 and c_2. Then the coloring obtained by applying the above extension procedure, on each layer L_i, on vertices $\{x_1^i, y_1^i, x_2^i, y_2^i\}$ is also a dynamic coloring of G. Hence it is enough to know the first and the last vertices in the ordering σ_i of L_i that are assigned the colors c_1 and c_2.

The number of colorings for a pair of colors c_1 and c_2 is at most $|L_i|^4$. Since there are three such pairs, the total number of colorings arising out of L_i is at most $3|L_i|^4$. Using these local colorings at the levels, we check for the existence of a dynamic coloring of G using a dynamic programming routine.

Let C_i be the set of colorings of L_i obtained from the above description. We call a triplet coloring (c_{i-1}, c_i, c_{i+1}), where $c_{i-1} \in C_{i-1}$, $c_i \in C_i$ and $c_{i+1} \in C_{i+1}$, as a *feasible coloring* (or simply *feasible*) for L_i, where $1 \leq i \leq p-1$, if every vertex of L_i admits dynamic coloring property when the colorings c_{i-1}, c_i, and c_{i+1} are assigned to L_{i-1}, L_i and L_{i+1} respectively. Similarly, we call a pair (c_0, c_1) (resp. (c_{p-1}, c_p)) as feasible for the layer L_0 (resp. L_p). Let T_i denote the set of all feasible colorings corresponding to the layer L_i.

We now use a dynamic programming approach to check if there exists a dynamic coloring of G using three colors. We have an entry $d[i, f_i]$ for each feasible coloring f_i at layer L_i that defines the existence of a coloring in $G[L_0 \cup L_1 \cup \cdots \cup L_{i+1}]$ such that all vertices in $L_0 \cup L_1 \cup \cdots \cup L_i$ satisfy dynamic coloring property given the feasible coloring f_i at L_i.

Let $f_i = (c_{i-1}, c_i, c_{i+1})$. We set the entry $d[i, f_i] = true$ if f_i is feasible and there exists a feasible coloring $f_{i-1} = (x, c_{i-1}, c_i)$ at L_{i-1} such that $d[i-1, f_{i-1}] = true$. Otherwise, we set $d[i, f_i] = false$. We initialize $d[0, f] = true$, for each feasible coloring $f = (c_0, c_1)$ of L_0 if the vertex $v_0 \in L_0$ satisfies dynamic coloring when the colorings c_0 and c_1 are assigned to the layers L_0 and L_1 respectively. Otherwise, we initialize $d[0, f] = false$. If there exists an entry $d[p, f_p]$, for some feasible coloring f_p of L_p, such that $d[p, f_p] = true$, we decide that there exists a dynamic coloring of G using three colors. If $d[p, f_p] = false$ for every feasible coloring f_p of L_p, then we decide that G does not have a dynamic coloring using three colors.

The correctness of the algorithm follows from the description of the algorithm. We now compute the running time of the algorithm which includes guessing the colorings at each layer and applying the dynamic programming routine. Computing the colorings for all the layers takes $O(p \cdot |L_i|^4) \leq O(n^5)$ time. The number of feasible colorings is at most $|L_{i-1}|^4 \cdot |L_{i+1}|^4 \cdot |L_{i+1}|^4 \leq n^{12}$. Computing if a coloring is feasible coloring can be done in $O(n^2)$ time. All the entries $d[i, \cdot]$ pertaining a layer can be computed in $O(n^{14})$ time. Since $i \leq p \leq n$, the total time taken is $O(n^{15})$. □

Theorem 4. DYNAMIC COLORING *can be solved in polynomial time on bipartite permutation graphs.*

Proof. Let $G = (V, E)$ be a connected bipartite permutation graph. Since bipartite permutation graphs are a sub-class of biconvex graphs, it follows from Lemma 2 that $\chi_d(G) \leq 4$. It is easy to see that (i) $\chi_d(G) = 1$ if and only if $G = K_1$, and (ii) $\chi_d(G) = 2$ if and only if $G = K_2$. If G does not belong to any of the above cases, then $\chi_d(G) \in \{3, 4\}$. We check whether $\chi_d(G) = 3$ using Lemma 1. If $\chi_d(G) \neq 3$, we decide that $\chi_d(G) = 4$. □

5 Biconvex Graphs

Definition 4 (Biconvex Graph). *An ordering σ of X in a bipartite graph $B = (X, Y, E)$ has the adjacency property if for every vertex $y \in Y$, the neighborhood $N(y)$ consists of vertices that are consecutive (an interval) in the ordering σ of X. A bipartite graph (X, Y, E) is biconvex if there are orderings of X (with respect to Y) and Y (with respect to X) that fulfills the adjacency property.*

Theorem 5. Dynamic Coloring *can be solved in polynomial time on biconvex graphs.*

Towards showing Theorem 5, we first show that the dynamic chromatic number of a biconvex graph is at most 4.

Lemma 2. *If G is a biconvex graph then $\chi_d(G) \leq 4$.*

Proof. Let $G = (X, Y, E)$ be a biconvex graph. We assume that G has at least five vertices, otherwise, trivially $\chi_d(G) \leq 4$.

We use the property that G is biconvex. Let $\sigma = x_1, x_2, \ldots, x_p$ be an enumeration of vertices of X and $\pi = y_1, y_2, \ldots, y_q$ be an enumeration of vertices of Y. For each $i \in [p]$, color x_i with 1 if i is odd, else color it with 2. For each $j \in [q]$, color y_j with 3 if i is odd, else color it with 4. Consider any vertex $x_i \in X$ with degree at least two. As G is convex over Y, the vertices adjacent to x_i are consecutive with respect to the ordering π. That is, if y_j is a neighbor of x_i, then at least one of y_{j+1} or y_{j-1} is a neighbor of x_i. Hence the neighborhood of x_i contains a vertex of color 3 and a vertex of color 4. Similarly, we can show that the neighborhood of every vertex $y_j \in Y$ contains a vertex of color 1 and a vertex of color 2. Hence, the above coloring is a dynamic coloring of G. □

We now proceed to the proof of Theorem 5 which is similar to the proof of bipartite permutation graphs in Theorem 4. Hence, we give a short proof highlighting the key differences.

Proof. (Short Proof of Theorem 5). Let G be a biconvex graph. We know that $\chi_d(G) \leq 4$, from Lemma 2. Similar to the proof of bipartite permutation graphs, it is sufficient to check whether $\chi_d(G) \in \{3, 4\}$. Since all connected biconvex graphs admit multi-chain ordering [6], it is possible to extend our algorithm in Theorem 4 to biconvex graphs.

The difference between bipartite permutation graphs and biconvex graphs is that the latter has two vertex orderings in a multi-chain ordering, say $\sigma_{i,1}$ and $\sigma_{i,2}$, for each layer L_i, one corresponding to L_{i-1} and the other corresponding to L_{i+1}. For each of the two orderings, we guess the first and last vertices in the respective ordering that are assigned colors based on how many colors are seen in each layer. However, we need to ensure that the guesses obtained in the orderings should complement each other in the sense that a vertex $v \in L_i$ cannot be assigned the color c_1 in one ordering and the color c_2 in the other ordering. The rest of the algorithm is similar to Theorem 4. The number of colorings at each layer is at most $|L_i|^8$. The number of feasible colorings is at most $|L_{i-1}|^8 \cdot |L_i|^8 \cdot |L_{i+1}|^8 \leq n^{24}$. Considering the number of entries in the dynamic programing table, the total time taken is $O(n^{27})$. □

6 Hardness Results on Sub-classes of Bipartite Graphs

Theorem 6 (⋆). Dynamic Coloring *is* NP*-complete on perfect elimination bipartite graphs, star-convex bipartite graphs and comb-convex graphs.*

7 Parameterization by Neighborhood Diversity

The graph parameter neighborhood diversity was introduced by Lampis [11], and it is a generalization of the parameter vertex cover. In this section, we show that Dynamic Coloring is fixed-parameter tractable parameterized by neighborhood diversity. Our main idea is to reduce our problem to the integer linear programming problem that is fixed-parameter tractable when parameterized by the number of variables.

Definition 5. (Neighborhood Diversity [11]). *Let $G = (V, E)$ be a graph. Two vertices $u, v \in V(G)$ are said to have the* same type *if and only if $N(u) \setminus \{v\} = N(v) \setminus \{u\}$. A graph G has neighborhood diversity at most t, if there exists a partition of $V(G)$ into at most t sets $V_1, V_2, \dots, V_t$ such that all the vertices in each set have the same type.*

Observe that each V_i either forms a clique or an independent set in G, for all $i \in [t]$. We call the set V_i as a *clique type* (resp. *independent type*) if $G[V_i]$ is a clique (resp. independent set). If $|V_i| = 1$, then we consider V_i as an independent type. For each $i, j \in [t]$, $i \neq j$, it is the case that either every vertex in V_i is adjacent to every vertex in V_j or no vertex in V_i is adjacent to any vertex in V_j.

For each $A \subseteq \{1, 2, \cdots, t\}$, we denote a *subset type* of G by $T_A = \{V_i : i \in A\}$. We denote the set of types neighboring the type V_i in G by $adj(V_i)$. That is, $V_j \in adj(V_i)$, if every vertex in V_i is adjacent to every vertex in V_j. Given a proper coloring $f : V(G) \rightarrow [k]$, we say V_i *admits dynamic coloring* with respect to f if for all $v \in V_i$, $|f(N(v))| \geq 2$.

Given a graph $G = (V, E)$, there exists an algorithm that runs in polynomial time [11] and finds a minimum sized neighborhood partition of $V(G)$. So, we assume that the types $V_1, V_2, \dots, V_t$ are given as input. If $t = 1$, then G is a complete graph and the problem can be solved easily. Hence, we assume $t \geq 2$.

Observation 7. *If there exists a proper coloring of G and V_i is a clique type, then V_i admits dynamic coloring.*

Proof. Recall that each clique type has at least two vertices. Let V_i be a clique type. Since $t \geq 2$ and G is connected, there exists a type $V_j \in adj(V_i)$ such that every vertex in V_i is part of a triangle (with two vertices from V_i and a vertex from V_j) that uses three distinct colors in any proper coloring. Thus each vertex in V_i has at least two differently colored neighbors which implies that V_i admits dynamic coloring. □

We now present the main theorem of the section.

Theorem 8. DYNAMIC COLORING *can be solved in* $O(q^{2.5q+o(q)}n)$ *time, where* $q = 2^t$ *and* t *is the neighborhood diversity of* G.

We use *Integer Linear Programming* (ILP) to show that DYNAMIC COLORING is FPT when parameterized by neighborhood diversity. The following result shows that ILP is FPT when parameterized by the number of variables.

Theorem 9. ([7,9,12]). An ILP feasibility instance of size n can be solved in $O(q^{2.5q+o(q)}n)$ time and $n^{O(1)}$ space, where q is the number of variables.

The crux of the proof is to distribute the colors across the type sets in a dynamic coloring of G (if one exists). Instead of looking at the list of colors featuring in the types of T_A in a dynamic coloring, where $A \subseteq [t]$, we are only interested in the number of colors that appear exclusively in each of the types of T_A in the coloring.

We now define variables and constraints for ILP. For each subset $A \subseteq [t]$, we have a variable n_A that denotes the number of colors used exclusively in all the types of T_A and not used in any of the types $\{V_1, V_2, \ldots, V_t\} \setminus T_A$. For example, if $A = \{4, 6\}$ (i.e., $T_A = \{V_4, V_6\}$) and $n_A = 3$, then there are three colors say c_1, c_2, c_3 (the exact values of which will be decided later) where each of the colors is used in both V_4 and V_6. Moreover, the colors c_1, c_2, c_3 are not assigned to any of the vertices in types $\{V_1, V_2, \ldots, V_t\} \setminus \{V_4, V_6\}$. Notice that the number of variables is at most 2^t. With this, we proceed to describe the constraints of ILP.

(C0) Consider only those subsets types T_A such that there do not exist types $V_i, V_j \in T_A$ such that $V_i \in adj(V_j)$.
We only consider those subset types T_A which do not have a pair of adjacent types. This constraint ensures that, if two types V_i and V_j are adjacent then the set of colors used in V_i is disjoint from the set of colors used in V_j.

(C1) The sum of all the variables is at most k. That is $\sum_A n_A \leq k$.
This constraint ensures that the number of colors used in any coloring is at most k.

(C2) For each clique type V_i, $1 \leq i \leq t$, the sum of the variables n_A for which $V_i \in T_A$ is equal to the number of vertices in V_i. That is $\sum_{A:V_i \in T_A} n_A = |V_i|$.
This constraint ensures that the number of colors used for coloring a clique type V_i in any coloring is equal to $|V_i|$.

(C3) For each independent type V_i, $1 \leq i \leq t$, the sum of the variables n_A for which $V_i \in T_A$ is at most the minimum of k and $|V_i|$. That is, $\sum_{A:V_i \in T_A} n_A \leq \min\{k, |V_i|\}$. Also the sum of variables n_A for which $V_i \in T_A$ is at least one. That is, $\sum_{A:V_i \in T_A} n_A \geq 1$.
This constraint ensures that, in any coloring the number of colors used for coloring an independent type V_i is (i) at most the minimum of k and $|V_i|$, and (ii) at least one.

(C4) For each independent type V_i, $1 \leq i \leq t$, if the degree of every vertex in V_i is at least two then the sum of variables n_A for which there exists $V_j \in adj(V_i) \cap T_A$ is at least 2. That is, $\sum_{A:\exists V_j \in adj(V_i) \cap T_A} n_A \geq 2$.
This constraint ensures that the number of colors used in the neighborhood of any vertex (with degree at least two) in an independent type is at least two.

(C5) For each $A \subseteq [t]$, $n_A \geq 0$.
The number of colors used exclusively in all the types in T_A is at least 0.

We use Theorem 9 to obtain a feasible assignment for ILP, if one exists. We claim the following: there is a feasible assignment of ILP if and only if there is a dynamic coloring of G using at most k colors.

Feasibility Implies Colorability: Using a feasible assignment of variable values returned by ILP, we construct a dynamic coloring $f : V(G) \rightarrow [k]$ that assigns colors greedily to the vertices of G, a pseudo-code is presented as Algorithm 1.

We now show that f is a dynamic coloring of G. To show this, we need to show that (a) every vertex is colored, (b) f is a proper coloring and (c) every vertex with degree at least two has two distinctly colored neighbors. Every vertex is assigned a color in Algorithm 1. The constraint (C0) ensures that no color is used in both V_i and V_j if they are adjacent. The constraint (C2) ensures that no two vertices in a clique type are assigned the same color. Hence f is a proper coloring of G.

We now show that for each vertex v with degree at least two, $|f(N(v))| \geq 2$. Since f is a proper coloring, from Observation 7, we have that each clique type V_i admits dynamic coloring. Consider an independent type V_j. Since the graph is connected, we have that $|adj(V_j)| \geq 1$. If there exists a clique type $V_\ell \in adj(V_j)$ then each vertex v in V_j has $|f(N(v))| \geq 2$. This is because the size of a clique type V_ℓ is at least two and f is a proper coloring. Otherwise, if all the types in $adj(V_j)$ are independent types, then there exists at least two distinctly colored vertices in the neighborhood of every vertex in V_j due to constraint (C4). Thus f is a dynamic coloring of G.

Algorithm 1: A dynamic coloring from a feasible assignment of ILP.

Input: T_A and n_A, for each $A \subseteq [t]$
Output: A dynamic coloring of G

```
1  C(T_A) = ∅, for each A ⊆ [t]          ▷ C(T_A): set of colors associated to T_A
2  c = 0                                  ▷ c: color counter
3  for each A ⊆ [t] do
4  |   C(T_A) = {c+1, c+2, ..., c+n_A}
5  |   c = c + n_A
6  C(V_i) = ∅, for each i ∈ [t]          ▷ C(V_i): set of colors associated to V_i
7  for each i ∈ [t] do
8  |   C(V_i) = ∪_{A:V_i ∈ T_A} C(T_A)
9  for each i ∈ [t] do
10 |   if V_i is a clique type then
11 |   |   Color the vertices of V_i from C(V_i) such that each vertex is
   |   |     assigned a distinct color
12 |   else
13 |   |   Color the vertices of V_i from C(V_i) such that each color of C(V_i) is
   |   |     used at least once
14 return (Coloring of G)
```

Colorability Implies Feasibility: Given a dynamic coloring $f : V(G) \rightarrow [k]$ of G, we find a feasible assignment to the ILP. For each $A \subseteq [t]$, we set n_A to be the number of colors that are assigned exclusively to each of the types in T_A, that is

$$n_A = |\bigcap_{V_i \in T_A} f(V_i) - \bigcup_{V_i \notin T_A} f(V_i)|.$$

We now show that such an assignment satisfies the constraints (C0)-(C5). As f is a proper coloring, no two adjacent types share the same color. Hence the constraint (C0) is satisfied.

Each color $c \in [k]$ is uniquely associated with a type T_A, where c is used in all the types of T_A and not used in any of the types $\{V_1, V_2, \ldots, V_t\} \setminus T_A$. The color c is therefore contributed to the variable n_A and not contributed to any other variable $n_{A'}$ where $A \neq A'$. Hence we get that the sum of variables is at most k. Hence the constraints (C1) and (C5).

Every vertex in a clique type V_i is assigned a distinct color in f and hence the sum of variables n_A such that $V_i \in T_A$ is equal to $|V_i|$. Hence the constraint (C2) is satisfied.

Consider an independent type V_j. Clearly $|f(V_j)| \leq \min\{k, |V_j|\}$. Hence the constraint (C3) is satisfied. For each vertex $v \in V_j$ of degree at least two, we have that $|f(N(v))| \geq 2$. That is there are two distinct colors in the neighboring types of V_j. Since these colors contribute to some variables, the constraint (C4) is satisfied.

Running Time: The running time of the algorithm is proportional to the time needed to (i) reduce dynamic coloring problem to the ILP problem, and (ii) find a feasible solution to the created ILP instance. The constraint (C0) considers the subset types from the 2^t subset types such that no two types in a subset type are adjacent. This can be done in $O(2^t t^2)$ time. The constraints (C1) and (C5) can be constructed in $O(2^t)$ time. The constraints (C2) and (C3) can be constructed in $O(t2^t)$ time. The constraint (C4) can be constructed in $O(t2^t)$ time. Finding a feasible assignment of ILP using Theorem 9 takes $O(q^{2.5q+o(q)}n)$ time, where $q = 2^t$. The latter part dominates the former and hence the overall running time is $O(q^{2.5q+o(q)}n)$ time, where $q = 2^t$.

This completes the proof of Theorem 8.

8 Parameterizations by Twin-Cover and Clique-Width

Theorem 10 ($\star$). Dynamic Coloring *can be solved in* $O(q^{2.5q+o(q)}n)$ *time, where* $q = 2^{t+2^t}$ *and* t *is the size of the twin-cover of* G.

Theorem 11 ($\star$). Dynamic Coloring *can be solved in* $O(3^{O(wk)}poly(n))$ *time where* w *is the clique-width of the graph* G *and* k *is the number of colors.*

9 Conclusion

In this paper, we study Dynamic Coloring on various restricted graph classes and from the viewpoint of parameterized complexity. We presented polynomial time algorithms for chordal graphs and biconvex graphs. We have strengthened the NP-completeness result for bipartite graphs by showing that the problem remains NP-complete for star-convex bipartite graphs, comb-convex bipartite graphs and perfect elimination bipartite graphs. We show that the problem is FPT when parameterized by neighborhood diversity, twin-cover or the combined parameters clique-width and the number of colors.

We conclude the paper with the following list of open problems.

1. What is the parameterized complexity of Dynamic Coloring parameterized by (a) distance to cluster, and (b) distance to co-cluster?
2. What is the complexity of Dynamic Coloring for (a) convex bipartite graphs, and (b) permutation graphs?
3. Will our results hold for a generalized version of the problem called r-Dynamic Coloring [14], where every vertex v of degree at least one, is adjacent to at least $\min\{r, d\}$ many colors, where d is the degree of v?.

Acknowledgments. We would like to thank anonymous referees for their helpful comments. The first author and the third author acknowledges SERB-DST for supporting this research via grants PDF/2021/003452 and SRG/2020/001162 respectively.

References

1. Alishahi, M.: Dynamic chromatic number of regular graphs. Discret. Appl. Math. **160**(15), 2098–2103 (2012)
2. Brandstädt, A., Lozin, V.V.: On the linear structure and clique-width of bipartite permutation graphs. Ars Comb. **67**, 273–281 (2003)
3. Chen, Y., Fan, S., Lai, H.-J., Song, H., Sun, L.: On dynamic coloring for planar graphs and graphs of higher genus. Discret. Appl. Math. **160**(7–8), 1064–1071 (2012)
4. Cygan, M., et al.: Parameterized Algorithms, vol. 4. Springer, Cham (2015). https://doi.org/10.1007/978-3-319-21275-3
5. Downey, R.G., Fellows, M.R.: Fundamentals of Parameterized Complexity, vol. 4. Springer, Cham (2013). https://doi.org/10.1007/978-1-4471-5559-1
6. Enright, J.A., Stewart, L., Tardos, G.: On list coloring and list homomorphism of permutation and interval graphs. SIAM J. Discret. Math. **28**(4), 1675–1685 (2014)
7. Frank, A., Tardos, É.: An application of simultaneous diophantine approximation in combinatorial optimization. Combinatorica **7**(1), 49–65 (1987)
8. Jahanbekam, S., Kim, J., Suil, O., West, D.B.: On r-dynamic coloring of graphs. Discrete Appl. Math. **206**, 65–72 (2016)
9. Kannan, R.: Minkowski's convex body theorem and integer programming. Math. Oper. Res. **12**(3), 415–440 (1987)
10. Kim, S.J., Lee, S.J., Park, W.J.: Dynamic coloring and list dynamic coloring of planar graphs. Discrete Appl. Math. **161**(13), 2207–2212 (2013)
11. Lampis, M.: Algorithmic meta-theorems for restrictions of treewidth. Algorithmica **64**(1), 19–37 (2012)
12. Lenstra, H.W.: Integer programming with a fixed number of variables. Math. Oper. Res. **8**(4), 538–548 (1983)
13. Li, X., Yao, X., Zhou, W., Broersma, H.: Complexity of conditional colorability of graphs. Appl. Math. Lett. **22**(3), 320–324 (2009)
14. Montgomery, B.: Dynamic Coloring of Graphs. West Virginia University, Morgantown (2001)
15. Rose, D.J., Tarjan, R.E., Lueker, G.S.: Algorithmic aspects of vertex elimination on graphs. SIAM J. Comput. **5**(2), 266–283 (1976)
16. Saqaeeyan, S., Mollaahamdi, E.: Dynamic chromatic number of bipartite graphs. Sci. Ann. Comput. Sci. **26**(2), 249 (2016)

Enumeration of Minimal Tropical Connected Sets

Ivan Bliznets[1(✉)], Danil Sagunov[2], and Eugene Tagin[3]

[1] Utrecht University, Utrecht, Netherlands
iabliznets@gmail.com, i.bliznets@uu.nl

[2] St. Petersburg Department of Steklov Institute of Mathematics of the RAS, Saint Petersburg, Russia

[3] St. Petersburg State University, Saint Petersburg, Russia

Abstract. A subset of vertices in a vertex-colored graph is called tropical if vertices of each color present in the subset. This paper is dedicated to the enumeration of all minimal tropical connected sets in various classes of graphs. We show that all minimal tropical connected sets can be enumerated in $\mathcal{O}(1.7142^n)$ time on n-vertex interval graph which improves previous $\mathcal{O}(1.8613^n)$ upper bound obtained by Kratsch et al. Moreover, for chordal and general class of graphs we present algorithms with running times in $\mathcal{O}(1.937^n)$ and $\mathcal{O}(1.999958^n)$, respectively. The last two algorithms answer question implicitly asked in the paper [Kratsch et al. SOFSEM 2017]: «Is the number of tropical sets significantly smaller than the trivial upper bound 2^n?».

Keywords: tropical sets · enumeration algorithms · graph motif · chordal graphs · beating brute-force

1 Introduction

Efficient enumeration of objects with special properties is an important problem in computer science. There are many problems in graph theory in which the answer is a list of subsets of vertices that have a certain property or the cardinality of this set. Most often one is looking for the inclusion minimal/maximal induced subgraphs with additional attributes. The most classical result is that all maximal independent sets can be enumerated in $\mathcal{O}^*(3^{\frac{n}{3}})$ time [32], moreover, the running time is tight since there are graphs that have $3^{\frac{n}{3}}$ maximal independent sets [32]. It is also known that all minimal dominating sets can be listed in $\mathcal{O}(1.7159^n)$ time [21]. If the input graph is restricted to a special type of graphs like trees, chordal, and interval graphs faster algorithms were designed. For example, in chordal graphs, all minimal dominating sets can be enumerated

Work of Ivan Bliznets is supported by the project CRACKNP that has received funding from the European Research Council (ERC) under the European Union's Horizon 2020 research and innovation program (grant agreement No 853234).

M. Mavronicolas (Ed.): CIAC 2023, LNCS 13898, pp. 127–141, 2023.
https://doi.org/10.1007/978-3-031-30448-4_10

in $\mathcal{O}(1.5048^n)$ time [24], in trees in $\mathcal{O}(1.4656^n)$ time [28] and in interval graphs in $\mathcal{O}^*(3^{\frac{n}{3}})$ time [11] ($\mathcal{O}^*()$ suppress polynomial factors in the same way as $\mathcal{O}$ suppress constant factors).

Listing all potential candidates might be essential in some applications and the enumeration algorithms perform exactly this task. Such algorithms sometimes are key ingredients of very efficient algorithms for certain problems. For example, Lawler's $\mathcal{O}^*((1+3^{\frac{1}{3}})^n)$ algorithm [30] for a chromatic number is based on the fact that all maximal independent sets can be enumerated within $\mathcal{O}^*(3^{\frac{n}{3}})$ running time. The same fact is also used in the fastest known algorithm for 4-coloring by Fomin, Gaspers, and Saurabh [18]. Construction of efficient enumeration algorithms also often leads to new combinatorial upper bounds on the number of objects with special properties.

In the paper, we consider a problem of enumeration of subsets in vertex-colored graphs. More precisely, we are interested in enumeration of tropical subsets with additional properties. A set of vertices is called tropical if vertices of each color are presented in the set. There are papers dedicated to the study of various variants of tropical sets. For example tropical dominating sets were studied in [16], tropical matchings in [10], tropical paths in [9], tropical vertex-disjoint cycles in [31]. However, it seems that tropical connected sets attract the greatest attention [7,8,17,27]. Most probably that can be explained by close connection of connected tropical sets with a Graph Motif problem that was motivated by applications in biological network analysis [29] and later found applications in social networks [2] and in the context of mass spectrometry [5].

Angles d'Auriac et al. [17] proved that finding a minimum tropical connected set is NP-complete even on trees of height three as well as on split and interval graphs. An exact exponential-time algorithm for Minimum Tropical Connected Set was presented by Chapelle et al. [8]. In the case of a general input graph, they provide a $\mathcal{O}(1.5359^n)$ algorithm while in the case of trees they give a $\mathcal{O}(1.2721^n)$ algorithm. Later focus was shifted to enumeration of all minimal tropical connected sets and Kratsch et al. [27] presented algorithms tailored to special types of input graphs. So for split graphs they constructed a $\mathcal{O}(1.6402^n)$ algorithm, for interval graphs $\mathcal{O}(1.8613^n)$ time algorithm, for co-bipartite graphs and block graphs a $\mathcal{O}^*(3^{\frac{n}{3}})$ algorithm was presented. Moreover, in the same paper, several lower bounds on the maximum number of minimal tropical connected sets were given: for co-bipartite, interval, and block graphs the lower bound was $3^{\frac{n}{3}}$, for split graphs it was 1.4766^n and for chordal graphs it was 1.4916^n. No algorithm was presented for the case of a general input graph or for the case when the input graph is known to be chordal. We present a quote from the paper by Kratsch et al. [27]: "Interestingly, the best known upper bound for the maximum number of minimal tropical connected sets in an arbitrary graph and even for chordal graphs is the trivial one which is 2^n". The main goal of our paper is to answer this implicit question and present the first non-trivial upper bounds on the maximum number of minimal connected tropical sets in chordal and general graphs. We note that these types of questions, i.e. whether there is an algorithm faster than naive brute-force search play a tremendous role in computer science, especially in areas like fine-grained complexity, parameterized algorithm, and exact exponential algorithms. For many problems, it is easy to come up with algorithms significantly faster than brute-force search. However, for some problems, even a tiny improvement over brute-force search is a highly non-trivial task [1,3,4,13–15,19,20,33]. Moreover, there are a lot of important problems for which we do not know algorithms faster than simple brute-force search. For example, Set Cover problem, Satisfiability, and Orthogonal Vectors. Moreover, it is conjectured that it is impossible to construct such algorithms: Orthogonal Vector Conjecture [35], Set Cover Conjecture [12], Strong Exponential Time Hypothesis [12].

As a result of our research, we present an algorithm that enumerates all minimal tropical connected sets in $\mathcal{O}(1.999958^n)$ time in general graphs and an algorithm that performs the same task on chordal graphs in $\mathcal{O}(1.937^n)$ time. Moreover, we present an algorithm for interval graphs that runs in $\mathcal{O}(1.7142^n)$ time, which improves the previous asymptotic upper bounds of $\mathcal{O}(1.8613^n)$.

2 Preliminaries

We consider finite undirected graphs without loops or multiple edges. For graph G, $V(G)$ is the set of vertices of G, E(G) is a set of edges of G and $n = |V(G)|$ if not stated otherwise. $N(v)$ is the set of neighbours of vertex $v \in V(G)$. $N[v] = N(v) \cup \{v\}$ is the set of neighbours of vertex v including itself. For a set of vertices $X \subseteq V(G)$, $N_G[X] = \cup_{v \in X} N_G[v]$ and $N_G(X) = N_G[X] \setminus X$. For a subset $X \subseteq V(G)$ of vertices, $G[X]$ denotes the subgraph of G induced by X. A clique is a subset of vertices $D \subseteq V(G)$ such that $G[D]$ is a complete graph. *Chordal graph* is a graph without induced cycles of length bigger than 3. Chordal graphs admit many equivalent definitions, more details can be found here [26]. *Interval graphs* is a subclass of chordal graphs in which each vertex can be assigned an interval on a line such that two vertices have a common edge if and only if the corresponding intervals overlap [6]. $c : V(G) \to \mathbb{N}$ is a coloring function (not necessary proper), which assigns to each vertex a certain color. Let $c(X) = \{c(v) : v \in X\}$ be a set of different colors assigned to vertices of $X \subseteq V(G)$. Let $\mathcal{C} = c(V(G))$ be a set of all colors of graph G. We assume that $\mathcal{C} = \{1, 2, \ldots, C\}$. *A tropical set* of graph G is a subset of vertices $X \subseteq V(G)$ such that $c(X) = c(V(G))$. *A tropical connected set* of graph G is a subset of vertices $X \subseteq V(G)$ such that X is tropical and $G[X]$ is a connected subgraph. Let $\gamma = \frac{|\mathcal{C}|}{n}$. *A rainbow set* is a tropical set of the smallest size $|\mathcal{C}|$, i.e. a set that contains each color exactly once. A subset of vertices $X \subseteq V(G)$ is called *minimal tropical connected set* if there is no $Y \subsetneq X$ such that Y is tropical and $G[Y]$ is a connected subgraph.

Let $n, \ell, C, n_1, n_2, \ldots n_C$ be positive integers such that $n_1 + n_2 + \cdots + n_C = n$. We denote by $P_{n,\ell,C}^{n_1,n_2,\ldots,n_C}$ the number of tuples $(a_1, a_2, \ldots, a_C) \in \mathbb{Z}_{>0}^C$ such that $a_1 + a_2 + \cdots + a_C = \ell$ and $1 \le a_i \le n_i$ for each $1 \le i \le C$. Let $P_{n,\ell,C} = \max_{n_1,\ldots,n_C} P_{n,\ell,C}^{n_1,n_2,\ldots,n_C}$.

We assume that G is connected. Since otherwise, we can simply run our algorithms on each connected component of G separately and output a union of the obtained results. In our algorithms we use upper bounds on the number of tropical, rainbow sets and binomial coefficients given in the lemmas below. Due to the space constraints, proofs of lemmas marked by ($\star$) are omitted.

Lemma 1. [22] *For any positive integer n and $0 \leq \alpha \leq 1$ we have $\binom{n}{\alpha n} \leq 2^{H(\alpha)n}$, where $H(\cdot)$ is the binary entropy function i.e. $H(x) = -x\log_2(x) - (1-x)\log_2(1-x)$.*

Lemma 2 ($\star$). *Let G be a colored graph with n vertices and a number of used colors is γn then:*

1. *the number of all rainbow sets is at most $(\frac{1}{\gamma})^{\gamma n}$;*
2. *the number of tropical sets is at most $(2^{\frac{1}{\gamma}} - 1)^{\gamma n}$.*

Moreover, all rainbow and tropical sets can be listed almost within the same running time i.e. within $\mathcal{O}^*((\frac{1}{\gamma})^{\gamma n})$ and $\mathcal{O}^*((2^{\frac{1}{\gamma}} - 1)^{\gamma n})$ running time.

Lemma 3 ($\star$). *If $n_1, n_2, \ldots, n_k$ are positive integer numbers such that $n_1 + n_2 + \cdots + n_k = n$ then $n_1 n_2 \ldots n_k \leq 3^{\frac{n}{3}}$.*

Lemma 4 ($\star$). *For any positive integers n, ℓ, C we have: (i) $P_{n,\ell,C} \leq (\frac{n}{C})^C$; (ii) $P_{n,\ell,C} \leq \binom{\ell-1}{C-1}$.*

Lemma 5 ($\star$). *Let (G, c) be a colored graph and $S \subseteq V(G)$. There is a polynomial time algorithm that tests whether S is a* Minimal Tropical Connected Set.

3 General Graphs

In this section we present an algorithm that enumerates all inclusion-minimal tropical connected sets. The running time of the algorithm is $\mathcal{O}(1.999958^n)$. Hence, the number of Minimal Tropical Connected Sets is at most $\mathcal{O}(1.999958^n)$. These results answers an implicit question from [27], where trivial upper bound 2^n was given. In order to present the algorithm with mentioned running time we construct two auxiliary algorithms for the problem. The first one is given in the lemma below.

Lemma 6. *Let G be a vertex-colored graph with n vertices colored with $C = \gamma n$ colors. There is an algorithm that enumerates all* Minimal Tropical Connected Sets *in $\mathcal{O}^*((2^{1/\gamma} - 1)^{\gamma n})$ time.*

Proof. From Lemma 2, it follows that the number of all tropical sets is at most $((2^{1/\gamma} - 1)^{\gamma n})$. It is straightforward to enumerate all of them within this running time. What is left is to delete all sets that are not minimal tropical connected. However, by Lemma 5 we can run such test for each candidate in polynomial time. Hence in $\mathcal{O}^*((2^{1/\gamma} - 1)^{\gamma n})$ time we can list all Minimal Tropical Connected Sets. □

Before we proceed to the second auxiliary algorithm we state the following definition and theorem from [34].

Definition 1. [34] *For a given subset of vertices T we call a superset S of T* T-connecting *if S induces a connected graph. Moreover, we call S a minimal T-connecting if no strict subset of S is T-connecting.*

Theorem 1. [34] *For an n vertex graph $G = (V, E)$ and a terminal set $T \subseteq V$ where $|T| \leq \frac{n}{3}$ there are at most $\binom{n-|T|}{|T|-2} \cdot 3^{(n-|T|)/3}$ minimal T-connecting vertex sets and they can be enumerated in time $\mathcal{O}^*(\binom{n-|T|}{|T|-2} \cdot 3^{(n-|T|)/3})$.*

Lemma 7 ($\star$). *For an n vertex graph $G = (V, E)$ and a terminal set $T \subseteq V$ there are at most $2^{n-|T|}$ minimal T-connecting vertex sets and they can be enumerated in time $\mathcal{O}^*(2^{n-|T|})$.*

Equipped with the previous theorem and lemma, we are ready to prove the following result.

Lemma 8. *Let (G, c) be a graph with n vertices colored in $C = \gamma n$ colors and $\gamma \leq \frac{1}{3}$. There is an algorithm that enumerates all* Minimal Tropical Connected Sets *in time*

$$\max\{\max_{\alpha:\gamma\leq\alpha\leq 1-2\gamma} 2^{H(\frac{\gamma}{\alpha})\cdot\alpha n} \cdot \min\{2^{H(\frac{\gamma}{1-\alpha})\cdot(1-\alpha)n} \cdot 3^{\frac{1-\alpha}{3}n}, 2^{(1-\alpha)n}\},$$

$$\max_{\alpha:1-2\gamma\leq\alpha\leq 1} 2^{H(\frac{\gamma}{\alpha})} \cdot 2^{(1-\alpha)n}\}$$

up to a polynomial factor.

Proof. Recall that our graph contains vertices of $C = \gamma n$ different colors and the number of vertices colored in the i-th color is exactly n_i, i.e. $n_1 + n_2 + \cdots + n_C = n$. Let $V_i = \{v_i^1, \ldots, v_i^{n_i}\}$ be a set of all vertices of the i-th color.

We know that any tropical set must contain a rainbow set. With each minimal tropical connected set X we associate a rainbow set R_X constructed in the following way: for each $i \in \{1, 2, 3, \ldots, C\}$ we put v_i^j in R_X if $v_i^j \in X$ and for each $p < j$ we have that $v_i^p \notin X$. We note that $X \setminus R_X$ is an inclusion-minimal set that connects vertices from R_X, otherwise X is not minimal tropical connected set.

Now we are ready to describe the algorithm. In the first step we list all potential candidates for the role of R_X. So, basically, we consider many branchings and each branch defines a corresponding R_X. So in branch with $R_X = \{v_1^{j_1}, v_2^{j_2}, \ldots, v_C^{j_C}\}$ we assume that R_X is part of a minimal tropical connected set, while vertices $v_i^{p_i}$ with $p_i < j_i$ are not, hence in this branch these vertices can be simply deleted from the graph. At this point at each branch we already decided about $\ell = j_1 + j_2 + \cdots + j_C$ vertices whether they belong to a minimal tropical connected set or not. There are $n - j_1 - j_2 - \cdots - j_C$ vertices that are left, let us call the set of these vertices W. Now it is enough to list all inclusion-minimal sets $Y' \subseteq W$ such that $R_X \cup Y'$ is connected and discard those sets that are not minimal tropical connected sets. Check whether a set is a minimal tropical connected set can be done in a polynomial time by Lemma 5. Moreover, by Theorem 1 and Lemma 7 we can list all Y' that connect R_X in time $\min\{\mathcal{O}^*(\binom{|W|}{|C|-2} \cdot 3^{\frac{|W|}{3}}), \mathcal{O}^*(2^{|W|})\}$ if $|W| \geq 2C$ or in time $\mathcal{O}^*(2^{|W|})$ otherwise. Denote by

$$f(w,c) = \begin{cases} \min(\binom{w}{c-2} \cdot 3^{\frac{w}{3}}, 2^w), & \text{if } w \geq 2c \\ 2^w, & \text{otherwise} \end{cases}$$

So the running time of the algorithm up to a polynomial factor is equal to:

$$\sum_{\substack{1 \leq j_1 \leq n_1 \\ \dots \\ 1 \leq j_C \leq n_C}} f(n - (j_1 + j_2 + \dots + j_C), C).$$

Recall that $P_{n,\ell,C} = \max_{n_1,n_2,\dots,n_C} P^{n_1,n_2,\dots,n_C}_{n,\ell,C}$ and $P^{n_1,n_2,\dots,n_C}_{n,\ell,C}$ is the number of tuples $(a_1, \dots, a_C)$ such that $a_1 + a_2 + \dots + a_C = \ell$ and $1 \leq a_i \leq n_i$. So, the running time can be rewritten (up to a polynomial factor) as $\sum_{C \leq \ell \leq n} P_{\ell,C} \cdot f(n-\ell, C)$. By Lemma 4 we know that $P_{n,\ell,C} \leq \binom{\ell-1}{C-1} \leq \binom{\ell}{C}$

So, the running time up to the polynomial factor is bounded by $\max_{C \leq \ell \leq n} \binom{\ell}{C} \cdot f(n-\ell, C)$. Since $\gamma \leq \frac{1}{3}$ we know that $C \leq n - 2C$. So we can split interval $[C, n]$ into two intervals $[C, n-2C]$ and $[n-2C, n]$. So, it is obvious that:

$$\max_{C \leq \ell \leq n} \binom{\ell}{C} \cdot f(n-\ell, C) =$$

$$\max\left\{\max_{C \leq \ell \leq n-2C} \binom{\ell}{C} \cdot f(n-\ell, C), \max_{n-2C \leq \ell \leq n} \binom{\ell}{C} \cdot f(n-\ell, C)\right\} =$$

$$\max\left\{\max_{C \leq \ell \leq n-2C} \binom{\ell}{C} \cdot \min\left\{\binom{n-\ell}{C-2} \cdot 3^{\frac{n-\ell}{3}}, 2^{n-\ell}\right\}, \max_{n-2C \leq \ell \leq n} \binom{\ell}{C} \cdot 2^{n-\ell}\right\}.$$

Let $\ell = \alpha n$, recall that $C = \gamma n$. Note that $\binom{w}{c-2} \leq w^2 \binom{w}{c}$ for any w, c and $\binom{n}{\beta n} \leq 2^{H(\beta)n}$ for arbitrary $0 \leq \beta \leq 1$. Keeping the above said in mind, the running time up to the polynomial factor is bounded by:

$$\max\{\max_{\alpha: \gamma \leq \alpha \leq 1-2\gamma} 2^{H(\frac{\gamma}{\alpha}) \cdot \alpha n} \cdot \min\{2^{H(\frac{\gamma}{1-\alpha}) \cdot (1-\alpha)n} \cdot 3^{\frac{1-\alpha}{3}n}, 2^{(1-\alpha)n}\},$$

$$\max_{\alpha: 1-2\gamma \leq \alpha \leq 1} 2^{H(\frac{\gamma}{\alpha})} \cdot 2^{(1-\alpha)n}\}$$

So, we obtain the desired result. □

Now, we have all tools to show the main result of this section.

Theorem 2. *Let G be a colored graph with n vertices. There is an algorithm that enumerates all* MINIMAL TROPICAL CONNECTED SETS *in time* $\mathcal{O}(1.999958^n)$. *Hence, the number of all* MINIMAL TROPICAL CONNECTED SETS *in a graph on n vertices is at most* $\mathcal{O}(1.999958^n)$.

Proof. In order to construct an algorithm with desired running time, we carefully choose the right algorithm from algorithms presented in Lemmas 6 and 8. Note that $(2^{1/\gamma} - 1)^\gamma$ is decreasing function for $\gamma \in [0, 1]$, see Fig. 2. So it is more reasonable to use the algorithm from Lemma 6 when γ is large enough, i.e. input graph G has a sufficiently large number of colors. In contrary, if we plot the function

$$\max\{\max_{\alpha:\gamma\leq\alpha\leq 1-2\gamma} 2^{H(\frac{\gamma}{\alpha})\cdot\alpha n} \cdot \min\{2^{H(\frac{\gamma}{1-\alpha})\cdot(1-\alpha)n} \cdot 3^{\frac{1-\alpha}{3}n}, 2^{(1-\alpha)n}\}, \max_{\alpha:1-2\gamma\leq\alpha\leq 1} 2^{H(\frac{\gamma}{\alpha})\cdot\alpha n} \cdot 2^{(1-\alpha)n}\},$$

we see that the function is non-decreasing for $\gamma \in [0, 0.1]$, Fig. 2, so the second algorithm shows its best performance when the number of different colors is small.

So if the number of different colors in graph G is bigger than $0.08369n$ we run the first algorithm, i.e. if $\gamma \geq 0.08369$ then we run the algorithm with running time $\mathcal{O}^*((2^{1/\gamma} - 1)^{\gamma n}) \leq \mathcal{O}(1.999958^n)$. Otherwise, we run the second algorithm with running time

$$\max\{\max_{\alpha:\gamma\leq\alpha\leq 1-2\gamma} 2^{H(\frac{\gamma}{\alpha})\cdot\alpha n} \cdot \min\{2^{H(\frac{\gamma}{1-\alpha})\cdot(1-\alpha)n} \cdot 3^{\frac{1-\alpha}{3}n}, 2^{(1-\alpha)n}\}, \max_{\alpha:1-2\gamma\leq\alpha\leq 1} 2^{H(\frac{\gamma}{\alpha})\cdot\alpha n} \cdot 2^{(1-\alpha)n}\} \leq \mathcal{O}^*(1.999958^n).$$

So, in any case we get the desired running time. □

4 Chordal Graphs

The objective of this section is to present an algorithm that enumerates all Minimal Tropical Connected Sets in chordal graphs within $\mathcal{O}(1.937^n)$ running time which is smaller than in the case of arbitrary graphs. As a consequence we get that the number of all Minimal Tropical Connected Sets in any colored chordal graph is at most $\mathcal{O}(1.937^n)$. We note that this answers an implicit question from [27] where even for chordal graphs, the trivial 2^n bound was the only given bound on the number of minimal tropical connected sets. In order to achieve this improvement compared to the case of general input graph we replace algorithm described in Lemma 2 with a more efficient one. Instead of enumerating T-connecting sets we will be interested in enumerating connected dominating sets in special chordal subgraphs.

Before we proceed we recall some properties of chordal graphs and tree-decomposition.

A tree decomposition of a graph G is a pair $(\{X_i \mid i \in I\}, T = (I, F))$ with $\{X_i \mid i \in I\}$ a collection of subsets of $V(G)$, called *bags*, and $T = (I, F)$ a tree, such that

1. For every $v \in V(G)$, there exists $i \in I$ with $v \in X_i$.

2. For every $\{v, w\} \in E$, there exists $i \in I$ with $v, w \in X_i$
3. For every $i, j, k \in I$, if j is contained in a path from i to k in T, then $X_i \cap X_k \subseteq X_j$.

The following lemma is folklore and easily follows from lemma 7.1 in [12].

Lemma 9. [12] *Let $\mathcal{T} = (T, \{X_t\}_{t \in V(T)})$ be a tree decomposition of non-complete graph G and let u, v, w be nodes in tree T with bags B_u, B_v, B_w such that shortest path from u to w in tree T goes through vertex v. If $x \in B_u, y \in B_w$ then there is no path from x to y in graph $G \setminus B_v$ (note the statement trivially holds if x or y belongs to B_v).*

The following lemma is well known [25].

Lemma 10. [25] *Let G be a chordal graph then there exists a tree decomposition of G in which all bags are cliques. Moreover, such decomposition can be constructed in polynomial time.*

Now we are ready to present relevant results about connected dominating sets.

Definition 2. *For a connected graph G a subset of vertices $X \subseteq V(G)$ is called* connected dominating set, *if X induces a connected subgraph and $N[X] = V(G)$.*

Theorem 3. [23] *Any chordal graph with n vertices has no more than 1.4736^n minimal connected dominating sets. And all of them can be enumerated within $\mathcal{O}(1.4736^n)$ running time.*

Before we proceed with the algorithm we prove several auxiliary lemmas.

Lemma 11. *Let X be a vertex subset in graph G. Let S be a minimal set that connects X, i.e. is an inclusion-minimal subset of $V(G)$ such that the induced subgraph $G[S \cup X]$ is connected. If $S \cup X$ is a dominating set then there is a minimal connected dominating subset $M \subseteq V(G)$ such that $S \subseteq M \subseteq S \cup X$.*

Proof. We know that $S \cup X$ is connected and a dominating set. So it must contain some minimal connected dominating set. Let us call this set M'. If $S \subseteq M'$ then we are done and can take $M = M'$. If this is not the case, consider $S' = M' \setminus X$. Since $M' \subseteq (S \cup X)$ and $S \nsubseteq M'$ we have that $S' \subsetneq S$. M' is connected and dominating so $M' \cup X$ is also connected. Since $S' \cup X = M' \cup X$ we have that S is not an inclusion-minimal subset of $V(G)$ such that $G[S \cup X]$ is connected. So we get a contradiction. Hence, S must be a subset of M'. And we can take $M = M'$. □

Definition 3. *Let G be a chordal graph. For a subset of vertices $X \subseteq V(G)$ we call an X-*restriction *a chordal graph G_X obtained in the following way:*

1. *Take the tree decomposition $\mathcal{T}$ of G where each bag is a clique, one that is described in Lemma 10*

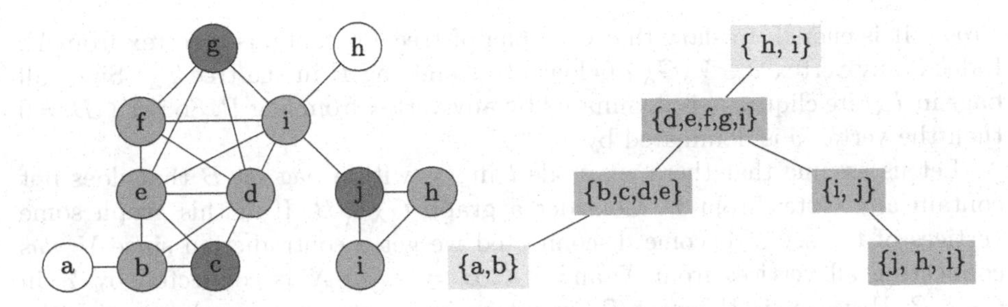

Fig. 1. The left part of figure shows graph G, set $X = \{c, g, j\}$ colored red and subgraph G_X that consist of orange and red vertices. The right part of figure shows tree-decomposition T of graph G, and an inclusion-minimal subtree T_X with all bags containing vertices from X.

2. *Find inclusion-minimal subtree $T_X \subseteq T$ such that T_X includes all nodes whose bags contain vertices from X*
3. *G_X is a graph, obtained by removing all vertices in G that are not contained in bags of T_X.*

Note that G_X is a chordal graph (as an induced subgraph of a chordal graph) and T_X is a tree-decomposition of G_X. Illustration of this definition is presented in Fig. 1.

Lemma 12. *Let G be a chordal graph and $X \subseteq V(G)$, then all minimal connected subgraphs of G containing X must be subgraphs of a restriction G_X.*

Proof. Let T be the tree decomposition of G (with corresponding tree T) constructed by the algorithm from Lemma 10 and T_X be a subtree in the decomposition that we used to construct an X-restriction graph G_X. Assume that there is a connected minimal subgraph H such that $X \subseteq V(H)$ and $V(H) \nsubseteq V(G_X)$. It can happen only if there are $x, y \in X$ such that there exists an vertex inclusion-minimal path p in $V(H)$ that connects x, y and has vertices outside G_X. Let us consider the shortest such path p and let z be a vertex on it that does not belong to G_X. Consider the shortest path $p_T = v_1, v_2, \ldots, v_q$ in T from the subtree T_X to the subtree T_z induced by bags containing z. Since, $z \notin G_X$ we have $T_X \cap T_z = \emptyset$. Hence, there is only one such path as otherwise T is not a tree. We note that $v_1 \in T_X$. Consider a subset of vertices S' from $V(G)$ that forms a bag of vertices for node v_1. S' is a separator and any path going from a vertex in G_X to z must pass through one of the vertices in S'. So it means that path $p = x, \ldots, z, \ldots y$ must contain vertices $u_1, u_2 \in S'$ on the subpaths from x to z and from z to y. However, $u_1, u_2 \in S'$ and S' is a clique as a bag of a node in a tree-decomposition for a chordal graph. So it means that the path p can be shorten as instead of going from u_1 to z and from z to u_2 we can go straight-ahead from u_1 to u_2. This leads to a desired contradiction. □

Lemma 13. *Let G be a chordal graph, $X \subseteq V(G)$, G_X be an X-restriction of G. If Y is connected in G_X and $X \subseteq Y$ then Y is a dominating set in G_X.*

Proof. It is enough to show that each bag of tree T_X contains a vertex from Y. Indeed, any vertex $v \in V(G_X)$ belongs to some bag B in the tree T_X. Since all bags in T_X are cliques, v is dominated by any vertex from bag B. So if $Y \cap B \neq \emptyset$ then the vertex v is dominated by Y.

Let us assume that there is a node t in T_X with a bag set B that does not contain any vertex from Y. Consider a graph $G_X \setminus B$. If in this graph some vertices of the set X become disconnected we get a contradiction since Y was connecting all vertices from X and $Y \cap B = \emptyset$. So X is connected by Y in $G_X \setminus B$. However, in this case T_X is not an inclusion-minimal subtree with the required property as some connected component of $T_X \setminus \{t\}$ will contain all bags with vertices from the set Y (this follows from Lemma 9). It is not possible that vertices $x_1, x_2 \in X$ belong to bags from different components of $T_X \setminus t$ since the path that connects x_1, x_2 in $G[Y]$ must go through some vertex from bag B and this contradict the fact that $B \cap Y = \emptyset$. □

Lemma 14. *Let G be a chordal graph with n vertices colored in $C = \gamma n$ colors. There is an algorithm that enumerates all* Minimal Tropical Connected Sets *within* $\max_{\alpha:\gamma \leq \alpha \leq 1} \min\{(\frac{1}{\gamma})^{\gamma n}, 2^{H(\frac{\gamma}{\alpha})\cdot \alpha n}\} \cdot 1.4736^{(1-\alpha+\gamma)n}$ *running time.*

Proof. First of all the algorithm in Lemma 10 constructs a tree-decomposition of graph G in which each bag is a clique. As in the case of general graph we list all potential candidates for the role of the rainbow set R_X. Recall that as before for each tropical set X we associated a rainbow set R_X. The R_X was constructed in the following way: for each $i \in \{1, 2, 3, \dots, C\}$ we put v_i^j in R_X if $v_i^j \in X$ and for each $p < j$ we have that $v_i^p \notin X$. Note that if R_X is a chosen rainbow set in the minimal tropical connected set then vertices v_i^p such that $p < j$ can be deleted from G. Denote by G' the obtained graph after such deletion.

After this, for the fixed rainbow set R_X, the algorithm constructs an R_X-restriction G'_{R_X}. On the next step the algorithm enumerates all minimal connected dominating sets of the graph G'_{R_X}. Let D be a minimal connected dominating set of G'_{R_X}. If $D \cup R_X$ is a minimal tropical connected set we output $D \cup R_X$ (we can test it by Lemma 5).

Let us prove that we output all Minimal Tropical Connected Sets. If Y is Minimal Tropical Connected Set then it contains the associated rainbow subset X' (here it might be the case that $Y = X'$, but it does not contradict anything). Recall that at some point we generate X' as a rainbow set in our algorithm. Since Y is minimal then $S' = Y \setminus X'$ is a minimal set that connects vertices X' (as otherwise there will be a tropical connected set that is a subset of Y). Note that any minimal set that connects X' lies inside $G'_{X'}$ by Lemma 12. For the graph $G'_{X'}$ and sets S', X' conditions of the Lemma 11 are true (take $G'_{X'}$ as G, S' as S and X' as X). Indeed, S' is a minimal set connecting X', $S' \cup X'$ is connected and that is why by Lemma 13 is a dominating set in $G'_{X'}$. Hence, at some point our algorithm considers minimal connected dominating set M' of $G'_{X'}$ such that $S' \subseteq M' \subseteq S' \cup X' = Y$ and outputs $M' \cup X'$ which is exactly Y.

It is left to prove the upper bound on the running time. Construction of the required tree-decomposition of the chordal graph takes polynomial time as well as vertex deletion and construction of R_X-restriction for fixed R_X and G'. So most of the time is consumed by the enumeration of all the rainbow sets and the enumeration of all the connected dominating sets in graph G'_{R_X} for fixed rainbow set R_X. So, as in the case with general graphs, the overall running time up to polynomial factor is:

$$\sum_{\substack{1 \le j_1 \le n_1 \\ \dots \\ 1 \le j_C \le n_C}} 1.4736^{n-(j_1+j_2+\dots+j_C)+C} = \sum_{\ell: C \le \ell \le n} P_{n,\ell,C} \cdot 1.4736^{n-\ell+C} \le$$

$$n \cdot \max_{\ell: C \le \ell \le n} P_{n,\ell,C} \cdot 1.4736^{n-\ell+C}.$$

By Lemma 4 we know that $P_{n,\ell,C} \le \min\{(\frac{n}{C})^C, \binom{\ell-1}{C-1}\} \le \min\{(\frac{n}{C})^C, \binom{\ell}{C}\}$. Making the substitution $C = \gamma n$ and $\ell = \alpha n$ we have that the running time is at most:

$$\max_{\alpha: \gamma \le \alpha \le 1} \left\{ \min\{(\frac{1}{\gamma})^{\gamma n}, 2^{H(\frac{\gamma}{\alpha}) \cdot \alpha n}\} \cdot 1.4736^{(1-\alpha+\gamma)n} \right\}.$$

□

Now we have all ingredients to prove the main result of this section.

Theorem 4. *All* Minimal Tropical Connected Sets *in a chordal graph on n vertices can be enumerated within $\mathcal{O}(1.937^n)$ running time. Hence, the maximum number of* Minimal Tropical Connected Sets *is at most $\mathcal{O}(1.937^n)$.*

Proof. Our algorithm for chordal graphs as well as the algorithm for the general case combines two algorithms and chooses between them depending on the number of colors in the input graph. However, instead of the algorithm from Lemma 8 we use the algorithm from Lemma 14. We recall that the running time of the algorithm from Lemma 6 is decreasing so it is more suitable for the case when the number of colors $C = \gamma n$ is large. In contrary, the function $\max_{\alpha: \gamma \le \alpha \le 1} \min\{(\frac{1}{\gamma})^{\gamma n}, 2^{H(\frac{\gamma}{\alpha}) \cdot \alpha n}\} \cdot 1.4736^{(1-\alpha+\gamma)n}$ is increasing for $\gamma \in [0, \frac{1}{3}]$, see Fig. 2, so the algorithm from Lemma 14 is preferable for small γ. So if $\gamma \le 0.3019$ then we run the algorithm from Lemma 14 and the running time will be bounded by $\mathcal{O}(1.937^n)$. If $\gamma > 0.3019$ then we run the algorithm from Lemma 6 and again the running time will be bounded by $\mathcal{O}(1.937^n)$. □

5 Interval Graphs

The main result of this section improves the previous known upper bound on the maximum number of minimal tropical connected sets in an interval graph on n vertices. Kratsch et al. [27] showed that the number is at most $\mathcal{O}(1.8613^n)$. Our upper bound is $\mathcal{O}(1.7142^n)$. Specifically, we prove the following:

Theorem 5. *There is an algorithm with running time $\mathcal{O}(1.7142^n)$ that enumerates all minimal tropical connected sets in a given interval graph on n vertices. Hence, the number of minimal tropical connected sets in any interval graph is at most $\mathcal{O}(1.7142^n)$.*

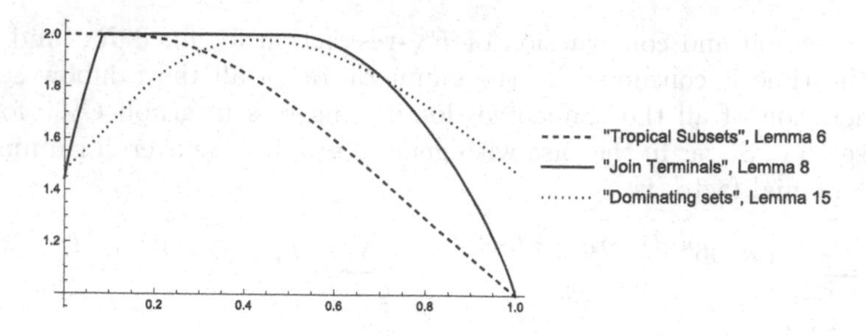

Fig. 2. The dependence of exponent on γ for the algorithms from Lemmas 6, 8, 14

In order to prove the theorem we use the following

Lemma 15. *Given an interval graph G of order n and a subset of vertices $U \subseteq V(G)$ such that $|U| = n'$ we can list all minimal sets Z that connects U in time $\mathcal{O}^*(3^{\frac{n-n'}{3}})$ (i.e. inclusion-minimal sets $Z \subseteq V(G) \setminus U$ such that $G[Z \cup U]$ is a connected graph).*

Proof. As our graph is interval we can in polynomial time construct a interval model such that each vertex has a corresponding interval on a line and:

- two vertices share an edge if and only if corresponding intervals intersect;
- no two intervals share endpoints.

Let us construct such model and fix it. Note that if $G[U]$ is a connected subgraph then the only set Z that satisfies the conditions is $\emptyset$. Denote by $U_1, U_2, \ldots, U_q$ connected components of $G[U]$. We enumerate them from left to right i.e. U_1 is the leftmost connected component in the interval model and U_q is the rightmost connected component. Denote by $\ell(U_i), r(U_i)$ the leftmost and the rightmost point of the connected component U_i on the fixed line model of G.

We must add a few vertices/intervals that join connected components $U_1, \ldots, U_q$. Hence, we must add some vertex whose corresponding interval starts to the left of $r(U_1)$ and ends to the right of $r(U_1)$ as otherwise the connected component U_1 will stay isolated from the other connected components. For a connected subgraph W denote by $N_r(W)$ vertices whose intervals start before $r(W)$ and end to the right of it. For any minimal U-connecting set Z we have $|Z \cap N_r(U_1)| \geq 1$. On the other hand $|Z \cap N_r(U_1)| < 2$, otherwise there are $v_1, v_2 \in Z \cap N_r(U_1)$. If $r(v_1)$ is to the left of $r(v_2)$ then $(Z \setminus v_1)$ connects all components $U_1, \ldots, U_q$ as Z was doing so. If $r(v_1)$ is to the right of $r(v_2)$ then $(Z \setminus v_2)$ connects all components $U_1, \ldots, U_q$. In any case, this contradicts to the fact that Z is inclusion-minimal. So it must be the case that $|Z \cap N_r(U_1)| = 1$.

Based on the above proved facts we suggest the following algorithm:

1. Branch on $|N_r(U_1)|$ possibilities to select a vertex from $N_r(U_1)$ that belongs to Z and discard from the graph the rest of vertices from $N_r(U_1)$. Assume that we pick vertex v' at this step.

2. Run the whole algorithm recursively on the new graph $(G \setminus N_r(U_1)) \cup \{v'\}$ and with a new subset (that needs to be connected) $U \cup \{v'\}$.

The correctness of the presented algorithm follows from the above observations. Since in the recurrence call each time we create i branchings and decrease n by i, we have that the running time of the algorithm is at most $\mathcal{O}^*(3^{\frac{n-n'}{3}})$ (as the maximum of $i^{\frac{1}{i}}$ is achieved when $i = 3$ in the set of natural numbers). □

Now we have all needed tools to present proof of Theorem 5:

Proof. (Proof of Theorem 5). As before we assume that our graph contains vertices of $C = \gamma n$ different colors and the number of vertices colored in the i-th color is exactly n_i, i.e. $n_1 + n_2 + \cdots + n_C = n$. Let $V_i = \{v_i^1, \ldots, v_i^{n_i}\}$ be a set of all vertices of the i-th color.

We know that any tropical set must contain a rainbow set. With each Minimal Tropical Connected Set X we associate a rainbow set R_X constructed as before (for each $i \in \{1, 2, 3, \ldots, C\}$ we put v_i^j in R_X if $v_i^j \in X$ and for each $p < j$ we have that $v_i^j \notin X$). We note that $X \setminus R_X$ is an inclusion-minimal set that connects vertices from R_X, otherwise X is not minimal tropical connected set.

Now we are ready to describe the algorithm. In the first step we list all potential candidates for the role of R_X. So, basically, we consider many branchings and each branch defines a corresponding R_X. So in branch in which $R_X = \{v_1^{j_1}, v_2^{j_2}, \ldots, v_C^{j_C}\}$ we assume that R_X is part of a minimal tropical connected set, while vertices $v_i^{p_i}$ with $p_i < j_i$ are not, hence in this branch these vertices can be simply deleted from the graph. At this point at each branch we already decided about $\ell = j_1 + j_2 + \cdots + j_C$ vertices whether they belong to a minimal tropical connected set or not. There are $n - j_1 - j_2 - \cdots - j_C$ vertices that are left, let us call the set of these vertices W. Now it is enough to list all inclusion-minimal sets $Y' \subseteq W$ such that $R_X \cup Y'$ is connected and discard those sets that are not minimal tropical connected sets. Check whether a set is a minimal tropical connected set can be done in a polynomial time by Lemma 5 and by Lemma 15 we can list all Y' that connect R_X in time $\mathcal{O}^*(3^{\frac{|W|}{3}})$. So the running time of the algorithm up to a polynomial factor is bounded by:

$$\sum_{\substack{1 \le j_1 \le n_1 \\ \cdots \\ 1 \le j_C \le n_C}} 3^{\frac{n-(j_1+j_2+\cdots+j_C)}{3}} \le \sum_{C \le \ell \le n} P_{n,\ell,C} \cdot 3^{\frac{n-\ell}{3}}.$$

Taking into account inequalities from Lemma 4 we have that the running time of our algorithm is at most $poly(n) \cdot \max_\ell[\min\{2^\ell, 3^{\frac{n}{3}}\} \cdot 3^{\frac{n-\ell}{3}}]$. The maximum of previous expression is achieved when $2^\ell = 3^{\frac{n}{3}}$ (since $2^\ell 3^{\frac{n-\ell}{3}}$ is an increasing function of ℓ when n is fixed and $3^{\frac{n}{3}} 3^{\frac{n-\ell}{3}}$ is decreasing). So in the worst case we have $\ell = \frac{n}{3} \cdot \log_2 3$ and the running time of our algorithm is bounded by $\mathcal{O}^*(3^{1/3 \cdot n \cdot (2 - \frac{1}{3} \cdot \frac{\log 3}{\log 2})}) = \mathcal{O}(1.7142^n)$. □

Acknowledgements. We would like to thank Lucas Meijer and anonymous reviewers for comments that helped to improve the paper.

References

1. Agrawal, A., Fomin, F.V., Lokshtanov, D., Saurabh, S., Tale, P.: Path contraction faster than 2^n. SIAM J. Discret. Math. **34**(2), 1302–1325 (2020)
2. Betzler, N., Van Bevern, R., Fellows, M.R., Komusiewicz, C., Niedermeier, R.: Parameterized algorithmics for finding connected motifs in biological networks. IEEE/ACM Trans. Comput. Biol. Bioinf. **8**(5), 1296–1308 (2011)
3. Bliznets, I., Fomin, F.V., Pilipczuk, M., Villanger, Y.: Largest chordal and interval subgraphs faster than 2^n. Algorithmica **76**(2), 569–594 (2016)
4. Bliznets, I., Sagunov, D.: Solving target set selection with bounded thresholds faster than 2^n. Algorithmica, 1–22 (2022)
5. Böcker, S., Rasche, F., Steijger, T.: Annotating fragmentation patterns. In: Salzberg, S.L., Warnow, T. (eds.) WABI 2009. LNCS, vol. 5724, pp. 13–24. Springer, Heidelberg (2009). https://doi.org/10.1007/978-3-642-04241-6_2
6. Brandstädt, A., Le, V.B., Spinrad, J.P.: Graph classes: a survey. SIAM (1999)
7. Chapelle, M., Cochefert, M., Kratsch, D., Letourneur, R., Liedloff, M.: Exact exponential algorithms to find a tropical connected set of minimum size. In: Cygan, M., Heggernes, P. (eds.) IPEC 2014. LNCS, vol. 8894, pp. 147–158. Springer, Cham (2014). https://doi.org/10.1007/978-3-319-13524-3_13
8. Chapelle, M., Cochefert, M., Kratsch, D., Letourneur, R., Liedloff, M.: Exact exponential algorithms to find tropical connected sets of minimum size. Theor. Comput. Sci. **676**, 33–41 (2017)
9. Cohen, J., Italiano, G.F., Manoussakis, Y., Thang, N.K., Pham, H.P.: Tropical paths in vertex-colored graphs. J. Comb. Optim. **42**(3), 476–498 (2021)
10. Cohen, J., Manoussakis, Y., Phong, H., Tuza, Z.: Tropical matchings in vertex-colored graphs. Electron. Notes Discrete Math. **62**, 219–224 (2017)
11. Couturier, J.F., Letourneur, R., Liedloff, M.: On the number of minimal dominating sets on some graph classes. Theoret. Comput. Sci. **562**, 634–642 (2015)
12. Cygan, M., et al.: Parameterized Algorithms. Springer, Cham (2015). https://doi.org/10.1007/978-3-319-21275-3
13. Cygan, M., Pilipczuk, M., Pilipczuk, M., Wojtaszczyk, J.O.: Solving the 2-disjoint connected subgraphs problem faster than 2^n. Algorithmica **70**(2), 195–207 (2014)
14. Cygan, M., Pilipczuk, M., Wojtaszczyk, J.O.: Irredundant set faster than $O(2^n)$. In: Calamoneri, T., Diaz, J. (eds.) CIAC 2010. LNCS, vol. 6078, pp. 288–298. Springer, Heidelberg (2010). https://doi.org/10.1007/978-3-642-13073-1_26
15. Cygan, M., Pilipczuk, M., Wojtaszczyk, J.O.: Capacitated domination faster than $o(2^n)$. Inf. Process. Lett. **111**(23–24), 1099–1103 (2011)
16. d'Auriac, J.A.A., et al.: Tropical dominating sets in vertex-coloured graphs. J. Discrete Algorithms **48**, 27–41 (2018)
17. d'Auriac, J.A.A., Cohen, N., El Mafthoui, H., Harutyunyan, A., Legay, S., Manoussakis, Y.: Connected tropical subgraphs in vertex-colored graphs. Discrete Math. Theor. Comput. Sci. **17**(3), 327–348 (2016)
18. Fomin, F.V., Gaspers, S., Saurabh, S.: Improved exact algorithms for counting 3- and 4-colorings. In: Lin, G. (ed.) COCOON 2007. LNCS, vol. 4598, pp. 65–74. Springer, Heidelberg (2007). https://doi.org/10.1007/978-3-540-73545-8_9
19. Fomin, F.V., Giannopoulou, A.C., Pilipczuk, M.: Computing tree-depth faster than 2^n. Algorithmica **73**(1), 202–216 (2015)
20. Fomin, F.V., Grandoni, F., Kratsch, D.: Solving connected dominating set faster than 2^n. Algorithmica **52**(2), 153–166 (2008)

21. Fomin, F.V., Grandoni, F., Pyatkin, A.V., Stepanov, A.A.: Combinatorial bounds via measure and conquer: bounding minimal dominating sets and applications. ACM Trans. Algorithms (TALG) **5**(1), 1–17 (2008)
22. Fomin, F.V., Kratsch, D.: Exact exponential algorithms (2010)
23. Golovach, P.A., Heggernes, P., Kratsch, D., Saei, R.: Enumeration of minimal connected dominating sets for chordal graphs. Discrete Appl. Math. **278**, 3–11 (2020). https://doi.org/10.1016/j.dam.2019.07.015
24. Golovach, P.A., Kratsch, D., Liedloff, M., Sayadi, M.Y.: Enumeration and maximum number of minimal dominating sets for chordal graphs. Theor. Comput. Sci. **783**, 41–52 (2019). https://doi.org/10.1016/j.tcs.2019.03.017
25. Golumbic, M.C.: Algorithmic Graph Theory and Perfect Graphs. Elsevier, Amsterdam (2004)
26. Heggernes, P.: Minimal triangulations of graphs: a survey. Discret. Math. **306**(3), 297–317 (2006)
27. Kratsch, D., Liedloff, M., Sayadi, M.Y.: Enumerating minimal tropical connected sets. In: Steffen, B., Baier, C., van den Brand, M., Eder, J., Hinchey, M., Margaria, T. (eds.) SOFSEM 2017. LNCS, vol. 10139, pp. 217–228. Springer, Cham (2017). https://doi.org/10.1007/978-3-319-51963-0_17
28. Krzywkowski, M.: Trees having many minimal dominating sets. Inf. Process. Lett. **113**(8), 276–279 (2013)
29. Lacroix, V., Fernandes, C.G., Sagot, M.F.: Motif search in graphs: application to metabolic networks. IEEE/ACM Trans. Comput. Biol. Bioinf. **3**(4), 360–368 (2006)
30. Lawer, E.L.: A note on the complexity of the chromatic number problem. Inf. Process. Lett. (1976)
31. Le, H., Highley, T.: Tropical vertex-disjoint cycles of a vertex-colored digraph: barter exchange with multiple items per agent. Discrete Math. Theor. Comput. Sci. **20** (2018)
32. Moon, J.W., Moser, L.: On cliques in graphs. Israel J. Math. **3**, 23–28 (1965)
33. Razgon, I.: Computing minimum directed feedback vertex set in $o*(1.9977^n)$. In: Theoretical Computer Science, pp. 70–81. World Scientific (2007)
34. Telle, J.A., Villanger, Y.: Connecting terminals and 2-disjoint connected subgraphs. In: Brandstädt, A., Jansen, K., Reischuk, R. (eds.) WG 2013. LNCS, vol. 8165, pp. 418–428. Springer, Heidelberg (2013). https://doi.org/10.1007/978-3-642-45043-3_36
35. Vassilevska Williams, V.: Hardness of easy problems: basing hardness on popular conjectures such as the strong exponential time hypothesis (invited talk). In: 10th International Symposium on Parameterized and Exact Computation (IPEC 2015). Schloss Dagstuhl-Leibniz-Zentrum fuer Informatik (2015)

Dynamic Flows with Time-Dependent Capacities

Thomas Bläsius, Adrian Feilhauer(✉), and Jannik Westenfelder

Karlsruhe Institute of Technology (KIT), 76131 Karlsruhe, Germany
{thomas.blaesius,adrian.feilhauer}@kit.edu

Abstract. Dynamic network flows, sometimes called flows over time, extend the notion of network flows to include a transit time for each edge. While Ford and Fulkerson showed that certain dynamic flow problems can be solved via a reduction to static flows, many advanced models considering congestion and time-dependent networks result in NP-hard problems. To increase understanding of these advanced dynamic flow settings we study the structural and computational complexity of the canonical extensions that have time-dependent capacities or time-dependent transit times.

If the considered time interval is finite, we show that already a single edge changing capacity or transit time once makes the dynamic flow problem weakly NP-hard. In case of infinite considered time, one change in transit time or two changes in capacity make the problem weakly NP-hard. For just one capacity change, we conjecture that the problem can be solved in polynomial time. Additionally, we show the structural property that dynamic cuts and flows can become exponentially complex in the above settings where the problem is NP-hard. We further show that, despite the duality between cuts and flows, their complexities can be exponentially far apart.

1 Introduction

Network flows are a well established way to model transportation of goods or data through systems representable as graphs. Dynamic flows (sometimes called flows over time) include the temporal component by considering the time to traverse an edge. They were introduced by Ford and Fulkerson [3], who showed that maximum dynamic flows in static networks can be found using *temporally repeated flows*, which send flow over paths of a static maximum flow as long as possible.

Since capacities in real-world networks tend to be more dynamic, several generalizations have been considered in the literature. One category here is congestion modeling networks, where transit times of edges can depend on the flow routed over them [6,7]. Other generalizations model changes in the network independently from the routed flow [4,9,11]. This makes it possible to model known physical changes to the network and allows for situations, where we have estimates of the overall congestion over time that is caused by external entities that are not part of the given flow problem. There are also efforts to include different

M. Mavronicolas (Ed.): CIAC 2023, LNCS 13898, pp. 142–156, 2023.
https://doi.org/10.1007/978-3-031-30448-4_11

objectives for the flow, e.g., for evacuation scenarios, it is beneficial for a flow to maximize arrival for all times, not just at the end of the considered time interval [2].

Most problems modeling congestion via flow-dependent transit times are NP-hard. If the transit time depends on the current load of the edge, the flow problems become strongly NP-hard and no ε approximation exists unless P = NP [6]. If the transit time of an edge instead only depends on its inflow rate while flow that entered the edge earlier is ignored the flow problems are also strongly NP-hard [7]. When allowing to store flow at vertices, pseudo-polynomial algorithms are possible if there are time-dependent capacities [4] and if there additionally are time-dependent transit times [9,11]. In the above mentioned evacuation scenario, one aims at finding the so-called earliest arrival flow (EAF). It is also NP-hard in the sense that it is hard to find the average arrival time of such a flow [2]. Moreover, all known algorithms to find EAFs have worst case exponential output size for all known encodings [2].

In this paper, we study natural generalizations of dynamic flows that have received little attention so far, allowing time-dependent capacities or time-dependent transit times. We prove that finding dynamic flows with time-dependent capacities or time-dependent transit times is weakly NP-hard, even if the graph is acyclic and only a single edge experiences a capacity change at a single point in time. This shows that a single change in capacity already increases the complexity of the – otherwise polynomially solvable – dynamic flow problem. It also implies that the dynamic flow problem with time-dependent capacities is not FTP in the number of capacity changes. The above results hold in the setting where the considered time interval is finite. If we instead consider infinite time, the results remain the same for time-dependent transit times. For time-dependent capacities, two capacity changes make the problem weakly NP-hard. We conjecture that it can be solved in polynomial when there is only one change.

Beyond these results on the computational complexity, we provide several structural insights. For static flows, one is usually not only interested in the flow value but wants to output a maximum flow or a minimum cut. The concept of flows translates more or less directly to the dynamic setting [3], we need to consider time-dependent flows and cuts if we have time-dependent capacities or transit times. In this case, instead of having just one flow value per edge, the flow is a function over time. Similarly, in a dynamic cut, the assignment of vertices to one of two partitions changes over time. The cut–flow duality, stating that the capacity of the minimum cut is the same as the value of the maximum flow also holds in this and many related settings [5,8,11]. Note that the output complexity can potentially be large if the flow on an edge or the partition of a vertex in a cut changes often. For dynamic flows on static graphs (no changes in capacities or transit times) vertices start in the target vertices' partition and at some point change to the source partition, but never the other way [10], which shows that cuts have linear complexity in this setting.

In case of time-dependent capacities or transit times, we show that flow and cut complexity are sometimes required to be exponential. Specifically, for all cases where we show weak NP-hardness, we also give instances for which every maximum flow and minimum cut have exponential complexity. Thus, even a single edge changing capacity or transit time once can jump the output complexity from linear to exponential. Moreover, we give examples where the flow complexity is exponential while there exists a cut of low complexity and vice versa.

We note that the scenario of time-dependent capacities has been claimed to be strongly NP-complete [9] before. However, we suspect the proof to be flawed as one can see that this scenario can be solved in pseudo-polynomial time. Moreover, the above mentioned results on the solution complexity make it unclear whether the problem is actually in NP.

Due to space constraints, many proofs are shortened or omitted here, they can be found in the full version of the paper [1].

2 Preliminaries

We consider dynamic networks $G = (V, E)$ with directed edges and designated source and target vertices $s, t \in V$. Edges $e = (v, w) \in E$ have a time-dependent non negative *capacity* $u_e \colon [0, T] \to \mathbb{R}_0^+$, specifying how much flow can enter e via v at each time. We allow u_e to be non-continuous but only for finitely many points in time. In addition, each edge $e = (v, w)$ also has a non negative *transit time* $\tau_e \in \mathbb{R}^+$, denoting how much time flow takes to move from v to w when traversing e. Note that the capacity is defined on $[0, T]$, i.e., time is considered from 0 up to a *time horizon* T.

Let f be a collection of measurable functions $f_e \colon [0, T - \tau_e] \to \mathbb{R}$, one for each edge $e \in E$, assigning every edge a flow value depending on the time. The restriction to the interval $[0, T - \tau_e]$ has the interpretation that no flow may be sent before time 0 and no flow should arrive after time T in a valid flow. To simplify notation, we allow time values beyond $[0, T - \tau_e]$ and implicitly assume $f_e(\Theta) = 0$ for $\Theta \notin [0, T - \tau_e]$. We call f a *dynamic flow* if it satisfies the *capacity constraints* $f_e(\Theta) \leq u_e(\Theta)$ for all $e \in E$ and $\Theta \in [0, T - \tau_e]$, and *strong flow conservation*, which we define in the following.

The *excess flow* $\mathrm{ex}_f(v, \Theta)$ of a vertex v at time Θ is the difference between flow sent to v and the flow sent from v up to time Θ, i.e.,

$$\mathrm{ex}_f(v, \Theta) := \int_0^{\Theta} \sum_{e=(u,v)\in E} f_e(\zeta - \tau_e) - \sum_{e=(v,u)\in E} f_e(\zeta)\, \mathrm{d}\zeta.$$

We have strong flow conservation if $\mathrm{ex}_f(v, \Theta) = 0$ for all $v \in V \backslash \{s, t\}$ and $\Theta \in [0, T]$.

The *value* of f is defined as the excess of the target vertex at the time horizon $|f| := \mathrm{ex}_f(t, T) = -\mathrm{ex}_f(s, T)$. The *maximum dynamic flow problem with time-dependent capacities* is to find a flow of maximum value. We refer to its input as *dynamic flow network*.

A cut-flow duality similar to the one of the static maximum flow problem holds for the maximum dynamic flow problem with the following cut definition.

A *dynamic cut* or *cut over time* is a partition of the vertices $(S, V \setminus S)$ for each point in time, where the source vertex s always belongs to S while the target t never belongs to S. Formally, each vertex $v \in V$ has a boolean function $S_v \colon [0,T] \to \{0,1\}$ assigning v to S at time Θ if $S_v(\Theta) = 1$. As for the flow, we extend S_v beyond $[0,T]$ and set $S_v(\Theta) = 1$ for $\Theta > T$ for all $v \in V$ (including t). The *capacity* $\text{cap}(S)$ of a dynamic cut S is the maximum flow that could be sent on edges from S to $V \setminus S$ during the considered time interval $[0,T]$, i.e.,

$$\text{cap}(S) = \int_0^T \sum_{\substack{(v,w)\in E \\ S_v(\Theta)=1 \\ S_w(\Theta+\tau_e)=0}} u_{(v,w)}(\Theta)\,\mathrm{d}\Theta.$$

An edge (v,w) *contributes* to the cut S at time Θ if it contributes to the above sum, so $S_v(\Theta) = 1 \wedge S_w(\Theta + \tau_e) = 0$. Note that this is similar to the static case, but in the dynamic variant the delay of the transit time needs to be considered. Thus, for the edge $e = (v,w)$ we consider v at time Θ and w at time $\Theta + \tau_e$. Moreover, setting $S_v(\Theta) = 1$ for all vertices v if $\Theta > T$ makes sure that no point in time beyond the time horizon contributes to $\text{cap}(S)$.

Theorem 1 (Min-Cut Max-Flow Theorem [8,11]). *For a maximum flow over time f and a minimum cut over time S it holds $|f| = cap(S)$.*

Proof. The theorem by Philpott [8, Theorem 1] is more general than the setting considered here. They in particular allow for time-dependent storage capacities of vertices. We obtain the here stated theorem by simply setting them to constant zero. The theorem by Tjandra [11, Theorem 3.4] is even more general and thus also covers the setting with time-dependent transit times. □

Though the general definition allows the capacity functions to be arbitrary, for our constructions it suffices to use piecewise constant capacities. We note that in this case, there always exists a maximum flow that is also piecewise constant, assigning flow values to a set of intervals of non-zero measure. The property that the intervals have non-zero measure lets us consider an individual point Θ in time and talk about the contribution of an edge to a cut or flow at time Θ, as Θ is guaranteed to be part of a non-empty interval with the same cut or flow. For the remainder of this paper, we assume that all flows have the above property.

We define the following additional useful notation. We use $S(\Theta) := \{v \in V \mid S_v(\Theta) = 1\}$ and $\bar{S}(\Theta) := \{v \in V \mid S_v(\Theta) = 0\}$ to denote the cut at time Θ. Moreover, a vertex v changes its partition at time Θ if $S_v(\Theta - \varepsilon) \neq S_v(\Theta + \varepsilon)$ for every sufficiently small $\varepsilon > 0$. We denote a change from S to $\bar{S}$ with $S_v \xrightarrow{\Theta} \bar{S}_v$ and a change in the other *direction* from $\bar{S}$ to S with $\bar{S}_v \xrightarrow{\Theta} S_v$. We denote the number of partition changes of a vertex v in a cut S with $\text{ch}_v(S)$. Moreover the total number of changes in S is the *complexity* of the cut S. For a flow f, we define changes on edges as well as the complexity of f analogously.

In the above definition of the maximum dynamic flow problem we allow time-dependent capacities but assume constant transit times. Most of our results translate to the complementary scenario where transit times are time-dependent while capacities are constant. In this setting $\tau_e(\Theta)$ denotes how much time flow takes to traverse e, if it enters at time Θ. Similarly to the above definition, we allow τ_e to be non-continuous for finitely many points in time.

Additionally we look at the scenarios where infinite time ($\Theta \in (-\infty, \infty)$) is considered instead of only considering times in $[0, T]$. This removes structural effects caused by the boundaries of the considered time interval. Intuitively, because we are working with piecewise constant functions with finitely many incontinuities, there exists a point in time Θ that is sufficiently late that all effects of capacity changes no longer play a role. From that time on, one can assume the maximum flow and minimum cut to be constant. The same holds true for a sufficiently early point in time. Thus, to compare flow values it suffices to look at a finite interval I. Formally, f is a maximum dynamic flow with *infinite considered time* if it is constant outside of I and maximum on I, such that for any larger interval $J \supset I$ there exists a large enough interval $K \supset J$ so that a maximum flow with considered time interval K can be f during J. Minimum cuts with infinite considered time are defined analogously. Such maximum flows and minimum cuts always exist as temporally repeated flows provide optimal solutions to dynamic flows and we only allow finitely many changes to capacity or traversal time.

We will need the set of all integers up to k and denote it $[k] := \{i \in \mathbb{N}^+ | i \leq k\}$.

3 Computational Complexity

In this section we study the computational complexity of the dynamic flow problem with time-dependent capacities or transit times. We consider finite and infinite time. For all cases except for a single capacity change with infinite considered time, we prove NP-hardness.

We start by showing hardness in the setting where we have time-dependent capacities with only one edge changing capacity once. Our construction directly translates to the setting of infinite considered time with one edge changing capacity twice. For the case of time-dependent transit times we prove hardness for one change even in the infinite considered time setting. This also implies hardness for one change when we have a finite time horizon.

We reduce from the *partition problem*, which is defined as follows. Given a set of positive integers $S = \{b_1, \ldots, b_k\}$ with $\sum_{i=1}^{k} b_i = 2L$, is there a subset $S' \subset S$ such that $\sum_{a \in S'} a = L$?

Theorem 2. *The dynamic flow problem with time-dependent capacities is weakly NP-hard, even for acyclic graphs with only one capacity change.*

Proof. Given an instance of the partition problem, we construct $G = (V, E)$ as shown in Fig. 1 and show that a solution to partition is equivalent to a flow of value 1 in G.

Every $b_i \in S$ corresponds to a vertex x_i which can be reached by x_{i-1} with one edge of transit time b_i and one bypass edge of transit time zero. The last of these vertices x_k is connected to the target t with an edge only allowing flow to pass during $[L+1, L+2]$, where the lower border is ensured by the capacity change of (x_k, t) and the upper border is given by the time horizon $T = L+3$. The source s is connected to x_0 with an edge of low capacity $\frac{1}{L+1}$, so that the single flow unit that can enter this edge in $[0, T-2]$ can pass (x_k, t) during one time unit.

Since a solution to the partition problem is equivalent to a path of transit time L through the x_i, we additionally provide paths of transit time $0, 1, \ldots, L-1$ bypassing the b_i edges via the y_i so that a solution for partition exists, if and only if flow of value 1 can reach t. To provide the bypass paths, we set $\ell \in \mathbb{N}_0$ so that $L = 2^{\ell+1} + r$, $r \in \mathbb{N}_0, r < 2^{\ell+1}$ and define vertices $y_i, i \in \mathbb{N}_0, i \leq \ell$. They create a path of transit time $L-1$ where the edges' transit times are powers of two and one edge of transit time r and all edges can be bypassed by an edge with transit time zero. This allows all integer transit times smaller than $L-1$. All edges except for (s, x_0) have unit capacity when they are active.

Given a solution S' to the partition problem, we can route flow leaving s during $[0,1]$ through the x_i along the non zero transit time edges if and only if the corresponding b_i is in S'. Flow leaving s in $[1, L+1]$ can trivially reach x_k during $[L+1, L+2]$ using the bypass paths, providing a maximum flow of 1.

Only one unit of flow can reach x_0 until $L+2$, considering the time horizon $T = L+3$ and the transit time of (x_k, t), the flow can have value at most 1. Given a flow that sends one unit of flow to t, we can see that the flow has to route all flow that can pass (s, x_0) during $[0, L+1]$ to t. Due to the integrality of transit times, the flow leaving s during $[0,1]$ has to take exactly time L to traverse from x_0 to x_k. The bypass paths via y_0 are too short for this. As such, this time is the sum of edge transit times taken from the partition instance and zeroes from bypass edges, and there exists a solution S' to the partition problem that consists of the elements corresponding to the non zero transit time edges taken by this flow. □

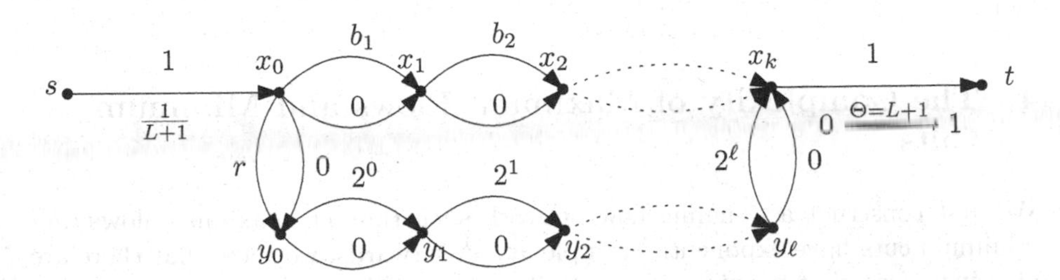

Fig. 1. Graph constructed for the reduction of the partition problem to dynamic flow with time-dependent capacities. Flow leaving s at time zero can only reach t if it takes exactly time L to traverse from x_0 to x_k, such choosing a partition. Black numbers are transit times, blue numbers indicate capacity, all unspecified capacities are 1, time horizon is $T = L+3$. (Color figure online)

Corollary 1. *The dynamic flow problem with time-dependent capacities and infinite considered time is weakly NP-hard, even for acyclic graphs with only two capacity changes.*

Proof. In the proof of Theorem 2, we restricted the flow on the edge from x_k to t to have non-zero capacity only at time $[L + 1, L + 2]$. For Theorem 2, we achieved the lower bound with one capacity change and the upper bound with the time horizon. Here, we can use the same construction but use a second capacity change for the upper bound. □

For the case of time-dependent transit times, we use a similar reduction. We start with the case of infinite considered time.

Theorem 3. *The dynamic flow problem with infinite considered time and time-dependent transit times is weakly NP-hard, even for acyclic graphs with only one transit time change.*

To translate this result to the case of a finite time horizon, note that we can use the above construction and choose the time horizon sufficiently large to obtain the following corollary.

Corollary 2. *The dynamic flow problem with time-dependent transit times is weakly NP-hard, even for acyclic graphs with only one transit time change.*

This leaves one remaining case: infinite considered time and a single capacity change. For this case, we can show that there always exists a minimum dynamic cut, where each vertex changes partition at most once and all partition changes are of the same direction. Furthermore, for given partitions before and after the changes, a linear program can be used to find the optimal transition as long as no vertex changes partition more than once and all partition changes are of the same direction. This motivates the following conjecture.

Conjecture 1. The minimum cut problem in a dynamic flow network with only a single change in capacity and infinite considered time can be solved in polynomial time.

4 The Complexity of Maximum Flows and Minimum Cuts

We first construct a dynamic flow network such that all maximum flows and minimum cuts have exponential complexity. Afterwards, we show that there are also instances that require exponentially complex flows but allow for cuts of linear size and vice versa. These results are initially proven for a single change in capacity and are then shown to also hold in the setting with time dependent transit times, likewise with only one change in transit time required.

4.1 Exponentially Complex Flows and Cuts

We initially focus on the complexity of cuts and only later show that it transfers to flows. Before we start the construction, note that the example in Fig. 2 shows how the partition change of two vertices a and b can force a single vertex v to change its partition back and forth. This type of enforced partition change of v is at the core of our construction.

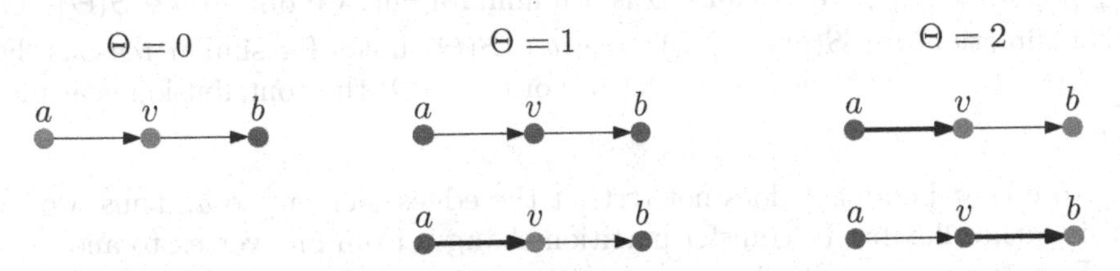

Fig. 2. Example where a changes from $\bar{S}$ (red) to S (blue) at time 1 and b changes from S to $\bar{S}$ at time 2. Only edges from S to $\bar{S}$ contribute to the cut (bold edges). Assuming v starts in $\bar{S}$ and $u_{(a,v)} < u_{(v,b)}$ as well as $\tau_{(a,v)} = \tau_{(v,b)} = 0$, v has to change to S at time 1 and back to $\bar{S}$ at time 2 in a minimum cut (top row). The bottom row illustrates the alternative (more expensive) behavior of v. (Color figure online)

More specifically, we first give a structure with which we can force vertices to mimic the partition changes of other vertices, potentially with fixed time delay.

The *mimicking gadget* links two non terminal vertices $a, b \in V \backslash \{s, t\}$ using edges $(a, b), (b, t) \in E$ with capacities $u_{(a,b)} = \alpha, u_{(b,t)} = \beta$. The following lemma shows what properties α and β need to have such that the mimicking gadget does its name credit, i.e., that b mimics a with delay $\tau_{(a,b)}$. A visualization of the mimicking gadget is shown in Fig. 3.

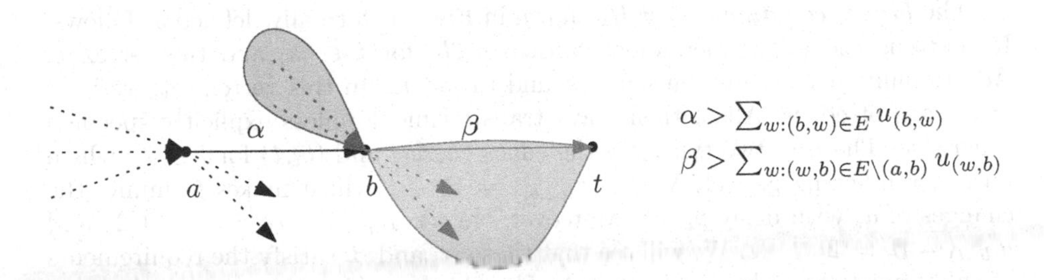

Fig. 3. Gadget linking the partitions of two vertices a and b, so that b mimics a with a delay of $\tau_{(a,b)}$; α, β are capacities.

Lemma 1. *Let G be a graph that contains the mimicking gadget as a sub-graph, such that*

$$\alpha > \sum_{w|(b,w)\in E} u_{(b,w)} \quad \text{and} \quad \beta > \sum_{w|(w,b)\in E\backslash(a,b)} u_{(w,b)}.$$

Then, $S_b(\Theta) = S_a(\Theta - \tau_{(a,b)})$ *for every minimum cut* S *and times* $\Theta \in (\tau_{(a,b)}, T - \tau_{(b,t)})$.

Proof. We first show $a \in S(\Theta - \tau_{(a,b)}) \implies b \in S(\Theta)$. With the partition of a fixed, we look at possible contribution to S of edges incident to b at time Θ. For $b \in \bar{S}(\Theta)$ the contribution is at least α, because $\Theta \in (\tau_{(a,b)}, T - \tau_{(b,t)})$ ensures that (a, b) can contribute to S. For $b \in S(\Theta)$ the contribution is at most $\sum_{w|(b,w)\in E} u_{(b,w)} < \alpha$. Because S is a minimum cut, we obtain $b \in S(\Theta)$. The other direction $a \in \bar{S}(\Theta - \tau_{(a,b)}) \implies b \in \bar{S}(\Theta)$ holds for similar reasons. For $b \in S(\Theta)$ the contribution is at least β. For $b \in \bar{S}(\Theta)$ the contribution is at most $\sum_{w|(w,b)\in E} u_{(w,b)} < \beta$. □

Note that Lemma 1 does not restrict the edges incident to a. Thus, we can use it rather flexibly to transfer partition changes from one vertex to another.

To enforce exponentially many partition changes, we next give a gadget that can double the number of partition changes of one vertex. To this end, we assume that, for every integer $i \in [k]$, we already have access to vertices a_i with *period* $p_i := 2^i$, i.e., a_i changes partition every p_i units of time. Note that a_1 is the vertex with the most changes. With this, we construct the so-called binary counting gadget that produces a vertex v with period $p_0 = 1$, which results in it having twice as many changes as a_1. Roughly speaking, the binary counting gadget, shown in Fig. 5, consists of the above mentioned vertices a_i together with additional vertices b_i such that b_i mimics a_i. Between the a_i and b_i lies the central vertex v with edges from the vertices a_i and edges to the b_i. Carefully chosen capacities and synchronization between the a_i and b_i results in v changing partition every step.

To iterate this process using v as vertex for the binary counting gadget of the next level, we need to ensure functionality with the additionally attached edges of the mimicking gadget.

The *binary counting gadget* H_k shown in Fig. 5 is formally defined as follows. It contains the above mentioned vertices a_i, b_i for $i \in [k]$ and the vertex v. Additionally, it contains the source s and target t. On this vertex set, we have five types of edges. All of them have transit time 1 unless explicitly specified otherwise. The first two types are the edges (a_i, b_i) and (b_i, t) for $i \in [k]$, which form a mimicking gadget. We set $\tau_{(a_i,b_i)} = p_i + 1$ which makes b_i mimic the changes of a_i with delay $p_i + 1$. Moreover, we set $u_{(a_i,b_i)} = \alpha_i := 2^{i-1} + 2\varepsilon$ and $u_{(b_i,t)} = \beta_i := 2^{i-1} + \varepsilon$. We will see that these α_i and β_i satisfy the requirements of the mimicking gadget in Lemma 1. The third and fourth types of edges are (a_i, v) and (v, b_i) for $i \in [k]$ with capacities $u_{(a_i,v)} = u_{(v,b_i)} = 2^{i-1}$. These edges have the purpose to force the partition changes of v, similar to the simple example in Fig. 2. Finally, we have the edge (s, v) with capacity $u_{(s,v)} = 1 - \varepsilon$. It has the purpose to fix the initial partition of v and introduce some asymmetry to ensure functionality even if additional edges are attached to v.

Our plan is to prove that the binary counting gadget H_k works as desired by induction over k. We start by defining the desired properties that will serve as induction hypothesis.

Definition 1. *Let G be a graph. We say that H_k is a* valid binary counting gadget *in G if H_k is a subgraph of G and every minimum cut S has the following properties.*

- *For $i \in [k-1]$, the vertex a_i has period p_i. It changes its partition 2^{k-i} times starting with a change from S to $\bar{S}$ at time 0 and ending with a change at time $2^k - 2^i$.*
- *For $i = k$, a_k changes from S to $\bar{S}$ at time 0 and additionally back to S at time 2^k.*

Note that in a valid binary counting gadget the a_i and v form a binary counter from 0 to $2^k - 1$ when regarding $\bar{S}$ as zero and S as 1, with v being the least significant bit (shifted back two time steps); see Fig. 4.

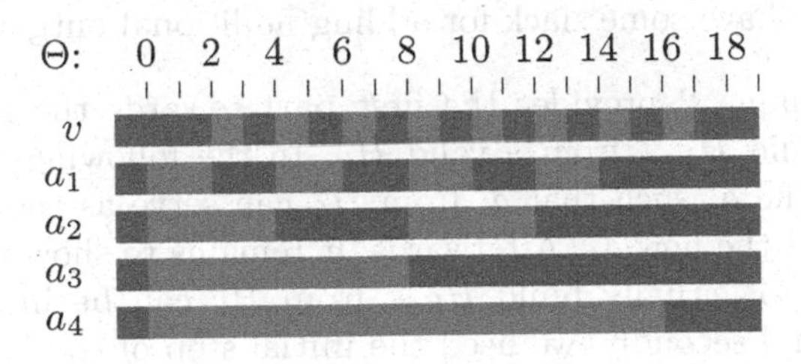

Fig. 4. Visualization of the partition change patterns of a valid binary counting gadget H_4. In blue sections the vertex is in S and in red sections it is in $\bar{S}$. (Color figure online)

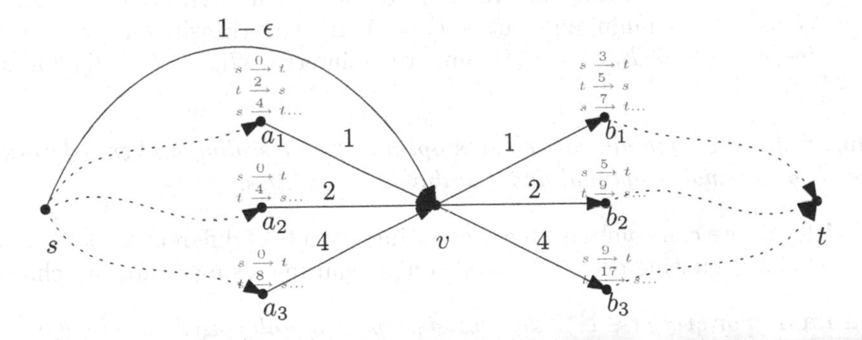

Fig. 5. Intuitive visualization of the binary counting gadget allowing to double the number of partition changes of a single vertex, ensuring 2^k changes of vertex v assuming that vertices $a_i, i < k$ are changing partition $\text{ch}_{a_i} = 2^{k-i}$ times each, with $\text{ch}_{a_k} = 2$ and correct timing. The shown black numbers are capacities. For improved readability mimicking gadgets are omitted and the partition changes of the a_i, b_i are denoted in blue by their terminal, i.e. s for S and t for $\bar{S}$.

Lemma 2. *Let G be a graph containing the valid binary counting gadget H_k such that the central vertex v has no additional incoming edges, the sum of the capacities of additional outgoing edges of v is less than $1 - \varepsilon$, and no additional edges are incident to the b_i. Then, for every minimum cut S, the central vertex v has period $p_0 = 1$ and changes 2^k times, starting with a change from S to $\bar{S}$ at time 2 and ending with a change at time $2^k + 1$.*

Proof (Proof sketch). The edge (s, v) sets the default partition of v to S. Beyond that, v is influenced by the edges from a_i and to b_i. Mimicking gadgets ensure that the b_i change partitions with the same frequency as the a_i with appropriate delay and in opposite directions. The changes of the b_i-side from S to $\bar{S}$ induce changes of v from S to $\bar{S}$, while changes on the a_i-side from $\bar{S}$ to S induce changes of v back to S. Moreover, the exponentially increasing capacities make sure that later changes essentially supersede earlier ones. We additionally need to ensure that we have some slack for adding additional outgoing edges to v. □

Note that Lemma 2 provides the first part towards the induction step of constructing a valid H_{k+1} from a valid H_k. In the following, we show how to scale periods of the a_i such that a_i from H_k can serve as the a_{i+1} from H_{k+1} and v can serve as the new a_1. Afterwards, it remains to show two things. First, additional edges to actually build H_{k+1} from H_k can be introduced without losing validity. And secondly, we need the initial step of the induction, i.e., the existence of a valid H_1 even in the presence of only one capacity change.

We say that a minimum dynamic cut S *remains optimal under scaling and translation of time* if S is a minimum cut on graph $G = (V, E)$ with transit times τ_e, capacities $u_e(\Theta)$ and time interval $[0, T]$ if and only if $\hat{S}$ with $\hat{S}_v(r \cdot \Theta + T_0) := S_v(\Theta)\ \forall \Theta \in [0, T]$ is a minimum cut on $\hat{G} = (V, E)$ with transit times $\hat{\tau}_e := r \cdot \tau_e$, capacities $\hat{u}_e(r \cdot \Theta + T_0) := u_e(\Theta)$ and time interval $[T_0, r \cdot T + T_0]$ for any $r \in \mathbb{R}^+, T_0 \in \mathbb{R}$.

Lemma 3. *Any dynamic cut remains optimal under scaling and translation of time. It also remains optimal under scaling of capacities.*

With this, we can combine binary counting gadgets of different sizes to create a large binary counting gadget H_ℓ while only requiring a single capacity change.

Lemma 4. *For every $\ell \in \mathbb{N}^+$, there exists a polynomially sized, acyclic dynamic flow network with only one capacity change that contains a valid binary counting gadget H_ℓ.*

Proof (Proof sketch). To create the necessary change patterns for the a_i of H_ℓ, we chain binary counting gadgets of increasing size, beginning with H_2 up to H_ℓ, together, as shown in Fig. 6. The coarse idea is to ensure the behavior of all $a_{k,i}$ by having them mimic the central vertex v_{k-i} of the correct smaller binary counting gadget. We use Lemma 3 to scale and shift the various counting gadgets as necessary. We note that ensuring that all requirements of the mimicking gadget (Lemma 1) and the counting gadget (Lemma 2) remain true in the whole

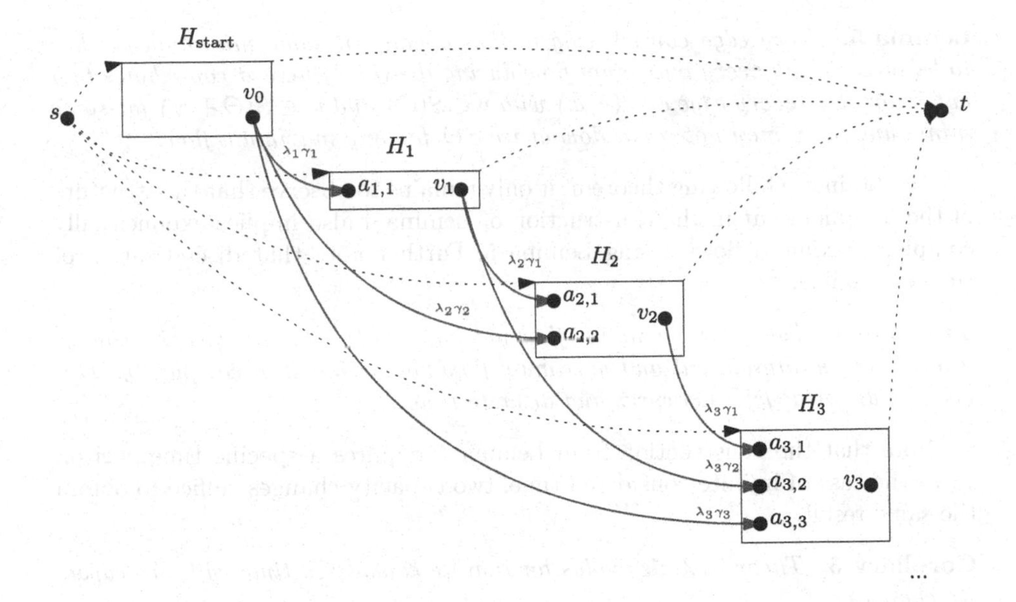

Fig. 6. Construction linking binary counting gadgets to ensure $\mathrm{ch}_{v_\ell} = 2^\ell$ partition changes at v_ℓ in a minimum cut; purple edges represent mimicking gadgets, numbers are capacities. (Color figure online)

construction is rather technical but possible. The more interesting part of the proof is starting the construction, i.e., the base case of the induction.

For the base case, we need to ensure that there is one vertex v_0 that changes twice, first from S to $\bar{S}$ and later to $\bar{S}$ to S, based on just one capacity change. We achieve this as illustrated in Fig. 7. The edge (v_0, t) changes capacity from 1 to 3 at time $\Theta = 2$ and has transit time $T - 3$. Note that v_0 starts in S as (s, v_0) has capacity 2 and (v_0, t) starts with capacity 1. The partition change $S \xrightarrow{2} \bar{S}$ directly follows from the capacity change of (v_0, t) at time $\Theta = 2$ increasing the potential contribution of $v_0 \in S(2)$. The other partition change $\bar{S} \xrightarrow{3} S$ is a result of the approaching time horizon T, which reduces the potential contribution of $v_0 \in S(3)$ to zero. □

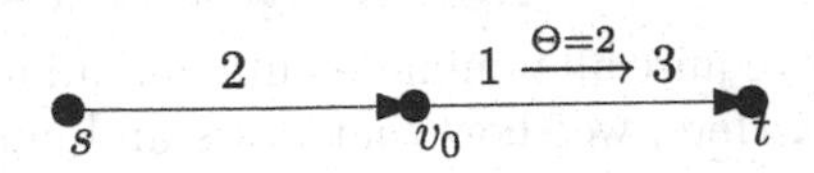

Fig. 7. Construction of H_{start}, providing the partition changes of v_0 needed for G_ℓ with one capacity change, numbers are capacities, $\tau_{(v_0,t)} = T - 3$.

To be able to use the complexity of minimum cuts to show complexity of maximum flows, we need the following lemma.

Lemma 5. *Every edge contributing to the capacity of some minimum cut has to be saturated by every maximum flow during the time where it contributes to a cut. Moreover, every edge* $e = (v, w)$ *with* $v \in \bar{S}(\Theta)$ *and* $w \in S(\Theta + \tau_e)$ *for some minimum cut* S *may not route flow at time* Θ *for any maximum flow.*

To obtain the following theorem, it only remains to observe that the structure of the minimum cut in the construction of Lemma 4 also implies exponentially complex maximum flows, using Lemma 5. Further note that discretization of time is possible.

Theorem 4. *There exist dynamic flow networks with only one capacity change where every minimum cut and maximum flow has exponential complexity. This even holds for acyclic networks and discrete time.*

Note that the construction from Lemma 4 requires a specific time horizon T. In the case of infinite considered time, two capacity changes suffice to obtain the same result.

Corollary 3. *Theorem 4 also holds for infinite considered time with two capacity changes.*

Note that if Conjecture 1 holds, two capacity changes are necessary in this setting.

As mentioned in the introduction, the above complexity results transfer to the setting where we have time-dependent transit times instead of time-dependent capacities. The result of Corollary 3 can even be strengthened to only require a single transit time change.

Corollary 4. *Theorem 4 also holds in the setting of static capacities and time-dependent transit times with a single change, with finite time horizon and with infinite considered time.*

Note that the construction of Theorem 4 causes every minimum cut and every maximum flow to have exponential complexity. In the following we show that exponentially many changes in cut or flow can occur independently. Specifically, we provide constructions that require exponentially complex flows but allow for cuts of low complexity and vice versa.

4.2 Complex Flows and Simple Cuts (and Vice Versa)

All above constructions require all minimum cuts *and* all maximum flows to have exponential complexity. Here, we show that flows and cuts can be independent in the sense that their required complexity can be exponentially far apart (in both directions).

Theorem 5. *There exist acyclic dynamic flow networks with only one capacity change where every maximum flow has exponential complexity, while there exists a minimum cut of constant complexity. The same is true for static capacities and time-dependent transit times.*

Proof. Figure 8 shows a graph with these properties. This is achieved by only allowing flow to enter v_0 during $[0, 1]$, but it has to leave v_k during $[0, 2^k]$ due to the reduced capacity of (v_k, t) for a time horizon $T \geq 2^k$. Apart from v_k all v_i are connected to the next v_{i+1} with a pair of edges, with transit times 2^{k-i-1} and zero, all these edges have capacity 1. So all 2^k paths of different transit time through the v_i have to be used to route flow for a maximum flow. Every second of those paths has an even transit time, so flow has to traverse the edge with transit time zero between v_{k-1} and v_k every second integer time interval, which results in exponentially many changes in flow over that edge. However assigning all v_i to $\bar{S}$ for all time is a minimum cut without partition changes. This generalizes to time-dependent transit times, as we can block the edge (s, v_0) at time $\Theta = 1$ by increasing its transit time to T at that time. □

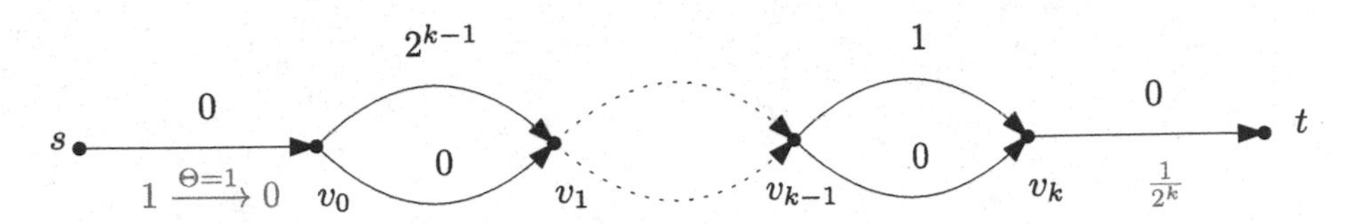

Fig. 8. Example of a graph where any maximum flow contains exponentially many changes, but there is a minimum cut with no changes, black numbers are transit times, blue numbers are capacities, unspecified capacities are 1. (Color figure online)

Theorem 6. *There exist acyclic dynamic flow networks with only one capacity change where every minimum cut has exponential complexity, while there exists a maximum flow of linear complexity. The same is true for static capacities and time-dependent transit times.*

References

1. Bläsius, T., Feilhauer, A., Westenfelder, J.: Dynamic flows with time-dependent capacities (2023). https://doi.org/10.48550/ARXIV.2302.07657, https://arxiv.org/abs/2302.07657
2. Disser, Y., Skutella, M.: The simplex algorithm is NP-mighty. ACM Trans. Algorithms (TALG) **15**(1), 1–19 (2018)
3. Ford Jr., L.R., Fulkerson, D.R.: Constructing maximal dynamic flows from static flows. Oper. Res. **6**(3), 419–433 (1958)
4. Halpern, J.: A generalized dynamic flows problem. Networks **9**(2), 133–167 (1979)
5. Koch, R., Nasrabadi, E., Skutella, M.: Continuous and discrete flows over time. Math. Methods Oper. Res. **73**(3), 301 (2011)
6. Köhler, E., Skutella, M.: Flows over time with load-dependent transit times. SIAM J. Optim. **15**(4), 1185–1202 (2005)
7. Langkau, K.: Flows over time with flow-dependent transit times (2003)
8. Philpott, A.B.: Continuous-time flows in networks. Math. Oper. Res. **15**(4), 640–661 (1990)

9. Sha, D., Cai, X., Wong, C.: The maximum flow in a time-varying network. In: Nguyen, V.H., Strodiot, J.J., Tossings, P. (eds.) Optimization. Lecture Notes in Economics and Mathematical Systems, vol. 481, pp. 437–456. Springer, Heidelberg (2000). https://doi.org/10.1007/978-3-642-57014-8_29
10. Skutella, M.: An introduction to network flows over time. In: Cook, W., Lovász, L., Vygen, J. (eds.) Research Trends in Combinatorial Optimization, pp. 451–482. Springer, Heidelberg (2009). https://doi.org/10.1007/978-3-540-76796-1_21
11. Tjandra, S.A.: Dynamic network optimization with application to the evacuation problem (2003)

On One-Sided Testing Affine Subspaces

Nader H. Bshouty(✉)

Department of Computer Science, Technion, Haifa, Israel
bshouty@cs.technion.ac.il

Abstract. We study the query complexity of one-sided ϵ-testing the class of Boolean functions $f : \mathcal{F}^n \to \{0,1\}$ that describe affine subspaces and Boolean functions that describe axis-parallel affine subspaces, where $\mathcal{F}$ is any finite field. We give a polynomial-time ϵ-testers that ask $\tilde{O}(1/\epsilon)$ queries. This improves the query complexity $\tilde{O}(|\mathcal{F}|/\epsilon)$ in [11].

We then show that any one-sided ϵ-tester with proximity parameter $\epsilon < 1/|\mathcal{F}|^d$ for the class of Boolean functions that describe $(n-d)$-dimensional affine subspaces and Boolean functions that describe axis-parallel $(n-d)$-dimensional affine subspaces must make at least $\Omega(1/\epsilon + |\mathcal{F}|^{d-1}\log n)$ and $\Omega(1/\epsilon + |\mathcal{F}|^{d-1}n)$ queries, respectively. This improves the lower bound $\Omega(\log n/\log\log n)$ that is proved in [11] for $\mathcal{F} = \mathrm{GF}(2)$. We also give testers for those classes with query complexity that almost match the lower bounds. (See the definitions of the classes in the introduction and many other results in Figs. 1 and 2).

1 Introduction

Property testing of Boolean function was first considered in the seminal works of Blum, Luby, and Rubinfeld [1] and Rubinfeld and Sudan [15] and has recently become a very active research area. See, for example, the works referenced in the surveys and books [8,10,13,14].

Let $\mathcal{F}$ be a finite field. A Boolean function $f : \mathcal{F}^n \to \{0,1\}$ describes a $(n-d)$-dimensional affine subspace if $f^{-1}(1) \subseteq \mathcal{F}^n$ is a $(n-d)$-dimensional affine subspace. We denote the class of all such functions by d-**AS**. The class **AS** $= \cup_k k$-**AS** and $(\le d)$-**AS** $= \cup_{k\le d} k$-**AS**. A Boolean function $f : \mathcal{F}^n \to \{0,1\}$ describes an axis-parallel $(n-d)$-dimensional affine subspace if $f^{-1}(1) \subseteq \mathcal{F}^n$ is an axis parallel $(n-d)$-dimensional affine subspace, i.e., there are d entries $1 \le i_1 < i_2 < \cdots < i_d \le n$ and constants $\lambda_i \in \mathcal{F}$, $i \in [d]$, such that $f^{-1}(1) = \{a \in \mathcal{F}^n | a_{i_1} = \lambda_1, \ldots, a_{i_d} = \lambda_d\}$. We denote the class of all such functions by d-**APAS**. In the same way, we define the class **APAS** and $(\le d)$-**APAS**. If in the above definitions, instead of "affine subspace" we have "linear subspace", then we get the classes d-**LS**, **LS**, $(\le d)$-**LS**, d-**APLS**, **APLS** and $(\le d)$-**APLS**. Those classes are studied in [9,11,12].

A related classes of Boolean functions $f : \{0,1\}^n \to \{0,1\}$ that are studied in the literature, [2,4–7,9,11,12], are d-**Monomial** (conjunction of d negated

Center for Theoretical Sciences, Guangdong Technion, (GTIIT), China.

M. Mavronicolas (Ed.): CIAC 2023, LNCS 13898, pp. 157–171, 2023.
https://doi.org/10.1007/978-3-031-30448-4_12

Boolean variables)[1], **Monomial** (conjunction of negated Boolean variables), ($\le d$)-**Monomial** (conjunction of at most d negated Boolean variables), d-**Term** (conjunction of d literals[2]), **Term** (conjunction of literals), ($\le d$)-**Term** (conjunction of at most d literals). Those are equivalent to the two family of classes **APLS** (for **Monomial**) and **APAS** (for **Term**) over the binary field GF(2).

In property testing a class C of Boolean functions, a *tester* for C is a randomized algorithm T that has access to a Boolean function f via a black-box oracle that returns $f(x)$ when a point x is queried. Given a proximity parameter, ϵ, if $f \in C$, the tester T accepts with probability at least 2/3, and if f is ϵ-far from C (i.e., for every $g \in C$, $\Pr_x[f(x) \neq g(x)] > \epsilon$) then it rejects with probability at least 2/3. We say that T is a *one-sided* tester if it always accepts when $f \in C$; otherwise, it is called a *two-sided* tester.

Testing the classes d-**AS**, **AS** and ($\le d$)-**AS** correspond to testing functions that describe $(n - d')$-dimensional affine subspaces where, in the class d-**AS**, d' is known (to the tester) and equal to d, in the class **AS**, d' is unknown, and in the class ($\le d$)-**AS**, d is an upper bound on d'. The same applies to the other classes.

Testers for the above classes were studied in [2,4,6,7,9,11,12]. In [12], Parnas et al. gave two-sided testers for the above classes that make $O(1/\epsilon)$ queries. See also [2,9]. The one-sided testers were studied by Goldreich and Ron in [11]. They gave a polynomial-time one-sided testers for the classes **AS**, **APAS**, **LS**, ($\le d$)-**LS**, **APLS** and ($\le d$)-**APLS** that make $\tilde{O}(|\mathcal{F}|/\epsilon)$ queries[3]. In this paper, we give a polynomial-time[4] testers for these classes that make $\tilde{O}(1/\epsilon)$ queries.

For the classes d-**AS** and d-**APAS**, Goldreich and Ron gave the lower bound $\Omega(1/\epsilon + \log n / \log\log n)$ for the query complexity of any tester when $\mathcal{F} = \mathrm{GF}(2)$ and $\epsilon \le 2^{-d}$. In this paper, we give the lower bounds $\Omega(1/\epsilon + |\mathcal{F}|^{d-1} n)$ and $\Omega(1/\epsilon + |\mathcal{F}|^{d-1} \log n)$, respectively, for the proximity parameter $\epsilon < 1/|\mathcal{F}|^d$. We also give testers for those classes with query complexity that almost match the lower bounds.

See other results in Figs. 1 and 2.

2 Overview of the Testers and the Lower Bounds

2.1 The Algorithm for Functions that Describe Linear Subspace

In this section, we give the one-sided testers for **LS** and ($\le d$)-**LS**.

First, we briefly outline the stages and then give more details.

[1] In the literature, this class is defined as conjunction of d (non-negated) variables. Testability of f for this class is equivalent to testability of $f(x+1^n)$ of d-**Monomial** as defined in this paper. The same applies to the classes ($\le d$)-**Monomial** and **Monomial**.

[2] A literal is a variable or its negation.

[3] They also gave a tester for ($\le d$)-**AS**$\cup\{z(x)\}$ and ($\le d$)-**APAS**$\cup\{z(x)\}$ with the same query complexity where $z(x)$ is the zero function.

[4] Goldreich and Ron algorithm and our algorithm run in time linear in the number of queries.

Class	Lower Bound	Upper Bound	ϵ
AS	$\Omega(1/\epsilon)$	$\tilde{O}(1/\epsilon)$	$< \frac{1}{2}$
	0	0	$> \frac{1}{2}$
$(\leq d)$-**AS**	$\Omega(1/\epsilon + \lvert\mathcal{F}\rvert^{d-1} n)$	$\tilde{O}(1/\epsilon) + \tilde{O}(\lvert\mathcal{F}\rvert^{d}) n$ †	$< \frac{1}{\lvert\mathcal{F}\rvert^d}$
	$\Omega(1/\epsilon)$	$\tilde{O}(1/\epsilon) + O(1/(\epsilon - 1/\lvert\mathcal{F}\rvert^d))$	$> \frac{1}{\lvert\mathcal{F}\rvert^d}$
	0	0	$> \frac{1}{2}$
d-**AS**	$\Omega(1/\epsilon + \lvert\mathcal{F}\rvert^{d-1} n)$	$\tilde{O}(1/\epsilon) + \tilde{O}(\lvert\mathcal{F}\rvert^{d}) n$ †	$< \frac{1}{\lvert\mathcal{F}\rvert^d}$
	$\Omega(1/\epsilon + n)$	$\tilde{O}(1/\epsilon) + \tilde{O}(\lvert\mathcal{F}\rvert^{d}) n$ †	$< 1 - \frac{1}{\lvert\mathcal{F}\rvert^d}$
	$\Omega(1)$	$O(1/(\epsilon - 1 + \lvert\mathcal{F}\rvert^{-d}))$	$> 1 - \frac{1}{\lvert\mathcal{F}\rvert^d}$
	0	0	$> 1 - \frac{1}{\lvert\mathcal{F}\rvert^d} +$ †
APAS	$\Omega(1/\epsilon)$	$\tilde{O}(1/\epsilon)$	$> \frac{1}{2}$
	0	0	$> \frac{1}{2}$
$(\leq d)$-**APAS**	$\Omega(1/\epsilon + \lvert\mathcal{F}\rvert^{d-1} \log n)$	$\tilde{O}(1/\epsilon) + \lvert\mathcal{F}\rvert^{d+o(d)} \log n$	$< \frac{1}{\lvert\mathcal{F}\rvert^d}$
	$\Omega(1/\epsilon)$	$\tilde{O}(1/\epsilon) + O(1/(\epsilon - 1/\lvert\mathcal{F}\rvert^d))$	$> \frac{1}{\lvert\mathcal{F}\rvert^d}$
	0	0	$> \frac{1}{2}$
d-**APAS**	$\Omega(1/\epsilon + \lvert\mathcal{F}\rvert^{d-1} \log n)$	$O(1/\epsilon) + \lvert\mathcal{F}\rvert^{d+o(d)} \log n$	$< \frac{1}{\lvert\mathcal{F}\rvert^d}$
	$\Omega(L)$‡	$O(1/\epsilon) + \lvert\mathcal{F}\rvert^{d+o(d)} \log n$	$< 1 - \frac{1}{\lvert\mathcal{F}\rvert^d}$
	$\Omega(1)$	$O(1/(\epsilon - 1 + \lvert\mathcal{F}\rvert^{-d}))$	$> 1 - \frac{1}{\lvert\mathcal{F}\rvert^d}$
	0	0	$> 1 - \frac{1}{\lvert\mathcal{F}\rvert^d} +$
LS	$\Omega(1/\epsilon)$	$\tilde{O}(1/\epsilon)$	$< \frac{1}{2}$
	0	0	$> \frac{1}{2}$
$(\leq d)$-**LS**	$\Omega(1/\epsilon)$	$\tilde{O}(1/\epsilon)$	$< \frac{1}{2}$
	0	0	$> \frac{1}{2}$
d-**LS**	$\Omega(1/\epsilon + n)$	$\tilde{O}(1/\epsilon) + O(\lvert\mathcal{F}\rvert^{d} n)$	$< 1 - \frac{1}{\lvert\mathcal{F}\rvert^d}$
	$\Omega(1)$	$O(1/(\epsilon - 1 + \lvert\mathcal{F}\rvert^{-d}))$	$> 1 - \frac{1}{\lvert\mathcal{F}\rvert^d}$
	0	0	$> 1 - \frac{1}{\lvert\mathcal{F}\rvert^d} +$
APLS	$\Omega(1/\epsilon)$	$\tilde{O}(1/\epsilon)$	$> \frac{1}{2}$
	0	0	$> \frac{1}{2}$
$(\leq d)$-**APLS**	$\Omega(1/\epsilon)$	$\tilde{O}(1/\epsilon)$	$> \frac{1}{2}$
	0	0	$> \frac{1}{2}$
d-**APLS**	$\Omega(L)$‡	$O(L)$‡	$< \frac{1}{\lvert\mathcal{F}\rvert} - \frac{1}{\lvert\mathcal{F}\rvert^d}$
	$\Omega(1)$	$O(1/(\epsilon - \lvert\mathcal{F}\rvert^{-1} + \lvert\mathcal{F}\rvert^{-d}) + \log n)$	$> \frac{1}{\lvert\mathcal{F}\rvert} - \frac{1}{\lvert\mathcal{F}\rvert^d}$
	$\Omega(1)$	$O(1/(\epsilon - 1 + \lvert\mathcal{F}\rvert^{-d}))$	$> 1 - \frac{1}{\lvert\mathcal{F}\rvert^d}$
	0	0	$> 1 - \frac{1}{\lvert\mathcal{F}\rvert^d} +$

Fig. 1. A table of the lower bounds and upper bounds achieved in this paper. Any upper bound (resp. lower bound) for the proximity parameter ϵ is also an upper bound for $\epsilon' \geq \epsilon$ (resp. $\epsilon' \leq \epsilon$). † Those testers are exponential time testers. $\lvert\mathcal{F}\rvert^{-d}+$ means $\lvert\mathcal{F}\rvert^{-d} + o(\lvert\mathcal{F}\rvert^{-d})$. See the full paper.‡ Here $L = \frac{1}{\epsilon} + \min\left(\frac{\log(1/\epsilon)}{\log\lvert\mathcal{F}\rvert}, d\right) \cdot \log \frac{n}{d}$.

Class	Lower Bound	Upper Bound	ϵ
Term	$\Omega(1/\epsilon)$	$\tilde{O}(1/\epsilon)$	$< \frac{1}{2}$
	0	0	$> \frac{1}{2}$
($\le d$)-**Term**	$\Omega(1/\epsilon + 2^d \log n)$	$\tilde{O}(1/\epsilon) + 2^{d+o(d)} \log n$	$< \frac{1}{2^d}$
	$\Omega(1/\epsilon)$	$\tilde{O}(1/\epsilon) + O(1/(\epsilon - 1/2^d))$	$> \frac{1}{2^d}$
	0	0	$> \frac{1}{2}$
d-**Term**	$\Omega(1/\epsilon + 2^d \log n)$	$O(1/\epsilon) + 2^{d+o(d)} \log n$	$< \frac{1}{2^d}$
		$O(1/\epsilon) + 2^{d+o(d)} \log n$	$< 1 - \frac{1}{2^d}$
	$\Omega(1)$	$O(1/(\epsilon - 1 + 2^{-d}))$	$> 1 - \frac{1}{2^d}$
	0	0	$> 1 - \frac{1}{2^d} +^{\dagger}$
Monomial	$\Omega(1/\epsilon)$	$\tilde{O}(1/\epsilon)$	$< \frac{1}{2}$
	0	0	$> \frac{1}{2}$
($\le d$)-**Monom.**	$\Omega(1/\epsilon)$	$\tilde{O}(1/\epsilon)$	$< \frac{1}{2}$
	$\Omega(1/\epsilon)$	$O(1/\epsilon) + \tilde{O}(2^{2d})$	$< \frac{1}{2}$
	0	0	$> \frac{1}{2}$
d-**Monomial**	$\Omega(L)^{\dagger}$	$O(L)$	$< \frac{1}{2} - \frac{1}{2^d}$
	$\Omega(1)$	$O(1/(\epsilon - 1/2 + 2^{-d}) + \log n)$	$> \frac{1}{2} - \frac{1}{2^d}$
	$\Omega(1)$	$O(1/(\epsilon - 1 + 2^{-d}))$	$> 1 - \frac{1}{2^d}$
	0	0	$> 1 - \frac{1}{2^d} +$

Fig. 2. A table of the lower bounds and upper bounds achieved in this paper for **Term** and **Monomial.** † $L = \frac{1}{\epsilon} + \min(\log(1/\epsilon), d) \cdot \log \frac{n}{d}$. ‡ $2^{-d}+$ means $2^{-d} + o(2^{-d})$.

High-Level Description of the Tester: We first give a tester for the class d-**WSLS**[5] (Well-Structured Linear Space[6]), a subclass of **LS** that contains functions f that satisfy $f^{-1}(1) = \{(a, b)|a \in \mathcal{F}^{n-d}, b = \phi(a)\}$ for some linear function $\phi : \mathcal{F}^{n-d} \to \mathcal{F}^d$. In other words, the class of functions f where the generator matrix[7] of $f^{-1}(1)$ is of rank $n - d$ and has full rank in the first $n - d$ columns. The tester for d-**WSLS** is built of three sub-testers that test the following three properties $P_1 - P_3$ of f.

- $P_1(d)$: For every $a \in \mathcal{F}^{n-d}$, there is at most one $b \in \mathcal{F}^d$ such that $f(a, b) = 1$.
- $P_2(d)$: For every $a \in \mathcal{F}^{n-d}$, there is at least one $b \in \mathcal{F}^d$ such that $f(a, b) = 1$.
- $P_3(d)$: There is a linear function $\phi : \mathcal{F}^{n-d} \to \mathcal{F}^d$ such that if $f(a, b) = 1$ then $b = \phi(a)$.

Obviously, $f \in d$-**WSLS** if and only if it satisfies the above three properties. We also show that if a function f is ϵ-far from d-**WSLS**, then it is $\Omega(\epsilon)$-far from one of the above three properties.

In the second stage, we test **LS**. Note here that in the above tester, we assume that the tester knows d. Here, the tester does not know d, so it first assumes that $d = 0$. This stage is an iterative one that is rooted in the observation that if

[5] d is known to the tester.

[6] In coding theory for $f \in d$-**WSLS**, $f^{-1}(1)$ is called systematic code.

[7] The generator matrix of a linear subspace $L \subseteq \mathcal{F}^n$, is a matrix with a minimum number of rows where the span of its rows is L.

$f \in d^*$-**LS**, then for every $d \le d^*$, we can find a non-singular matrix M such that if $d < d^*$, then $f_d(x) = f(xM)$ satisfies properties $P_1(d)$ and $P_2(d)$ but not $P_3(d)$ and if $d = d^*$ then $f_d \in d$-**WSLS**. The tester runs the tester of d-**WSLS** on f_d. In this case, either f_d satisfies $P_3(d)$, in which case it is in d-**WSLS**, and then the tester accepts[8], or, it is far from property $P_3(d)$, in which case the tester receives $a \in \mathcal{F}^{n-d}$ such that no $b \in \mathcal{F}^d$ satisfies $f(a, b) = 1$. Then we show how to use M and a to construct a matrix M' such that $f_{d+1} = f(xM')$ satisfies properties $P_1(d+1)$ and $P_2(d+1)$ but not $P_3(d+1)$ if $d+1 < d^*$ and $f_{d+1} \in d^*$-**WSLS** if $d+1 = d^*$. The tester then moves to the next iteration. The tester accepts if the tester of d-**WSLS** accepts or $d = D := (2 + \log(1/\epsilon)/\log|\mathcal{F}|)$. In the latter case, f_d (and therefore f) is ϵ-close to the zero function[9], and therefore it is ϵ-close to **LS**.

If f is ϵ-far from **LS**, then it is ϵ-far from every d-**WSLS**. So, with high probability, in one of the iterations, f_d is either $\Omega(\epsilon)$-far from[10] $P_1(d)$ or $P_3(d)$ and the tester rejects.

A Detailed Description of the Tester: We now give a detailed outline of the stages.

For testing **LS** we have two main stages. The first is composed of the following three substages that test whether the function describes a *well-structured* $(n-d)$-*dimensional subspace*. A function describes a well-structured $(n-d)$-dimensional subspace if $f^{-1}(1) = \{(a, \phi(a)) | a \in \mathcal{F}^{n-d}\}$, where $\phi : \mathcal{F}^{n-d} \to \mathcal{F}^d$ is a linear function. Then, in the second stage, we show how to test whether a function describes a linear subspace using the first three substages.

In the first substage, we give a tester that tests whether f is a function that *describes a well-structured* $(n-d)$-*dimensional injective relation* (property $P_1(d)$). That is, it satisfies: For every $a \in \mathcal{F}^{n-d}$, there is *at most one* $b \in \mathcal{F}^d$ such that $f(a, b) = 1$. The class of such functions is denoted by d-**R**. We show that if f is ϵ-far from d-**R**, then there are $\alpha, \beta < 1$ such that $\alpha\beta = O(\epsilon/\log(1/\epsilon))$ and $\Pr_{a \in \mathcal{F}^{n-d}}[\Pr_{b \in \mathcal{F}^d}[f(a, b) = 1] \ge \beta] \ge \alpha$. Then with a proper double search, the tester, with high probability, can find $a, b^{(1)} \ne b^{(2)}$ such that $f(a, b^{(1)}) \ne f(a, b^{(2)})$ and rejects. If $f \in d$-**R**, then no such $a, b^{(1)} \ne b^{(2)}$ can be found. Therefore, this is a one-sided tester. The query complexity of this stage is $\tilde{O}(\log^2(1/\epsilon)/\epsilon) = \tilde{O}(1/\epsilon)$.

In the second substage, we give a tester that tests whether f *describes a well-structured* $(n-d)$-*dimensional bijection* (properties $P_1(d)$ and $P_2(d)$). That is: For every $a \in \mathcal{F}^{n-d}$, there is *exactly one* $b \in \mathcal{F}^d$ such that $f(a, b) = 1$. The class of such functions is denoted by d-**F**. The tester for d-**F** first runs the above tester for d-**R** with proximity parameter $\epsilon/2$ and rejects if it rejects. So, we may assume that f is $\epsilon/2$-close to d-**R**. Define the function $R_f : \mathcal{F}^{n-d} \to \mathcal{F}^d \cup \{\perp\}$

---

[8] It is easy to show that if $f_d(x) = f(xM)$ is ϵ-close to d^*-**LS** if and only if $f(x)$ is ϵ-close to d^*-**LS**.

[9] If $d = D$, then by property $P_1(d)$, it follows that f is ϵ-close to the zero function.

[10] If in every iteration f_d is not ϵ-far from $P_1(d)$ and $P_3(d)$, then it gets to iteration D, and therefore f is ϵ-close to **LS**. A contradiction.

where $R_f(a)$ is equal to the first $b \in \mathcal{F}^d$ (in some total order) that satisfies $f(a,b) = 1$ and $\perp$ if no such b exists. We show that if f is $\epsilon/2$-close to d-**R** and ϵ-far from d-**F** then[11] $\Pr[R_f(a) = \perp] \geq \epsilon|\mathcal{F}|^d/2$. See details in Sect. 3. Since computing $R_f(a)$ takes $|\mathcal{F}|^d$ queries, the query complexity of testing whether $\Pr[R_f(a) = \perp] \geq \epsilon|\mathcal{F}|^d/2$ is $O(1/\epsilon)$. This is also a one-sided tester because when $f \in d$-**F**, $\Pr[R_f(a) = \perp] = 0$. The query complexity of this stage is $\tilde{O}(1/\epsilon)$.

In the third substage, we give a tester that tests whether a function f describes a well-structured $(n-d)$-dimensional linear subspace (properties $P_1(d)$, $P_2(d)$ and $P_3(d)$). The class of such functions is denoted by d-**WSLS**. First, the tester runs the tester for d-**F** with proximity parameter $\epsilon/2$ and rejects if it rejects. Now define a function $F_f : \mathcal{F}^{n-d} \to \mathcal{F}^d$ where $F_f(a) = R_f(a)$ if $R_F(a) \neq \perp$ and $f(a,b) = 0^d$ otherwise. We show that if f is ϵ-far from d-**WSLS** and $\epsilon/2$-close to d-**F**, then F_f is $(|\mathcal{F}|^d\epsilon/2)$-far from linear functions. See details in Sect. 3. The tester then uses the testers in [1,15] to test if F_f is $(|\mathcal{F}|^d\epsilon/2)$-far from linear functions. Since computing $F_f(a)$ takes $|\mathcal{F}|^d$ queries and the testers in [1,15] make $O(2/(\epsilon|\mathcal{F}|^d)$ queries, the query complexity of this test is $O(1/\epsilon)$. Since the testers in [1,15] are one-sided, this tester is also one-sided. The query complexity of this tester is $\tilde{O}(1/\epsilon)$.

Now, in the second stage, we give a tester that tests whether f describes a linear subspace. Recall that the class of such functions is denoted by **LS**. The tester at the $(d+1)$-th iteration uses a non-singular $n \times n$ matrix M such that $f_d(x) := f(xM)$ satisfies

1. If f is ϵ-far from **LS** then f_d is ϵ-far from **LS**.
2. If $f \in$ **LS** then $f_d \in$ **LS**.
3. If $f \in$ **LS** then $f_d^{-1}(1) = \{(a, \phi(a)) | a \in L\}$ for some linear subspace $L \subseteq \mathcal{F}^{n-d}$ and linear function $\phi : \mathcal{F}^{n-d} \to \mathcal{F}^d$ (satisfies properties $P_1(d)$ and $P_3(d)$. It satisfies property $P_2(d)$ if and only if $L = \mathcal{F}^{n-d}$).

Items 1 and 2 are true for any non-singular matrix M. At the $(d+1)$-th iteration, the tester runs the tester that tests whether $f_d \in d$-**WSLA** with proximity parameter $\epsilon/2$ and accepts if it accepts. We show that if $f_d \in$ **LS** and the tester rejects, then it is because some $a' \in \mathcal{F}^{n-d}$ has no $b \in \mathcal{F}^d$, such that $f_d(a', b) = 1$ (property $P_2(d)$ is not true - i.e., $L \neq \mathcal{F}^{n-d}$). In that case, the tester does not reject and uses the point $a' \in \mathcal{F}^{n-d}$ and M to construct a new non-singular matrix M' such that $f_{d+1} = f(xM')$ satisfies the above items 1–3. Items 1 and 2 hold for f_{d+1} because M' is non-singular. For item 3, we will have, if $f \in$ **LS**, then $f_{d+1}^{-1}(1) = \{(a, \phi'(a)) | a \in L'\}$ for some linear subspace $L' \subseteq \mathcal{F}^{n-d-1}$ and a linear function $\phi' : \mathcal{F}^{n-d-1} \to \mathcal{F}^{d+1}$. The tester then continues to the $(d+2)$-th iteration if $d < (2 + \log(1/\epsilon)/\log|\mathcal{F}|)$; otherwise, it accepts.

If $f \in$ **LS**, then at each iteration, the tester either accepts or moves to the next iteration. Also, when $d = (2 + \log(1/\epsilon)/\log|\mathcal{F}|)$, the tester accepts. So, this tester is one-sided.

[11] if $|\mathcal{F}|^d\epsilon > 2$, the tester accepts. This is because any function in d-**R** is $|\mathcal{F}|^{n-d}/|\mathcal{F}|^n \leq 1/|\mathcal{F}|^d \leq \epsilon/2$ close to any function in d-**F**. Therefore, if f is $\epsilon/2$-close to d-**R**, and $|\mathcal{F}|^d\epsilon > 2$ then it is ϵ-close to d-**F**.

On the other hand, if f is ϵ-far from **LS**, then it is ϵ-far from the function h that satisfies $h^{-1}(1) = \{0^n\}$ (which is in **LS**). Therefore,[12], $\Pr[f \neq 0] \geq \epsilon/2$. Now since for $d = 2+\log(1/\epsilon)/\log|\mathcal{F}|$, every function g in d-**R** satisfies $\Pr[g(x) = 1] \leq |\mathcal{F}|^{-d} \leq \epsilon/4$, the tester of d-**WSLA**, with high probability, rejects when it calls the tester of d-**R**.

Therefore, this tester is one-sided, and its query complexity is $\tilde{O}(1/\epsilon)$. This completes the description of the tester of the class **LS**.

The above tester also works for testing the class $(\leq k)$-**LS**. The only change is that the tester rejects if $d > k$.

2.2 Comparison with Goldreich and Ron Algorithm

Our method and the method of Goldreich and Ron [11] are rooted in different characterization of $(n-d)$-dimensional subspaces. Our method is based on the existence of a change of basis for the subspace $f^{-1}(1)$ that makes it "well-structured", then moving from linear space of dimension $n-i$ to dimension $n-i-1$. In contrast, the method in [11] is based on the existence of a basis for the d-dimensional subspace orthogonal to $f^{-1}(1)$ and moving from subspace of dimension i to dimension $i+1$. Both are iterative, but construct and test different subspaces. Our approach improves the query complexity of testing from $\tilde{O}(|\mathcal{F}|/\epsilon)$ (in [11]) to $\tilde{O}(1/\epsilon)$.

The issue with [11] is that it is centered on the function $g_{H,V}$, where H is the tested space and V is the basis of the linear subspace that complements H. The value of $g_{H,V}(x)$ identifies the unique coset in which x resides and $\perp$ otherwise. That is, $g_{H,V}(x) = u$ if $x+uV \in H$ and $\perp$ if such value does not exist or is not unique. In the ith iteration, evaluating $g_{H,V}$ requires $|\mathcal{F}|^i$ queries where $i = \dim(V)$. Also, for the test, their algorithm needs to reach iteration i such that $|\mathcal{F}|^i \geq 1/\epsilon$. Now, if ϵ is slightly smaller than $1/|\mathcal{F}|^i$, their algorithm enters iteration $i+1$ and makes at least $|\mathcal{F}|/\epsilon$ queries. This is the source of the extra factor of $|\mathcal{F}|$ in the query complexity of [11], and it seems that the approach they use cannot get around it.

2.3 The Algorithm for Functions that Describe Affine Subspace

Our tester that tests whether a function describes an affine subspace, **AS**, is built on the reduction of Goldreich and Ron's [11] that reduces one-sided testing **AS** to one-sided testing **LS** and then the above two main stages for testing **LS**. For completeness, we present Goldreich and Ron's reduction.

They show that testing whether a function $f(x)$ describes an affine subspace (resp. axis-parallel affine subspace) can be randomly reduced to testing whether $h(x) = f(x+a)$ describes a linear subspace (resp. axis-parallel linear subspace) where $a \in f^{-1}(1)$. This follows from the fact that if $f^{-1}(1) = u + L$ for some

[12] This is true since $\Pr[f \neq 0] \geq \Pr[f \neq h] - \Pr[h \neq 0] \geq \epsilon - 1/|\mathcal{F}|^n$. Now we may assume that $\epsilon \geq 2/|\mathcal{F}|^n$ because, otherwise, we can query f in all the points using $O(|\mathcal{F}|^n) = O(1/\epsilon)$ queries.

linear subspace $L \subseteq \mathcal{F}^n$, then for any $a \in f^{-1}(1)$, $f^{-1}(1) = a + L$ and, therefore, $h^{-1}(1) = L$.

Thus, in the reduction, the tester accepts if f is evaluated to 0 on uniformly at random $O(1/\epsilon)$ points[13]. Otherwise, let a be a point such that $f(a) = 1$. Then run the tester for functions that describe linear subspaces to test $f(x + a)$.

See more details in [11] Sect. 4.

2.4 The Algorithm for Functions that Describe Axis-Parallel Affine Subspace

The class of functions that describe axis-parallel affine subspace and the class of functions that describe axis-parallel linear subspace are denoted by **APAS** and **APLS**, respectively. Then d-**APAS**, d-**APLS**, $(\leq d)$-**APAS**, and $(\leq d)$-**APLS** are defined similarly to those in the previous subsection. When the field is $\mathcal{F} = \mathrm{GF}(2)$, those classes are equivalent to **Term**, **Monomial**, d-**Term**, d-**Monomial**, $(\leq d)$-**Term**, and $(\leq d)$-**Monomial**, respectively.

We first give an overview of the testers for **APAS** and **APLS**. As in the previous section, the reduction of Goldreich and Ron reduces the problem of testing whether the function describes an axis-parallel affine subspace (**APAS**) to testing whether the function describes an axis-parallel linear subspace (**APLS**).

The tester for testing whether the function describes an axis-parallel linear subspace, first runs the tester for **LS** with proximity parameter $\epsilon/100$ and rejects if it rejects. Then it draws uniformly at random $x, y, z \in f^{-1}(1)$ and tests if $f(w^{x,y} + z) = 1$ where for every $i \in [n]$, $w_i^{x,y} = 0$ if $x_i = y_i = 0$ and $w_i^{x,y} \in \{0, 1\}$ drawn uniformly at random, otherwise. If $f(w^{x,y} + z) = 1$, then the tester accepts; otherwise, it rejects.

We show that if $f \in$ **APLS**, then with probability 1, $f(w^{x,y} + z) = 1$. This fact is obvious. We also show that if f is ϵ-far from **APLS** and $\epsilon/100$-close to **LS**, then with constant probability $f(w^{x,y} + z) \neq 1$. Obviously, this tester is one-sided and makes $\tilde{O}(1/\epsilon)$ queries.

We provide some intuition for why this is true. Let f be ϵ-far from **APLS** and $\epsilon/100$-close to **LS**. If $f^{-1}(1)$ is very close to a linear subspace L, then, for a uniformly random $x, y, z \in f^{-1}(1)$, with high probability, x, y, z are in L. Since $f^{-1}(1)$ is ϵ-far from **APLS**, it follows that L is also $\Omega(\epsilon)$-far from **APLS**. Assuming $x, y \in L$, with high probability, $w^{x,y}$ is not in L. This is due to the fact that, if $L \in$ **LS****APLS**, then one of the entries in the points in L is a *non-zero* linear combination of the other entries; therefore this entry is uniformly at random in $w^{x,y}$. So, whp, $w^{x,y} \notin L$, but not necessarily (whp) not in $f^{-1}(1)$ because $w^{x,y}$ is not a uniformly random point. To add some randomness to $w^{x,y}$, we add a random z to $w^{x,y}$. Then, if $x, y, z \in L$, whp, $w^{x,y} + z$ is not L. Because z is almost random uniform in L and $w^{x,y}$ is not in L, whp, $w^{x,y} + z$ is an almost random uniform point in some coset outside L. Then again, since $f^{-1}(1)$

[13] If f is ϵ-far from **AS**, then it is ϵ-far from the function $h(x)$ that satisfies $h^{-1}(1) = \{0^n\}$. Therefore, whp, some point a satisfies $f(a) = 1$. This is not true for $(\leq d)$-**AS** because $h \notin (\leq d)$-**AS**.

is very close to L, it follows that, whp, $w^{x,y} + z \notin f^{-1}(1)$. This implies that, whp, $f(w^{x,y} + z) \neq 1$. For more details, refer to the full paper.

Now for testing the class ($\leq d$)-**APLS**, we prove that if f is ($\epsilon/100$)-close to **APLS** and ($\epsilon/100$)-close to ($\leq d$)-**LS**, then it is ϵ-close to ($\leq d$)-**APLS**. So we run the tester for **APLS** and ($\leq d$)-**LS**, with proximity parameter $\epsilon/100$, and accept if both accept.

2.5 Lower Bound for Testing Classes with Fixed/Bounded Dimension

For the class of Boolean functions that describe $(n-d)$-dimensional affine/linear subspaces (d-**AS** and d-**LS**) and Boolean functions that describe axis-parallel $(n-d)$-dimensional affine/linear subspaces (d-**APAS** and d-**APLS**), we give lower bounds that depend on n, the number of variables. See Figs. 1 and 2 and the proofs in Sect. 6.

Here we will give the technique used to prove the lower bound for the class d-**APLS**. For this class, we give the lower bound

$$\Omega\left(\frac{1}{\epsilon} + \min\left(\frac{\log(1/\epsilon)}{\log|\mathcal{F}|}, d\right) \cdot \log\frac{n}{d}\right)$$

for the query complexity.

First, the lower bound $\Omega(1/\epsilon)$ follows from [3]. Then any tester for the above classes can distinguish between functions in the class d-**APLS** and d'-**APLS** for $d' = \min(\log(1/\epsilon)/\log|\mathcal{F}|, d) - 1$. This is because the distance between any function in d'-**APLS** and a function in d-**APLS** is at least ϵ. Since the tester is one-sided, using Yao's principle, we show that there is a deterministic algorithm that can distinguish between all the functions in d-**APLS** and a subclass $C \subseteq d'$-**APLS** of size $|C| \geq (2/3)|d'\text{-}\mathbf{APLS}|$. We then show that for any $f \in C$, this algorithm asks queries that eliminate all possible entries in the points of $f^{-1}(1)$ that are not identically zero, except for at most d entries. Therefore, with d more queries, we get an exact learning algorithm for C. Thus, the number of queries of the tester must be at least the information-theoretic lower bound for learning C minus d, which is $\log|C| - d$. This gives the lower bound.

3 Definitions and Preliminary Results

Let $\mathcal{F}$ be a finite field of $q = |\mathcal{F}|$ elements, and $B(\mathcal{F})$ be the set of all Boolean functions $f : \mathcal{F}^n \to \{0,1\}$. We say that $f \in B(\mathcal{F})$ *describes a well-structured $(n-d)$-dimensional injective relation* if for every $a \in \mathcal{F}^{n-d}$, there is at most one element $b \in \mathcal{F}^d$ such that[14] $f(a,b) = 1$. The class of such functions is denoted by d-**R**. Here $\mathcal{F}^0 = \{()\}$, so every Boolean function describes a well-structured n-dimensional injective relation. That is 0-**R**$= B(\mathcal{F})$.

[14] By $f(a,b)$, we mean the following: If $a = (a_1, \ldots, a_{n-d})$ and $b = (b_1, \ldots, b_d)$, then $f(a,b) = f(a_1, \ldots, a_{n-d}, b_1, \ldots, b_d)$.

For a class $C \subseteq B(\mathcal{F})$ and functions $f, g \in B(\mathcal{F})$ we define $\mathrm{dist}(f,g) = \Pr[f(x) \neq g(x)]$ and $\mathrm{dist}(f,C) = \min_{h\in C} \mathrm{dist}(f,h)$. For any $f \in B(\mathcal{F})$ define the function $R_f : \mathcal{F}^{n-d} \to \mathcal{F}^d \cup \{\perp\}$, $\perp \notin \mathcal{F}^d$, where $R_f(a)$ is equal to the minimum $b \in \mathcal{F}^d$ (in some total order over $\mathcal{F}^d$) that satisfies $f(a,b) = 1$ and $R_f(a) = \perp$ if no such b exists. If $d = 0$, we have $R_f : \mathcal{F}^n \to \{(), \perp\}$, where $R_f(a) = ()$ if $f(a) = 1$ and $R_f(a) = \perp$ if $f(a) = 0$. For any $f \in B(\mathcal{F})$ define $f_{\mathbf{R}} \in B(\mathcal{F})$ as $f_{\mathbf{R}}(a,b) = 1$ if $b = R_f(a)$ and $f_{\mathbf{R}}(a,b) = 0$ otherwise.

In the full paper we show

Lemma 1. *Let* $q = |\mathcal{F}|$ *and* $r = \max(0, d\log q - \log(2/\epsilon))$. *If* f *is* ϵ*-far from* d-**R** *then there is* ℓ_0, $r+1 \le \ell_0 \le d\log q$ *such that for*

$$\alpha = \frac{\epsilon q^d}{2^{\ell_0+1}\min(d\log q - 1, \log(1/\epsilon))}\ ,\quad \beta = \frac{2^{\ell_0-1}}{q^d}$$

we have $\Pr_{a\in\mathcal{F}^{n-d}}[\Pr_{b\in\mathcal{F}^d}[f(a,b) \neq f_{\mathbf{R}}(a,b)] \ge \beta] \ge \alpha$.

In particular, $\alpha\beta \ge \epsilon/(4\log(1/\epsilon))$.

We say that $f \in B(\mathcal{F})$ *describes a well-structured* $(n-d)$*-dimensional bijection*, if for every $a \in \mathcal{F}^{n-d}$, there is exactly one $b \in \mathcal{F}^d$ such that $f(a,b) = 1$. This class is denoted by d-**F**. In particular, $f \in 0$-**F** if it is the constant 1 function.

We define $F_f : \mathcal{F}^{n-d} \to \mathcal{F}^d$ where $F_f(a) = R_f(a)$ if $R_f(a) \neq \perp$ and $F_f(a) = 0^d$ otherwise. Define $f_{\mathbf{F}} \in B(\mathcal{F})$ as $f_{\mathbf{F}}(a,b) = 1$ if $b = F_f(a)$ and $f_{\mathbf{F}}(a,b) = 0$ otherwise.

In the full paper we prove

Lemma 2. *If* f *is* $\epsilon/2$*-close to* d-**R** *and* ϵ*-far from* d-**F**, *then* $\Pr[R_f(a) = \perp] \ge \epsilon q^d/2$.

We say that $L \subseteq \mathcal{F}^n$ is a *well-structured* $(n-d)$*-dimensional linear subspace* if there is a linear function $\phi : \mathcal{F}^{n-d} \to \mathcal{F}^d$ such that $L = \{(a, \phi(a)) \mid a \in \mathcal{F}^{n-d}\}$.

We say that $f \in B(\mathcal{F})$ *describes a well-structured* $(n-d)$*-dimensional linear subspace* if $f^{-1}(1)$ is a well-structured $(n-d)$-dimensional linear subspace. We denote by d-**WSLS** the class of Boolean functions that describes a well-structured $(n-d)$-dimensional linear subspace. Consider the class **Linear** of linear functions $\Lambda : \mathcal{F}^{n-d} \to \mathcal{F}^d$. In the full paper, we prove

Lemma 3. *If* f *is* ϵ*-far from* d-**WSLS** *and* $\epsilon/2$*-close to* d-**F**, *then* F_f *is* $(q^d\epsilon/2)$*-far from* **Linear**.

In the full paper we prove.

Lemma 4. *For any function* $f \in B(\mathcal{F})$ *and any nonsinglar* $n \times n$*-matrix* M *we have*

1. *If* $f \in$ **LS** *and* $h(x) = f(xM)$, *then* $h \in$ **LS** *and* $\dim(h^{-1}(1)) = \dim(f^{-1}(1))$
2. $\mathrm{dist}(f(x), \mathbf{LS}) = \mathrm{dist}(f(xM), \mathbf{LS})$ *(Fig. 3).*

Test-d-R(f, ϵ)
Input: Oracle that accesses a Boolean function $f : \mathcal{F}^n \rightarrow \{0, 1\}$.
Output: Either "Accept" or "Reject"

1. For $\ell = \max(1, d \log q - \log(1/\epsilon))$ to $d \log q$
2. Let $\alpha(\ell) = \frac{\epsilon q^d}{2^{\ell+1} \min(d \log q - 1, \log(1/\epsilon))}$, $\beta(\ell) = \frac{2^{\ell-1}}{q^d}$
3. Draw uniformly at random $r = 10/\alpha(\ell)$ assignments $a^{(1)}, \ldots, a^{(r)} \in \mathcal{F}^{n-d}$
   Draw uniformly at random $s = 10/\beta(\ell)$ assignments $b^{(1)}, \ldots, b^{(s)} \in \mathcal{F}^d$
   If there is $a^{(i)}$ and two $b^{(j_1)} \neq b^{(j_2)}$ such that
   $f(a^{(i)}, b^{(j_1)}) = f(a^{(i)}, b^{(j_2)}) = 1$ then Reject
4. Accept.

Fig. 3. A tester for d-**R**.

4 Three Testers

In this section, we give testers for d-**R**, d-**F** and d-**WSLS**.

The proof of the following is in the full paper.

Lemma 5. *There is a polynomial-time one-sided tester for* d-**R** *that makes*

$$O(\min(\log(1/\epsilon), d \log q)^2/\epsilon) = O(\log^2(1/\epsilon)/\epsilon) = \tilde{O}(1/\epsilon).$$

queries.

Test-d-F(f, ϵ)
Input: Oracle that accesses a Boolean function $f : \mathcal{F}^n \rightarrow \{0, 1\}$.
Output: Either "Accept" or "Reject" with $v =$"empty" or $v \in \mathcal{F}^{n-d}$ s.t. $R_f(v) = \perp$

1. If **Test-d-R**$(f, \epsilon/2)$ =Reject then Reject; Return $v =$ "empty".
2. For $i = 1$ to $\lfloor 10/(q^d \epsilon) \rfloor$
3. Draw uniformly at random $a \in \mathcal{F}^{n-d}$
4. If $R_f(a) = \perp$ then Reject: Return $v = a$
5. Accept

Fig. 4. A Tester for d-**F**.

We now give a tester for d-**F**. See Fig. 4. Notice that when the tester rejects, it also returns $v \in \{\text{"empty"}\} \cup \mathcal{F}^d$. We will use this in the next section. So, we can ignore that for this section. In the full paper we prove.

Lemma 6. *There is a polynomial-time one-sided tester for d-***F** *that makes* $\tilde{O}(1/\epsilon)$ *queries.*

Test-d**-WSLS**(f,ϵ)
Input: Oracle that accesses a Boolean function $f:\mathcal{F}^n\to\{0,1\}$.
Output: Either "Accept" or "Reject" with $v=$"empty" or $v\in\mathcal{F}^{n-d}$ s.t. $R_f(v)=\perp$
Test-Linear(F,ϵ) tests whether $F:\mathcal{F}^{n-d}\to\mathcal{F}^d$ is linear or ϵ-far from linear

1. If **Test-**d**-F**$(f,\epsilon/2)$ =Reject then Reject; Return v (that **Test-**d**-F** returns).
2. If **Test-Linear**$(F_f,q^d\epsilon/2)$ =Reject then
   If for some query a that **Test-Linear** asks $R_f(a)=\perp$ then Reject; Return $v=a$
   Otherwise, Reject; Return $v=$"empty".
3. Accept

Fig. 5. A Tester for d-**WSLS**.

Lemma 7. *There is a polynomial-time one-sided tester for d-***WSLS** *that makes* $\tilde{O}(1/\epsilon)$ *queries.*

5 A Tester for AS

Test-LS(f,ϵ)
Input: Oracle that accesses a Boolean function $f:\mathcal{F}^n\to\{0,1\}$.
Output: Either "Accept" or "Reject".

1. If $f(0^n)=0$ then Reject.
2. $k\leftarrow 0; N=I_n$;
3. While $k\le m:=\log(1/\epsilon)/\log(q)+2$ do
4. $\quad v\leftarrow$ **Test-**k**-WSLS**$(f(xN),\epsilon,\delta=1-1/(10m))$; If Accept, then Accept.
5. $\quad$ If Reject and $v=$"empty" then Reject.
6. $\quad$ If Reject and $v\ne$"empty" $(R_{f_k}(v)=\perp)$ then
7. $\quad\quad$ Find a non-singular $(n-k)\times(n-k)$ matrix M s.t. $v=e_{n-k}M$
   $\quad\quad$ $N\leftarrow N\cdot diag(M,I_k)$.
   $\quad\quad$ $k\leftarrow k+1$
8. Accept

Fig. 6. A Tester for **LS**.

Consider the tester in Fig. 6. In the full paper, we prove.

Theorem 1. *There are polynomial-time one-sided testers for* **AS** *and* **LS** *that make* $\tilde{O}(1/\epsilon)$ *queries.*

Theorem 2. *There is a polynomial-time one-sided tester for* $(\le d)$-**LS** *that makes* $\tilde{O}(1/\epsilon)$ *queries.*

6 Lower Bounds

In this section, we give the lower bound for d-APLS. The other lower bounds are in the full paper.

Theorem 3. *Any one-sided tester for d-**APLS** and d-**APAS** with proximity parameter $\epsilon \le q^{-1} - q^{-d}$ must make at least*

$$\Omega\left(\frac{1}{\epsilon} + \min\left(\frac{\log(1/\epsilon)}{\log|\mathcal{F}|}, d\right) \cdot \log\frac{n}{d}\right)$$

queries.

In particular,

Corollary 1. *Any one-sided tester for d-**Monomial** and d-**Term** with proximity parameter $\epsilon \le 1/2 - 2^{-d}$ must make at least*

$$\Omega\left(\frac{1}{\epsilon} + \min\left(\log(1/\epsilon), d\right) \cdot \log\frac{n}{d}\right)$$

queries.

The following is an information-theoretic lower bound.

Lemma 8. *Any deterministic algorithm that exactly learns*[15] *a class C of Boolean functions $f : \mathcal{F}^n \to \{0,1\}$ must ask at least $\log|C|$ black-box queries.*

The following lemma is proved in the full paper.

Lemma 9. *Let $d' < d$. Then* $\mathrm{dist}(d\text{-}\mathbf{AS}, d'\text{-}\mathbf{AS}) = \mathrm{dist}(d\text{-}\mathbf{APLS}, d'\text{-}\mathbf{APLS}) = q^{-d'} - q^{-d}$.

We are now ready to prove Theorem 3.

Proof. The lower bound $\Omega(1/\epsilon)$ follows from [3].

Let T be a tester for d-**APLS** (resp. d-**APAS**) with proximity parameter $\epsilon \le 1 - q^{-d}$, which makes Q queries. Consider the class d'-**APLS** where $d' = \min(\lfloor \log(1/(\epsilon + q^{-d}))/\log q) \rfloor, d-1)$. Then, by Lemma 9, dist(d-**APLS**, d'-**APLS**) $= q^{-d'} - q^{-d} \ge \epsilon$ (resp. dist(d-**APAS**, d'-**APLS**) $\ge \epsilon$). Therefore

1. If $f \in d$-**APLS** then $T(f) =$Accept.
2. If $f \in d'$-**APLS** then with probability at least 2/3, $T(f) =$ Reject.

Using Yao's principle[16], there is a deterministic algorithm A that has query complexity Q (as T) and a class $C \subseteq d'$-**APLS** such that $|C| \ge (2/3)|d'$-**APLS**$|$ and

---

[15] For $f \in C$ and access to a black-box to f, the algorithm returns a function equivalent to f.

[16] For a random uniform $g \in d'$-**APLS**, we have $\mathbf{E}_s[\mathbf{E}_g[T(g)]] = \mathbf{E}_g[\mathbf{E}_s[T(g)]] \ge 2/3$ where s is the random seed of T. Then there is s_0 such that $\mathbf{E}_g[T(g)] \ge 2/3$.

1. If $f \in d$-**APLS** then $A(f) = $ Accept.
2. If $f \in C$ then $A(f) = $ Reject.

We will show in the following how to change A to an exact learning algorithm for C that makes $Q + d$ queries, and then, by Lemma 8, the query complexity of T is at least[17]

$$\log |C| - d \geq \log \left(\frac{2}{3}|d'\text{-}\mathbf{APLS}|\right) - d = \log \left(\frac{2}{3}\binom{n}{d'}\right) - d$$

$$= \Omega\left(\min\left(\frac{\log(1/\epsilon)}{\log |\mathcal{F}|}, d\right) \cdot \log \frac{n}{d}\right).$$

It remains to show how to change A to an exact learning algorithm for C that makes $Q + d$ queries. To this end, consider the following algorithm (e_i is the point that contains 1 in the i-th coordinate and zero elsewhere)

1. Given access to a black-box for $f \in C$.
2. Let $X = [n]$.
3. Run A and for every query b that A asks such that $f(b) = 1$, define $X \leftarrow X \backslash \{i | b_i = 1\}$.
4. For every $i \in X$ if $f(e_i) = 1$ then remove i from X.
5. Return the function h that satisfies $h^{-1}(1) = \{a \in \mathcal{F}^n | (\forall i \in X) a_i = 0\}$.

Now, suppose $f^{-1}(1) = \{a \in \mathcal{F}^n | a_{i_1} = \cdots = a_{i_{d'}} = 0\}$. We now show

Claim. After step 3, we have $|X| \leq d$ and $\{i_1, \ldots, i_{d'}\} \subseteq X$.

Proof. If, on the contrary, some $j \in [d']$, $i_j \notin X$, there is b such that $b_{i_j} = 1$ and $f(b) = 1$. Then $b \in f^{-1}(1)$ and therefore $b_{i_j} = 0$. A contradiction.

Suppose, on the contrary, X contains more than d elements. Let $i_{d'+1}, \ldots, i_d \in X$ be distinct and distinct from $i_1, \ldots, i_{d'}$. Consider the function g such that $g^{-1}(1) = \{a \in \mathcal{F}^b | a_{i_1} = \cdots = a_{i_d} = 0\}$. Since A accepts $g \in d$-**APLS** and rejects $f \in C$, there must be a query b that A makes such that $g(b) \neq f(b)$. Since $g^{-1}(1) \subset f^{-1}(1)$, we have $b \in f^{-1}(1) \backslash h^{-1}(1)$, and then for some $j > d'$, we have $b_{i_j} = 1$ and $f(b) = 1$. Therefore, $i_j \notin X$ after step 3. A contradiction. This finishes the proof of the claim. □

By the above claim, step 5 makes at most d queries; therefore, the query complexity of the learning algorithm is $Q + d$. If, after step 3, $i \in \{i_1, \ldots, i_{d'}\}$, then $f(e_i) = 0$, and then i is not removed from X after step 4. If after step 3, $i \notin \{i_1, \ldots, i_{d'}\}$ and $i \in X$, then the query e_i satisfies $f(e_i) = 1$, and then i is removed from X after step 4. So, after step 4, we have $X = \{i_1, \ldots, i_{d'}\}$ and hence $h = f$. □

---

[17] Here, we assume that $d \ll n$. For large d, we can replace step 4 in the learning algorithm that makes at most d queries with the algorithm in [16] that makes $d' \log(d/d') - O(d')$ queries. This changes $\log |C| - d$ to $\log |C| - d' \log(d/d') - O(d')$, and we get the lower bound for any d.

References

1. Blum, M., Luby, M., Rubinfeld, R.: Self-testing/correcting with applications to numerical problems. J. Comput. Syst. Sci. **47**(3), 549–595 (1993)
2. Bshouty, N.H.: Almost optimal testers for concise representations. In: Approximation, Randomization, and Combinatorial Optimization. Algorithms and Techniques, APPROX/RANDOM 2020, 17–19 August 2020, Virtual Conference, pp. 5:1–5:20 (2020)
3. Bshouty, N.H., Goldreich, O.: On properties that are non-trivial to test. In: Electronic Colloquium on Computational Complexity (ECCC), vol. 13 (2022)
4. Chakraborty, S., García-Soriano, D., Matsliah, A.: Efficient sample extractors for juntas with applications. In: Aceto, L., Henzinger, M., Sgall, J. (eds.) ICALP 2011. LNCS, vol. 6755, pp. 545–556. Springer, Heidelberg (2011). https://doi.org/10.1007/978-3-642-22006-7_46
5. Chen, X., Xie, J.: Tight bounds for the distribution-free testing of monotone conjunctions. In: Proceedings of the Twenty-Seventh Annual ACM-SIAM Symposium on Discrete Algorithms, SODA 2016, Arlington, VA, USA, 10–12 January 2016, pp. 54–71 (2016)
6. Dolev, E., Ron, D.: Distribution-free testing for monomials with a sublinear number of queries. Theory Comput. **7**(1), 155–176 (2011)
7. Glasner, D., Servedio, R.A.: Distribution-free testing lower bound for basic Boolean functions. Theory of Comput. **5**(1), 191–216 (2009)
8. Goldreich, O. (ed.): Property Testing - Current Research and Surveys. Lecture Notes in Computer Science, vol. 6390. Springer, Heidelberg (2010). https://doi.org/10.1007/978-3-642-16367-8
9. Goldreich, O.: Reducing testing affine spaces to testing linearity. Electron. Colloquium Comput. Complex 80 (2016)
10. Goldreich, O.: Introduction to Property Testing. Cambridge University Press, Cambridge (2017)
11. Goldreich, O., Ron, D.: One-sided error testing of monomials and affine subspaces. Electron. Colloquium Comput. Complex 68 (2020)
12. Parnas, M., Ron, D., Samorodnitsky, A.: Testing basic Boolean formulae. SIAM J. Discrete Math. **16**(1), 20–46 (2002)
13. Ron, D.: Property testing: a learning theory perspective. Found. Trends Mach. Learn. **1**(3), 307–402 (2008)
14. Ron, D.: Algorithmic and analysis techniques in property testing. Found. Trends Theor. Comput. Sci. **5**(2), 73–205 (2009)
15. Rubinfeld, R., Sudan, M.: Robust characterizations of polynomials with applications to program testing. SIAM J. Comput. **25**(2), 252–271 (1996)
16. Jun, W., Cheng, Y., Ding-Zhu, D.: An improved zig zag approach for competitive group testing. Discret. Optim. **43**, 100687 (2022)

Stable Scheduling in Transactional Memory

Costas Busch[1], Bogdan S. Chlebus[1(✉)], Dariusz R. Kowalski[1], and Pavan Poudel[2]

[1] Augusta University, Augusta, GA, USA
{kbusch,bchlebus,dkowalski}@augusta.edu
[2] ATGWORK, Norcross, GA, USA

Abstract. We study computer systems with transactions executed on a set of shared objects. Transactions arrive continually subjects to constrains that are framed as an adversarial model and impose limits on the average rate of transaction generation and the number of objects that transactions use. We show that no deterministic distributed scheduler in the queue-free model of transaction autonomy can provide stability for any positive rate of transaction generation. Let a system consist of m shared objects and an adversary be constrained such that each transaction may access at most k shared objects. We prove that no scheduler can be stable if a generation rate is greater than $\max\{\frac{2}{k+1}, \frac{2}{\lfloor\sqrt{2m}\rfloor}\}$. We develop a centralized scheduler that is stable if a transaction generation rate is at most $\max\{\frac{1}{4k}, \frac{1}{4\lceil\sqrt{m}\rceil}\}$. We design a distributed scheduler in the queue-based model of transaction autonomy, in which a transaction is assigned to an individual processor, that guarantees stability if the rate of transaction generation is less than $\max\{\frac{1}{6k}, \frac{1}{6\lceil\sqrt{m}\rceil}\}$. For each of the schedulers we give upper bounds on the queue size and transaction latency in the range of rates of transaction generation for which the scheduler is stable.

Keywords: Transactional memory · shared object · dynamic transaction generation · adversarial model · stability · latency

1 Introduction

Threads that execute concurrently need to synchronize access to shared objects to avoid conflicts. Traditional low-level thread synchronization mechanisms such as locks and barriers are prone to deadlock and priority inversion, among multiple vulnerabilities. The concept of transactional memory has emerged as a high-level abstraction of the functionality of multiprocessor systems, see Herlihy and Moss [17] and Shavit and Touitou [21]. The idea is to designate blocks of program code as 'transactions' to be executed atomically. Transactions are executed speculatively, in the sense that if a transaction aborts due to synchronization conflicts or failures, then the transaction's execution is rolled back to be

M. Mavronicolas (Ed.): CIAC 2023, LNCS 13898, pp. 172–186, 2023.
https://doi.org/10.1007/978-3-031-30448-4_13

restarted later. A transaction commits if there are no conflicts or failures, and its effects become visible to all processes. If multiple transactions concurrently attempt to access the same object, then this creates a conflict for access and could trigger aborting some of the involved transactions. The synchronization conflict between the transactions is handled by contention managers, also known as schedulers, see Hendler and Suissa-Peleg [16] and Spear et al. [22]. Schedulers determine an execution schedule for transactions striving to avoid conflicts for access to shared objects.

The adversarial models of generating transactions that we use are inspired by the adversarial queueing theory, which has been applied to study stability of routing algorithms with packets injected continually. Routing of packets in communication networks is constrained by properties of networks, like their topology and capacities of links or channels. In the case of transactional memory, executing multiple transactions concurrently is constrained by the requirement that a transaction needs to have an exclusive access to each object it wants to interact with in order to be executed succesfully.

A computer system includes a fixed set of shared objects. Transactions are spawned continually. The system is synchronous in that an execution of an algorithm scheduling transactions is structured into rounds. It takes one round to execute a transaction successfully. Multiple transactions can be invoked concurrently, but a transaction requires exclusive access to each object that it needs to interact with in order to be executed successfully. If multiple transactions accessing the same object are invoked at a round then all of them are aborted. The arrival of threads with transactions is governed by an adversarial model with parameters bounding the average generation rate and the number of transactions that can be generated at one round. Processed transactions may be additionally constrained by imposing an upper bound on the number of objects a transaction needs to access.

The task for such a computer system is to eventually execute each generated transaction, while striving to minimize the number of pending transactions at any round and the time a pending transaction spends waiting for execution. Once a transaction is generated, it may need to wait to be invoked. It is a scheduling algorithm that manages the timings of invocations of pending transactions. We consider both centralized and distributed schedulers.

There are two models of generating transactions which specify the autonomy of individual transactions. In the queue-free case, for each transaction there is a corresponding autonomous processor responsible for its execution. A distributed scheduler in the queue-free model is executed by the processors that attempt to invoke transactions on shared objects. In the queue-based model, there is a fixed set of processors, and each thread with a transaction is assigned to a processor. A distributed scheduler in the queue-based model is executed by the processors that communicate through the shared objects by performing transactions on them. A centralized scheduler is not affected by constraints on autonomy of each transaction, since all pending transactions are managed en masse. The schedulers we consider are deterministic, in that they do not resort to randomization.

Table 1. A summary of the ranges of rates of transaction generation for which deterministic schedulers are stable. The used notations are as follows: m is the number of shared objects, k is the maximum number of shared objects accessed by a transaction, and ρ is the rate of transaction generation. Upper bounds limit transaction generation rates for which stability is achievable. Lower bounds limit transaction generation rates for which stability is not possible. A lower bound for centralized schedulers holds a priori for distributed queue-based schedulers.

Scheduler	Lower bound	Upper bound
distributed queue-free	stability impossible	
centralized	$\rho > \max\{\frac{2}{k+1}, \frac{2}{\lfloor\sqrt{2m}\rfloor}\}$	$\rho \le \max\{\frac{1}{4k}, \frac{1}{4\lceil\sqrt{m}\rceil}\}$
distributed queue-based		$\rho < \max\{\frac{1}{6k}, \frac{1}{6\lceil\sqrt{m}\rceil}\}$

The Contributions. We show first that no deterministic distributed scheduler in the queue-free model of transaction autonomy can provide stability for any positive rate of transaction generation. Let a computer system consist of m shared objects and the adversary be constrained such that each transaction needs to access at most k of the shared objects. We show that no scheduler can be stable if a generation rate is greater than $\max\{\frac{2}{k+1}, \frac{2}{\lfloor\sqrt{2m}\rfloor}\}$. We develop a centralized scheduler that is stable if the transaction generate rate is at most $\max\{\frac{1}{4k}, \frac{1}{4\lceil\sqrt{m}\rceil}\}$. We design a distributed scheduler, in the queue-based model of transaction autonomy in which a transaction is assigned to an individual processor, that guarantees stability if the rate of transaction generation is less than $\max\{\frac{1}{6k}, \frac{1}{6\lceil\sqrt{m}\rceil}\}$. For each of the two schedulers we develop, we give upper bounds on the queue size and transaction latency in the range of rates of transaction generation for which the scheduler is stable. Table 1 gives a summary of the ranges of rates of transaction generation for which deterministic schedulers are stable.

Related Work. Scheduling transactions has been studied for both shared memory multi-core and distributed systems. Most of the previous work on scheduling transactions considered an offline case where all transactions are known at the outset. Some previous work considered online scheduling where a batch of transactions arrives one by one and the performance of an online scheduler is compared to a scheduler processing the batch offline. No previous work known to the authors of this paper addressed dynamic transaction arrivals with potentially infinitely many transactions to be scheduled in a never-ending execution.

Attiya et al. [4] and Sharma and Busch [19,20] considered transaction scheduling in distributed systems with provable performance bounds on communication cost. Transaction scheduling in a distributed system with the goal of minimizing execution time was first considered by Zhang et al. [25]. Busch et al. [8] considered minimizing both the execution time and communication cost simultaneously. They showed that it is impossible to simultaneously minimize execution time and communication cost for all the scheduling problem instances in arbitrary graphs even in the offline setting. Specifically, Busch et al. [8] demonstrated a tradeoff between minimizing execution time and communication cost and provided offline algorithms that separately optimizw execution time and communication cost. Busch et al. [9] considered transaction scheduling tailored to specific popular topologies and provided offline algorithms that minimize simultaneously execution time and communication cost. Distributed directory protocols have been designed by Herlihy and Sun [18], Sharma and Busch [19], and Zhang et al. [25], with the goal to optimize communication cost in scheduling transactions. Zhang and Ravindran [23] provided a distributed dependency-aware model for scheduling transactions in a distributed system that manages dependencies between conflicting and uncommitted transactions such that they can commit safely. This model has the inherent tradeoff between concurrency and communication cost. Zhang and Ravindran [24] provided cache-coherence protocols for distributed transactional memory based on a distributed queuing protocol. Attiya et al. [3] and Attiya and Milani [5] studied competitive performance of schedulers compared to the clairvoyant one. Busch et al. [10] studied online algorithms to schedule transactions for distributed transactional memory systems where transactions residing at nodes of a communication graph operate on shared, mobile objects.

Adversarial queuing is a methodology to capture stability of processing incoming tasks without any statistical assumptions about task generation. It provides a framework to develop worst-case bounds on performance of deterministic distributed algorithms in a dynamic setting. This approach to study routing algorithms in store-and-forward networks was proposed by Borodin et al. [7], and continued by Andrews et al. [2]. Adversarial queuing has been applied to other dynamic tasks in communication networks. Bender et al. [6] considered broadcasting in multiple-access channels with queue-free stations in the framework of adversarial queuing. Chlebus et al. [13] proposed to investigate deterministic distributed broadcast in multiple access channels performed by stations with queues in the adversarial setting. This direction was continued by Chlebus et al. [12] who studied the maximum throughput in such a setting. Anantharamu et al. [1] considered packet latency of deterministic broadcast algorithms with injection rates less than 1. Chlebus et al. [11] studied adversarial routing in multiple-access channels subject to energy constraints. Garncarek et al. [14] investigated adversarial stability of memoryless packet scheduling policies in multiple access channels. Garncarek et al. [15] studied adversarial communication through channels with collisions between communicating agents represented as graphs.

2 Technical Preliminaries

A distributed system consists of processors and a fixed set of m shared objects. The system executes an algorithm. An execution of the algorithm is synchronous in that it is partitioned into time steps, which we call *rounds*. Intuitively, the executed algorithm spawns threads and each thread generates and executes transactions. To simplify the model of transaction generation and scheduling, we assign transactions directly to processors and disregard threads entirely.

We consider two frameworks of generating transactions. In the *queue-based* model, we assume a fixed number of processors in the system, each with a unique name. Each new transaction is generated at a specific round and assigned to one such a processor. All the transactions at a processor pending at a round make its *queue* at the round. In a *queue-free* model of transaction autonomy, each new transaction generated at a round is associated with an anonymous virtual processor that exists only for the purpose to execute this transaction and disappears after the transaction's successful execution.

The *type* of a transaction is the set of objects it may need to access during execution. To determine the type of a transaction, it suffices to read it to list all the mentioned objects. The number of objects in a transaction's type is the *weight* of this transaction and also of the type. If the types of two transactions share an object, then we say that this creates a *conflict for access* to this object, and that the transactions involved in a conflict for access to an object *collide* at this object. A set of transactions with the property that no two different transactions in the set collide at some shared object is called *conflict free*.

Scheduling Transactions. Transactions are managed by a *scheduler*. This is an algorithm that determines for each round which pending transactions are invoked at this round. A transaction invoked at a round that gets executed successfully is no longer pending, while an aborted transaction stays pending at the next round. Scheduling transactions is constrained by whether this is a queue-free or queue-based model. In the queue-free model, transactions are managed en-masse and only conflicts for access to objects determine feasibility of concurrently performing a set of transactions. This means that if a pending transaction invoked at a round is not involved in conflict with any object it needs to access, for any of the transactions invoked at this round concurrently, then this transaction is executed successfully. In particular, if a set of transactions is conflict-free then all the transactions in this set can be executed together at one round. We assume conservatively that if a pending transaction invoked at a round is involved in conflict with some other transaction invoked concurrently, for an object they need to access, then all such transactions get aborted at this round and stay pending at the next round. The queue-based model is more restricted, in that the queue-free model's constraints on concurrent execution of transactions do apply, but additionally, for each processor, at most one transaction in this processor's queue can be performed at a round.

A *centralized scheduler* is an sequential algorithm that knows all the transactions pending at a round and receives instantaneous feedback from each object about committing to an invoked transaction or aborting it. Such a scheduler can invoke concurrently any set of pending transactions at a round in the queue-free model, but at most one transaction in the queue of a processor in the queue-based model.

A *distributed scheduler* is a distributed algorithm executed by all the involved processors. The processors communicate among themselves through shared objects. These are the processors that determine the distributed system in the queue-based case, and anonymous processors in queue-free case, one dedicated processor per each transaction. If a processor invokes a transaction at a round, it receives an instantaneous feedback from each object of the type of the transaction about committing to an invoked transaction or aborting it.

Adversaries Generate Transactions. We consider a setting in which new transactions arrive continually to the system. The process of generating transactions is represented quantitatively by adversarial models. We study two types of adversaries corresponding to the queue-free and queue-based models. In the queue-free model, a transaction generated at a round contributes a unit to the *congestion* at the round at each object the transaction includes in its type. This is the *queue-free adversary*. In the queue-based model, a transaction generated at a round at a processor contributes a unit to the *congestion* at the round at each object the transaction includes in its type and also to the processor the transaction is generated at. This is the *queue-based adversary*.

Quantitative restrictions imposed on adversaries are expressed in terms of bounds on congestion. A queue-free adversary generates transactions with *generation rate* ρ and *burstiness component* b if, in each contiguous time interval τ of length t and for each shared object, the amount of congestion created for the object at all the rounds in τ together is at most $\rho t + b$. A queue-based adversary generates transactions with *generation rate* ρ and *burstiness component* b if, in each contiguous time interval τ of length t and for each shared object and for each processor, the amount of congestion created for the object at all the rounds in τ together is at most $\rho t + b$ and the amount of congestion created for the processor at all the rounds in τ together is at most $\rho t + b$. For these adversarial models, we assume that $\rho > 0$ is a real number and $b > 0$ is an integer. Given the parameters ρ and b, such an adversary is said to be of *type* (ρ, b). The *burstiness* of an adversary is the maximum number of transactions the adversary can generate in one round.

Performance of Scheduling. A scheduler is *stable*, against a given type of adversary, if the number of pending transactions stays bounded in the course of any execution in which transactions are generated by the adversary of this type. For an object and a round number r, at most r transactions that contributed to congestion at this object can get executed in the first r rounds. It follows that no scheduler can be stable if its injection rate is greater than 1. In view of this, we consider only adversaries of types (ρ, b) in which $0 < \rho \leq 1$. A transaction's *delay* is the number of rounds between its generation and successful execution. The *latency* of a scheduler in an execution is the maximum delay of a transaction generated in the execution.

Proposition 1. *No deterministic distributed scheduler for a system with one shared object can be stable against a queue-free adversary of type* $(\rho, 2)$, *for any constant* $\rho > 0$.

In view of Proposition 1, we will consider a centralized deterministic scheduler for the queue-free model.

3 A Lower Bound

We show that no scheduler can handle dynamic transactions if a generation rate is sufficiently high with respect to the number of shared objects m and an upper bound k on the weight of a transaction. If a and b are integers where $a \leq b$ then let $[a, b]$ denote the set of integers $\{a, a+1, \ldots, b\}$.

Lemma 1. *For an integer $n > 0$, there is a family of sets $A_1, A_2, \ldots, A_{n+1}$, each a subset of $[1, \frac{n(n+1)}{2}]$, such that every set A_i has n elements, any two sets A_i and A_j, for $i \neq j$, share an element, and each element of $[1, \frac{n(n+1)}{2}]$ belongs to exactly two sets A_i and A_j, for $i \neq j$.*

We give a lower bound on generation rate to keep scheduling stable.

Theorem 1. *A queue-free adversary of type (ρ, b) generating transactions for a system of m objects such that each transaction is of weight at most k can make a scheduling algorithm unstable if injection rate ρ satisfies $\rho > \max\{\frac{2}{k+1}, \frac{2}{\lfloor\sqrt{2m}\rfloor}\}$.*

Proof. Let the m objects be denoted as $o_1, o_2, \ldots, o_m$. Suppose first that $\frac{k(k+1)}{2} \leq m$. The transactions to be generated will use only the objects $o_1, o_2, \ldots, o_s$, where $s = \frac{k(k+1)}{2}$. Let us take the family of sets $A_1, A_2, \ldots, A_{k+1}$ as in Lemma 1, in which n is set to k. We will use a fixed set of transactions $T_1, T_2, \ldots, T_{k+1}$ defined such that transaction T_i uses object o_j if and only if $j \in A_i$. In particular, each transaction uses k objects. The adversary generates these transactions listed in order $L_0, L_1, L_2, \ldots$, where L_{i-1} is the ith transaction generated and L_i is a transaction identical to $T_{1+i \bmod (k+1)}$. Consider a round $r+1$. Let i be the highest index of a transaction L_i generated by round r. Then in round $r+1$ the adversary generates transactions that make a maximal prefix of the sequence $L_{i+1}, L_{i+2}, \ldots$ such that the total number of transactions generated by round $r+1$ satisfies the constraints on objects' congestion of type (ρ, b). The adversary may generate no transaction at a round and it may generate multiple transaction at a round. For example, the adversary generates exactly the transactions $L_0, \ldots, L_{b-1}$ simultaneously in the first round.

By Lemma 1, at most one transaction can be executed at a round. The $k+1$ transactions $T_1, T_2, \ldots, T_{k+1}$ require $k+1$ rounds to have each one executed, one transaction per round. Discounting for the burstiness of generation, which is possible due to the burstiness component b in the type (ρ, b), these transactions can be generated with a frequency of at most one new transaction generated per round if the execution is to stay stable.

The group of transactions $T_1, \ldots, T_{k+1}$ together contribute 2 to the congestion of each used object, by Lemma 1. If an execution is stable then the inequality $\rho(k+1) \leq 2$ holds. This gives a bound $\rho \leq \frac{2}{k+1}$ on the generation rate of an adversary if the execution is stable. In case $\rho > \frac{2}{k+1}$, the adversary can generate at least one transaction at every round, and for each round r it can generate two transactions at some round after r. Such an execution is unstable, because at most one transaction among $T_1, \ldots, T_{k+1}$ can be executed in one round.

Next, consider the case $\frac{k(k+1)}{2} > m$. Let n be the greatest positive integer such that $\frac{n(n+1)}{2} \leq m$. We use a similar reasoning as in the case $\frac{k(k+1)}{2} \leq m$, with the family of sets $A_1, A_2, \ldots, A_{n+1}$ as in Lemma 1. In particular, we use a set of transactions $T_1, T_2, \ldots, T_{n+1}$ defined such that transaction T_i uses an object o_j if and only if $j \in A_i$. The rules of generating new transactions by the adversary are similar. We obtain the inequality $\rho \leq \frac{2}{n+1}$ by the same argument. The inequality $\frac{n(n+1)}{2} \leq m$ implies $n+1 = \lfloor \frac{1}{2}(1+\sqrt{1+8m}) \rfloor$, by algebra. We have the estimates $\frac{2}{n+1} = \frac{2}{\lfloor \frac{1}{2}(1+\sqrt{1+8m}) \rfloor} \leq \frac{2}{\lfloor \sqrt{2m} \rfloor}$. If $\rho > \frac{2}{\lfloor \sqrt{2m} \rfloor}$ then also $\rho > \frac{2}{n+1}$. It follows that if the adversary is of a type (ρ, b) such that $\rho > \frac{2}{\lfloor \sqrt{2m} \rfloor}$, then this adversary generating transactions at full power can generate at least one transaction at every round, and for each round r it can generate two transactions at some round after r. This makes the queue of transactions grow unbounded.

4 A Centralized Scheduler

We present a scheduling algorithm that processes all transactions pending at a round. The algorithm is centralized in that it is aware of all the pending transactions while selecting the ones to be executed at a round. Throughout this Section we assume the queue-free model of autonomy of individual transactions, and the corresponding queue-free adversarial model of transaction generation.

The centralized scheduler identifies a conflict-free set of transactions pending execution that is maximal with respect to inclusion among all pending transactions. This is accomplished by first ordering all pending transaction on the time of generation and then processing them greedily one by one in this order. The word 'greedily' in this context means passing over a transaction only when its type includes an object that belongs to the type of a transaction already selected for execution at the current round.

The algorithm is called Centralized-Scheduler, its pseudocode is given in Fig. 1. The variable `Pending` denotes a list of all pending transactions. At the beginning of a round, all newly generated transactions are appended to the tail of this list. The list is processed in the order from head to tail, which prioritizes transactions on their arrival time, such that those generated earlier get processed before these generated later. The transactions already selected for execution are stored in the set `Execute`. If a transaction in `Pending` is processed, it is checked for conflicts with transactions already placed in the set `Execute`. If a processed transaction does not collide with any transaction already in `Execute` then it is removed from `Pending` and added to `Execute`, and otherwise it is passed over. After the whole list `Pending` have been scanned, all the transactions in `Execute` get executed concurrently. No invoked transaction is aborted in the resulting execution, because conflicts of transactions are avoided by the process to add transactions to the set `Execute`.

To assess the efficiency of executing transactions, let us partition an execution of the algorithm Centralized-Scheduler into contiguous *milestone intervals* of rounds, denoted $I_1, I_2, I_3, \ldots$, such that the length of each interval equals $4b \cdot \min\{k, \lceil \sqrt{m} \rceil\}$ rounds.

```
Algorithm CENTRALIZED-SCHEDULER

initialize Pending ← an empty list
for each round do
    append all transactions generated in the previous round at the tail of list Pending
    initialize Execute ← an empty set
    if Pending is nonempty then
        repeat
        (a) entry ← first unprocessed list item on Pending, starting from head to-
            wards tail
        (b) if entry is conflict-free with all the transactions in Execute then
                remove entry from Pending and add it to set Execute
        until entry points at the tail of list Pending
    execute all the transactions in Execute
```

Fig. 1. A pseudocode of the algorithm scheduling all pending transactions en masse. Transactions pending execution are stored in a list Pending in the order of generation, with the oldest at the head. The set Execute includes transactions to execute at a round. It is selected in a greedy manner, prioritizing older transactions over newer and avoiding conflicts for access to shared objects.

The following invariant holds for all milestone intervals of an execution.

Lemma 2. *If a generation rate satisfies* $\rho \leq \max\{\frac{1}{4k}, \frac{1}{4\lceil\sqrt{m}\rceil}\}$, *then there are at most* $2bm$ *pending transactions at the first round of a milestone interval, and all these transactions get executed by the end of the interval.*

Algorithm CENTRALIZED-SCHEDULER is stable and has bounded transaction latency for suitably low transaction generation rates.

Theorem 2. *If algorithm* CENTRALIZED-SCHEDULER *is executed against an adversary of type* (ρ, b), *such that each generated transaction accesses at most* k *objects out of* m *shared objects available and transaction-generation rate* ρ *satisfies* $\rho \leq \max\{\frac{1}{4k}, \frac{1}{4\lceil\sqrt{m}\rceil}\}$, *then the number of pending transactions at a round is at most* $4bm$ *and transaction latency is at most* $8b \cdot \min\{k, \lceil\sqrt{m}\rceil\}$.

Proof. To estimate the number of transactions pending at a round, let this round belong to a milestone interval I_k. The number of old transactions at any round of milestone interval I_k is at most $2mb$, by the centralized milestone invariant formulated as Lemma 2. During the interval I_k, at most $2mb$ new transactions can be generated. So $2mb + 2mb = 4mb$ is an upper bound on the number of pending transactions at the round.

To estimate transaction latency, we use the property that a transaction generated in a milestone interval gets executed by the end of the next interval, again by the centralized milestone invariant formulated as Lemma 2. This means that transaction latency is at most twice the length of a milestone interval, which is $2 \cdot 4b \cdot \min\{k, \lceil\sqrt{m}\rceil\} = 8b \cdot \min\{k, \lceil\sqrt{m}\rceil\}$.

5 A Distributed Scheduler

We now consider distributed scheduling. Let a distributed system consist of n processors. The processors communicate among themselves through some m shared objects by invoking transactions and receiving instantaneous feedback from each involved object. Every transaction type includes at most k objects.

Each generated transaction is assigned to a specific processor and resides in its local queue while pending execution. This means we consider the queue-based model of autonomy of individual transactions, and the corresponding queue-based adversarial model of transaction generation.

We employ a specific communication mechanism between a pair of processors. One of the processors, say s, is a sender and the other processor, say r, is a receiver. The two processors s and r communicate through a designated object o. Communication occurs at a given round. All the processors are aware that this particular round is a round of communication from s to r. Each of the participants s and r may invoke a transaction involving object o at the round, while at the same time all the remaining processors pause and do not invoke any transactions at this round.

Assume first that both s and r have pending transactions that access object o. At a round of communication, the recipient processor r invokes a transaction t_r that uses object o. If the sender processor s wants to convey bit 1 then s also invokes a transaction t_s that uses object o. In this case, both transactions t_r and t_s get aborted, so that the processor r receives the respective feedback from the system and interprets it as receiving 1. If the sender processor s wants to convey bit 0 then s does not invoke any transactions using object o at this round. In this case, transaction t_r gets executed successfully, so that r receives the respective feedback from the system and interprets it as receiving 0. This is how one bit can be transmitted successfully from a sender s to a recipient r.

That was an example of a perfect cooperation between a sender and receiver, but alternative scenarios are possible as well. Suppose that the sender s has a pending transaction using object o and wants to communicate with r but the recipient r either does not want to communicate or does not have a pending transaction using object o. What occurs is that s invokes a suitable transaction t_s which gets executed but r does not receive any information. Alternatively, suppose that the receiver r has a pending transaction using object o and wants to communicate while the sender s either does not want to communicate or does not have a pending transaction using object o. What occurs is that the receiver r invokes a suitable transaction t_r which gets executed, which the receiver r interprets as receiving the bit 0.

That communication mechanism can be extended to transmit the whole type of any transaction in the following way. The type identifies a subset of all m objects. Having a fixed ordering of the objects, the type can be represented as a sequence of m bits, in which 1 at position i represents that the ith object belongs to the type, and 0 represents that the ith object does not belong. A processor s can transmit a transaction type to recipient r by transmitting m bits representing the type in m successive rounds while using some designated object o. We say that by this operation processor s *sends the transaction type to processor* r *via object* o. This operation works as desired assuming each of the processors has at least m transactions involving object o. If at least one of these processors either does not have m transactions involving object o or does not want to participate, then either no bits are transmitted, or the receiver r possibly receives a sequence of 0s only, which it interprets as no type of transaction successfully transmitted.

Pending transactions at a processor are grouped by their types. All pending transactions of the same type at a processor make a *block of transactions* of this type. The *weight of a block* is defined to be the weight of its type. If there are sufficiently many transactions in a block then the block and the type are said to be *large*. A boundary number defining sizes to be large is denoted by L and equals $L = (n-1)^2n^2m^2$. If the number of transactions of some type in a queue at a processor is at least kL but less than $(k+1)L$, for a positive integer k, then we treat these transactions as contributing k large blocks.

An execution of the scheduling algorithm is partitioned into epochs, and each consecutive epoch consists of three phases, labeled Phase 1, Phase 2, and Phase 3. Each phase is executed the same number of $L = (n-1)^2n^2m^2$ rounds. The algorithm is called DISTRIBUTED-SCHEDULER and its pseudocode is given in Fig. 2.

In the beginning of Phase 1, each processor v that has a large block of transactions of some type, selects one such a block, and this type then is *active* at the processor in the epoch. A processor that starts Phase 1 with an active type is called *active* in this phase. Processors store large blocks in the order of generation of their last-added transaction. Each processor chooses as active a large block that comes first in this order. The purpose of Phase 1 is to spread the information of active types of all the active processors as widely as possible. Each active processor uses transactions of its active type for communication. Such communication involves executing transactions, so a block of transactions of a given type may gradually get smaller. Once a type of a large block becomes active in the beginning of Phase 1, it stays considered as active for the durations of an epoch, even if the number of transactions in the block becomes less than L. Phase 1 assigns segments of $(n-1)n^2m^2$ rounds for each pair of processors s and r and each object o to spend with s acting as sender to r acting as receiver with communication performed via object o.

Phase 2 is spent on executing transactions in some active blocks selected such that they do not create conflicts for access to shared objects. In the beginning of Phase 2, each processor computes a selection of active large blocks of transactions to execute in Phase 2 among those learned in Phase 1. This common selection is computed greedily as follows. The active types learned in Phase 1 are ordered by the owners' names. There is a working set of active types selected for execution, which is initialized empty. The active types are considered one by one. If a processed active type can be added to the working set without creating a conflict for access to an object, then the type is added to the set, and otherwise it is passed over. This computation is performed locally by each active processor at the beginning of the first round of Phase 2 and each active processor obtains the

---

`Algorithm` DISTRIBUTED-SCHEDULER

```
Phase 1 : sharing information about large active blocks during L rounds
  repeat n − 1 times
    for each sender processor s and recipient processor r and object o do
      in a segment of rounds assigned for this selection of s, r, and o:
        if v is active and this is a round when s = v then
          act as sender to transmit all relevant information to r via object o
        elseif v is active and this is a round when r = v then
          act as recipient to receive information from s via object o
Phase 2 : executing large blocks of transactions during L rounds
  if v is active then
    select active blocks for execution among those learned in Phase 1
  if v is active and its active block got selected then
    for each among L consecutive rounds do
      if there is a transaction of the active type in the queue then
        invoke such a transaction
Phase 3 : executing remaining transactions by solo processors in L rounds
  for L consecutive rounds
    if this is a round among L/n ones assigned to v then
      if the queue is nonempty then invoke a transaction
```

Fig. 2. A pseudocode of an epoch for a processor v. Pending transactions are dispersed among the processors. Number $L = (n-1)^2n^2m^2$ is the duration of each phase. In Phase 1, processors s and r use transactions from their active large blocks to implement communication. A sender processor s transmits the active type for each processor it knows about. In Phase 2, large active blocks are selected for execution in a greedy manner, with blocks ordered by the processors' names. In Phase 3, each processor gets assigned a unique exclusive contiguous segment of L/n rounds, in which to execute up to L/n transactions from its queue in a first-in first-out manner.

same output. The rounds of Phase 2 are spent on executing the transactions of the active blocks selected for execution. An active processor whose active large block has been selected executes pending transactions in its selected active block as long as some transactions from the block are still available in the queue or Phase 2 is over, whichever happens earlier.

Phase 3 is spent by each processor executing solo its pending transactions, those that have never been included in large blocks. Each processor is assigned a unique exclusive contiguous segment of $L/n = (n-1)^2nm^2$ rounds to execute such transactions. Transactions are performed in the order of their adding to the queue, with those waiting longest executed before those generated later.

Let $P = \sum_{i=1}^{k} \binom{m}{i}$ be the number of possible different transaction types in a system of m shared objects such that a type includes at most k objects.

We will use the estimate $P \leq 2^{\mathrm{H}(\frac{k}{m})\,m}$, for $k \leq \frac{m}{2}$, where $\mathrm{H}(x)$ is the binary entropy function $\mathrm{H}(x) = x \lg x + (1-x)\lg(1-x)$ for $0 < x < 1$.

An execution of algorithm DISTRIBUTED-SCHEDULER is partitioned into contiguous *milestone intervals* denoted $I_1, I_2, \ldots$. Each milestone interval consists of $2bnP \cdot \min\{k, \lceil\sqrt{m}\rceil\}$ epochs. Alternatively, a milestone interval consists of $6bnLP \cdot \min\{k, \lceil\sqrt{m}\rceil\}$ rounds, after translating the lengths of epochs into rounds.

The following Lemma 3 gives and invariant that holds for all milestone intervals of an execution of algorithm DISTRIBUTED-SCHEDULER.

Lemma 3. *For a generation rate $\rho < \max\{\frac{1}{6k}, \frac{1}{6\lceil\sqrt{m}\rceil}\}$, and assuming the bulk of the system is sufficiently large with respect to ρ, there are at most bn^5m^3P pending transactions at a first round of every milestone interval, and all these transactions get executed by the end of the interval.*

Algorithm DISTRIBUTED-SCHEDULER is stable and has bounded transaction latency for suitably low transaction generation rates.

Theorem 3. *If algorithm* DISTRIBUTED-SCHEDULER *is executed against an adversary of type (ρ, b), such that each generated transaction accesses at most $k \leq \frac{m}{2}$ objects out of m shared objects available, and the generation rate ρ satisfies $\rho < \max\{\frac{1}{6k}, \frac{1}{6\lceil\sqrt{m}\rceil}\}$, and the bulk of the system is sufficiently large with respect to ρ, then the number of pending transactions at a round is at most $2bn^5m^3\,2^{H(\frac{k}{m})m}$ and latency is at most $12bn^5m^2\,2^{H(\frac{k}{m})m}\min\{k, \lceil\sqrt{m}\rceil\}$.*

Proof. To estimate the number of transactions pending at a round, let this round belong to a milestone interval I_k. The number of old transactions at any round of the interval I_k is at most bn^5m^3P, by the distributed milestone invariant formulated as Lemma 3. During the interval I_k, at most bn^5m^3P new transactions can be generated, again by Lemma 3, because they will become old when the next interval begins. So $2bn^5m^3P \leq 2bn^5m^3\,2^{\mathrm{H}(\frac{k}{m})m}$ is an upper bound on the number of pending transactions at any round, since $P = \sum_{i=1}^{k}\binom{m}{i} \leq 2^{\mathrm{H}(\frac{k}{m})m}$, for $k \leq \frac{m}{2}$.

To estimate the transaction latency, we use the property that a transaction generated in an interval gets executed by the end of the next interval, again by the distributed milestone invariant formulated as Lemma 3. This means that transaction latency is at most twice the length of an interval, which is $2 \cdot 6bnLP\min\{k, \lceil\sqrt{m}\rceil\}$, where $L = (n-1)^2n^2m^2$. We obtain that the latency is at most $12bn^5m^2\,2^{\mathrm{H}(\frac{k}{m})m}\min\{k, \lceil\sqrt{m}\rceil\}$.

6 Conclusion

We propose to study transactional memory systems with continual generation of transactions. The critical measure of quality of such systems is stability understood as having the number of pending transactions bounded from above at all times, for a given generation rate. Transactions are modeled as sets of accesses to shared objects, and it is assumed that conflicting transactions cannot be executed concurrently. We identify a lower bound on generation rate that makes stability impossible and also develop centralized and distributed optimal scheduling algorithms that handle generation rates asymptotically equal to the lower bound.

The quality of schedulers, on a range of generation rates that guarantee stability, is further assessed by the queue size and latency. The centralized scheduler has these bounds polynomial in the parameters of the system and the adversary's type, but the distributed scheduler has the bounds exponential. It is an open question if it is possible to develop distributed scheduling with polynomial queues and latency for the region of generation rates for which stability is feasible.

Acknowledgements. This work was partly supported by the National Science Foundation grant No. 2131538.

References

1. Anantharamu, L., Chlebus, B.S., Kowalski, D.R., Rokicki, M.A.: Packet latency of deterministic broadcasting in adversarial multiple access channels. J. Comput. Syst. Sci. **99**, 27–52 (2019)
2. Andrews, M., Awerbuch, B., Fernández, A., Leighton, F.T., Liu, Z., Kleinberg, J.M.: Universal-stability results and performance bounds for greedy contention-resolution protocols. J. ACM **48**(1), 39–69 (2001)
3. Attiya, H., Epstein, L., Shachnai, H., Tamir, T.: Transactional contention management as a non-clairvoyant scheduling problem. Algorithmica **57**(1), 44–61 (2010)
4. Attiya, H., Gramoli, V., Milani, A.: Directory protocols for distributed transactional memory. In: Guerraoui, R., Romano, P. (eds.) Transactional Memory. Foundations, Algorithms, Tools, and Applications. LNCS, vol. 8913, pp. 367–391. Springer, Cham (2015). https://doi.org/10.1007/978-3-319-14720-8_17
5. Attiya, H., Milani, A.: Transactional scheduling for read-dominated workloads. J. Parallel Distrib. Comput. **72**(10), 1386–1396 (2012)
6. Bender, M.A., Farach-Colton, M., He, S., Kuszmaul, B.C., Leiserson, C.E.: Adversarial contention resolution for simple channels. In: Proceedings of the 17th ACM Symposium on Parallel Algorithms and Architectures (SPAA), pp. 325–332 (2005)
7. Borodin, A., Kleinberg, J.M., Raghavan, P., Sudan, M., Williamson, D.P.: Adversarial queuing theory. J. ACM **48**(1), 13–38 (2001)
8. Busch, C., Herlihy, M., Popovic, M., Sharma, G.: Time-communication impossibility results for distributed transactional memory. Distrib. Comput. **31**(6), 471–487 (2018)
9. Busch, C., Herlihy, M., Popovic, M., Sharma, G.: Fast scheduling in distributed transactional memory. Theory Comput. Syst. **65**(2), 296–322 (2021)
10. Busch, C., Herlihy, M., Popovic, M., Sharma, G.: Dynamic scheduling in distributed transactional memory. Distrib. Comput. **35**(1), 19–36 (2022)
11. Chlebus, B.S., Hradovich, E., Jurdziński, T., Klonowski, M., Kowalski, D.R.: Energy efficient adversarial routing in shared channels. In: Proceedings of the 31st ACM Symposium on Parallelism in Algorithms and Architectures (SPAA), pp. 191–200. ACM (2019)
12. Chlebus, B.S., Kowalski, D.R., Rokicki, M.A.: Maximum throughput of multiple access channels in adversarial environments. Distrib. Comput. **22**(2), 93–116 (2009)
13. Chlebus, B.S., Kowalski, D.R., Rokicki, M.A.: Adversarial queuing on the multiple access channel. ACM Trans. Algorithms **8**(1), 5:1–5:31 (2012)
14. Garncarek, P., Jurdzinski, T., Kowalski, D.R.: Stable memoryless queuing under contention. In: Proceedings of the 33rd International Symposium on Distributed Computing (DISC). LIPIcs, vol. 146, pp. 17:1–17:16. Schloss Dagstuhl - Leibniz-Zentrum für Informatik (2019)

15. Garncarek, P., Jurdzinski, T., Kowalski, D.R.: Efficient local medium access. In: Proceedings of the 32nd ACM Symposium on Parallelism in Algorithms and Architectures (SPAA), pp. 247–257. ACM (2020)
16. Hendler, D., Suissa-Peleg, A.: Scheduling-based contention management techniques for transactional memory. In: Guerraoui, R., Romano, P. (eds.) Transactional Memory. Foundations, Algorithms, Tools, and Applications. LNCS, vol. 8913, pp. 213–227. Springer, Cham (2015). https://doi.org/10.1007/978-3-319-14720-8_10
17. Herlihy, M., Moss, J.E.B.: Transactional memory: architectural support for lock-free data structures. In: Proceedings of the 20th Annual International Symposium on Computer Architecture, pp. 289–300. ACM (1993)
18. Herlihy, M., Sun, Y.: Distributed transactional memory for metric-space networks. Distrib. Comput. **20**(3), 195–208 (2007)
19. Sharma, G., Busch, C.: Distributed transactional memory for general networks. Distrib. Comput. **27**(5), 329–362 (2014). https://doi.org/10.1007/s00446-014-0214-7
20. Sharma, G., Busch, C.: A load balanced directory for distributed shared memory objects. J. Parallel Distrib. Comput. **78**, 6–24 (2015)
21. Shavit, N., Touitou, D.: Software transactional memory. Distrib. Comput. **10**(2), 99–116 (1997)
22. Spear, M.F., Dalessandro, L., Marathe, V.J., Scott, M.L.: A comprehensive strategy for contention management in software transactional memory. In: Proceedings of the 14th ACM SIGPLAN Symposium on Principles and Practice of Parallel Programming (PPOPP), pp. 141–150. ACM (2009)
23. Zhang, B., Ravindran, B.: Brief announcement: On enhancing concurrency in distributed transactional memory. In: Proceedings of the 29th Annual ACM Symposium on Principles of Distributed Computing (PODC), pp. 73–74. ACM (2010)
24. Zhang, B., Ravindran, B.: Brief announcement: queuing or priority queuing? On the design of cache-coherence protocols for distributed transactional memory. In: Proceedings of the 29th Annual ACM Symposium on Principles of Distributed Computing (PODC), pp. 75–76. ACM (2010)
25. Zhang, B., Ravindran, B., Palmieri, R.: Distributed transactional contention management as the traveling salesman problem. In: Halldórsson, M.M. (ed.) SIROCCO 2014. LNCS, vol. 8576, pp. 54–67. Springer, Cham (2014). https://doi.org/10.1007/978-3-319-09620-9_6

Parameterizing Path Partitions

Henning Fernau[1(✉)], Florent Foucaud[2], Kevin Mann[1], Utkarsh Padariya[3], and K. N. Rajath Rao[3]

[1] Universität Trier, Fachbereich IV, Informatikwissenschaften, Trier, Germany
{fernau,mann}@uni-trier.de
[2] Université Clermont-Auvergne, CNRS, Mines de Saint-Étienne, Clermont-Auvergne-INP, LIMOS, 63000 Clermont-Ferrand, France
florent.foucaud@uca.fr
[3] International Institute of Information Technology Bangalore, Bengaluru, India
{utkarsh.prafulchandra,rajath.rao}@iiitb.ac.in

Abstract. We study the algorithmic complexity of partitioning the vertex set of a given (di)graph into a small number of paths. The Path Partition problem (PP) has been studied extensively, as it includes Hamiltonian Path as a special case. The natural variants where the paths are required to be either *induced* (Induced Path Partition, IPP) or *shortest* (Shortest Path Partition, SPP), have received much less attention. Both problems are known to be NP-complete on undirected graphs; we strengthen this by showing that they remain so even on planar bipartite directed acyclic graphs (DAGs), and that SPP remains NP-hard on undirected bipartite graphs. When parameterized by the natural parameter "number of paths", both problems are shown to be W[1]-hard on DAGs. We also show that SPP is in XP both for DAGs and undirected graphs for the same parameter (IPP is known to be NP-hard on undirected graphs, even for two paths). On the positive side, we show that for undirected graphs, both problems are in FPT, parameterized by neighborhood diversity. When considering the dual parameterization (graph order minus number of paths), all three variants, IPP, SPP and PP, are shown to be in FPT for undirected graphs.

Keywords: Path Partitions · NP-hardness · Parameterized Complexity · Neighborhood Diversity · Vertex Cover Parameterization

1 Introduction

Graph partitioning and graph covering problems are among the most studied problems in graph theory and algorithms. There are several types of graph partitioning and covering problems including covering the vertex set by stars (Dominating Set), covering the vertex set by cliques (Clique Covering),

F. Foucaud—Research financed by the French government IDEX-ISITE initiative 16-IDEX-0001 (CAP 20-25) and by the ANR project GRALMECO (ANR-21-CE48-0004).
K. N. R. Rao—Research funded by a DAAD-WISE Scholarship 2022.

M. Mavronicolas (Ed.): CIAC 2023, LNCS 13898, pp. 187–201, 2023.
https://doi.org/10.1007/978-3-031-30448-4_14

partitioning the vertex set by independent sets (Coloring), and covering the vertex set by paths or cycles [22]. In recent years, partitioning and covering problems by paths have received considerable attention in the literature because of their connections with well-known graph-theoretic theorems and conjectures like the Gallai-Milgram theorem and Berge's path partition conjecture. Also, these studies are motivated by applications in diverse areas such as code optimization [1], machine learning/AI [32], transportation networks [33], parallel computing [26], and program testing [25]. There are several types of paths that can be considered: unrestricted paths, induced paths, shortest paths, or directed paths (in a directed graph). A path P is an *induced* path in G if the subgraph induced by the vertices of P is a path. An induced path is also called a *chordless* path; *isometric* path and *geodesic* are other names for shortest path. Various questions related to the complexity of these path problems, even in standard graph classes, remained open for a long time (even though they have uses in various fields), a good motivation for this project.

In this paper, we mainly study the problem of partitioning the vertex set of a graph (undirected or directed) into the minimum number of disjoint *paths*, focussing on three problems, Path Partition (PP), Induced Path Partition (IPP) and Shortest Path Partition (SPP)—formal definitions are given in Sect. 2. A *path partition* (pp) of a graph G is a partitioning of the vertex set into unrestricted paths. The *path partition number* of G is the smallest size of a pp of G. Similar definitions apply to ipp and spp. PP is studied extensively under the names of Path Cover and also Hamiltonian Completion on many graph classes [4,6,16]. We give a wide range of results in this paper including complexity (NP- or W[1]-hardness) and algorithms (polynomial time or FPT); see Table 1. The types of graphs we consider are general directed, directed acyclic (DAG), general undirected and bipartite undirected graphs. We also consider some structural parameters like neighborhood diversity and vertex cover number.

The three problems considered are all NP-hard. PP can be seen as an extension of Hamiltonian Path and is thus NP-hard, even for one path. IPP is NP-hard, even for two paths [20]. Recently, SPP was proved to be NP-hard [23]. On trees, the three problems are equivalent, and are solvable in polynomial time. For a detailed survey on these types of problems (both partitioning and covering versions), see [22]. The covering version of SPP (where the paths need not necessarily be disjoint) was recently studied, see [8] for an XP algorithm, [2] for NP-hardness on chordal graphs and approximation algorithms for chordal graphs and other classes, and [32] for a $\log n$-factor approximation algorithm.

The versions of these problems where covering is not required (and the endpoints of the solution paths are prescribed in the input) are studied as Disjoint Paths (DP) [27], Disjoint Induced Paths (DIP) and Disjoint Shortest Paths (DSP) [21]. DP in particular has been extensively studied, due to its connections to the Graph Minor theorem [27]. Robertson and Seymour showed that DP is in FPT, parameterized by the number of paths [27], contrasting (D)PP. Recently, DSP was shown to have an XP algorithm and to be W[1]-hard when parameterized by the number of paths [21], solving a 40-year-old open problem.

Table 1. Summary of known results concerning path partitioning problems. The abbreviations c. and h. refer to completeness and hardness, respectively. In parentheses, we put further input specifications, with UG referring to undirected graphs. Our results are highlighted in bold face.

parameter	PP	SPP	IPP
none (UG)	NP-c. [15]	NP-c. [23]	NP-c. [20]
none (bipartite UG)	NP-c. [18,20]	**NP-c.**	open
solution size k (UG)	paraNP-h. [15]	**in XP**	paraNP-h. [20]
solution size k (DAG)	polynomial [3, Problem 26-2]	**NP-c. W[1]-h. in XP**	**NP-c. W[1]-h.**
neighborhood diversity (UG)	FPT [14]	**FPT**	**FPT**
dual $n-k$ (UG)	**FPT**	**FPT**	**FPT**

Our Contribution. Table 1 summarizes known results about the three problems, with our results are highlighted in bold. We fill in most of the hiterto open questions concerning variations of PP, SPP and IPP, e.g., we show that SPP has a poly-time algorithm for a fixed number of paths (in undirected, DAGs, and planar-directed graphs). This is surprising as both PP and IPP are NP-hard for $k = 1$ and for $k = 2$, respectively, see [20]. Many of our results concern our problems restricted to DAGs. Notice that PP has a polynomial time algorithm when restricted to DAGs (using Maximum Matching), but the complexity for IPP and SPP was open for such inputs. We show that IPP and SPP are NP-hard even when restricted to planar DAGs whose underlying graph is bipartite. We strengthen this result using a similar construction as in the classic proof of DP being W[1]-hard on DAGs by Slivkins [29], to show that IPP and SPP are W[1]-hard on DAGs. The complexity of these problems has not been studied when parameterized by structural parameters. We show that IPP and SPP both belong to FPT when parameterized by standard structural parameters like vertex cover and neighborhood diversity using ILP-techniques. Moreover, when considering the dual parameterization (graph order minus number of paths), all three variants, IPP, SPP and PP, are shown to be in FPT for undirected graphs.

It is interesting to note the differences in the complexities of the three problems on different input classes. For example, SPP can be solved in XP-time on undirected graphs, while this is not possible for the other two problems. For DAGs, PP is polynomial-time solvable, but the other two problems are NP-hard. In a way, IPP can be seen as an intermediate problem between PP and SPP. Thus it is not too surprising that, when the complexity of the three problems differs, IPP sometimes behaves like PP, and sometimes, like SPP.

The following combinatorial properties are helpful to connect the problems: (1) Every shortest path is also an induced path. (2) Every induced path of length at most two is also a shortest path. (3) In bipartite graphs, an induced path of length three is a shortest path. (4) If $G = (V, E)$ is a subgraph of H and p is a shortest path in H with only vertices of V, then p is also a shortest path in G. Proofs of statements marked with (*) are omitted due to space constraints; [10].

2 Definitions

We are using standard terminology concerning graphs, classical and parameterized complexity and we will not iterate this standard terminology here. In particular, a *path* P can be described by a sequence of non-repeated vertices such that there is an edge between vertices that are neighbors in this sequence. Sometimes, it is convenient to consider P as a set of vertices. We are next defining the problems considered in this paper. All problems can be considered on undirected or directed graphs, or also on directed acyclic graphs (DAG). We will specify this by prefixing U, D, or DAG to our problem name abbreviations.

Disjoint Paths (DP for short)

Input: A graph G, pairs of terminal vertices $\{(s_1, t_1), \dots, (s_k, t_k)\}$
Problem: Are there pairwise vertex-disjoint paths $P_1, \dots, P_k$ such that, for $1 \leq i \leq k$, the end-points of P_i are s_i and t_i?

A path P is an *induced path* in G if the induced graph $G[P]$ is a path; here, P is considered as a vertex set. Analogously to DP, we can define the problem Disjoint Induced Paths or DIP for short. A *shortest path* is a path with end-points u, v that is shortest among all paths from u to v. The greatest length of any shortest path in a graph G is also known as its *diameter*, written as diam(G). Hence, we can define the problem Disjoint Shortest Paths or DSP for short.

Remark 2.1 (∗). *The problems* DIP *and* DP *are* equivalent *when the inputs are undirected graphs or DAGs, but not for general directed graphs.*

Next, we define the corresponding partition problems. The induced and non-induced versions do no longer coincide as seen above for the set of DP problems. We say that a sub-graph G' of $G = (V, E)$ *spans* G if its vertex set is V.

Path Partition (PP for short)

Input: A graph G, a non-negative integer k
Problem: Are there pairwise vertex-disjoint paths $P_1, \dots, P_{k'}$, with $k' \leq k$, such that, together, these paths span G?

Asking for shortest or induced paths in the partition similarly gives the problems Shortest Path Partition, or SPP, and Induced Path Partition, or IPP, respectively. Contrasting Remark 2.1, the complexities of DAGPP and DAGIPP differ drastically, see Table 1. As also sketched in [12], these problems are tightly linked to Hamiltonian Completion, asking to add at most k (directed) edges to a (di)graph to guarantee the existence of a Hamiltonian path.

Omitting the vertex-disjointness condition, we arrive at *covers* instead:

Path Cover (PC for short)

Input: A graph G, a non-negative integer k
Problem: Are there paths $P_1, \dots, P_{k'}$, with $k' \leq k$, such that, together, these paths span G?

Similarly, we can define the problems SHORTEST PATH COVER (SPC) and INDUCED PATH COVER (IPC).

The *neighborhood diversity* [19] of a graph G (or just $\mathsf{nd}(G)$) is the number of equivalency classes of the following equivalence: two vertices $u, v \in V$ are equivalent (we also say that they have the *same type*) if they have the same neighborhoods except for possibly themselves, i.e., if $N(v) \setminus \{u\} = N(u) \setminus \{v\}$. The equivalence classes are all cliques, possibly of size one, but one class that collects all isolated vertices.

3 NP-Hardness Results

A well-known result related to PC in DAGs is Dilworth's theorem: the minimal size of PC equals the maximal cardinality of an anti-chain [7]. Fulkerson [13] gave a constructive proof of this theorem. Hence, the PC problem in DAGs can be reduced to a maximum matching problem in a bipartite graph. Even PP can be solved in polynomial time by reducing it to a matching problem in a bipartite graph [3, Problem 26-2]. We show that DAGSPP, DAGSPC, DAGIPP and DAGIPC are NP-hard even when restricted to planar bipartite DAGs.

Theorem 3.1. DAGSPP *is NP-hard even when the inputs are restricted to planar bipartite DAGs of maximum degree 3.*

Proof (sketch). Our reduction is adapted from [24,30]. We reduce from the PLANAR 3-DIMENSIONAL MATCHING problem, or PLANAR 3-DM, which is NP-complete (see [9]), even when each element occurs in either two or three triples. A 3-DM instance consists of three disjoint sets X, Y, Z of equal cardinality p and a set T of triples from $X \times Y \times Z$. Let $q = |T|$. The question is if there are p triples which contain all elements of X, Y and Z. We associate a bipartite graph with this instance. We assume that the four sets T, X, Y and Z are pairwise disjoint. We also assume that each element of $X \cup Y \cup Z$ belongs to at most three triples. We have a vertex for each element in X, Y, Z and each triple in T. There is an edge connecting triples to elements if and only if the element belongs to the triple. This graph G is bipartite with vertex bipartition of $T, X \cup Y \cup Z$, and has maximum degree 3. We say the instance is planar if G is planar. Given an instance of PLANAR 3-DM, $G = (T, X \cup Y \cup Z, E)$, and a planar embedding of it, we build an instance $G' = (V', E')$ of DAGSPP.

Construction: We replace each $v_i = (x, y, z) \in T$, where $x \in X$, $y \in Y$, $z \in Z$, with a gadget $H(v_i)$ that consists of 9 vertices named l^i_{jk} where $1 \leq j, k \leq 3$ and with edges as shown in Fig. 1; if the planar embedding has x, y, z in clockwise order seen as neighbors of v_i, then we add the arcs (l^i_{12}, x), (l^i_{22}, z) and (l^i_{32}, y), otherwise, we add the arcs (l^i_{12}, x), (l^i_{22}, y) and (l^i_{32}, z).

We observe the following two properties of G'.

Claim 3.2 (*). *G' is a planar DAG with maximum degree 3 in which every shortest/induced path is of length at most* 3, *and the underlying undirected graph of G' is bipartite.*

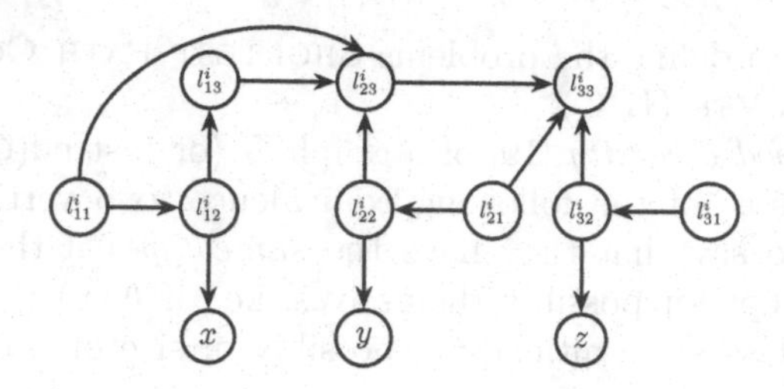

Fig. 1. The vertex gadget as defined in the proof of Theorem 3.1

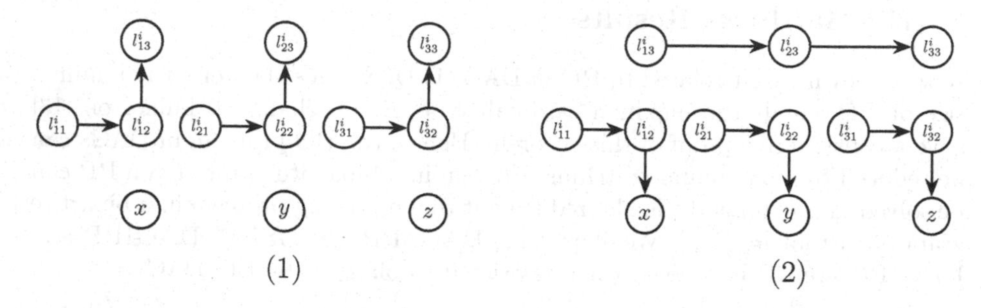

Fig. 2. Two different vertex partitions of a $H(v_i)$ gadget into 3-vertex paths, corresponding to different triple selections in the construction of Theorem 3.1.

Claim 3.3 (∗). *The* PLANAR 3-DM *instance has a solution if and only if* G' *can be partitioned into* $p + 3q$ *shortest paths.*

The intuition behind the proof of Claim 3.3 is that each shortest or induced path in a solution must contain exactly three vertices, and each gadget $H(v_i)$ is partitioned into P_3-paths in one of the two ways shown in Fig. 2. □

The proof above can also be adapted to DAGSPC, DAGIPP and DAGIPC.

Corollary 3.4. DAGSPC, DAGIPP, *and* DAGIPC *are* NP*-hard even when restricted to planar bipartite DAGs of maximum degree 3.*

Next, we prove that SPP is NP-hard even when the input graph is restricted to bipartite 5-degenerate graphs with diameter at most 4. To prove this, we reduce from 4-SPP on bipartite graphs to SPP on bipartite graphs. 4-SPP asks, given $G = (V, E)$, if there exists a partition $\mathbb{P}$ of V such that each set in $\mathbb{P}$ induces a shortest path of length 3 in G. First, we show that 4-SPP is NP-hard on bipartite graphs (Lemma 3.6) by a reduction from 4-IPP (also known as INDUCED P_4-PARTITION) on bipartite graphs. 4-IPP asks if there exists a partition $\mathbb{P}$ of V such that each set in $\mathbb{P}$ induces a path of length 3 in G.

Lemma 3.5. [24] 4-IPP *is* NP*-hard for bipartite graphs of maximum degree* 3.

Lemma 3.6 (∗). 4-SPP *is* NP*-hard for bipartite graphs of maximum degree* 3.

Theorem 3.7. SPP *is NP-hard, even for bipartite 5-degenerate graphs with diameter 4.*

Proof. To prove this claim, we use Lemma 3.6. Given an instance of 4-SPP, say, $G = (V, E)$, with bipartition $V = A \cup B$ and $|V| = 4k$, as the number of vertices must be divisible by 4, we create an instance $G' = (V', E')$ of SPP.

Construction: We add 10 new vertices to G, getting

$$V' = V \cup \{x_1, x_2, x_3, x_4, x_5, y_1, y_2, y_3, y_4, y_5\}.$$

We add edges from x_2 and x_4 to all vertices of $B \cup \{y_2, y_4\}$. Also, add edges from y_2 and y_4 to all vertices of A, add further edges to form paths $x_1x_2x_3x_4x_5$ and $y_1y_2y_3y_4y_5$. The remaining edges all stem from G. This describes E' of G'.

We have the following observations, due to the construction of G'.

Claim 3.8 (*). *$G' = (V', E')$ is bipartite and 5-degenerate.*

We can make the claimed bipartition explicit by writing $V' = A' \cup B'$, where $A' = A \cup \{x_2, x_4, y_1, y_3, y_5\}$ and $B' = B \cup \{x_1, x_3, x_5, y_2, y_4\}$.

Claim 3.9 (*). *Any shortest path of G', except for $x_1x_2x_3x_4x_5$, $x_1x_2y_2x_4x_5$, $x_1x_2y_4x_4x_5$ and $y_1y_2y_3y_4y_5$, $y_1y_2x_2y_4y_5$, $y_1y_2x_4y_4y_5$, contains at most four vertices. Hence, G' has a diameter at most 4.*

The arguments leading to the previous claim also give raise to the following one.

Claim 3.10. *The only two shortest paths in G' that have five vertices and that can simultaneously exist in a path partition are $x_1x_2x_3x_4x_5$ and $y_1y_2y_3y_4y_5$.* ◇

Notice that Claim 3.8 and Claim 3.9 together guarantee the additional properties of the constructed graph G' that have been claimed in Theorem 3.7.

Claim 3.11 (*). *Let $u, v \in V$ have distance $d < 3$ in G. Then, they also have distance d in G'. Hence, if p is a shortest path on at most three vertices in G, then p is also a shortest path in G'.*

Now, we claim that G is a Yes-instance of 4-SPP if and only if G' has a shortest path partitioning of cardinality $k' = k + 2$, where $|V| = 4k$. For the forward direction, let D be any solution of 4-SPP for G, containing k shortest paths. To construct a solution D' of SPP for the instance (G', k'), we just need to add the two paths $x_1x_2x_3x_4x_5$ and $y_1y_2y_3y_4y_5$ to D. By Claim 3.11, every path $p \in D$ is in fact a shortest path in G'. Hence, D' is a set of shortest paths with cardinality $k' = k + 2$ that covers all vertices of G'.

For the backward direction, assume G' has a solution D', where $|D'| = k' = k + 2$. As $|V'| = 4k + 10$ by construction, we know by Claim 3.9 that D' contains k paths of length three and two paths of length four. By Claim 3.10, $\{x_1x_2x_3x_4x_5, y_1y_2y_3y_4y_5\} \subseteq D'$ and the rest of the paths of D' are of length three and consists of vertices from V only. Let $D = D' \setminus \{x_1x_2x_3x_4x_5, y_1y_2y_3y_4y_5\}$. As the k paths of D are each of length three also in G (by Table 1) and as they cover V completely, D provides a solution to the 4-SPP instance G. □

The reduction above can also be used for proving the following result. (A graph is *d-degenerate* if every induced subgraph has a vertex of degree at most d).

Corollary 3.12. SPC *is NP-hard, even for bipartite 5-degenerate graphs with diameter 4.*

4 W[1]-Hardness Results

The natural or standard parameterization of a parameterized problem stemming from an optimization problem is its solution size. We will study this type of parameterization in this section for path partitioning problems. More technically speaking, we are parameterizing these problems by an upper bound on the number of paths in the partitioning. Unfortunately, our results show that for none of the variations that we consider, we can expect FPT-results.

Theorem 4.1. SPP *(parameterized by solution size) is W[1]-hard on DAGs.*

The following reduction is non-trivially adapted from [28].

Proof. We define a parameterized reduction from Clique to SPP on DAGs (both parameterized by solution size). Let (G, k) be an instance of Clique, where $k \in \mathbb{N}$ and $G = (V, E)$. We construct an equivalent instance (G', k') of the SPP problem, where G' is a DAG and $k' = \frac{k \cdot (k-1)}{2} + 3k$. Let $V = [n]$.

Overview of the Construction: We create an array of $k \times n$ identical gadgets for the construction, with each gadget representing a vertex in the original graph. We can visualize this array as having k rows and n columns. If the SPP instance (G', k') is a Yes-instance, then we show that each row has a so-called *selector* in the solution. Here, a selector is a path that traverses all but one gadget in a row, hence skipping exactly one of the gadgets. The vertices in G corresponding to the skipped gadgets form a clique of size k in G. To ensure that all selected vertices form a clique in G, we have so-called *verifiers* in SPP. Verifiers are the paths that are used for each pair of rows to ensure that the vertices corresponding to the selected gadgets in these rows are adjacent in G. This way, we also do not have to check separately that the selected vertices are distinct.

Construction Details: The array of gadgets is drawn with row numbers increasing downward and column numbers increasing to the right. Arcs between columns go down and arcs within the same row go to the right. To each row i, with $i \in [k]$, we add a *start terminal*, an arc (s_i, s_i'), and an *end terminal*, an arc (t_i', t_i). Next, add arcs starting from s_i, s_i' to t_l and t_l', with $l > i$. Also, for each row, we have $k - i$ *column start terminals*, arcs $(s_{i,j}, s_{i,j}')$ with $i < j$, and $i - 1$ *column end terminals*, arcs $(t_{j,i}, t_{j,i}')$ with $j \in [i-1]$. Also, add arcs from vertices $s_{i,j}$ and $s_{i,j}'$ to all vertices $t_{j,p}'$ with $p < j$ if $j > i$, and to t_l, t_l' if $l \geq i$.

Gadgets are denoted by $G_{i,u}$, $i \in [k]$, corresponding to $u \in V$. Each gadget $G_{i,u}$ consists of $k-1$ arcs $(a_r^{i,u}, b_r^{i,u})$, with $r \in [k] \setminus \{i\}$. To each row $i \in [k]$, we also add *dummy gadgets* $G_{i,0}$ and $G_{i,n+1}$. $G_{i,0}$ consists of two arcs, $(a_1^{i,0}, a_2^{i,0})$ and $(b_1^{i,0}, b_2^{i,0})$, also add arcs from $b_1^{i,0}$ and $b_2^{i,0}$ to t_m, t_m' and $t_{m,h}$ where $m < h$ and $i \leq$

h. $G_{i,n+1}$ consists of a directed P_{k+2} and an arc $(b_1^{i,n+1}, b_2^{i,n+1})$; the P_{k+2}-vertices are named $a_j^{i,n+1}$, with $j \in [k+2]$, where $a_{k+2}^{i,n+1}$ has out-degree zero. We can speak of $T^{i,u} := \{a_r^{i,u} \mid r \in [k] \setminus \{i\}\} \cup \{a_1^{i,0}, a_2^{i,0}, a_j^{i,n+1} \mid j \in [k+2]\}$ as being on the *top level*, while the vertices $B^{i,u} := \{b_r^{i,u} \mid r \in [k] \setminus \{i\}\} \cup \{b_1^{i,0}, b_2^{i,0}, b_1^{i,n+1}, b_2^{i,n+1}\}$ form the *bottom level* of gadget $G_{i,u}$.

Due to the natural ordering of the wires within a gadget, we can also speak of the first vertex on the top level of a gadget or that last vertex on the bottom level of a gadget. On both levels, the vertices are connected following their natural wire ordering. More technically speaking, this means that within gadget $G_{i,u}$, with $u \in V$, there is an arc from the first vertex of the upper level to the second vertex of the upper level, from the second vertex of the upper level to the third vertex of the upper level etc., up to an arc from the penultimate vertex of the upper level to the last vertex of the upper level. Moreover, there is an arc from the last vertex of the upper level of $G_{i,u-1}$ to the first vertex of the upper level of $G_{i,u}$ and from the last vertex of the upper level of $G_{i,u}$ to the first vertex of the upper level of $G_{i,u+1}$. Analogously, the vertices of the lower level of the gadgets are connected. These notions are illustrated in Fig. 3. In the figure, we omit the superscripts $(a_r, b_r) = (a_r^{i,u}, b_r^{i,u})$, where $r \in [k-i]$.

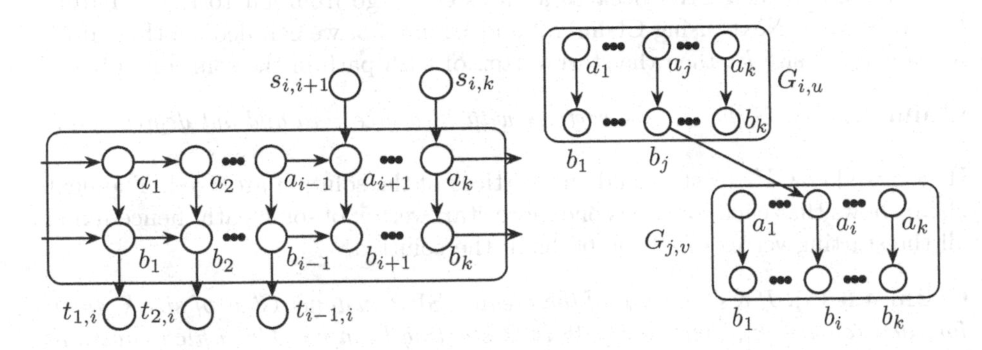

Fig. 3. Gadget $G_{i,u}$, $0 < u \leq n$, $i \in [k]$

Fig. 4. $G_{i,u}$ connected to $G_{j,v}$, $i < j$

A **selector** is a path that starts at s_i, enters its row at the top level, and exits it at the bottom level, ending at t_i and skipping exactly one gadget $G_{i,u}$, with $i \in [k]$ and $u \in V$. In order to implement this, we add the arcs $(s_i', a_1^{i,0})$ and $(b_2^{i,n+1}, t_i')$ for row i, as well as *skipping arcs* that allow to skip a gadget $G_{i,u}$ for $u \in V$. These connect the top level of the last wire of $G_{i,u-1}$ to the bottom level of the first wire of $G_{i,u+1}$.

A **verifier** is a path that routes through one of the wires of the skipped gadget and connects column terminals $s_{i,j}$ to $t_{i,j}$. In order to implement this, we add the arcs $(s_{i,j}', a_j^{i,u})$ and $(b_i^{i,u}, t_{i,j}')$ to every gadget in rows i and j. To connect the gadget in row i and j, for every edge uv in G, we add an arc between the wires $(b_j^{i,u}, a_i^{j,v})$ for each $i < j$, see Fig. 4.

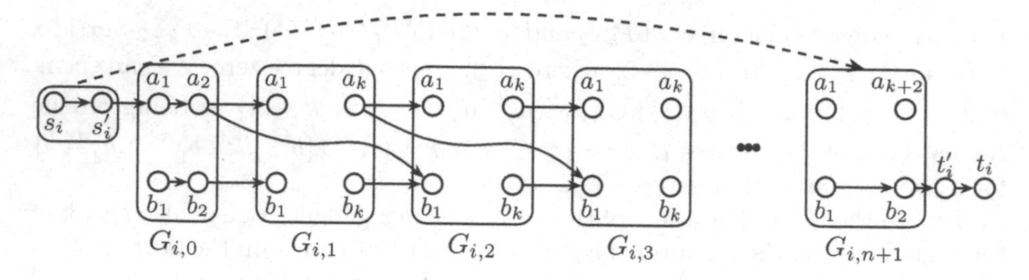

Fig. 5. The ith row (only the first two skipping arcs are shown), the dashed line indicates arcs from s_i, s_i' to $a_j^{i,n+1}$, with $j \in [k+2]$

To force the start and end vertices of paths in the spp of G' of size k', we add the following arcs, called *shortest paths enforcers*. For every row i, from all the vertices of gadget $G_{j,u}$, where $j < i$ and $0 \leq u \leq n$, from every vertex s_l and s_l' of a start terminal and from every vertex $s_{l,m}$, $s_{l,m}'$ of a column start terminal, where $l \leq i$ and $l < m$, add arcs to $a_j^{i,n+1}$, with $j \in [k+2]$ (Fig. 5).

We are now going to show a number of properties of our construction.

Observe that G' is a DAG because all arcs either go from left to right or from top to bottom. Next, using Claim 4.2 and Claim 4.3, we can deduce that, if G' has an spp of size k', then the start vertex of each path in the solution is fixed.

Claim 4.2 (∗). *G' has $k' - k$ vertices with in-degree zero and out-degree zero.*

Hence, we know $k' - k$ start and end vertices of the solution are fixed. The next claim shows that each row i has one more start vertex of some path, hence fixing all the starting vertex of all the paths in the solution.

Claim 4.3 (∗). *If a solution $\mathbb{P}$ of the created* SPP *instance G' is of size k' then, for each row $i \in [k]$, there is a path in $\mathbb{P}$ starting from $v \in T^{i,u}$ which covers at least the vertex $a_1^{i,n+1}$.*

Hence, if G' has an spp of size k', then the paths in the solution must start at: for $i \in [k]$, s_i, $b_1^{i,0}$, $v, v \in T^{i,u}$ and $s_{i,j}$, with $i < j$. Next, we will show observations about these paths' ending vertices.

Claim 4.4 (∗). *If G' has an spp of size k', then a path (in this* SPP *solution) starting at s_i has to end at t_i, with $i \in [k]$.*

With arguments along similar lines, we can show:

Claim 4.5. *If G' has an spp of size k', then any path in this partition starting at $s_{i,j}$ has to end at $t_{i,j}, i < j$.*

Claim 4.6. *If G' has an spp of size k', then any path in this partition starting at $v \in T^{i,u}$ has to end at $a_i^{i,n+1}, i \in [k]$.*

Also, if G' has an spp of size k', then for each row i, with $i \in [k]$, any path starting at s_i has to skip exactly one gadget and end at t_i. If we do not skip any gadget, $s_{i,j}$ cannot be connected to $t_{i,j}, j > i$. Once we skip a gadget, we are at the bottom level of the row and hence we cannot skip again. To conclude if G has a shortest path partition of size k' (then in the solution) the path starting at s_i has to end at t_i and this path has to skip exactly one gadget. Through this skipped gadget, $s_{i,j}$ is connected to $t_{i,j}$. The bottom of the row is covered by a path starting from $b^{i,0}$ and the top of the row is covered by a path starting at $v \in T^{i,u}$ after the skipped gadget and ending at $a_{k+2}^{i,n+1}$.

Finally, we have the following claim about G' that follows by construction.

Claim 4.7. *There exists a path between $s_{i,j}$ and $t_{i,j}$ through two gadgets $G_{i,u}$ and $G_{j,v}$ where $i > j$ if and only if there is an edge uv in the graph G.*

Claim 4.8 (*). *G has a k-clique iff G' has a shortest path partition of size k'.*

As the construction is polynomial-time, W[1]-hardness follows. □

5 XP Algorithms

We now present our XP algorithms to prove the following result. We note that the result for USPP also follows from [8] (with a different proof).

Proposition 5.1. USPP *and* DAGSPP *are in XP.*

For our XP algorithms, the following combinatorial result is crucial.

Lemma 5.2 (*). *Let $G = (V, E)$ be a directed graph. Then, V can be partitioned into k vertex-disjoint shortest paths if and only if there are k vertex-disjoint paths between some s_i and t_i, for $1 \leq i \leq k$, such that $\sum_{i=1}^{k} d(s_i, t_i) = |V| - k$. A similar characterization is true for the undirected case.*

This lemma allows us to prove Proposition 5.1 by cycling through all possible $X := \{(s_1, t_1), \ldots, (s_k, t_k)\}$, resulting in an instance of DP. Now, either apply [11, Theorem 3] for DAGs or [17,27] for undirected graphs to get the XP-result. Notice that these algorithmic results rule out paraNP-hardness results.

6 Neighborhood Diversity Parameterization

One of the standard structural parameters studied within parameterized complexity is the *vertex cover number*. As graphs with bounded vertex cover number are highly restricted, less restrictive parameters that generalize vertex cover are interesting, as *neighborhood diversity* is, introduced by Lampis [19].

Crucial to our FPT-results is the following interesting combinatorial fact.

Proposition 6.1 (*). *If G is connected, then* $\mathrm{diam}(G) \leq nd(G)$.

Similar arguments show that the number of neighborhood diversity equivalence classes that a shortest path P intersects equals its number of vertices if P contains at least four vertices. In shortest paths on two or three vertices, however, their endpoints might have the same type. Now, call two shortest paths P_i and P_j (viewed as sets of vertices) *equivalent*, denoted by $P_i \equiv P_j$, in a graph G with $d = \mathsf{nd}(G)$ if $|P_i \cap C_l| = |P_j \cap C_l|$ for all l with $1 \leq l \leq d$, where $C_1, \ldots, C_d$ denote the nd-equivalence classes. Any two equivalent shortest paths have the same length. Proposition 6.1 and our discussions imply that there are $\mathcal{O}(2^{\mathsf{nd}(G)})$ many equivalence classes of shortest paths (+).

Theorem 6.2. USPP *is FPT when parameterized by neighborhood diversity.*

Proof. As we can solve SPP separately on each connected component, we can assume in the following that the input graph is connected.

Given that the neighborhood diversity of a graph $G = (V, E)$ is bounded by an integer d, then there exists a partition of V into d nd-classes $C_1, \ldots, C_d$. Hence, each C_i for $i \in [d]$ either induces a clique or an independent set. Such a partition can be found in linear time with a fast modular decomposition algorithm [31].

Compute the set of all shortest path equivalence classes; this can be represented by a set $\mathcal{P}$ of shortest paths, in which any two shortest paths are not equivalent. By (+), we know that $|\mathcal{P}| \in \mathcal{O}(2^{\mathsf{nd}(G)})$. Construct and solve the following Integer Linear Program (ILP), with variables $z_p \geq 0$ corresponding to $p \in \mathcal{P}$. Each $p \in \mathcal{P}$ is characterized by a vector $(p^1, \ldots, p^d)$ with $p^j = |C_j \cap p|$.

$$\text{minimize} \sum_{p \in \mathcal{P}} z_p$$

$$\text{subject to} \sum_{p \in \mathcal{P}} z_p \cdot p^j = |C_j| \text{ for all } j \in [d]$$

The variable z_p encodes how many shortest paths equivalent to p are taken in the solution. Hence, the objective function expresses minimizing the number of shortest paths used in the partition. The constraints ensure the path partitioning.

Claim 6.3 (*). *There exists a spp of G with k shortest paths if and only if the objective function attains the value k in the ILP described above.*

Notice that the number of variables of the ILP is exactly $|\mathcal{P}|$, which is $\mathcal{O}(2^d)$, as observed above. Now, we apply [5, Theorem 6.5] to prove our FPT claim. □

The arguments leading to Proposition 6.1 and the subsequent discussions are all proofs by contradiction that show shortcuts in shortest paths that are 'too long'. This also works for induced paths instead of shortest paths, leading to:

Proposition 6.4. *The length of a longest induced path in a graph G is upper-bounded by the neighborhood diversity* $\mathsf{nd}(G)$.

Theorem 6.5. UIPP *is FPT when parameterized by neighborhood diversity.*

By using results of Lampis [19], we can immediately infer the following.

Corollary 6.6. USPP, UIPP *are FPT, parameterized by vertex cover number.*

We have a similar result for UPP, either by a more general approach in [14], or based on a direct reasoning. It is open if one can get a polynomial-size kernel.

Proposition 6.7 (*). UPP *is FPT when parameterized by vertex cover number, as it possesses a single-exponential size kernel.*

We can also obtain a direct FPT algorithm with running time $\mathcal{O}^*\left(\mathsf{vc}(G)! \cdot 2^{\mathsf{vc}(G)}\right)$.

7 Duals and Distance to Triviality

Given a typical graph problem that is (as a standard) parameterized by solution size k, it takes as input a graph G of order n and k, then its *dual parameter* is $k_d = n - k$. This applies in particular to our problems UPP, UIPP and USPP. As these problems always have as a trivial solution the number n of vertices (i.e., n trivial paths), we can also interpret this dual parameterization as a parameterization led by the idea of *distance from triviality.* Moreover, all our problems can be algorithmically solved for each connected component separately, so that we can assume, w.l.o.g., that we are dealing with connected graphs. Namely, if (G, k) with $G = (V, E)$ is a graph and $C \subsetneq V$ describes a connected component, then we can solve $(G[C], k')$ and $(G - C, k - k')$ independently for all $1 \leq k' < k$. We now prove that our problems, with dual parameterizations, are in FPT by providing a kernelization algorithm. The following claims are crucial.

Lemma 7.1 (*). *Let G be a graph of order n. If G has a matching that covers $2k$ vertices, then $(G, n - k)$ is a Yes-instance of* UPP, UIPP *and* USPP.

The preceding lemma has the following interesting consequence.

Corollary 7.2 (*). *If $G = (V, E)$ is a graph that possesses some $X \subseteq V$ with $|X| \geq 2k$ such that $\deg(v) \geq 2k$ for every $v \in X$, then G has a matching of size k and hence $(G, n - k)$ is a Yes-instance of* UPP, UIPP *and* USPP.

This consequence, as well as the following combinatorial observation, has no direct bearing on our algorithmic result, but may be of independent interest.

Lemma 7.3 (*). *If G is a connected graph with $\mathrm{diam}(G) > k$, then $(G, n - k)$ is a Yes-Instance of* UPP, UIPP *and* USPP.

Our combinatorial thoughts, along with Corollary 6.6 and Proposition 6.7, allow us to show the following algorithmic result, the main result of this section.

Theorem 7.4 (*). UPP, UIPP *and* USPP *can be solved in FPT time with dual parameterization.*

8 Conclusion

We have explored the algorithmic complexity of the three problems PP, IPP and SPP, and as witnessed by Table 1, our results show some interesting algorithmic differences between these three problems.

Many interesting questions remain to be investigated. For example, what is the parameterized complexity of SPP on undirected graphs, parameterized by the number of paths? Is it W[1]-hard, like for DAGs? This was asked in [8], and our W[1]-hardness result for DAGs can be seen as a first step towards an answer.

We have seen that PP, IPP and SPP admit FPT algorithms when parameterized by neighborhood diversity. Can we obtain such algorithms for other (e.g., more general) parameters (as was done for PP and modular-width in [14])?

Moreover, in the light of our FPT algorithms for the dual parameterizations of PP, IPP and SPP, we can ask whether they admit a polynomial kernel.

References

1. Boesch, F.T., Gimpel, J.F.: Covering points of a digraph with point-disjoint paths and its application to code optimization. J. ACM **24**(2), 192–198 (1977)
2. Chakraborty, D., Dailly, A., Das, S., Foucaud, F., Gahlawat, H., Ghosh, S.K.: Complexity and algorithms for isometric path cover on chordal graphs and beyond. In: Proceedings of the 33rd International Symposium on Algorithms and Computation, ISAAC 2022. LIPIcs, vol. 248, pp. 12:1–12:17. Schloss Dagstuhl - Leibniz-Zentrum für Informatik (2022)
3. Cormen, T.H., Leiserson, C.E., Rivest, R.L., Stein, C.: Introduction to Algorithms, 3rd edn. The MIT Press, Cambridge (2009)
4. Corneil, D.G., Dalton, B., Habib, M.: LDFS-based certifying algorithm for the minimum path cover problem on cocomparability graphs. SIAM J. Comput. **42**(3), 792–807 (2013)
5. Cygan, M., et al.: Parameterized Algorithms. Springer, Cham (2015). https://doi.org/10.1007/978-3-319-21275-3
6. Damaschke, P., Deogun, J.S., Kratsch, D., Steiner, G.: Finding Hamiltonian paths in cocomparability graphs using the bump number algorithm. Order **8**(4), 383–391 (1992). https://doi.org/10.1007/BF00571188
7. Dilworth, R.P.: A decomposition theorem for partially ordered sets. In: Gessel, I., Rota, G.C. (eds.) Classic Papers in Combinatorics. Modern Birkhäuser Classics, pp. 139–144. Springer, Boston (2009). https://doi.org/10.1007/978-0-8176-4842-8_10
8. Dumas, M., Foucaud, F., Perez, A., Todinca, I.: On graphs coverable by k shortest paths. In: Proceedings of the 33rd International Symposium on Algorithms and Computation, ISAAC 2022. LIPIcs, vol. 248, pp. 40:1–40:15. Schloss Dagstuhl - Leibniz-Zentrum für Informatik (2022)
9. Dyer, M.E., Frieze, A.M.: Planar 3DM is NP-complete. J. Algorithms **7**(2), 174–184 (1986)
10. Fernau, H., Foucaud, F., Mann, K., Padariya, U., Rao, K.N.R.: Parameterizing path partitions. CoRR ArXiv preprint arXiv:2212.11653 (2022)
11. Fortune, S., Hopcroft, J.E., Wyllie, J.: The directed subgraph homeomorphism problem. Theoret. Comput. Sci. **10**, 111–121 (1980)
12. Franzblau, D.S., Raychaudhuri, A.: Optimal Hamiltonian completions and path covers for trees, and a reduction to maximum flow. ANZIAM J. **44**, 193–204 (2002)

13. Fulkerson, D.R.: Note on Dilworth's decomposition theorem for partially ordered sets. Proc. AMS **7**(4), 701–702 (1956)
14. Gajarský, J., Lampis, M., Ordyniak, S.: Parameterized algorithms for modular-width. In: Gutin, G., Szeider, S. (eds.) IPEC 2013. LNCS, vol. 8246, pp. 163–176. Springer, Cham (2013). https://doi.org/10.1007/978-3-319-03898-8_15
15. Garey, M.R., Johnson, D.S.: Computers and Intractability. Freeman, Dallas (1979)
16. Goodman, S., Hedetniemi, S.: On the Hamiltonian completion problem. In: Bari, R.A., Harary, F. (eds.) Graphs and Combinatorics. Lecture Notes in Mathematics, vol. 406, pp. 262–272. Springer, Heidelberg (1974). https://doi.org/10.1007/BFb0066448
17. Kawarabayashi, K.I., Kobayashi, Y., Reed, B.: The disjoint paths problem in quadratic time. J. Combin. Theory Ser. B **102**(2), 424–435 (2012)
18. Krishnamoorthy, M.S.: An NP-hard problem in bipartite graphs. ACM SIGACT News **7**(1), 26 (1975)
19. Lampis, M.: Algorithmic meta-theorems for restrictions of treewidth. Algorithmica **64**(1), 19–37 (2012). https://doi.org/10.1007/s00453-011-9554-x
20. Le, H.O., Le, V.B., Müller, H.: Splitting a graph into disjoint induced paths or cycles. Discret. Appl. Math. **131**(1), 199–212 (2003)
21. Lochet, W.: A polynomial time algorithm for the k-disjoint shortest paths problem. In: Proceedings of the 2021 ACM-SIAM Symposium on Discrete Algorithms (SODA), pp. 169–178. SIAM (2021)
22. Manuel, P.D.: Revisiting path-type covering and partitioning problems. CoRR ArXiv preprint arXiv:1807.10613 (2018)
23. Manuel, P.D.: On the isometric path partition problem. Discuss. Math. Graph Theory **41**(4), 1077–1089 (2021)
24. Monnot, J., Toulouse, S.: The path partition problem and related problems in bipartite graphs. Oper. Res. Lett. **35**(5), 677–684 (2007)
25. Ntafos, S., Hakimi, S.: On path cover problems in digraphs and applications to program testing. IEEE Trans. Softw. Eng. **SE-5**(5), 520–529 (1979)
26. Pinter, S.S., Wolfstahl, Y.: On mapping processes to processors in distributed systems. Int. J. Parallel Prog. **16**(1), 1–15 (1987). https://doi.org/10.1007/BF01408172
27. Robertson, N., Seymour, P.: Graph minors XIII. The disjoint paths problem. J. Combin. Theory Ser. B **63**(1), 65–110 (1995)
28. Schrijver, A.: Finding k disjoint paths in a directed planar graph. SIAM J. Comput. **23**(4), 780–788 (1994)
29. Slivkins, A.: Parameterized tractability of edge-disjoint paths on directed acyclic graphs. SIAM J. Discrete Math. **24**(1), 146–157 (2010)
30. Steiner, G.: On the k-path partition of graphs. Theoret. Comput. Sci. **290**(3), 2147–2155 (2003)
31. Tedder, M., Corneil, D., Habib, M., Paul, C.: Simpler linear-time modular decomposition via recursive factorizing permutations. In: Aceto, L., Damgård, I., Goldberg, L.A., Halldórsson, M.M., Ingólfsdóttir, A., Walukiewicz, I. (eds.) ICALP 2008, Part I. LNCS, vol. 5125, pp. 634–645. Springer, Heidelberg (2008). https://doi.org/10.1007/978-3-540-70575-8_52
32. Thiessen, M., Gaertner, T.: Active learning of convex halfspaces on graphs. In: Ranzato, M., Beygelzimer, A., Dauphin, Y., Liang, P., Vaughan, J.W. (eds.) Proceedings of the 35th Conference on Neural Information Processing Systems, NeurIPS 2021, vol. 34, pp. 23413–23425. Curran Associates, Inc. (2021)
33. Wang, C., et al.: Optimizing cross-line dispatching for minimum electric bus fleet. IEEE Trans. Mob. Comput. **22**(4) (2023)

Maintaining Triconnected Components Under Node Expansion

Simon D. Fink(✉) and Ignaz Rutter

Faculty of Informatics and Mathematics, University of Passau, Passau, Germany
{finksim,rutter}@fim.uni-passau.de

Abstract. SPQR-trees model the decomposition of a graph into triconnected components. In this paper, we study the problem of dynamically maintaining an SPQR-tree while expanding vertices into arbitrary biconnected graphs. This allows us to efficiently merge two SPQR-trees by identifying the edges incident to two vertices with each other. We do this working along an axiomatic definition lifting the SPQR-tree to a stand-alone data structure that can be modified independently from the graph it might have been derived from. Making changes to this structure, we can now observe how the graph represented by the SPQR-tree changes, instead of having to reason which updates to the SPQR-tree are necessary after a change to the represented graph.

Using efficient expansions and merges allows us to improve the runtime of the Synchronized Planarity algorithm by Bläsius et al. [2] from $O(m^2)$ to $O(m \cdot \Delta)$, where Δ is the maximum pipe degree. This also reduces the time for solving several constrained planarity problems, e.g. for Clustered Planarity from $O((n+d)^2)$ to $O(n+d \cdot \Delta)$, where d is the total number of crossings between cluster borders and edges and Δ is the maximum number of edge crossings on a single cluster border.

Keywords: SPQR-Tree · Dynamic Algorithm · Cluster Planarity

1 Introduction

The SPQR-tree is a data structure that represents the decomposition of a graph at its *separation pairs*, that is the pairs of vertices whose removal disconnects the graph. The components obtained by this decomposition are called *skeletons*. SPQR-trees form a central component of many graph visualization techniques and are used for, e.g., planarity testing and variations thereof [6,11] and for computing embeddings and layouts [10,13]. Initially, SPQR-trees were devised by Di Battista and Tamassia for incremental planarity testing [6]. Their use was quickly expanded to other on-line problems [5] and to the fully-dynamic setting, that is allowing insertion and deletion of vertices and edges in $O(\sqrt{n})$ time [7], where n is the number of vertices in the graph.

Funded by DFG-grant RU-1903/3-1.

M. Mavronicolas (Ed.): CIAC 2023, LNCS 13898, pp. 202–216, 2023.
https://doi.org/10.1007/978-3-031-30448-4_15

In this paper, we consider an incremental setting where we allow a single operation that expands a vertex v into an arbitrary biconnected graph G_ν. The approach of Eppstein et al. [7] allows this in $O((\deg(v)+|G_\nu|)\cdot\sqrt{n})$ time by only representing parts of triconnected components.[1] We improve this to $O(\deg(v)+|G_\nu|)$ using an algorithm that is much simpler and explicitly yields full triconnected components, which will become important for our applications later. In addition, our approach also allows to efficiently merge two SPQR-trees as follows. Given two biconnected graphs G_1, G_2 containing vertices v_1, v_2, respectively, together with a bijection between their incident edges, we construct a new graph G by replacing v_1 with $G_2 - v_2$ in G_1, identifying edges using the given bijection. Given the SPQR-trees of G_1 and G_2, we show that the SPQR-tree of G can be found in $O(\deg(v_1))$ time. More specifically, we present a data structure that supports the following operations: $\texttt{InsertGraph}_\texttt{SPQR}$ expands a single vertex in time linear in the size of the expanded subgraph, $\texttt{Merge}_\texttt{SPQR}$ merges two SPQR-trees in time linear in the degree of the replaced vertices, `IsPlanar` indicates whether the currently represented graph is planar in constant time, and `Rotation` yields one of the two possible planar rotations of a vertex in a triconnected skeleton in constant time. Furthermore, our data structure can be adapted to yield consistent planar embeddings for all triconnected skeletons and to test for the existence of three distinct paths between two arbitrary vertices with an additional factor of $\alpha(n)$ for all operations, where α is the inverse Ackermann function.

The main idea of our approach is that the subtree of the SPQR-tree affected by expanding a vertex v has size linear in the degree of v, but may contain arbitrarily large skeletons. In a "non-normalized" version of an SPQR-tree, the affected cycle ('S') skeletons can easily be split to have a constant size, while we develop a custom splitting operation to limit the size of triconnected 'R' skeletons. This limits the size of the affected structure to be linear in the degree of v and allows us to perform the expansion efficiently.

In addition to the description of this data structure, the technical contribution of this paper is twofold: First, we develop an axiomatic definition of the decomposition at separation pairs, putting the SPQR-tree as "mechanical" data structure into focus instead of relying on and working along a given graph structure. As a result, we can deduce the represented graph from the data structure instead of computing the data structure from the graph. This allows us to make more or less arbitrary changes to the data structure (respecting its consistency criteria) and observe how the graph changes, instead of having to reason which changes to the graph require which updates to the data structure.

Second, we explain how our data structure can be used to improve the runtime of the algorithm by Bläsius et al. [2] for solving SYNCHRONIZED PLANARITY from $O(m^2)$ to $O(m\cdot\Delta)$, where Δ is the maximum pipe degree (i.e. the maximum degree of a vertex with synchronization constraints that enforce its rotation to be the same as that of another vertex). SYNCHRONIZED PLANARITY can be used

[1] Unfortunately, the recent improvements by Holm and Rotenberg are not applicable here, as they maintain triconnectivity in an only incremental setting [12], while maintaining only planarity information in the fully-dynamic setting [11].

to model and solve a vast class of different kinds of constrained planarity. Among them is the notorious CLUSTERED PLANARITY, whose complexity was open for 30 years before Fulek and Tóth gave an algorithm with runtime $O((n+d)^8)$ in 2019 [9], where d is the total number of crossings between cluster borders and edges. Shortly thereafter, Bläsius et al. [2] gave a solution in $O((n+d)^2)$ time. We improve this to $O(n + d \cdot \Delta)$, where Δ is the maximum number of edge crossings on a single cluster border.

This work is structured as follows. After preliminaries in Sect. 2, we describe the skeleton decomposition and show how it relates to the SPQR-tree in Sect. 3. Section 4 extends this data structure by the capability of splitting triconnected components. In Sect. 5, we use this to ensure the affected part of the SPQR-tree is small when we replace a vertex with a new graph. Section 6 shows how we reduce the runtime for solving SYNCHRONIZED and CLUSTERED PLANARITY. Omitted proofs and more details on SYNCHRONIZED and CLUSTERED PLANARITY can be found in the full version [8].

2 Preliminaries

In the context of this work, $G = (V, E)$ is a (usually biconnected and loop-free) multi-graph with n vertices V and m (possibly parallel) edges E. For a vertex v, we denote its open neighborhood (excluding v itself) by $N(v)$. For a bijection or matching ϕ we call $\phi(x)$ the *partner* of an element x. We use $A \uplus B$ to denote the union of two disjoint sets A, B. A separating k-set is a set of k vertices whose removal increases the number of connected components. Separating 1-sets are called *cutvertices*, while separating 2-sets are called *separation pairs*. A connected graph is *biconnected* if it does not have a cutvertex. A biconnected graph is *triconnected* if it does not have a separation pair. Maximal biconnected subgraphs are called *blocks*. Each separation pair divides the graph into *bridges*, the maximal subgraphs which cannot be disconnected by removing or splitting the vertices of the separation pair. A *bond* is a graph that consists solely of two *pole* vertices connected by multiple parallel edges, a *polygon* is a simple cycle, while a *rigid* is any simple triconnected graph. A *wheel* is a cycle with an additional central vertex connected to all other vertices.

Finally, the *expansion* that is central to this work is formally defined as follows. Let G_α, G_β be two graphs where G_α contains a vertex u and G_β contains $|N(u)|$ marked vertices, together with a bijection ϕ between the neighbors of u and the marked vertices in G_β. With $G_\alpha[u \rightarrow_\phi G_\beta]$ we denote the graph that is obtained from the disjoint union of G_α, G_β by identifying each neighbor x of u with its respective marked vertex $\phi(x)$ in G_β and removing u, i.e. the graph G_α where the vertex u was expanded into G_β; see Fig. 3 for an example.

3 Skeleton Decompositions

A *skeleton structure* $\mathcal{S} = (\mathcal{G}, \text{origV}, \text{origE}, \text{twinE})$ that *represents* a graph $G_\mathcal{S} = (V, E)$ consists of a set $\mathcal{G}$ of disjoint *skeleton* graphs together with three total, surjective mappings twinE, origE, and origV that satisfy the following conditions:

- Each skeleton $G_\mu = (V_\mu, E_\mu^{\text{real}} \cup E_\mu^{\text{virt}})$ in $\mathcal{G}$ is a multi-graph where each edge is either in E_μ^{real} and thus called *real* or in E_μ^{virt} and thus called *virtual*.
- Bijection twinE : $E^{\text{virt}} \to E^{\text{virt}}$ matches all virtual edges $E^{\text{virt}} = \bigcup_\mu E_\mu^{\text{virt}}$ such that $\text{twinE}(e) \neq e$ and $\text{twinE}^2 = \text{id}$.
- Surjection origV : $\bigcup_\mu V_\mu \to V$ maps all skeleton vertices to graph vertices.
- Bijection origE : $\bigcup_\mu E_\mu^{\text{real}} \to E$ maps all real edges to the graph edge set E.

Note that each vertex and each edge of each skeleton is in the domain of exactly one of the three mappings. As the mappings are surjective, V and E are exactly the images of origV and origE. For each vertex $v \in G_\mathcal{S}$, the skeletons that contain an *allocation vertex* v' with $\text{origV}(v') = v$ are called the *allocation skeletons* of v. Furthermore, let $T_\mathcal{S}$ be the graph where each node μ corresponds to a skeleton G_μ of $\mathcal{G}$. Two nodes of $T_\mathcal{S}$ are adjacent if their skeletons contain a pair of virtual edges matched with each other. We call a skeleton structure a *skeleton decomposition* if it satisfies the following conditions:

1 (bicon) Each skeleton is biconnected.
2 (tree) Graph $T_\mathcal{S}$ is simple, loop-free, connected and acyclic, i.e., a tree.
3 (orig-inj) For each skeleton G_μ, the restriction $\text{origV}\,|_{V_\mu}$ is injective.
4 (orig-real) For each real edge uv, the endpoints of $\text{origE}(uv)$ are $\text{origV}(u)$ and $\text{origV}(v)$.
5 (orig-virt) Let uv and $u'v'$ be two virtual edges with $uv = \text{twinE}(u'v')$. For their respective skeletons G_μ and G'_μ (where μ and μ' are adjacent in $T_\mathcal{S}$), it is $\text{origV}(V_\mu) \cap \text{origV}(V_{\mu'}) = \text{origV}(\{u, v\}) = \text{origV}(\{u', v'\})$.
6 (subgraph) The allocation skeletons of any vertex of $G_\mathcal{S}$ form a connected subgraph of $T_\mathcal{S}$.

Figure 1 shows an example of $\mathcal{S}$, $G_\mathcal{S}$, and $T_\mathcal{S}$. We call a skeleton decomposition with only one skeleton G_μ *trivial*. In this case, G_μ is isomorphic to $G_\mathcal{S}$, and origE and origV are actually bijections between the edges and vertices of both graphs.

To model the decomposition into triconnected components, we define the operations `SplitSeparationPair` and its converse, `JoinSeparationPair`, on a skeleton decomposition $\mathcal{S} = (\mathcal{G}, \text{origV}, \text{origE}, \text{twinE})$. For `SplitSeparationPair`, let u, v be a separation pair of skeleton G_μ and let (A, B) be a non-trivial bipartition of the bridges between u and v.[2] Applying `SplitSeparationPair`$(\mathcal{S}, (u, v), (A, B))$ yields skeleton decomposition $\mathcal{S}' = (\mathcal{G}', \text{origV}', \text{origE}', \text{twinE}')$ as follows. In $\mathcal{G}'$, we replace G_μ by two skeletons G_α, G_β, where G_α is obtained from $G_\mu[A]$ by adding a new virtual edge e_α between u and v. The same respectively applies to G_β with $G_\mu[B]$ and e_β. We set $\text{twinE}'(e_\alpha) = e_\beta$ and $\text{twinE}'(e_\beta) = e_\alpha$. Note that origV maps the endpoints of e_α and e_β to the same vertices. All other skeletons and their mappings remain unchanged.

[2] Note that a bridge might consist out of a single edge between u and v and that each bridge includes the vertices u and v.

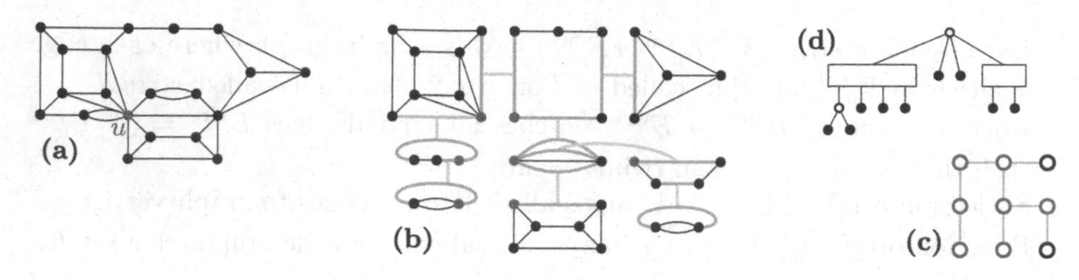

Fig. 1. Different views on the skeleton decomposition $\mathcal{S}$. **(a)** The graph $G_{\mathcal{S}}$ with a vertex u marked in blue. **(b)** The skeletons of $\mathcal{G}$. Virtual edges are drawn in gray with their matching twinE being shown in orange. The allocation vertices of u are marked in blue. **(c)** The tree $T_{\mathcal{S}}$. The allocation skeletons of u are marked in blue. **(d)** The embedding tree of vertex u as described in Sect. 6. P-nodes are shown as white disks, Q-nodes are shown as large rectangles. The leaves of the embedding tree correspond to the edges incident to u.

For `JoinSeparationPair`, consider virtual edges e_α, e_β with $\text{twinE}(e_\alpha) = e_\beta$ and let $G_\beta \neq G_\alpha$ be their respective skeletons. Applying `JoinSeparationPair` $(\mathcal{S}, e_\alpha)$ yields a skeleton decomposition $\mathcal{S}' = (\mathcal{G}', \text{origV}', \text{origE}', \text{twinE}')$ as follows. In $\mathcal{G}'$, we merge G_α with G_β to form a new skeleton G_μ by identifying the endpoints of e_α and e_β that map to the same vertex of $G_{\mathcal{S}}$. Additionally, we remove e_α and e_β. All other skeletons and their mappings remain unchanged.

The main feature of both operations is that they leave the graph represented by the skeleton decomposition unaffected while splitting a node or contracting and edge in $T_{\mathcal{S}}$, which can be verified by checking the individual conditions.

Lemma 1 (*). *Applying* `SplitSeparationPair` *or* `JoinSeparationPair` *on a skeleton decomposition* $\mathcal{S}$ *yields a skeleton decomposition* $\mathcal{S}'$ *with an unchanged represented graph* $G_{\mathcal{S}'} = G_{\mathcal{S}}$.

Note that this gives us a second way of finding the represented graph by exhaustively joining all skeletons until there is only one left, the unique trivial skeleton decomposition. A key point about the skeleton decomposition and especially the operation `SplitSeparationPair` is that they model the decomposition of a graph at separation pairs. This decomposition was formalized as *SPQR-tree* by Di Battista and Tamassia [5,6] and is unique for a given graph [10]. Angelini et al. [1] describe a decomposition tree that is conceptually equivalent to our skeleton decomposition. They also present an alternative definition for the SPQR-tree as a decomposition tree satisfying further properties. We adopt this definition as follows, not requiring planarity of triconnected components and allowing virtual edges and real edges to appear within one skeleton (i.e., having leaf Q-nodes merged into their parents).

Definition 1. *A skeleton decomposition* $\mathcal{S}$ *where any skeleton in* $\mathcal{G}$ *is either a polygon, a bond, or triconnected ("rigid"), and two skeletons adjacent in* $T_{\mathcal{S}}$ *are never both polygons or both bonds, is the unique SPQR-tree of* $G_{\mathcal{S}}$.

The main difference between the well-known ideas behind decomposition trees and our skeleton decomposition is that the latter allow an axiomatic access to the decomposition at separation pairs. For the skeleton decomposition, we employ a purely functional, "mechanical" data structure instead of relying on and working along a given graph structure. In our case, the represented graph is deduced from the data structure (i.e. SPQR-tree) instead of computing the data structure from the graph.

4 Extended Skeleton Decompositions

Note that most skeletons, especially polygons and bonds, can easily be decomposed into smaller parts. The only exception to this are triconnected skeletons which cannot be split further using the operations we defined up to now. This is a problem when modifying a vertex that occurs in triconnected skeletons that may be much bigger than the direct neighborhood of the vertex. To fix this, we define a further set of operations which allow us to isolate vertices out of arbitrary triconnected components by replacing them with a ("virtual") placeholder vertex. This placeholder then points to a smaller component that contains the actual vertex, see Fig. 2. Modification of the edges incident to the placeholder is disallowed, which is why we call them "occupied".

Formally, the structures needed to keep track of the components split in this way in an *extended* skeleton decomposition $\mathcal{S} = (\mathcal{G}, \text{origV}, \text{origE}, \text{twinE}, \text{twinV})$ are defined as follows. Skeletons now have the form $G_\mu = (V_\mu \uplus V_\mu^{\text{virt}},\ E_\mu^{\text{real}} \uplus E_\mu^{\text{virt}} \uplus E_\mu^{\text{occ}})$. Bijection $\text{twinV} : V^{\text{virt}} \to V^{\text{virt}}$ matches all *virtual vertices* $V^{\text{virt}} = \bigcup_\mu V_\mu^{\text{virt}}$, such that $\text{twinV}(v) \neq v$, $\text{twinV}^2 = \text{id}$. The edges incident to virtual vertices are contained in E_μ^{occ} and thus considered *occupied*; see Fig. 2b. Similar to the virtual edges matched by twinE, any two virtual vertices matched by twinV induce an edge between their skeletons in $T_\mathcal{S}$. Condition 2 (tree) also equally applies to those edges induced by twinV, which in particular ensures that there are no parallel twinE and twinV tree edges in $T_\mathcal{S}$. Similarly, the connected subgraphs of condition 6 (subgraph) can also contain tree edges induced by twinV. All other conditions remain unchanged, but we add two further conditions to ensure that twinV is consistent:

7 (stars) For each v_α, v_β with $\text{twinV}(v_\alpha) = v_\beta$, it is $\deg(v_\alpha) = \deg(v_\beta)$. All edges incident to v_α and v_β are occupied and have distinct endpoints (except for v_α and v_β). Each occupied edge is adjacent to exactly one virtual vertex.

8 (orig-stars) Let v_α and v_β again be two virtual vertices matched with each other by twinV. For their respective skeletons G_α and G_β (where α and β are adjacent in $T_\mathcal{S}$), it is $\text{origV}(V_\alpha) \cap \text{origV}(V_\beta) = \text{origV}(N(v_\alpha)) = \text{origV}(N(v_\beta))$.

Both conditions together yield a bijection $\gamma_{v_\alpha v_\beta}$ between the neighbors of v_α and v_β, as origV is injective when restricted to a single skeleton (condition 3 (orig-inj)) and $\deg(v_\alpha) = \deg(v_\beta)$. Operations `SplitSeparationPair` and `JoinSeparationPair` can also be applied to an extended skeleton decomposition, yielding an extended skeleton decomposition without modifying twinV. To ensure that conditions 7 (stars) and 8 (orig-stars) remain unaffected by both operations, `SplitSeparationPair` can only be applied to non-virtual vertices.

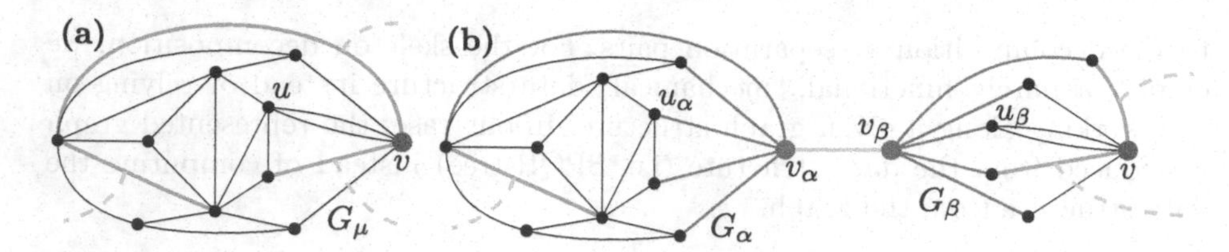

Fig. 2. **(a)** A triconnected skeleton G_μ with a highlighted vertex v incident to two gray virtual edges. **(b)** The result of applying `IsolateVertex` to isolate v out of the skeleton. The red occupied edges in the old skeleton G_α form a star with center v_α, while the red occupied edges in G_β connect all neighbors of v to form a star with center $v_\beta \neq v$. The centers v_α and v_β are virtual and matched with each other. Neighbor u of v was split into vertices u_α and u_β. (Color figure online)

The operations `IsolateVertex` and `Integrate` now allow us to isolate vertices out of triconnected components and integrate them back in, respectively. For `IsolateVertex`, let v be a non-virtual vertex of skeleton G_μ, such that v has no incident occupied edges. Applying `IsolateVertex`$(\mathcal{S}, v)$ on an extended skeleton decomposition $\mathcal{S}$ yields an extended skeleton decomposition $\mathcal{S}' = (\mathcal{G}', \text{origV}', \text{origE}', \text{twinE}', \text{twinV}')$ as follows. Each neighbor u of v is split into two non-adjacent vertices u_α and u_β, where u_β is incident to all edges connecting u with v, while u_α keeps all other edges of u. We set $\text{origV}'(u_\alpha) = \text{origV}'(u_\beta) = \text{origV}(u)$. This creates an independent, star-shaped component with center v, which we move to skeleton G_β, while we rename skeleton G_μ to G_α. We connect all u_α to a single new virtual vertex $v_\alpha \in V_\alpha^{\text{virt}}$ using occupied edges, and all u_β to a single new virtual vertex $v_\beta \in V_\beta^{\text{virt}}$ using occupied edges; see Fig. 2. Finally, we set $\text{twinV}'(v_\alpha) = v_\beta$, $\text{twinV}'(v_\beta) = v_\alpha$, and add G_β to $\mathcal{G}'$. All other mappings and skeletons remain unchanged.

For `Integrate`, consider two virtual vertices v_α, v_β with $\text{twinV}(v_\alpha) = v_\beta$ and the bijection $\gamma_{v_\alpha v_\beta}$ between the neighbors of v_α and v_β. An application of `Integrate`$(\mathcal{S}, (v_\alpha, v_\beta))$ yields an extended skeleton decomposition $\mathcal{S}' = (\mathcal{G}', \text{origV}', \text{origE}', \text{twinE}', \text{twinV}')$ as follows. We merge both skeletons into a skeleton G_μ (also replacing both in $\mathcal{G}'$) by identifying the neighbors of v_α and v_β according to $\gamma_{v_\alpha v_\beta}$. Furthermore, we remove v_α and v_β together with their incident occupied edges. All other mappings and skeletons remain unchanged.

Lemma 2 (*). *Applying* `IsolateVertex` *or* `Integrate` *on an extended skeleton decomposition* $\mathcal{S} = (\mathcal{G}, \text{origV}, \text{origE}, \text{twinE}, \text{twinV})$ *yields an extended skeleton decomposition* $\mathcal{S}' = (\mathcal{G}', \text{origV}', \text{origE}', \text{twinE}', \text{twinV}')$ *with* $G_{\mathcal{S}'} = G_{\mathcal{S}}$.

Furthermore, as `Integrate` is the converse of `IsolateVertex` and has no preconditions, any changes made by `IsolateVertex` can be undone at any time to obtain a (non-extended) skeleton decomposition, and thus possibly the SPQR-tree of the represented graph.

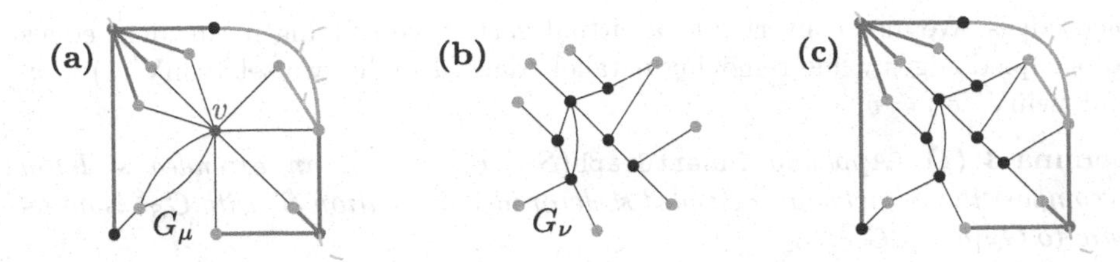

Fig. 3. Expanding a skeleton vertex v into a graph G_ν in the SPQR-tree of Fig. 4b. **(a)** The single allocation skeleton G_μ of u with the single allocation vertex v of u from Fig. 4b. The neighbors of v are marked in orange. **(b)** The inserted graph G_ν with orange marked vertices. Note that the graph is biconnected when all marked vertices are collapsed into a single vertex. **(c)** The result of applying `InsertGraph`($\mathcal{S}, u, G_\nu, \phi$) followed by an application of `Integrate` on the generated virtual vertices v and v'.

5 Node Expansion in Extended Skeleton Decompositions

We now introduce the dynamic operation that changes the represented graph by expanding a single vertex u into an arbitrary connected graph G_ν. This is done by identifying $|N(u)|$ marked vertices in G_ν with the neighbors of u via a bijection ϕ and then removing u and its incident edges. We use the "occupied stars" from the previous section to model the identification of these vertices, allowing us to defer the actual insertion to an application of `Integrate`. We need to ensure that the inserted graph makes the same "guarantees" to the surrounding graph in terms of connectivity as the vertex it replaces, that is all neighbors of u (i.e. all marked vertices in G_ν) need to be pairwise connected via paths in G_ν not using any other neighbor of u (i.e. any other marked vertex). Without this requirement, a single vertex could e.g. also be split into two non-adjacent halves, which could break a triconnected component apart. Thus, we require G_ν to be biconnected when all marked vertices are collapsed into a single vertex. Note that this also ensures that the old graph can be restored by contracting the vertices of the inserted graph. For the sake of simplicity, we require vertex u from the represented graph to have a single allocation vertex $v \in G_\mu$ with $\text{origV}^{-1}(u) = \{v\}$ so that we only need to change a single allocation skeleton G_μ in the skeleton decomposition. As we will make clear later on, this condition can be satisfied easily.

Formally, let $u \in G_\mathcal{S}$ be a vertex that only has a single allocation vertex $v \in G_\mu$ (and thus only a single allocation skeleton G_μ). Let G_ν be an arbitrary, new graph containing $|N(u)|$ marked vertices, together with a bijection ϕ between the marked vertices in G_ν and the neighbors of v in G_μ. We require G_ν to be biconnected when all marked vertices are collapsed into a single node. Operation `InsertGraph`($\mathcal{S}, u, G_\nu, \phi$) yields an extended skeleton decomposition $\mathcal{S}' = (\mathcal{G}', \text{origV}', \text{origE}', \text{twinE}', \text{twinV}')$ as follows, see also Fig. 3. We interpret G_ν as skeleton and add it to $\mathcal{G}'$. For each marked vertex x in G_ν, we set $\text{origV}'(x) = \text{origV}(\phi(x))$. For all other vertices and edges in G_ν, we set origV$'$ and origE$'$ to point to new vertices and edges forming a copy of G_ν in $\mathcal{G}_{\mathcal{S}'}$. We connect every marked vertex in G_ν to a new virtual vertex $v' \in G_\nu$ using occupied edges. We also convert v to a virtual vertex, converting its incident edges to occupied edges while removing parallel edges. Finally, we set $\text{twinV}'(v) = v'$ and $\text{twinV}'(v') = v$.

Lemma 3 (*). *Applying* `InsertGraph`$(\mathcal{S}, u, G_\nu, \phi)$ *on an extended skeleton decomposition* $\mathcal{S}$ *yields an extended skeleton decomposition* $\mathcal{S}'$ *with* $G_{\mathcal{S}'}$ *isomorphic to* $G_\mathcal{S}[u \rightarrow_\phi G_\nu]$.

On its own, this operation is not of much use though, as graph vertices only rarely have a single allocation skeleton. Furthermore, our goal is to dynamically maintain SPQR-trees, while this operation on its own will in most cases not yield an SPQR-tree. To fix this, we introduce the full procedure $\texttt{InsertGraph}_\texttt{SPQR}(\mathcal{S}, u, G_\nu, \phi)$ that can be applied to any graph vertex u and that, given an SPQR-tree $\mathcal{S}$, yields the SPQR-tree of $G_\mathcal{S}[u \rightarrow_\phi G_\nu]$. It consists of three preparations steps, the insertion of G_ν, and two further clean-up steps:

1. We apply `SplitSeparationPair` to each polygon allocation skeleton of u with more than three vertices, using the neighbors of the allocation vertex of u as separation pair.
2. For each rigid allocation skeleton of u, we move the contained allocation vertex v of u to its own skeleton by applying `IsolateVertex`$(\mathcal{S}, v)$.
3. We exhaustively apply `JoinSeparationPair` to any pair of allocation skeletons of u that are adjacent in $T_\mathcal{S}$. Due to condition 6 (subgraph), this yields a single component G_μ that is the sole allocation skeleton of u with the single allocation vertex v of u. Furthermore, the size of G_μ is linear in $\deg(u)$.
4. We apply `InsertGraph` to insert G_ν as skeleton, followed by an application of `Integrate` to the virtual vertices $\{v, v'\}$ introduced by the insertion, thus integrating G_ν into G_μ.
5. We apply `SplitSeparationPair` to all separation pairs in G_μ that do not involve a virtual vertex. These pairs can be found in linear time, e.g. by temporarily duplicating all virtual vertices and their incident edges and then computing the SPQR-tree.[3]
6. Finally, we exhaustively apply `Integrate` and also apply `JoinSeparationPair` to any two adjacent polygons and to any two adjacent bonds to obtain the SPQR-tree of the updated graph.

The basic idea behind the correctness of this procedure is that splitting the newly inserted component according to its SPQR-tree in step 5 yields biconnected components that are each either a polygon, a bond, or "almost" triconnected. The latter (and only those) might still contain virtual vertices and all their remaining separation pairs, which were not split in step 5, contain one of these virtual vertices. This, together with the fact that there still may be pairs of adjacent skeletons where both are polygons or both are bonds, prevents the instance from being an SPQR-tree. Both issues are resolved in step 6: The adjacent skeletons are obviously fixed by the `JoinSeparationPair` applications. To show that the virtual vertices are removed by the `Integrate` applications, making the remaining components triconnected, we need the following lemma.

[3] The wheels replacing virtual vertices in the proof of Theorem 2 also ensure this.

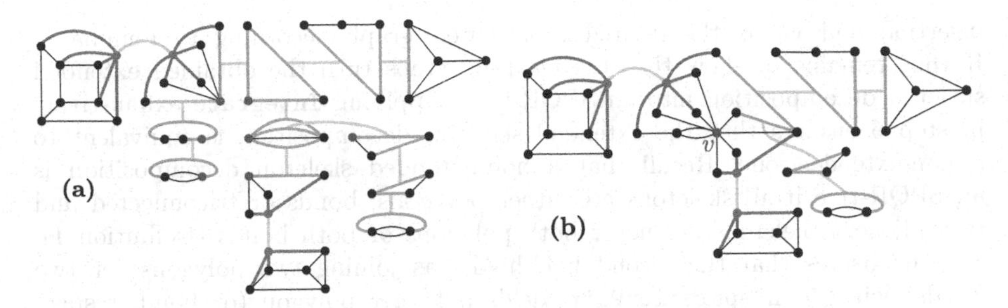

Fig. 4. The preprocessing steps of $\texttt{InsertGraph}_{\texttt{SPQR}}$ being applied to the SPQR-tree of Fig. 1b. **(a)** The state after step 2, after all allocation skeletons of u have been split. **(b)** The state after step 3, after all allocation skeletons of u have been merged into a single one.

Lemma 4. *Let G_α be a triconnected skeleton containing a virtual vertex v_α matched with a virtual vertex v_β of a biconnected skeleton G_β. Furthermore, let $P \subseteq \binom{V(G_\beta)}{2}$ be the set of all separation pairs in G_β. An application of* $\texttt{Integrate}(\mathcal{S}, (v_\alpha, v_\beta))$ *yields a biconnected skeleton G_μ with separation pairs $P' = \{\{u, v\} \in P \mid v_\beta \notin \{u, v\}\}$.*

Proof. We partition the vertices of G_μ into sets A, B, and N depending on whether the vertex stems from G_α, G_β, or both, respectively. The set N thus contains the neighbors of v_α, which were identified with the neighbors of v_β. We will show by contradiction that G_μ contains no separation pairs except for those in P'. Thus, consider a separation pair $u, v \in G_\mu$ not in P'. First, consider the case where $u, v \in A \cup N$. Observe that removing u, v in this case leaves B connected. Thus, we can contract all vertices of B into a single vertex, reobtain G_α and see that u, v is a separation pair in G_α. This contradicts the precondition that G_α is triconnected. Now consider the case where $u, v \in B \cup N$. Analogously to above, we find that u, v is a separation pair in G_β that does not contain v_β, a contradiction to $\{u, v\} \notin P'$. Finally, consider the remaining case where, without loss of generality, $u \in A, v \in B$. Since $\{u, v\}$ is a separation pair, u has two neighbors x, y that lie in different connected components of $G_\mu - \{u, v\}$ and therefore also in different components of $(G_\mu - \{u, v\}) - B$ which is isomorphic to $G_\alpha - \{u, v_\alpha\}$. This again contradicts G_α being triconnected. □

Theorem 1. *Applying* $\texttt{InsertGraph}_{\texttt{SPQR}}(\mathcal{S}, u, G_\nu, \phi)$ *to an SPQR-tree $\mathcal{S}$ yields an SPQR-tree $\mathcal{S}'$ in $O(|G_\nu|)$ time with $G_{\mathcal{S}'}$ isomorphic to $G_\mathcal{S}[u \rightarrow_\phi G_\nu]$.*

Proof. As all applied operations leave the extended skeleton decomposition valid, the final extended skeleton decomposition $\mathcal{S}'$ is also valid. The purpose of the preprocessing steps 1 to 3 is to ensure that the preconditions of `InsertGraph` are satisfied and the affected component is not too large. All rigids split in step 2 remain structurally unmodified in the sense that edges only changed their type, but the graph and especially its triconnectedness remains unchanged. Step 4 performs the actual insertion and yields the desired represented graph according to Lemma 3. It thus remains to show that the clean-up steps turn the obtained extended skeleton decomposition into an SPQR-tree. Applying `Integrate` exhaustively in step 6 ensures that the extended skeleton decomposition is equivalent to a non-extended one. Recall that a non-extended skeleton decomposition is an SPQR-tree if all skeletons are either polygons, bonds or triconnected and two adjacent skeletons are never both polygons or both bonds (Definition 1). Step 6 ensures that the second half holds, as joining two polygons (or two bonds) with `JoinSeparationPair` yields a bigger polygon (or bond, respectively). Before step 6, all skeletons that are not an allocation skeleton of u are still unmodified and thus already have a suitable structure, i.e., they are either polygons, bonds or triconnected. Furthermore, the allocation skeletons of u not containing virtual vertices also have a suitable structure, as their splits were made according to the SPQR-tree in step 5. It remains to show that the remaining skeletons, that is those resulting from the `Integrate` applications in step 6, are triconnected. Note that in these skeletons, step 5 ensures that every separation pair consists of at least one virtual vertex, as otherwise the computed SPQR-tree would have split the skeleton further. Further note that, for each of these virtual vertices, the matched partner vertex is part of a structurally unmodified triconnected skeleton that was split in step 2. Lemma 4 shows that applying `Integrate` does not introduce new separation pairs while removing two virtual vertices if one of the two sides is triconnected. We can thus exhaustively apply `Integrate` and thereby remove all virtual vertices and thus also all separation pairs, obtaining triconnected components. This shows that the criteria for being an SPQR-tree are satisfied and, as `InsertGraph` expanded u to G_ν in the represented graph, we now have the unique SPQR-tree of $G_{\mathcal{S}}[u \rightarrow_\phi G_\nu]$.

All operations we used can be performed in time linear in the degree of the vertices they are applied on. For the bipartition of bridges input to `SplitSeparationPair`, it is sufficient to describe each bridge via its edges incident to the separation pair instead of explicitly enumerating all vertices in the bridge. Thus, the applications of `SplitSeparationPair` and `IsolateVertex` in steps 1 and 2 touch every edge incident to u at most once and thus take $O(\deg(u))$ time. Furthermore, they yield skeletons that have a size linear in the degree of their respective allocation vertex of u. As the subtree of u's allocation skeletons has size at most $\deg(u)$, the `JoinSeparationPair` applications of step 3 also take at most $O(\deg(u))$ time. It follows that the resulting single allocation skeleton of u has size $O(\deg(u))$. The applications of `InsertGraph` and `Integrate` in step 4 take time linear in the number of identified neighbors, which is $O(\deg(u))$. Generating the SPQR-tree of the inserted graph in step 5 (where all virtual vertices where replaced by wheels) can be done in time linear in the size of the inserted graph [10], that is $O(|G_\nu|)$. Applying `SplitSeparationPair` according to all separation pairs identified by this SPQR-tree can also be done in $O(|G_\nu|)$ time in total. Note that there are at most $\deg(u)$ edges between the skeletons that existed before step 4 and those that were created or modified in steps 4 and 5, and these are the only edges that might now connect two polygons or two bonds.

As these tree edges have one endpoint in the single allocation skeleton of u, the applications of `Integrate` and `JoinSeparationPair` in step 6 run in $O(\deg(u))$ time in total. Furthermore, they remove all pairs of adjacent polygons and all pairs of adjacent bonds. This shows that all steps take $O(\deg(u))$ time, except for step 5, which takes $O(|G_\nu|)$ time. As the inserted graph contains at least one vertex for each neighbor of u, the total runtime is in $O(|G_\nu|)$. □

Corollary 1. *Let $\mathcal{S}_1, \mathcal{S}_2$ be two SPQR-trees together with vertices $u_1 \in G_{\mathcal{S}_1}$, $u_2 \in G_{\mathcal{S}_2}$, and let ϕ be a bijection between the edges incident to u_1 and the edges incident to u_2. Operation $\texttt{Merge}_{\texttt{SPQR}}(\mathcal{S}_1, \mathcal{S}_2, u_1, u_2, \phi)$ yields the SPQR-tree of the graph $G_{\mathcal{S}_1}[u_1 \rightarrow_\phi G_{\mathcal{S}_2} - u_2]$, i.e. the union of both graphs where the edges incident to u_1, u_2 were identified according to ϕ and u_1, u_2 removed, in time $O(\deg(u_1)) = O(\deg(u_2))$.*

Proof. Operation $\texttt{Merge}_{\texttt{SPQR}}$ works similar to the more general $\texttt{InsertGraph}_{\texttt{SPQR}}$, although the running time is better because we already know the SPQR-tree for the graph being inserted. We apply steps 1 to 3 to ensure that both u_1 and u_2 have sole allocation vertices v_1 and v_2, respectively. To properly handle parallel edges, we subdivide all edges incident to u_1, u_2 (and thus also the corresponding real edges incident to v_1, v_2) and then identify the subdivision vertices of each pair of edges matched by ϕ. By deleting vertices v_1 and v_2 and suppressing the subdivision vertices (that is, removing them and identifying each pair of incident edges) we obtain a skeleton G_μ that has size $O(\deg(u_1)) = O(\deg(u_2))$. Finally, we apply steps 5 and 6 to G_μ to obtain the final SPQR-tree. Again, as the partner vertex of every virtual vertex in the allocation skeletons of u is part of a triconnected skeleton, applying `Integrate` exhaustively in step 6 yields triconnected skeletons. As previously discussed, the preprocessing and clean-up steps run in time linear in degree of the affected vertices, thus the overall runtime is $O(\deg(u_1)) = O(\deg(u_2))$ in this case. □

5.1 Maintaining Planarity and Vertex Rotations

Note that expanding a vertex of a planar graph using another planar graph using $\texttt{InsertGraph}_{\texttt{SPQR}}$ (or merging two SPQR-trees of planar graphs using Corollary 1) might actually yield a non-planar graph. This is, e.g., because the rigids of both graphs might require incompatible orders for the neighbors of the replaced vertex. The aim of this section is to efficiently detect this case, that is a planar graph turning non-planar. To check a general graph for planarity, it suffices to check the rigids in its SPQR-tree for planarity and each rigid allows exactly two planar embeddings, where one is the reverse of the other [6]. Thus, if a graph becomes non-planar through an application of $\texttt{InsertGraph}_{\texttt{SPQR}}$, this will be noticeable from the triconnected allocation skeletons of the replaced vertex. To be able to immediately report if the instance became non-planar, we need to maintain a rotation, that is a cyclic order of all incident edges, for each vertex in any triconnected skeleton. Note that we do not track the direction of the orders, that is we only store the order up to reversal. As discussed later, the exact orders can also be maintained with a slight overhead.

Theorem 2. *SPQR-trees support the following operations:*

- $\texttt{InsertGraph}_{\texttt{SPQR}}(\mathcal{S}, u, G_\nu, \phi)$*: expansion of a single vertex u in time* $O(|G_\nu|)$,
- $\texttt{Merge}_{\texttt{SPQR}}(\mathcal{S}_1, \mathcal{S}_2, u_1, u_2, \phi)$*: merging of two SPQR-trees in time* $O(\deg(u_1))$,
- `IsPlanar`*: queries whether the represented graph is planar in time* $O(1)$*, and*
- `Rotation`(u)*: queries for one of the two possible rotations of vertices u in planar triconnected skeletons in time* $O(1)$.

Proof. Note that the flag `IsPlanar` together with the `Rotation` information can be computed in linear time when creating a new SPQR-tree and that expanding a vertex or merging two SPQR-trees cannot turn a non-planar graph planar. We make the following changes to the operations $\texttt{InsertGraph}_{\texttt{SPQR}}$ and $\texttt{Merge}_{\texttt{SPQR}}$ to maintain the new information. After a triconnected component is split in step 2 we now introduce further structure to ensure that the embedding is maintained on both sides. The occupied edges generated around the split-off vertex v (and those around its copy v') are subdivided and connected cyclically according to `Rotation`(v). Instead of "stars", we thus now generate occupied "wheels" that encode the edge ordering in the embedding of the triconnected component. When generating the SPQR-tree of the modified subgraph in step 5, we also generate a planar embedding for all its triconnected skeletons. If no planar embedding can be found for at least one skeleton, we report that the resulting instance is non-planar by setting `IsPlanar` to false. Otherwise, after performing all splits indicated by the SPQR-tree, we assign `Rotation` by generating embeddings for all new rigids. Note that for all skeletons with virtual vertices, the generated embedding will be compatible with the one of the neighboring triconnected component, that is, the rotation of each virtual vertex will line up with that of its matched partner vertex, thanks to the inserted wheel. Finally, before applying `Integrate` in step 6, we contract each occupied wheel into a single vertex to re-obtain occupied stars. The creation and contraction of wheels adds an overhead that is at most linear in the degree of the expanded vertex and the generation of embeddings for the rigids can be done in time linear in the size of the rigid. Thus, this does not affect the asymptotic runtime of both operations. □

Corollary 2 (∗). *The data structure from Theorem 2 can be adapted to also provide the exact rotations with matching direction for every vertex in a rigid. Furthermore, it can support queries whether two vertices* v_1, v_2 *are connected by at least 3 different vertex-disjoint paths via* `3Paths`(v_1, v_2) *in* $O((\deg(v_1) + \deg(v_2)) \cdot \alpha(n))$ *time. These adaptions change the runtime of* $\texttt{InsertGraph}_{\texttt{SPQR}}$ *to* $O(\deg(u) \cdot \alpha(n) + |G_\nu|)$*, that of* $\texttt{Merge}_{\texttt{SPQR}}$ *to* $O(\deg(u_1) \cdot \alpha(n))$*, and that of* `Rotation`$(u)$ *to* $O(\alpha(n))$.

6 Application to Synchronized Planarity

We show how extended skeleton decompositions and their dynamic operation $\texttt{InsertGraph}_{\texttt{SPQR}}$ can be used to improve the runtime of the algorithm for solving SYNCHRONIZED PLANARITY by Bläsius et al. [2] from $O(m^2)$ to $O(m \cdot \Delta)$, where Δ is the maximum pipe degree. The algorithm spends a major part of its runtime on computing so-called embedding trees, which describe all possible rotations of a single vertex in a planar graph and are used to communicate embedding restrictions between vertices with synchronized rotation. Once the embedding trees are available, the at most $O(m)$ executed operations run in time linear in the degree of the pipe/vertex they are applied on, that is in $O(\Delta)$ [2]. Thus, being able to generate these embedding trees efficiently by maintaining the SPQR-trees they are derived from is our main contribution towards the speedup of the SYNCHRONIZED PLANARITY algorithm. See the full version [8] for more details on the problems SYNCHRONIZED and CLUSTERED PLANARITY. There, we also give a short overview over the operations Bläsius et al. [2] use for solving SYNCHRONIZED PLANARITY, which we improve in the proof of Theorem 3.

An *embedding tree* $\mathcal{T}_v$ for a vertex v of a biconnected graph G describes the possible cyclic orderings or *rotations* of the edges incident to v in all planar embeddings of G [4]. The leaves of $\mathcal{T}_v$ are the edges incident to v, while its inner nodes are partitioned into two categories: *Q-nodes* define an up-to-reversal fixed rotation of their incident tree edges, while *P-nodes* allow arbitrary rotation; see Fig. 1d. To generate the embedding tree we use the observation about the relationship of SPQR-trees and embedding trees described by Bläsius and Rutter [3, Section 2.5]: there is a bijection between the P- and Q-nodes in the embedding tree of v and the bond and triconnected allocation skeletons of v in the SPQR-tree of G, respectively. Note that the detailed constructions for the following statements are given in the respective proofs the full version [8].

Lemma 5 (∗). *Let $\mathcal{S}$ be an SPQR-tree with a planar represented graph $G_\mathcal{S}$. The embedding tree for a vertex $v \in G_\mathcal{S}$ can be found in time $O(\deg(v))$.*

This can now be used to reduce the runtime of solving SYNCHRONIZED PLANARITY by generating an SPQR-tree upfront, maintaining it throughout all applied operations, and deriving any needed embedding tree from the SPQR-tree.

Theorem 3 (∗). SYNCHRONIZED PLANARITY *can be solved in time in $O(m \cdot \Delta)$, where m is the number of edges and Δ is the maximum degree of a pipe.*

Proof (Sketch). See the full version [8] for more background on the SYNCHRONIZED PLANARITY operations modified in the following. Operation `PropagatePQ` expands a vertex into a tree corresponding to the embedding tree of its partner vertex with synchronized rotation. This expansion can also be done in the SPQR-tree without a runtime overhead, while some care needs to be taken when expanding cut-vertices, as different parts of the tree need to be expanded in different blocks. Operation `EncapsulateAndJoin` generates a new bipartite component linear in size to the pipe it removes. Thus, the SPQR-tree for this new component can be computed without a runtime overhead. All other operations do not affect the SPQR-tree and once embedding trees are available, of the at most $O(m)$ applied operations, each takes $O(\Delta)$ time [2]. □

Corollary 3 (*). Clustered Planarity *can be solved in time in* $O(n+d\cdot\Delta)$, *where* d *is the total number of crossings between cluster borders and edges and* Δ *is the maximum number of edge crossings on a single cluster border.*

References

1. Angelini, P., Bläsius, T., Rutter, I.: Testing mutual duality of planar graphs. Int. J. Comput. Geom. Appl. **24**(4), 325–346 (2014). https://doi.org/10.1142/S0218195914600103
2. Bläsius, T., Fink, S.D., Rutter, I.: Synchronized planarity with applications to constrained planarity problems. In: Proceedings of the 29th Annual European Symposium on Algorithms (ESA'21). LIPIcs, vol. 204, pp. 19:1–19:14 (2021). https://doi.org/10.4230/LIPIcs.ESA.2021.19
3. Bläsius, T., Rutter, I.: Simultaneous PQ-ordering with applications to constrained embedding problems. ACM Trans. Algorithms **12**(2), 16:1–16:46 (2016). https://doi.org/10.1145/2738054
4. Booth, K.S., Lueker, G.S.: Testing for the consecutive ones property, interval graphs, and graph planarity using PQ-tree algorithms. J. Comput. Syst. Sci. **13**(3), 335–379 (1976). https://doi.org/10.1016/s0022-0000(76)80045-1
5. Di Battista, G., Tamassia, R.: On-line maintenance of triconnected components with SPQR-trees. Algorithmica **15**(4), 302–318 (1996). https://doi.org/10.1007/bf01961541
6. Di Battista, G., Tamassia, R.: On-line planarity testing. SIAM J. Comput. **25**(5), 956–997 (1996). https://doi.org/10.1137/s0097539794280736
7. Eppstein, D., Galil, Z., Italiano, G.F., Spencer, T.H.: Separator based sparsification. J. Comput. Syst. Sci. **52**(1), 3–27 (1996). https://doi.org/10.1006/jcss.1996.0002
8. Fink, S.D., Rutter, I.: Maintaining triconnected components under node expansion (2023). https://arxiv.org/abs/2301.03972
9. Fulek, R., Tóth, C.D.: Atomic embeddability, clustered planarity, and thickenability. J. ACM **69**(2), 13:1–13:34 (2022). https://doi.org/10.1145/3502264
10. Gutwenger, C.: Application of SPQR-trees in the planarization approach for drawing graphs. Ph.D. thesis (2010). https://eldorado.tu-dortmund.de/bitstream/2003/27430/1/diss_gutwenger.pdf
11. Holm, J., Rotenberg, E.: Fully-dynamic planarity testing in polylogarithmic time. In: Makarychev, K., Makarychev, Y., Tulsiani, M., Kamath, G., Chuzhoy, J. (eds.) Proceedings of the 52nd Annual ACM SIGACT Symposium on Theory of Computing (STOC 2020). vol. abs/1911.03449, pp. 167–180. ACM (2020). https://doi.org/10.1145/3357713.3384249
12. Holm, J., Rotenberg, E.: Worst-case polylog incremental SPQR-trees: embeddings, planarity, and triconnectivity. In: Proceedings of the 31st Annual ACM-SIAM Symposium on Discrete Algorithms (SODA 2020), pp. 2378–2397. SIAM (2020). https://doi.org/10.1137/1.9781611975994.146
13. Weiskircher, R.: New applications of SPQR-trees in graph drawing. Ph.D. thesis, Universität des Saarlandes (2002). https://doi.org/10.22028/D291-25752

Approximating Power Node-Deletion Problems

Toshihiro Fujito(✉), Kneto Mukae, and Junya Tsuzuki

Department of Computer Science and Engineering, Toyohashi University of Technology, Toyohashi 441-8580, Japan
fujito@tut.jp

Abstract. In the Power Vertex Cover *(PVC)* problem introduced in [1] as a generalization of the well-known Vertex Cover, we are allowed to specify costs for covering edges in a graph individually. Namely, two weights, $w(u, v)$ and $w(v, u)$, are associated with each "edge" $\{u, v\} \in E$ of an input graph $G = (V, E)$, and to cover an edge $\{u, v\}$, it is required to assign "power" $p \in \mathbb{R}^V$ on vertices of G s.t. either $p(u) \geq w(u, v)$ or $p(v) \geq w(v, u)$. The objective is to minimize the total power assigned on V, $\sum_{v \in V} p(v)$, while covering all the edges of G by p.

The *node-deletion* problem for a graph property π is the problem of computing a vertex subset $C \subseteq V$ of minimum weight, given a graph $G = (V, E)$, s.t. the graph satisfies π when all the vertices in C are removed from G. In this paper we consider node-deletion problems extended with the "covering-by-power" condition as in PVC, and present a unified approach for effectively approximating them. The node-deletion problems considered are Partial Vertex Cover (PartVC), Bounded Degree Deletion (BDD), and Feedback Vertex Set (FVS), each corresponding to graph properties π = "the graph has at most $|E| - k$ edges", π = "vertex degree of v is no larger than $b(v)$", and π = "the graph is acyclic", respectively. After reducing these problems to the Submodular Set Cover (SSC) problem, we conduct an extended analysis of the approximability of these problems in the new setting of power covering by applying some of the existing techniques for approximating SSC. It will be shown that 1) PPartVC can be approximated within a factor of 2, 2) PBDD for $b \in \mathbb{Z}_+^V$ within $\max\{2, 1 + b_{\max}\}$, where $b_{\max} = \max_{v \in V} b(v)$, or within $2 + \log b_{\max}$ (for $b_{\max} \geq 1$) by a combination of the greedy SSC algorithm and the local ratio method extended for power node-deletion problems, and 3) PFVS within 2, resulting in each of these bounds matching the best one known for the corresponding original problem.

Keywords: power cover · vertex cover · node-deletion problems · submodular set cover

1 Introduction

The Vertex Cover (VC) problem is one of the most well-known NP-hard graph problems. Given as an input is an undirected graph $G = (V, E)$, and it

Supported in part by JSPS KAKENHI under Grant Number 20K11676.

M. Mavronicolas (Ed.): CIAC 2023, LNCS 13898, pp. 217–231, 2023.
https://doi.org/10.1007/978-3-031-30448-4_16

is required to compute a minimum vertex subset $C \subseteq V$ s.t. every edge of G is "covered" by some vertex in C. In the "weighted" version, the vertices of G are associated with weights $w \in \mathbb{R}^V$, and subject to minimize is the sum of weights of vertices in a vertex cover. The decision version is one of Karp's NP-complete problems [25], and various generalizations have been introduced and subjected for further study, e.g., [12,20,22] just to mention a few. In such a basic model as VERTEX COVER, no distinction is made among the edges $e \in \delta(v) \subseteq E$ all incident to one vertex $v \in V$ as far as covering e by v is concerned (Note: $\delta(v)$ is the set of edges incident to v); that is, only one of two is possible, whether v covers all in $\delta(v)$ (by having v in C) or none (by not having v in C). In a more realistic setting though, the edges in $\delta(v)$ may have different characteristics so that some are harder to cover from v than others. POWER VERTEX COVER *(PVC)* introduced in [1], is such a model reflecting more refined conditions: here, each edge $\{u, v\}$ is a priori associated with a weight $w(\{u, v\})$, and it can be covered "from u" if and only if $u \in V$ has enough power $p(u)$, that is, $p(u) \geq w(\{u, v\})$, and can be "from v" if and only if $v \in V$ has enough power $p(v)$ of at least $w(\{u, v\})$. Furthermore, it could be the case that covering $\{u, v\}$ from u is costlier than doing so from v, or vice versa. To deal with such a more general case, (u, v) and (v, u) for $\{u, v\} \in E$ are distinguished so that it costs $w(u, v)$ to cover $\{u, v\}$ from u while it costs $w(v, u)$ when covering from v. POWER VERTEX COVER is such a problem of computing a "power assignment" $p : V \rightarrow \mathbb{R}$ minimizing $\sum_{v \in V} p(v)$, given $G = (V, E)$ and $w(u, v), w(v, u)$ for each $\{u, v\} \in E$, such that every edge in G is covered by p; that is, either $p(u) \geq w(u, v)$ or $p(v) \geq w(v, u)$ holds for all $\{u, v\} \in E$. It is easy to see that PVC generalizes vertex weighted VC since the latter is a special case of PVC where $w(u, v) =$ "weight of u" for all $\{u, v\} \in \delta(u)$.

Consider, for instance, setting up surveillance cameras at intersections within a certain district for detecting anomaly throughout all the streets running in the district. In a simplest model we just want to locate a minimum number of intersections for camera installation such that every road segment $\{u, v\}$ between intersections u and v can be "covered" by at least one camera, set up at one of its end-intersections u and v, and this is the minimum VC problem. In a more realistic model, we use the minimum "weight" VC problem where a vertex weight represents the cost for installing such a camera being able to monitor "all" the road segments incident to the vertex. However, road segments usually have different characteristics such as length, width, darkness, straightness, etc.; accordingly, the camera capability required for monitoring them differs from one road segment to another. Thus, while a vertex weight in the minimum weight VC model must represent the cost of a camera capable of monitoring the most demanding road segment among the incident ones so that it can cover "all" the incident road segments at once, a more flexible way of camera installation becomes possible in the power VC model, where the use of low-performance but less expensive cameras covering only some of incident road segments is allowed. It is thus more reasonable and realistic to model the problem by "power" VC representing the cost of a camera good enough for monitoring a road segment by the "edge" weight.

Angel et al. showed that PVC is approximable within 2 in general and it becomes polynomially solvable when graphs are bipartite [1]. As for the parameterized complexity of PVC, the following algorithms were presented in [2]; $O^*(1.325^P)$-time algorithm parameterized by the total value P of power, and $O^*(k^k)$-time algorithm parameterized by the number of vertices k that receive positive power, whereas no $n^{O(t)}$-time algorithm was shown to exist, unless the ETH is false, when parameterized by the graph's treewidth t.

The *node-deletion problem for a graph property* π (denoted ND(π)) is a typical graph optimization problem; that is, given a node-weighted graph G, find a vertex set of the minimum weight sum whose deletion (along with all the incident edges) from G leaves a subgraph satisfying the property π. A graph property π is *hereditary* if every subgraph of a graph satisfying π also satisfies π. A number of well-studied graph properties are hereditary such as independent set, planar, bipartite, acyclic, degree-constrained, circular-arc, circle graph, chordal, comparability, permutation, perfect. Naturally, many well known graph problems fall into this class of problems when desired graph properties are specified appropriately. Lewis and Yannakakis proved, however, that ND(π) is NP-hard whenever π is nontrivial and hereditary on induced subgraphs [30], using VERTEX COVER as a "core" problem in the class of node-deletion problems.

In this paper we consider extending some of the node-deletion problems other than VERTEX COVER adopting the "covering-by-power" condition and present a unified approach for effectively approximating them. The node-deletion problems considered are PARTIAL VERTEX COVER (PARTVC), BOUNDED DEGREE DELETION (BDD), and FEEDBACK VERTEX SET (FVS). PARTIAL VERTEX COVER is ND(π) for π = "the graph has at most $|E| - k$ edges", where k can range from 0 to $|E|$, and it is required to cover only a specified number k of edges (instead of all) in a given graph. BOUNDED DEGREE DELETION with degree bound of $b \in \mathbb{Z}_+^V$ (or b-BDD for short) is ND(π) such that π = "vertex degree of v is no larger than $b(v)$". It should be noted that VC is a special case of BDD coinciding with $\mathbf{0}$-BDD. FEEDBACK VERTEX SET (FVS) corresponds to ND(π) for π = "the graph is acyclic", and all of these are basic graph optimization problems. We call the power cover extension of these problems as POWER PARTIAL VERTEX COVER (PPartVC), POWER BOUNDED DEGREE DELETION (PBDD), and POWER FEEDBACK VERTEX SET (PFVS), respectively, and as POWER NODE DELETION for π (PND(π)) in a generic term.

Let us introduce the set $\overleftrightarrow{E}$ of directed edges derived from E s.t. $\overleftrightarrow{E} = \{(u,v),(v,u) \mid \{u,v\} \in E\}$, to distinguish edge weights $w(u,v)$ from $w(v,u)$ in PND(π), and let w now denote a weight function defined on $\overleftrightarrow{E}$. Formally, given in PND(π) is an undirected graph $G = (V,E)$ with edge weights $w : \overleftrightarrow{E} \to \mathbb{R}$, and it is required to compute a power assignment $p \in \mathbb{R}^V$ minimizing $\sum_{v \in V} p(v)$ s.t. $G - E(p) = (V, E - E(p))$ satisfies π, where $E(p)$ is the set of edges covered by p, i.e., $E(p) = \{\{u,v\} \in E \mid \text{either } w(u,v) \le p(u) \text{ or } w(v,u) \le p(v)\}$. What we do first is to reduce PND(π) to SUBMODULAR SET COVER (SSC) and we call obtained instances of SSC as the *SSC formulations* of PND(π). By applying some of the existing techniques for approximating SSC we conduct an extended analysis of the approximability of these problems in the new setting of power covering. After observing the factor 2 approximation of POWER VERTEX COVER as a starter when the current approach is applied, it will be shown that PND(π) can be approximated as good as ND(π) is known to be for those π's designated above. More specifically, it will be shown that 1) PPartVC can be approximated within a factor of 2, 2) PBDD for $b \in \mathbb{Z}_+^V$ within $\max\{2, 1+b_{\max}\}$, where where $b_{\max} = \max_{v \in V} b(v)$, and 3) PFVS within 2. It will be additionally shown that the approximation bound for PBDD can be further reduced to $2 + \log b_{\max}$ (for $b_{\max} \geq 1$) by a combination of the greedy SSC algorithm and the local ratio technique extended for power node-deletion problems.

1.1 Known Results on VC, PartVC, BDD, and FVS

It has been long known that VERTEX COVER (VC) can be approximated within a factor of 2 (achievable by a simple maximal matching heuristic [21] for the unweighted case), and a better approximation has been a subject of extensive research over the years. Yet the best constant bound has remained the same at 2 while the best known algorithm can accomplish only slightly better, within a factor of $2 - \Theta\left(1/\sqrt{\log |V|}\right)$ [24]. On the other hand, VERTEX COVER, known to be APX-complete, cannot be approximated within $2 - \epsilon$ for any $\epsilon > 0$ [29] if the unique games conjecture [26] holds, and it is currently known impossible to approximate it in polynomial time within $\sqrt{2} - \epsilon$ for any $\epsilon > 0$ unless P=NP [13, 27,28]. In PARTIAL VERTEX COVER (PartVC) it is required to cover only a specified number k of edges (instead of all) in a given graph. This "truncated" version of VC was shown to remain approximable with a factor of 2, the best constant bound known for VC, in [8,16]. The time complexity of these algorithms was later improved to $O\left(|E||V| \log |V| \log \frac{|V|^2}{|E|}\right)$ [23], and then to $O(|V|^2)$ [5].

BOUNDED DEGREE DELETION (BDD) has an application in computational biology [14] as well as in the area of property testing [33], whereas its "dual problem" of finding maximum s-plexes [35] has applications in social network analysis [4,32]. For b-BDD with $b \in \mathbb{Z}_+$, where the degree bound is uniformly equal to $b \in \mathbb{Z}_+$ over all the vertices, the approximation bound of $(b+2)$ implied by the hitting set formulation was first improved to $\max\{2, b+1\}$ by the local ratio method [17], and then to $\max\{2, b/2+1\}$ [19]. Okun and Barak considered general b-BDD where $b \in \mathbb{Z}_+^V$ is an arbitrary function, and obtained an approximation bound of $2 + \log b_{\max}$ by combination of the local ratio method and the greedy multicovering method [34]. More recently, b-BDD has been extensively studied for its parameterized complexity. It has been shown that, when parameterized by the size k of the deletion set, the problem is $W[2]$-hard for unbounded b and FPT for each fixed $b \geq 0$ [14], whereas, when parameterized by treewidth t, it is FPT with parameters k and t, and $W[2]$-hard with only parameter t [7]. A linear vertex kernel of b-BDD has been developed by generalizing the Nemhauser-Trotter theorem for VERTEX COVER to b-BDD [11,14,37].

Besides, 2-BDD has been recently highlighted under the name of *Co-Path/Cycle Packing* [10,11,15], mostly from the viewpoint of parameterized complexity, due to its important applications in bioinformatics.

FEEDBACK VERTEX SET (FVS), being the problem of hitting all the cycles existent in a given graph, is another fundamental problem in graph theory, and the decision version is one of Karp's NP-complete problems [25]. It is easy to see that FVS generalizes VC and inherits not only the NP-hardness but also the approximation hardness from it. While it was shown approximable within 2 [3,6] matching the best constant factor for VC, much of the interest has shifted more recently to the parameterized complexity and we now have enormous amount of results on parameterization of FVS. In fact it was even stated in [9] that the number of parameterized algorithms for FVS published then in the literature exceeds the number of parameterized algorithms for any other single problem.

1.2 Notation and Definitions

For any graph G let $V(G)$ and $E(G)$ denote the vertex set and the edge set of G, respectively. For an edge subset $X \subseteq E$ of $G = (V, E)$ the subgraph of which edge set limited to X is denoted by $G[X] = (V, X)$ and the one obtained by deleting all the edges in X from G by $G - X$. The set of edges incident to some vertex in X is denoted by $\delta(X)$ and $\delta(u)$ means $\delta(\{u\})$. The degree of a vertex u in G is denoted by $d(u) = |\delta(u)|$. The set of neighboring vertices of u, i.e., $\{v \in V \mid \{u, v\} \in E\}$, is denoted by $\Gamma(u)$.

2 Power Node-Deletion and Submodular Set Cover

Let N be a finite set and $f : 2^N \to \mathbb{R}$ a real valued function defined on the subsets of N. The set-function f is called *submodular* if the following inequality holds for any two subsets X and Y of N:

$$f(X \cap Y) + f(X \cup Y) \le f(X) + f(Y).$$

The formal framework of the SUBMODULAR SET COVER *(SSC)* problem is described as follows. A problem instance consists of a finite set N, a nonnegative cost c_j associated with each element $j \in N$, and a nondecreasing submodular function $f : 2^N \to \mathbb{Z}_+$ (for simplicity f is assumed to be nonnegative integer valued). The problem is then that of finding a spanning set of minimum cost, that is,

$$\text{(SSC)} \quad \min_{S \subseteq N} \left\{ \sum_{j \in S} c_j \,\middle|\, f(S) = f(N) \right\}$$

Algorithm 1. Primal-dual algorithm PD for SSC

Initialize $S = \emptyset, y = 0, l = 0$
while S is not a solution of SSC (i.e., $f(S) < f(N)$) **do**
  $l \leftarrow l + 1$
  $$j_l \leftarrow \arg\min_{j \in N-S} \left\{ \frac{w_j - \sum_{X: j \notin X, X \neq S} f_X(j) y_X}{f_S(j)} \left(= \frac{w_j - \sum_{1 \leq k \leq l-1} f_{S_k}(j) y_{S_k}}{f_S(j)} \right) \right\}$$
  ▷ Increase y_S until the dual constraint corresponding to j becomes tight for some $j \notin S$
  $$y_S \leftarrow \frac{w_{j_l} - \sum_{X: j_l \notin X, X \neq S} f_X(j_l) y_X}{f_S(j_l)} \left(= \frac{w_{j_l} - \sum_{1 \leq k < l-1} f_{S_k}(j_l) y_{S_k}}{f_S(j_l)} \right)$$
  Add j_l into S (and let $S_l \leftarrow S$)
for $k = l$ downto 1 **do**
  if $S - \{j_k\}$ is a solution of SSC **then** remove j_k from S
Output S

A primal-dual heuristic based on the following IP formulation of SSC is known effective among others for a certain type of SSC:

$$\begin{array}{llll} & \min \sum_{j \in N} w_j x_j & & \\ & \text{subject to:} & & \\ \text{(IP)} & & \sum_{j \in N-S} f_S(j) x_j \geq f_S(N-S) & S \subseteq N \\ & & x_j \in \{0,1\} & j \in N \end{array}$$

and the dual of its linear relaxation:

$$\begin{array}{llll} & \max \sum_{S \subseteq N} f_S(N-S) y_S & & \\ & \text{subject to:} & & \\ \text{(D)} & & \sum_{S: j \notin S} f_S(j) y_S \leq w_j & j \in N \\ & & y_S \geq 0 & S \subseteq N \end{array}$$

where f_S is the *contraction* of f onto $N - S$ for any $S \subseteq N$, that is the function defined on 2^{N-S} s.t. $f_S(X) = f(X \cup S) - f(S)$.

Definition 1. *We say that* $X \subseteq N$ *is a* minimal *solution for a SSC instance* (N, f, c) *iff*

- $f(X) = f(N)$*, i.e.,* X *is a solution for* (N, f, c)*, and*
- $f(X - \{x\}) < f(N)$*,* $\forall x \in X$*.*

The resulting primal-dual algorithm called PD is given in Algorithm 1, and its performance can be estimated by the following combinatorial bound:

Theorem 1 ([18]). *The algorithm PD computes a SSC solution for* (N, f, c) *approximating within a factor of* $\max \left\{ \frac{\sum_{j \in X} f_S(\{j\})}{f_S(N-S)} \right\}$, *where* max *is taken over any* $S \subseteq N$ *and any minimal solution* X *for* $(N-S, f_S)$.

2.1 Submodular Set Cover Formulation

Let us consider reducing PND(π) to SSC, starting with a simple observation. Let $W_u = \{w(u, v) \mid \{u, v\} \in \delta(u)\}$ be the set of weights of edges "out-going" from a vertex u. Then, we may assume w.l.o.g. that $p(u) \in W_u \cup \{0\}$ for any solution $p \in \mathbb{R}^V$ of PND(π) since an assignment of any other value to $p(u)$ yields some redundancies. This observation allows us to consider power covering of a graph to be the problem of covering edges by the weighted edge subsets of the following form: Define $\delta_w(u, v) \subseteq \delta(u)$ for $\{u, v\} \in \delta(u)$ to be the set of edges $\{u, x\}$ incident to u of which (u, x)-weight $w(u, x)$ is no larger than $w(u, v)$, i.e.,

$$\delta_w(u, v) = \{\{u, x\} \in \delta(u) \mid w(u, x) \leq w(u, v)\}.$$

Extending the edge weight function $w \in \mathbb{R}^{\overleftrightarrow{E}}$ to the edge "subset" weight function, the weight $w(\delta_w(u, v))$ of $\delta_w(u, v)$ is set equal to $w(u, v)$.

Proposition 1. *For* $G = (V, E)$ *and* $w : \overleftrightarrow{E} \to \mathbb{R}$ *let* $N = \{\delta_w(u, v) \mid (u, v) \in \overleftrightarrow{E}\}$ *and* $c(\delta_w(u, v)) = w(\delta_w(u, v)) = w(u, v)$, $\forall (u, v) \in \overleftrightarrow{E}$. *Suppose we have a nondecreasing submodular function* $f : 2^N \to \mathbb{Z}$ *s.t.*

$$G - \bigcup S = (V, E - \bigcup S) \text{ satisfies } \pi \text{ if and only if } f(S) = f(N).$$

for any G *and* $S \subseteq N$, *where* $\bigcup S$ *denotes* $\bigcup_{x \in S} x = \bigcup_{\delta_w(u,v) \in S} \delta_w(u, v)$ *for* $S \subseteq N$. *Then, PND(π) on an instance of* (G, w) *is equivalent to SSC on an instance of* (N, f, c). □

We say that (N, f, c) thus obtained from a PND(π) instance of (G, w) is a *SSC formulation* of PND(π). Throughout the paper, N and c will be set as in Proposition 1 from (G, w) regardless of π, whereas f needs to be suitably chosen depending on specifics of π.

Let us consider a SSC formulation (N, f, c) of PND(π) and applying Theorem 1 to it. To do so, we introduce the following "consistency condition" for a submodular system (N, f):

Definition 2. *For* $G = (V, E), N = \{\delta_w(u, v) \mid (u, v) \in \overleftrightarrow{E}\}$, $S \subseteq N$, *and* $X \subseteq N - S$, *let* $X_{\setminus S}$ *denote the set* $\{x \setminus \bigcup S \mid x \in X\}$ *and* $f_{\setminus S}$ *be* f *defined on* $G - \bigcup S = (V, E - \bigcup S)$ *by the same formulation. We say that a SSC formulation* (N, f, c) *of PND(π) is* consistent *if* $f_S(X) = f_{\setminus S}(X_{\setminus S})$ *for any* $S \subseteq N$ *and* $X \subseteq N - S$.

Corollary 1 (of Theorem 1). *If* (N, f, c) *is a consistent SSC formulation of PND(π), PND(π) can be approximated within a factor of* $\max \left\{ \frac{\sum_{j \in X} f(j)}{f(N)} \right\}$ *by PD, where* max *is taken over any graph* G *and any minimal solution* $X \subseteq N$ *in* (N, f).

Proof. Due to Theorem 1 we can say that the algorithm PD approximates PND(π) within a factor of $\max\left\{\frac{\sum_{j\in X} f_S(j)}{f_S(N-S)}\right\}$, where max is taken over any graph G, any $S \subseteq N$ and any minimal solution X in $(N-S, f_S)$. Since (N, f) here is consistent, the SSC system $(N-S, f_S)$ coincides with the SSC formulation $((N-S)_{\setminus S}, f_{\setminus S})$ of $G - \bigcup S = (V, E - \bigcup S)$. Hence,

$$\frac{\sum_{j\in X} f_S(j)}{f_S(N-S)} = \frac{\sum_{j\setminus\bigcup S\in X_{\setminus S}} f_{\setminus S}(j \setminus \bigcup S)}{f_{\setminus S}((N-S)_{\setminus S})}$$

and $\max\left\{\frac{\sum_{j\in X} f_S(j)}{f_S(N-S)}\right\}$ can be upper bounded by $\max\left\{\frac{\sum_{j\in X} f(j)}{f(N)}\right\}$ by taking the latter max over any graph G and any minimal solution X for (N, f). □

The minimality of a solution often plays a crucial role in the analysis of the performance of PD, and the next is an easy but useful observation for minimal solutions; any two members of N cannot together belong to any minimal solution if both are subsets of $\delta(v)$ for some $v \in V$.

Observation 1. Suppose $\{u, v\}$ and $\{u, v'\}$ are two distinct edges in $\delta(u)$. Then, $\delta_w(u, v) \notin X$ or $\delta_w(u, v') \notin X$ if X is a minimal solution for (N, f) (since either $\delta_w(u, v) \subseteq \delta_w(u, v')$ or $\delta_w(u, v') \subseteq \delta_w(u, v)$).

It follows from Observation 1 that a minimal solution X contains at most one subset of $\delta(u)$ for each $u \in V$.

3 Power (Partial) Vertex Cover

Let us start with the most basic problem of power node-deletion, Power Vertex Cover (PVC). Angel et al. designed a 2-approximation algorithm for it using the matching techniques [1]. We will show here that the same approximation bound follows from the current approach by simply setting $f(S) = |\bigcup S|$ for $S \subseteq N$. Since f is clearly a nondecreasing submodular function and $f(S) = f(N) = |E|$ iff $G - \bigcup S$ has no edge in it, (N, f, c) is a SSC formulation of PVC. Moreover, since

$$f_S(X) = f(X \cup S) - f(S) = \left|\bigcup(X \cup S)\right| - \left|\bigcup S\right| = \left|\bigcup X - \bigcup S\right|$$

for $X \subseteq N - S$, f_S coincides with f on $G - \bigcup S$.

Theorem 2. *PVC can be approximated within* 2 *by PD.*

Proof. Since f is consistent, it suffices to show, due to Corollary 1, that

$$\sum_{x\in X} |x| \leq 2 \cdot f(N) = 2 \cdot |E|$$

for any minimal solution X for $G = (V, E)$. Since any edge of E can belong to at most two members of X if X is minimal (see Observation 1), every edge of E can be counted at most twice in $\sum_{x\in X} |x|$. □

Consider next the partial vertex cover problem, in which a minimum weight set of vertices covering k edges is sought for some specified integer k. We define $f : N \to \mathbb{Z}$ for PPartVC s.t. $f(S) = \min\{|\bigcup S|, k\}$. Then, f is clearly nondecreasing and submodular. Besides,

$$f_S(X) = \min\left\{\left|\bigcup(X \cup S)\right|, k\right\} - \min\left\{\left|\bigcup S\right|, k\right\}$$
$$= \begin{cases} 0 & \text{if } |\bigcup S| \geq k \\ \min\{|\bigcup X - \bigcup S|, k - |\bigcup S|\} & \text{otherwise} \end{cases}$$

for $X \subseteq N - S$. So, f is consistent and the problem reduces to SSC on (N, f).

Lemma 1. *Let $N = \{\delta_w(u,v) \mid (u,v) \in \overleftrightarrow{E}\}$, and $f(S) = \min\{|\bigcup S|, k\}$ for $S \subseteq N$. Then, $\sum_{x \in X} f(\{x\}) \leq 2 \cdot f(N)$ for any minimal solution $X \subseteq N$ for SSC (N, f, w).*

Proof. Since $\bigcup N = E$, $f(N) = k$. Suppose there exists $x' \in X$ with $|x'| \geq k$. Then, minimal X can contain no other element, and $\sum_{x \in X} f(x) = \min\{|x'|, k\} = k \leq 2k$.

Assuming that $|X| \geq 2$ and $f(\{x\}) = \min\{|x|, k\} = |x|$, $\forall x \in X$, let $x \setminus X$ denote $x - \bigcup(X - x)$, that is, the set of edges covered only by x. Since X is a minimal solution, $X - x$ is not a solution for all $x \in X$, and hence, $|\bigcup(X - x)| = |\bigcup X| - |x \setminus X| < k$, $\forall x \in X$. Summing over all x's in X,

$$\sum_{x \in X}\left(\left|\bigcup X\right| - |x \setminus X|\right) = |X|\left|\bigcup X\right| - \sum_{x \in X}|x \setminus X|$$
$$= |X|\left(\left|\bigcup X\right| - \sum_{x \in X}|x \setminus X|\right) + (|X| - 1)\sum_{x \in X}|x \setminus X|$$
$$< k|X|$$

Multiplying both sides of the inequality by $2/|X|$, we have

$$2k > 2\left(\left|\bigcup X\right| - \sum_{x \in X}|x \setminus X|\right) + \frac{2(|X| - 1)}{|X|}\sum_{x \in X}|x \setminus X|$$
$$\geq 2\left(\left|\bigcup X\right| - \sum_{x \in X}|x \setminus X|\right) + \sum_{x \in X}|x \setminus X|$$

since $|X| \geq 2$. Observe that $x \setminus X$ and $x' \setminus X$ are disjoint for $x, x' \in X$ if $x \neq x'$, and that $\bigcup_{x \in X}(x \setminus X)$ is the set of edges covered only once by X while $\bigcup X - \bigcup_{x \in X}(x \setminus X)$ is the set of those covered twice by X. Thus,

$$\sum_{x \in X} f(\{x\}) = \sum_{x \in X}|x| = 2\left(\left|\bigcup X\right| - \sum_{x \in X}|x \setminus X|\right) + \sum_{x \in X}|x \setminus X|,$$

and it follows that $\sum_{x \in X} f(\{x\}) \leq 2k$. □

It is immediate that

Theorem 3. *PPartVC can be approximated within 2 by PD.*

4 Power Bounded Degree Deletion

For b-PBDD with π = "the degree of vertex v is bounded by $b(v)$", let us define $f : 2^N \to \mathbb{Z}_+$ s.t.

$$f(S) = \sum_{v \in V} \min\{d(v; S), d(v) \dot{-} b(v)\}$$

where

$$d(x; S) = \left|\delta(x) \cap \bigcup S\right|$$

is the vertex degree of $x \in V$ in $G[\bigcup S] = (V, \bigcup S)$, and $x \dot{-} y = \max\{x - y, 0\}$.

Easily, every vertex degree is $\leq b(v)$ in $G - \bigcup S = (V, E - \bigcup S)$ if and only if every vertex degree of $v \in V$ in $G[\bigcup S] = (V, \bigcup S)$ is $\geq d(v) \dot{-} b(v)$, which in turn holds if and only if

$$f(S) = \sum_{v \in V} (d(v) \dot{-} b(v)) = f(N).$$

To show that $\min_{S \subseteq N} \left\{\sum_{s \in S} w(s) \middle| f(S) = f(N)\right\}$ is actually an instance of SSC for N and f thus defined, it remains to prove that f is a nondecreasing submodular function on 2^N.

Lemma 2. *The set function $f : 2^N \to \mathbb{Z}_+$ is nondecreasing and submodular.*

Proof. It suffices to show that $d(x; S)$ at any fixed vertex $x \in V$, when seen as a function on 2^N, is nondecreasing and submodular. Clearly, $d(x; S)$ is nondecreasing at any x. To verify submodularity of $d(x; S)$, the following characterization of submodularity is helpful:

Proposition 2 (Lovász [31]). *Let g be a set-function defined on all subsets of N. Then g is submodular if and only if the derived set-functions*

$$g_a(X) = g(X \cup \{a\}) - g(X) \quad (X \subseteq N - \{a\})$$

are monotone decreasing for all $a \in N$. □

Letting $a = \delta_w(y, z)$ for any $(y, z) \in \overleftrightarrow{E}$ and $S \subseteq T \subseteq N - \{a\}$, consider $f_a(S)$ and $f_a(T)$ for $f = d(x; *)$. The values of $f_a(S)$ and $f_a(T)$ are increments of the f-value when a is added to S and T, respectively; we may write

$$\begin{aligned} f_a(S) &= d(x; S \cup \{\delta_w(y, z)\}) - d(x; S) \\ &= \left|\delta(x) \cap \left(\bigcup S \cup \delta_w(y, z)\right)\right| - \left|\delta(x) \cap \bigcup S\right| \\ &= \left|\delta_w(y, z) \cap \left(\delta(x) - \bigcup S\right)\right| \end{aligned}$$

and

$$f_a(T) = \left|\delta_w(y, z) \cap \left(\delta(x) - \bigcup T\right)\right|$$

Since $S \subseteq T$, $\bigcup S \subseteq \bigcup T$, and hence, $f_a(S) \geq f_a(T)$ for all $a \in N$. This shows that $d(x; *)$ is submodular at any $x \in V$.

Proposition 3 (see, e.g., [31]). *Let* g *and* h *be submodular set-functions s.t.* $g - h$ *is nondecreasing (or nonincreasing). Then,* $\min\{g, h\}$ *is also submodular.* □

It follows that $\min\{d(x; *), d(x) \dot{-} b(x)\}$ is submodular since $d(x; *)$ is nondecreasing, and so is $d(x; *) - (d(x) \dot{-} b(x))$.

Finally, $f(S) = \sum_{v \in V} \min\{d(v; S), d(v) \dot{-} b(v)\}$, the sum of submodular $\min\{d(x; *),\ d(x) \dot{-} b(x)\}$'s, is submodular since the sum of submodular functions is submodular. □

Therefore,

Proposition 4. *b-PBDD on* $(G = (V, E), w)$ *can be reduced to SSC on* $(N, f : 2^N \to \mathbb{Z}_+, c)$*, where* $N = \{\delta_w(u, v) \mid (u, v) \in \overleftrightarrow{E}\}$*,* $c(\delta_w(u, v)) = w(u, v)$*,* $\forall (u, v) \in \overleftrightarrow{E}$*, and* $f(S) = \sum_{v \in V} \min\{d(v; S), d(v) \dot{-} b(v)\}$*,* $\forall S \subseteq N$*.* □

To apply Corollary 1 to b-PBDD we need the following lemma:

Lemma 3. *Let* $N = \{\delta_w(u, v) \mid (u, v) \in \overleftrightarrow{E}\}$*, and define* $f : 2^N \to \mathbb{Z}_+$ *s.t.* $f(S) = \sum_{v \in V} \min\{d(v; S), d(v) \dot{-} b(v)\}$ *for* $S \subseteq N$*. If* $b_{\max} \geq 1$*,*

$$\sum_{x \in X} f(\{x\}) \leq (b_{\max} + 1)\, f(N)$$

for any minimal solution $X \subseteq N$ *for the SSC instance* (N, f, c)*.*

Proof. Omitted due to space constraints. □

Once this key lemma is proven, it is straightforward to obtain the approximation bound of PD for b-PBDD:

Theorem 4. *b-PBDD can be approximated within* $\max\{2, b_{\max} + 1\}$ *by PD.*

Proof. To apply Corollary 1 to b-PBDD, we need to verify that f defined as above is consistent (of which proof is omitted). The approximation bound guaranteed by Corollary 1 refers to the ratio of $\sum_{j \in X} f_S(\{j\})$ and $f_S(N - S)$, and $(N - S, f_S, c')$ is a SSC instance corresponding to the b-PBDD instance on the subgraph $G - \bigcup S$ of G. This ratio is no larger than $(b_{\max} + 1)$ on an arbitrary graph according to Lemma 3 as long as $b_{\max} \geq 1$. Besides, it can be verified, by the essentially same analysis as for PVC, that this ratio is no larger than 2 when $b_{\max} = 0$. Hence, the approximation bound of PD for b-PBDD is as claimed. □

4.1 Combination of Greedy and Local Ratio

It is well-known that the greedy algorithm as given in Algorithm 2 is quite effective for approximating SSC.

Algorithm 2. Greedy algorithm Greedy for SSC

Initialize $S = \emptyset$
while S is not a solution of SSC (i.e., $f(S) < f(N)$) **do**
  Add $\arg\min_{j \in N-S} \left\{ \frac{c(j)}{f(S \cup \{j\}) - f(S)} \right\}$ to S
Output S

Theorem 5 (Wolsey [36]). *When f is integer-valued with $f(\emptyset) = 0$, the algorithm Greedy approximates SSC (N, f, w) within a factor of $H\left(\max_{j \in N} f(\{j\})\right)$, where $H(k) = \sum_{i=1}^{k} \frac{1}{i}$ for a positive integer k.*

The *local ratio* method is another standard technique for approximating general ND(π). For a graph $G = (V, E)$ and a vertex $u \in V$, let S_u denote the subgraph u-star of G centered at u; that is, $S_u = (V(S_u), E(S_u)) = (\{u\} \cup \Gamma(u), \{\{u, v\} \mid v \in \Gamma(u)\})$. If $d(u) = |\Gamma(u)| > b(u)$ then S_u is a forbidden subgraph for π = "degree of every vertex v is bounded by $b(v)$". Consider the edge weight function w' on $\overleftrightarrow{E}$ for such a forbidden graph S_u, corresponding to the assignment of

$$p(v) = \begin{cases} \epsilon(d(u) - b(u)) & \text{if } v = u, \\ \epsilon & \text{if } v \in \Gamma(u), \\ 0 & \text{otherwise,} \end{cases}$$

s.t.

$$w'(\boldsymbol{e}) = \begin{cases} \epsilon(d(u) - b(u)) & \text{if } \boldsymbol{e} = (u, v) \in \overleftrightarrow{E}(S_u), \\ \epsilon & \text{if } \boldsymbol{e} = (v, u) \in \overleftrightarrow{E}(S_u), \\ \min\{\epsilon, w(v, x)\} & \text{if } \boldsymbol{e} = (v, x) \notin \overleftrightarrow{E}(S_u), v \in \Gamma(u), \\ 0 & \text{otherwise.} \end{cases} \tag{1}$$

where a positive constant ϵ is taken as large as possible without $w'(\boldsymbol{e})$ exceeding $w(\boldsymbol{e})$ at any $\boldsymbol{e} \in \overleftrightarrow{E}$.

Following the approach of Okun and Barak [34], we apply the local ratio approximation to forbidden u-stars S_u, as long as there remains such S_u with relatively large $d(u) - b(u)$. When such S_u's are exhausted, we switch to the greedy method Greedy and complete our approximation. The point at which we switch from the local ratio approximation to the greedy one is where $d(u)/b(u) = 1 + 1/\log b(u)$. In the first half where $d(u)/b(u) > 1 + 1/\log b(u)$ at some u-star S_u, we apply the local ratio algorithm to S_u. Then, in the second half where $d(v)/b(u) \leq 1 + 1/\log b(u)$ at every vertex v, we initiate running Greedy. While all the details are omitted here due to space constraints, the resulting algorithm Comb as given in Algorithm 3 can be shown to deliver improved approximation:

Theorem 6. *b-PBDD can be approximated by Comb within $2 + \log b_{\max}$ when $b_{\max} > 1$, and within $7/3$ when $b_{\max} = 1$.*

Algorithm 3. Combined algorithm Comb for b-PBDD

Input: $(G = (V, E), w)$
  Initialize: $\boldsymbol{F} = \emptyset$
  while $\exists u \in V$ s.t. $(d(u)/b(u) > 1 + 1/\log b(u)$ with $b(u) \geq 2)$ or $(d(u) > 3$ with $b(u) = 1)$ or $(d(u) > 0$ with $b(u) = 0)$ **do**
    $\epsilon \leftarrow \min\left\{\bigcup_{v \in \Gamma(u)}\{w(u,v), w(v,u)\}\right\}$
    Define w' according to eq. (1)
    $\boldsymbol{F}' \leftarrow \{(u,v) \in \overleftrightarrow{E} \mid w'(u,v) = w(u,v)\}$
    $w \leftarrow w - w'$
    $\boldsymbol{F} \leftarrow \boldsymbol{F} \cup \boldsymbol{F}'$
    $E \leftarrow E - F'$
  while $\exists u \in V$ with $d(u) > b(u)$ **do**
    Run Greedy

5 Power Feedback Vertex Set

Letting again $N = \{\delta_w(u,v) \mid (u,v) \in \overleftrightarrow{E}\}$ for $(G = (V,E), w)$, we now turn to PFVS, that is PND(π) with π = "the graph is acyclic". Let $M(G) = (E, r)$ be the cycle matroid of $G = (V, E)$ and $M^d(G) = (E, r^d)$ be the dual matroid of $M(G)$, where $r : 2^E \to \mathbb{Z}_+$ and $r^d : 2^E \to \mathbb{Z}_+$ are the rank functions of $M(G)$ and $M^d(G)$, respectively.

Proposition 5. *Define $f : 2^N \to \mathbb{Z}_+$ s.t. $f(S) = r^d(\bigcup S)$, $\forall S \subseteq N$. Then, PFVS on $(G = (V, E), w)$ can be formulated by SSC (N, f, c), where $c(\delta_w(u,v)) = w(u,v), \forall (u,v) \in \overleftrightarrow{E}$.*

Proof. Since r^d is nondecreasing and submodular, so is f. The set $\bigcup S$ of edges covered by S is spanning in the dual matroid $M^d(G)$ iff $f(S) = r^d(\bigcup S) = r^d(E) = f(N)$, and $\bigcup S$ is spanning in $M^d(G)$ iff $E - \bigcup S$ is independent in $M(G)$, the cycle matroid of G, that is, the graph $G - \bigcup S = (V, E - \bigcup S)$ is acyclic. Therefore, (N, f, c) is a SSC formulation of PFVS on (G, w) when $c(\delta_w(u,v)) = w(u,v)$, $\forall (u,v) \in \overleftrightarrow{E}$. □

Lemma 4. *Let $f(S) = r^d(\bigcup S)$ for $S \subseteq N$. Then,*

$$\sum_{x \in X} f(\{x\}) = \sum_{\delta_w(u,v) \in X} r^d(\delta_w(u,v)) \leq 2 \cdot f(N) = r^d(\bigcup N) - r^d(E)$$

for any minimal solution $X \subseteq N$ for the SSC instance (N, f, c).

Proof. Omitted due to space constraints. □

Theorem 7. *PFVS can be approximated within 2 by PD.*

Proof. To apply Corollary 1 to PFVS, we need to verify that f defined as above is consistent (of which proof is omitted). It follows from Corollary 1 and Lemma 4 that PFVS can be approximated within 2 by PD. □

References

1. Angel, E., Bampis, E., Chau, V., Kononov, A.: Min-power covering problems. In: Elbassioni, K., Makino, K. (eds.) ISAAC 2015. LNCS, vol. 9472, pp. 367–377. Springer, Heidelberg (2015). https://doi.org/10.1007/978-3-662-48971-0_32
2. Angel, E., Bampis, E., Escoffier, B., Lampis, M.: Parameterized power vertex cover. In: Heggernes, P. (ed.) WG 2016. LNCS, vol. 9941, pp. 97–108. Springer, Heidelberg (2016). https://doi.org/10.1007/978-3-662-53536-3_9
3. Bafna, V., Berman, P., Fujito, T.: A 2-approximation algorithm for the undirected feedback vertex set problem. SIAM J. Discrete Math. **12**(3), 289–297 (1999)
4. Balasundaram, B., Butenko, S., Hicks, I.V.: Clique relaxations in social network analysis: the maximum k-plex problem. Oper. Res. **59**(1), 133–142 (2011)
5. Bar-Yehuda, R.: Using homogeneous weights for approximating the partial cover problem. J. Algorithms **39**(2), 137–144 (2001)
6. Becker, A., Geiger, D.: Optimization of Pearl's method of conditioning and greedy-like approximation algorithms for the vertex feedback set problem. Artif. Intell. **83**(1), 167–188 (1996)
7. Betzler, N., Bredereck, R., Niedermeier, R., Uhlmann, J.: On bounded-degree vertex deletion parameterized by treewidth. Discrete Appl. Math. **160**(1–2), 53–60 (2012)
8. Bshouty, N.H., Burroughs, L.: Massaging a linear programming solution to give a 2-approximation for a generalization of the vertex cover problem. In: Morvan, M., Meinel, C., Krob, D. (eds.) STACS 1998. LNCS, vol. 1373, pp. 298–308. Springer, Heidelberg (1998). https://doi.org/10.1007/BFb0028569
9. Cao, Y.: A naive algorithm for feedback vertex set. In: 1st Symposium on Simplicity in Algorithms, volume 61 of OASIcs Open Access Series Informatics, Article no. 1, p. 9. Schloss Dagstuhl. Leibniz-Zent. Inform., Wadern (2018)
10. Chauve, C., Tannier, E.: A methodological framework for the reconstruction of contiguous regions of ancestral genomes and its application to mammalian genomes. PLoS Comput. Biol. **4**(11), e1000234 (2008)
11. Chen, Z.-Z., Fellows, M., Fu, B., Jiang, H., Liu, Y., Wang, L., Zhu, B.: A linear kernel for co-path/cycle packing. In: Chen, B. (ed.) AAIM 2010. LNCS, vol. 6124, pp. 90–102. Springer, Heidelberg (2010). https://doi.org/10.1007/978-3-642-14355-7_10
12. Chuzhoy, J., Naor, J.: Covering problems with hard capacities. SIAM J. Comput. **36**(2), 498–515 (2006)
13. Dinur, I., Khot, S., Kindler, G., Minzer, D., Safra, M.: Towards a proof of the 2-to-1 games conjecture? In: STOC'18–Proceedings of the 50th Annual ACM SIGACT Symposium on Theory of Computing, pp. 376–389. ACM, New York (2018)
14. Fellows, M.R., Guo, J., Moser, H., Niedermeier, R.: A generalization of Nemhauser and Trotter's local optimization theorem. J. Comput. System Sci. **77**(6), 1141–1158 (2011)
15. Feng, Q., Wang, J., Li, S., Chen, J.: Randomized parameterized algorithms for P_2-packing and co-path packing problems. J. Comb. Optim. **29**(1), 125–140 (2015)
16. Fujito, T.: A unified local ratio approximation of node-deletion problems. In: Proceedings of the ESA'96, pp. 167–178 (1996)
17. Fujito, T.: A unified approximation algorithm for node-deletion problems. Discrete Appl. Math. **86**(2–3), 213–231 (1998)
18. Fujito, T.: On approximation of the submodular set cover problem. Oper. Res. Lett. **25**(4), 169–174 (1999)

19. Fujito, T.: Approximating bounded degree deletion via matroid matching. In: Fotakis, D., Pagourtzis, A., Paschos, V.T. (eds.) CIAC 2017. LNCS, vol. 10236, pp. 234–246. Springer, Cham (2017). https://doi.org/10.1007/978-3-319-57586-5_20
20. Garey, M.R., Johnson, D.S.: The rectilinear steiner tree problem is NP-complete. SIAM J. Appl. Math. **32**(4), 826–834 (1977)
21. Gavril, F.: Cited in [20, page 134] (1974)
22. Hassin, R., Levin, A.: The minimum generalized vertex cover problem. ACM Trans. Algorithms **2**(1), 66–78 (2006)
23. Hochbaum, D.S.: Approximating clique and biclique problems. J. Algorithms **29**(1), 174–200 (1998)
24. Karakostas, G.: A better approximation ratio for the vertex cover problem. ACM Trans. Algorithms **5**(4), 8 (2009)
25. Richard, M.K.: Reducibility among combinatorial problems. In: Complexity of computer computations (Proceeding Symposium, IBM Thomas J. Watson Research Center, Yorktown Heights, New York, 1972), pp. 85–103 (1972)
26. Khot, S.: On the power of unique 2-prover 1-round games. In: Proceedings of the 34th Annual ACM Symposium on Theory of Computing, pp. 767–775. ACM, New York (2002)
27. Khot, S., Minzer, D., Safra,M.: On independent sets, 2-to-2 games, and Grassmann graphs. In STOC'17–Proceedings of the 49th Annual ACM SIGACT Symposium on Theory of Computing, pp. 576–589. ACM, New York (2017)
28. Khot, S., Minzer, D., Safra, M.: Pseudorandom sets in Grassmann graph have near-perfect expansion. In: 59th Annual IEEE Symposium on Foundations of Computer Science–FOCS 2018, pp. 592–601. IEEE Computer Soc., Los Alamitos, CA (2018)
29. Khot, S., Regev, O.: Vertex cover might be hard to approximate to within $2 - \epsilon$. J. Comput. System Sci. **74**(3), 335–349 (2008)
30. Lewis, J.M., Yannakakis, M.: The node-deletion problem for hereditary properties is NP-complete. J. Comput. System Sci. **20**(2), 219–230 (1980)
31. Lovász, L.: Submodular functions and convexity. In: Bachem, A., Korte, B., Grëtschel, M. (eds.) Mathematical Programming: The State of the Art, pp. 235–257. Springer, Berlin (1983)
32. Moser, H., Niedermeier, R., Sorge, M.: Exact combinatorial algorithms and experiments for finding maximum k-plexes. J. Comb. Optim. **24**(3), 347–373 (2012)
33. Newman, I., Sohler, C.: Every property of hyperfinite graphs is testable. SIAM J. Comput. **42**(3), 1095–1112 (2013)
34. Okun, M., Barak, A.: A new approach for approximating node deletion problems. Inform. Process. Lett. **88**(5), 231–236 (2003)
35. Seidman, S.B., Foster, B.L.: A graph-theoretic generalization of the clique concept. J. Math. Sociol. **6**(1), 139–154 (1978)
36. Wolsey, L.A.: An analysis of the greedy algorithm for the submodular set covering problem. Combinatorica **2**(4), 385–393 (1982)
37. Xiao, M.: On a generalization of Nemhauser and Trotter's local optimization theorem. J. Comput. System Sci. **84**, 97–106 (2017)

Phase Transition in Count Approximation by Count-Min Sketch with Conservative Updates

Éric Fusy and Gregory Kucherov(✉)

LIGM, CNRS, Univ. Gustave Eiffel, Marne-la-Vallée, France
{Eric.Fusy,Gregory.Kucherov}@univ-eiffel.fr

Abstract. Count-Min sketch is a hashing-based data structure to represent a dynamically changing associative array of counters. We analyse the counting version of Count-Min under a stronger update rule known as *conservative update*, assuming the uniform distribution of input keys. We show that the accuracy of conservative update strategy undergoes a phase transition, depending on the number of distinct keys in the input as a fraction of the size of the Count-Min array. We prove that below the threshold, the relative error is asymptotically $o(1)$ (as opposed to the regular Count-Min strategy), whereas above the threshold, the relative error is $\Theta(1)$. The threshold corresponds to the peelability threshold of random k-uniform hypergraphs. We demonstrate that even for small number of keys, peelability of the underlying hypergraph is a crucial property to ensure the $o(1)$ error. To our knowledge, this relationship has not been observed previously. Finally, we provide experimental data on the behavior of the average error for Zipf's distribution compared with the uniform one.

1 Introduction

Count-Min sketch is a hash-based data structure to represent a dynamically changing associative array $\boldsymbol{a}$ of counters in an approximate way. The array $\boldsymbol{a}$ can be seen as a mapping from some set K of keys to $\mathbb{N}$, where K is drawn from a (large) universe U. The goal is to support *point queries* about the (approximate) current value of $\boldsymbol{a}(p)$ for a key p. Count-Min is especially suitable for the streaming framework, when counters associated to keys are updated dynamically. That is, *updates* are (key,value) pairs (p, ℓ) with the meaning that $\boldsymbol{a}(p)$ is updated to $\boldsymbol{a}(p) + \ell$.

Count-Min sketch was proposed in [13], see e.g. [11] for a survey. A similar data structure was introduced earlier in [10] named *Spectral Bloom filter*, itself closely related to *Counting Bloom filters* [20]. The difference between Count-Min sketch and Spectral Bloom filter is marginal: while a Count-Min sketch requires hash functions to have disjoint codomains (rows of Count-Min matrix), a Spectral Bloom filter has all hash functions mapping to the same array. This difference is the same as between partitioned [2] and regular Bloom filters. In this paper, we will deal with the Spectral Bloom filter version but will keep the term Count-Min sketch as more common in the literature.

M. Mavronicolas (Ed.): CIAC 2023, LNCS 13898, pp. 232–246, 2023.
https://doi.org/10.1007/978-3-031-30448-4_17

Count-Min sketch supports negative update values ℓ provided that at each moment, each counter $\boldsymbol{a}(p)$ remains non-negative (so-called *strict turnstile model* [27]). When updates are positive, the Count-Min update algorithm can be modified to a stronger version leading to smaller errors in queries. This modification, introduced in [18] as *conservative update*, is mentioned in [11], without any formal analysis given in those papers. This variant is also discussed in [10] under the name *minimal increase*, where it is claimed that it decreases the probability of a positive error by a factor of the number of hash functions, but no proof is given. We discuss this claim in the concluding part of this paper.

The case of positive updates is widespread in practice. In particular, a very common instance is *counting* where all update values are 1. This task occurs in different scenarios in network traffic monitoring, as well as other applications related to data stream mining [18]. In bioinformatics, we may want to maintain, on the fly, multiplicities of fixed-length words occurring in a big sequence dataset [3,29,33]. We refer to [16] for more examples of applications.

While it is easily seen that the error in conservative update can only be smaller than in Count-Min, obtaining more precise bounds is a challenging problem. Count-Min guarantees, with high probability, that the additive error can be bounded by $\varepsilon\|\boldsymbol{a}\|_1$ for any ε, where $\|\boldsymbol{a}\|_1$ is the $L1$-norm of $\boldsymbol{a}$ [13]. In the counting setting, $\|\boldsymbol{a}\|_1$ is the length of the input stream which can be very large, and therefore this bound provides a weak guarantee in practice, unless the distribution of keys is very skewed and queries are made on frequent keys (*heavy hitters*) [8,12,27]. It is therefore an important practical question to analyse the improvement provided by the conservative update strategy compared to the original Count-Min sketch.

To our knowledge, the first attempt towards this goal was made in [7], under assumption that all $\binom{n}{k}$ counter combinations are equally likely at each step (n size of the Count-Min array, k number of hash functions) which amounts to assuming uniform distribution on $\binom{n}{k}$ input keys, each hashed to a distinct combination of counters. Thus, the number of distinct keys in the input is assumed to be much larger than the sketch size n. It was observed that the behavior of this model with uniformly distributed keys has important implications to non-uniformly distributed input. Another approach to bounding the error proposed in [16] is based on a simulation of spectral Bloom filters by a hierarchy of ordinary Bloom filters. However, the bounds provided are not explicit but are expressed via a recursive relation based on false positive rates of involved Bloom filters. Recent works [5,6] propose formulas for computing error bounds depending on key probabilities assumed independent but not necessarily uniform, in particular leading to an improved precision bounds for detecting heavy hitters.

In this paper, we provide a probabilistic analysis of the conservative update scheme for counting under the assumption of *uniform distribution of keys* in the input. Our main result is a demonstration that the error in count estimates undergoes a phase transition when the number of distinct keys grows relative to the size of the Count-Min array. We show that the phase transition threshold corresponds to the *peelability threshold* for random k-uniform hypergraphs. For the *subcritical regime*, when the number of distinct keys is below the threshold, we show that the relative error for a randomly chosen key tends to 0 asymptotically, with high probability. This contrasts with the regular Count-Min algorithm producing a relative error shown to be at least 1 with constant probability.

For the *supercritical regime*, we show that the average relative error is lower-bounded by a constant (depending on the number of distinct keys), with high probability. We prove this result for $k = 2$ and conjecture that it holds for arbitrary k as well. We provide computer simulations showing the growth of the expected relative error after the threshold, with a distribution showing a peculiar multi-modal shape. In particular, keys with small (or zero) error still occur after the threshold, but their fraction quickly decreases when the number of distinct keys grows.

After defining Count-Min sketch and conservative update strategy in Sect. 2 and introducing hash hypergraphs in Sect. 3, we formulate the conservative update algorithm (or regular Count-Min, for that matter) in terms of a hypergraph augmented with counters associated to vertices. In Sect. 4, we state our main results and illustrate them with a series of computer simulations. In Sect. 5 we outline the proof of our main result for the subcritical regime and provide a proof for the supercritical regime.

In addition, in Sect. 6, we study a specific family of 2-regular k-hypergraphs that are sparse but not peelable. For such graphs we show that while the relative error of every key is 1 with the regular Count-Min strategy, it is $1/k + o(1)$ for conservative update. While this result is mainly of theoretical interest, it illustrates that the peelability property is crucial for the error to be asymptotically vanishing. Finally, in Sect. 7, we turn to non-uniform distributions and provide a brief experimental analysis of the behavior of the average error for Zipf's distribution compared with the uniform one. Missing full proofs and additional experimental data can be found in [22].

2 Count-Min and Conservative Update

We consider a (counting version of) Count-Min sketch to be an array A of size n of counters initially set to 0, together with hash functions $h_1, \ldots, h_k$ mapping keys from a given universe to $[1..n]$. To count key occurrences in a stream of keys, regular Count-Min proceeds as follows. To process a key p, each of the counters $A[h_i(p)]$, $1 \leq i \leq k$, is incremented by 1. Querying the occurrence number $\boldsymbol{a}(p)$ of a key p returns the estimate $\hat{\boldsymbol{a}}_{CM}(p) = \min_{1 \leq i \leq k}\{A[h_i(p)]\}$. It is easily seen that $\hat{\boldsymbol{a}}_{CM}(p) \geq \boldsymbol{a}(p)$. A bound on the overestimate of $\boldsymbol{a}(p)$ is given by the following result adapted from [13].

Theorem 1 ([13]). *For $\varepsilon > 0$, $\delta > 0$, consider a Count-Min sketch with $k = \lceil \ln(\frac{1}{\delta}) \rceil$ and size $n = k\frac{e}{\varepsilon}$. Then $\hat{\boldsymbol{a}}_{CM}(p) - \boldsymbol{a}(p) \leq \varepsilon N$ with probability at least $1 - \delta$, where N is the size of the input stream.*

While Theorem 1 is useful in some situations, it has a limited utility as it bounds the error with respect to the stream size which can be very large.

Conservative update strengthens Count-Min by increasing only the smallest counters among $A[h_i(p)]$. Formally, for $1 \leq i \leq k$, $A[h_i(p)]$ is incremented by 1 if and only if $A[h_i(p)] = \min_{1\leq j\leq k}\{A[h_j(p)]\}$ and is left unchanged otherwise. The estimate of $\boldsymbol{a}(p)$, denoted $\hat{\boldsymbol{a}}_{CU}(p)$, is computed as before: $\hat{\boldsymbol{a}}_{CU}(p) = \min_{1\leq i\leq k}\{A[h_i(p)]\}$. It can be seen that $\hat{\boldsymbol{a}}_{CU}(p) \geq \boldsymbol{a}(p)$ still holds, and that $\hat{\boldsymbol{a}}_{CU}(p) \leq \hat{\boldsymbol{a}}_{CM}(p)$. The latter follows from the observation that on the same input, an entry of counter array A under conservative update can never get larger than the same entry under Count-Min.

3 Hash Hypergraphs and CU Process

With a counter array $A[1..n]$ and hash functions $h_1, ..., h_k$ we associate a k-uniform *hash hypergraph* $H = (V, E)$ with vertex-set $V = \{1, ..., n\}$ and edge-set $E = \{\{h_1(p), ..., h_k(p)\}\}$ for all distinct keys p. Let $\mathcal{H}^k_{n,m}$ be the set of k-uniform hypergraphs with n vertices and m edges. We assume that the hash hypergraph is a uniformly random Erdős-Rényi hypergraph in $\mathcal{H}^k_{n,m}$, which we denote by $H^k_{n,m}$, where m is the number of distinct keys in the input (for $k = 2$, we use the notation $G_{n,m} = H^2_{n,m}$). Even if this property is not granted by hash functions used in practice, it is a reasonable and commonly used hypothesis to conduct the analysis of sketch algorithms.

Below we show that the behavior of a sketching scheme depends on the properties of the associated hash hypergraph. It is well-known that depending on the m/n ratio, many properties of Erdős-Rényi (hyper)graphs follow a phase transition phenomenon [21]. For example, the emergence of a giant component, of size $O(n)$, occurs with high probability (hereafter, *w.h.p.*) at the threshold $\frac{m}{n} = \frac{1}{k(k-1)}$ [26].

Particularly relevant to us is the *peelability* property. Let $H = (V, E)$ be a hypergraph. The peeling process on H is as follows. We define $H_0 = H$, and iteratively for $i \geq 0$, we define V_i to be the set of leaves (vertices of degree 1) or isolated vertices in H_i, E_i to be the set of edges of H_i incident to vertices in V_i, and H_{i+1} to be the hypergraph obtained from H_i by deleting the vertices of V_i and the edges of E_i. A vertex in V_i is said to have *peeling level* i. The process stabilizes from some step I, and the hypergraph H_I is called the *core* of H, which is the largest induced sub-hypergraph whose vertices all have degree at least 2. If H_I is empty, then H is called *peelable*.

It is known [30] that peelability undergoes a phase transition. For $k \geq 3$, there exists a positive constant λ_k such that, for $\lambda < \lambda_k$, the random hypergraph $H^k_{n,\lambda n}$ is w.h.p. peelable as $n \to \infty$, while for $\lambda > \lambda_k$, the core of $H^k_{n,\lambda n}$ has w.h.p. a size concentrated around αn for some $\alpha > 0$ that depends on λ. The first peelability thresholds are $\lambda_3 \approx 0.818$, $\lambda_4 \approx 0.772$, etc., λ_3 being the largest.

For $k = 2$, for $\lambda < 1/2$, w.h.p. a proportion $1 - o(1)$ of vertices are in trees of size $O(1)$, (and a proportion $o(1)$ of the vertices are in the core), while for $\lambda > 1/2$, the core size is w.h.p. concentrated around αn for $\alpha > 0$ that depends on λ [32].

We note that properties of hash hypergraphs determine the behavior of some other hash-based data structures, such as Cuckoo hash tables [31] and Cuckoo filters [19], Minimal Perfect Hash Functions and Static Functions [28], Invertible Bloom filters [24], and others. We refer to [34] for an extended study of relationships between properties of hash hypergraphs and some of those data structures. In particular, peelability is directly relevant to certain constructions of Minimal Perfect Hash Functions as well as to good functioning of Invertible Bloom filters. However, its relation to Count-Min sketches is less direct and has not been observed earlier.

The connection to hash hypergraphs allows us to reformulate the Count-Min algorithm with conservative updates as a process, which we call CU-process, on a random hypergraph $H^k_{n,m}$, where n, m, k correspond to counter array length, number of distinct keys, and number of hash functions, respectively. Let $H = (V, E)$ be a hypergraph. To each vertex v we associate a counter c_v initially set to 0. At each step $t \geq 1$, a *CU-process* on H chooses an edge $e = \{v_1, \ldots, v_k\} \in E$ in H, and increments by 1 those c_{v_i} which verify $c_{v_i} = \min_{1 \leq j \leq k} c_{v_j}$. For $t \geq 0$ and $v \in V$, $c_v(t)$ will denote the value of the counter c_v after t steps, and $o_e(t)$ the number of times edge $e \in E$ has been drawn in the first t steps. The counter $c_e(t)$ of an edge $e = \{v_1, \ldots, v_k\}$ is defined as $c_e(t) = \min_{1 \leq i \leq k} c_{v_i}(t)$. Clearly, for each t and each e, $o_e(t) \leq c_e(t)$. The *relative error* of e at time t is defined as $R_e(t) = \frac{c_e(t) - o_e(t)}{o_e(t)}$. The following lemma can be easily proved by induction on t.

Lemma 1. *Let $H = (V, E)$ be a hypergraph on which a CU-process is run. At every step t, for each vertex v, there is at least one edge e incident to v such that $c_e(t) = c_v(t)$.*

Observe that, when H is a graph ($k = 2$), Lemma 1 is equivalent to the property that vertex counters cannot have a strict local maximum, i.e., at every step t, each vertex v has at least one neighbour u such that $c_u(t) \geq c_v(t)$.

4 Phase Transition of the Relative Error

4.1 Main Results

Let $H = (V, E)$ be a hypergraph, $|V| = n$, $|E| = m$. Let $N \geq 1$. We consider two closely related models of input to perform the CU-process. In the *N-uniform model*, the CU process is performed on a random sequence of keys (edges in E) of length $N \cdot m$, each key being drawn independently and uniformly in E. In the *N-balanced model*, the CU-process is performed on a random sequence of length $N \cdot m$, such that each $e \in E$ occurs exactly N times, and the order of keys is random. In other words, the input sequence of keys is a random permutation of the multiset made of N copies of each key of E. Clearly, both models are very close, since the number of occurrences of any key in the N-uniform model is concentrated around N (with Gaussian fluctuations of order $\sqrt{N}$) as N gets large. For both models, we use the notation $c_v^{(N)} = c_v(Nm)$ for the resulting counter of $v \in V$, $o_e^{(N)} = o_e(Nm)$ for the resulting number of occurrences of

$e \in E$, $c_e^{(N)} = c_e(Nm)$ for the resulting counter of $e \in E$, and $R_e^{(N)} = R_e(Nm) = (c_e^{(N)} - o_e^{(N)})/o_e^{(N)}$ for the resulting relative error of e. In the N-balanced model, since each key $e \in E$ occurs N times, we have $R_e^{(N)} = (c_e^{(N)} - N)/N$.

Our main result is the following.

Theorem 2 (subcritical regime). *Let $k \geq 2$, and let $\lambda < \lambda_k$, where $\lambda_2 = 1/2$, and for $k \geq 3$, λ_k is the peelability threshold as defined in Sect. 3. Consider a CU-process on a random hypergraph $H_{n,\lambda n}^k$ under either N-uniform or N-balanced model, and consider the relative error $R_e^{(N)}$ of a random edge in $H_{n,\lambda n}^k$. Then $R_e^{(N)} = o(1)$ w.h.p., as both n and N grow*[1].

Note that with the regular Count-Min algorithm (see Sect. 2), in the N-balanced model, the counter value of a node v is $\tilde{c}_v^{(N)} = N \cdot \deg(v)$, and the relative error $\tilde{R}_e^{(N)}$ of an edge $e = (v_1, \ldots, v_k)$ is always (whatever $N \geq 1$) equal to $\min(\deg(v_1), \ldots, \deg(v_k)) - 1$, and is thus always a non-negative integer. For fixed $k \geq 2$ and $\lambda > 0$, and for a random edge e in $H_{n,\lambda n}^k$, the probability that all k vertices belonging to e have at least one incident edge apart from e converges to a positive constant $c(\lambda, k) = (1 - e^{-k\lambda})^k$. Therefore, $\tilde{R}_e$ is a nonnegative integer whose probability to be non-zero converges to $c(\lambda, k)$. Thus, Theorem 2 ensures that, for $\lambda < \lambda_k$, conservative updates lead to a drastic decrease of the error, from $\Theta(1)$ to $o(1)$.

For a given hypergraph $H = (V, E)$ with m edges, we define $\mathrm{err}_N(H) = \frac{1}{m} \sum_{e \in E} R_e^{(N)}$ the *average* error over the edges of H. Formally, Theorem 2 does not imply that $\mathrm{err}_N(H)$ is $o(1)$, as it might possibly happen that a small fraction of edges have very large errors, yielding $\mathrm{err}_N(H)$ larger than $o(1)$. However, we believe that this is not the case. From the previous remark, it follows that the error of an edge $e = (v_1, \ldots, v_k)$ is upper-bounded by $\min(\deg(v_1), \ldots, \deg(v_k)) - 1$. Since the expected maximal degree in $H_{n,\lambda n}^k$ grows very slowly with n, one can expect that any set of $o(n)$ edges should have a contribution $o(1)$ w.h.p.. This is also supported by experiments given in the next section.

Based on Theorem 2 and the above discussion, we propound that a phase transition occurs for the average error, in the sense that it is $o(1)$ in the subcritical regime $\lambda < \lambda_k$, and $\Theta(1)$ in the supercritical regime $\lambda > \lambda_k$, w.h.p.. Regarding the supercritical regime, we are able to show that this indeed holds for $k = 2$ in the N-balanced model.

Theorem 3 (supercritical regime, case $k = 2$). *Let $\lambda > 1/2$. Then there exists a positive constant $f(\lambda)$ such that, in the N-balanced model, $\mathrm{err}_N(G_{n,\lambda n}) \geq f(\lambda)$ w.h.p., as n grows*[2].

[1] Formally, for any $\epsilon > 0$, there exists M such that $\mathbb{P}(R_e^{(N)} \leq \epsilon) \geq 1 - \epsilon$ if $n \geq M$ and $N \geq M$.

[2] Formally, for any $\epsilon > 0$, there exists M such that $\mathbb{P}(\mathrm{err}_N(G_{n,m}) \geq f(\lambda)) \geq 1 - \epsilon$ if $N \geq 1$ and $n \geq M$.

Our proof of Theorem 3 extends to $k \geq 3$ for $\lambda > \widetilde{\lambda_k}$, where $\widetilde{\lambda_k}$ is the threshold beyond which the giant component of $H^k_{n,\lambda n}$ has w.h.p. more edges than vertices. The analysis given in [4] ensures that $\widetilde{\lambda_k}$ exists and is explicitly computable, $\widetilde{\lambda_3} \approx 0.94, \widetilde{\lambda_4} \approx 0.98$. We believe however that the peelability threshold λ_k constitutes the right critical value in Theorem 3 for $k \geq 3$ as well, which is supported by simulations presented below. Proving this would require a different kind of argument than we use in our proof though.

4.2 Simulations

Here we provide several experimental results illustrating the phase transition stated in Theorems 2 and 3. Figure 1 shows plots for the average relative error $\mathrm{err}_N(H^k_{n,m})$ as a function of $\lambda = m/n$, for $k \in \{2, 3, 4\}$ for regular Count-Min and the conservative update strategies. Experiments were run for $n = 1,000$ with the N-uniform model (each edge drawn independently with probability $1/m$) and $N = 50,000$ (number of steps $N \cdot |E|$). For each λ, an average is taken over 15 random graphs.

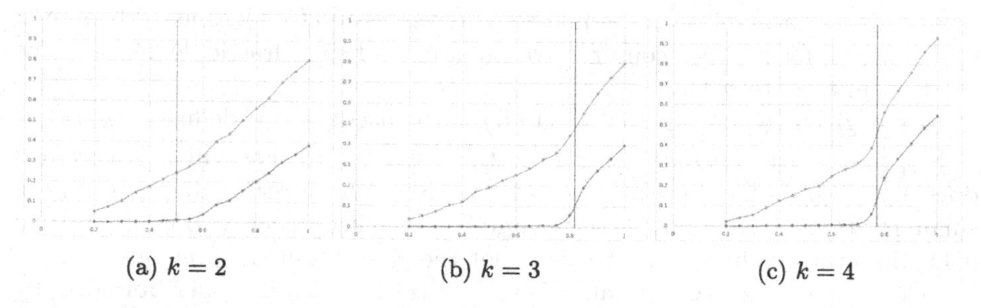

(a) $k = 2$ (b) $k = 3$ (c) $k = 4$

Fig. 1. Average relative error as a function of $\lambda = m/n$ for regular Count-Min (orange) and conservative update (blue), for $k \in \{2, 3, 4\}$. Vertical line shows the peelability threshold. (Color figure online)

The phase transitions are clearly seen to correspond to the critical threshold 0.5 for $k = 2$, and, for $k \in \{3, 4\}$, to the peelability thresholds $\lambda_3 \approx 0.818$, $\lambda_4 \approx 0.772$. Observe that the transition looks sharper for $k \geq 3$, which may be explained by the fact that the core size undergoes a discontinuous phase transition for $k \geq 3$, as shown in [30] (e.g. for $k = 3$, the fraction of vertices in the core jumps from 0 to about 0.13).

For the supercritical regime, we experimentally studied the empirical distribution of individual relative errors, which turns out to have an interesting multimodal shape for intermediate values of λ. Typical distributions for $k \in \{2, 3\}$ are illustrated in Fig. 2 where each point corresponds to an edge, and the edges are randomly ordered along the x-axis. Each plot corresponds to an individual random graph.

When λ grows beyond the peelability threshold, a fraction of edges with small errors still remains but vanishes quickly: these include edges incident to at least one leaf (these have error 0) and peelable edges (these have error $o(1)$), as follows from our proof of Theorem 2. For intermediate values of λ, the distribution presents several modes: besides the main mode (largest concentration on plots of Fig. 2), we observe a few other concentration values which are typically integers. While this phenomenon is still to be analysed, we explain it by the presence of certain structural graph motifs that involve disparities in node degrees. Note that the fraction of values concentrated around the main mode is dominant: for example, for $k = 3, \lambda = 3$ (Fig. 2d), about 90% of values correspond to the main mode (≈ 3.22). Finally, when λ becomes larger, these "secondary modes" disappear, and the distribution becomes concentrated around a single value. More data about concentration is given in the full version [22].

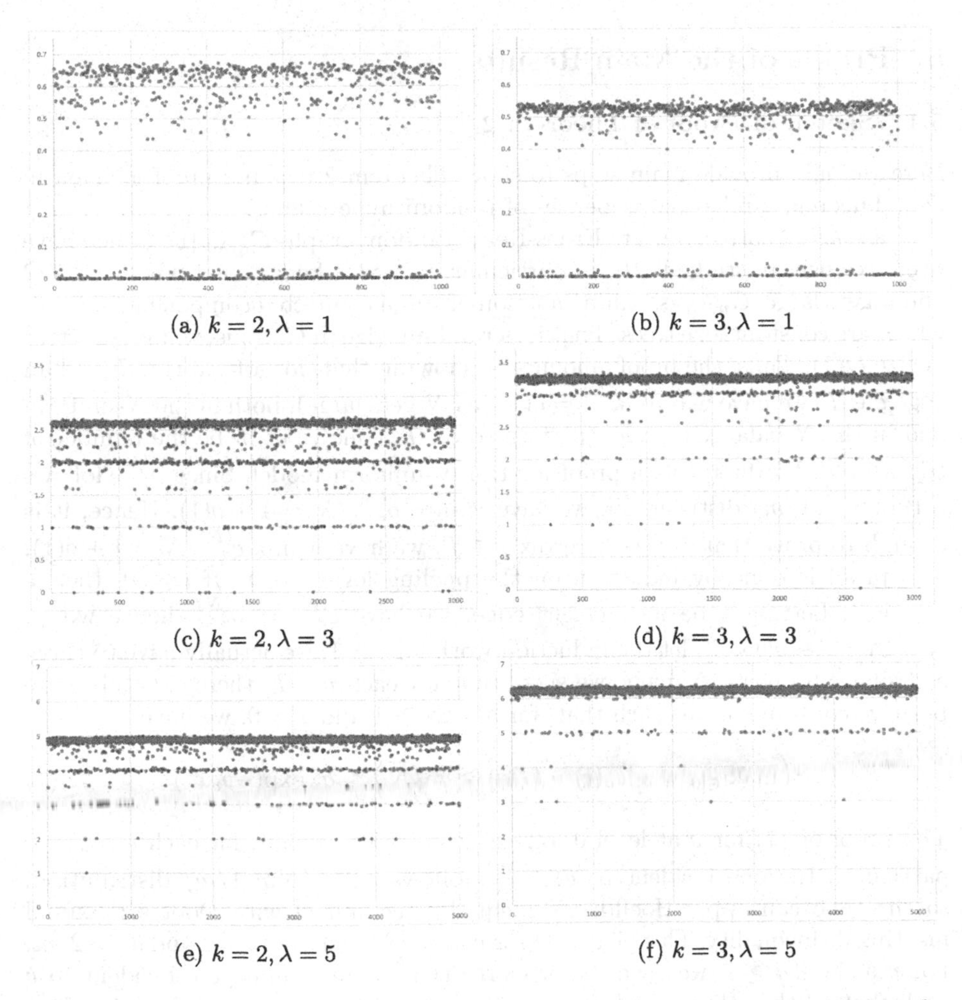

(a) $k = 2, \lambda = 1$ (b) $k = 3, \lambda = 1$

(c) $k = 2, \lambda = 3$ (d) $k = 3, \lambda = 3$

(e) $k = 2, \lambda = 5$ (f) $k = 3, \lambda = 5$

Fig. 2. Distribution of relative errors of individual edges shown in a random order along x-axis.

Our analysis suggests that a positive average error in the supercritical regime is due to a large core — a non-peelable subgraph with $\Theta(n)$ nodes — which exists in this regime. To test this claim (for $k = 3$), we simulated the CU-process on sparse random non-peelable 3-hypergraphs, namely 2-regular 3-hypergraphs with $2n$ edges and $3n$ vertices (n parameter). These are sparsest possible non-peelable 3-hypergraphs, with degree 2 of each vertex. In a separate experiment, we observed that the average error for such graphs is concentrated around a constant value of ≈ 0.217. While these graphs fall to the subcritical regime ($\lambda = 2/3 < \lambda_3 \approx 0.818$), they still generate an average error bounded away from 0. Along with our results of Sect. 6, this supports that peelability is crucial for the error to be $o(1)$, and the presence of large non-peelable subgraphs results in a $\Theta(1)$ error.

5 Proofs of the Main Results

5.1 Sketch of Proof of Theorem 2

Here we only provide main steps to show Theorem 2, the full proof is given in [22]. Theorem 2 relies on properties of random hypergraphs.

Case $k = 2$ corresponds to Erdős-Rényi random graphs $G_{n,\lambda n}$ [17] which have been extensively studied [21]. In particular, it is well known that when $\lambda < 1/2$ and n gets large, $G_{n,\lambda n}$ is, w.h.p., a union of small connected components most of which are constant-size trees. That is, a random edge in $G_{n,\lambda n}$ is, w.h.p., in a tree of size $O(1)$. Thus, the proof amounts to showing that, for a fixed tree T and an edge $e \in T$, we have w.h.p. $R_e^{(N)} = o(1)$ (as N gets large), both in the N-uniform and in the N-balanced model (performed on T alone). Let m be the number of edges in T. We first give a proof for the N-uniform model. Since $o_e^{(N)}$ follows a $\mathrm{Bin}(Nm, 1/m)$ distribution, we have w.h.p. $o_e^{(N)}/N = 1 + o(1)$. Hence, it is enough to prove that, for each vertex $v \in T$, we have w.h.p. $c_v^{(N)}/N = 1 + o(1)$. The proof is done by induction on the peeling level i of v. If $i = 0$, then v is a leaf. Letting e be its incident edge, we have $c_v^{(N)} = o_e^{(N)}$, hence w.h.p. $c_v^{(N)}/N = 1 + o(1)$. To let the induction work for $i \geq 1$, we actually have to carry a stronger property. Namely, we show that for each $v \in T$, there exist absolute positive constants a_v, b_v such that, for any $N \geq 1$ and $x > 0$, we have

$$\mathbb{P}\Big(\max_{t\in[0..Nm]}|c_v(t) - t/m| \geq x\sqrt{N}\Big) \leq a_v \exp(-b_v x^2). \tag{1}$$

The proof of (1) for v at level 0 follows from the fact that, for each $e \in T$ (in particular, the one incident to v), $o_e^{(N)}$ follows a $\mathrm{Bin}(Nm, 1/m)$ distribution, so that one can apply Hoeffding's inequality combined with Doob's maximal martingale inequality. This yields (1) for v at level 0, where $a_v = 2$ and $b_v = 2/m$. For v at level $i \geq 1$, we have the property that there is an edge e incident to v such that all the other neighbors $v_1, \ldots, v_h$ of v (i.e., the neighbors not incident to e) have level smaller than i, and thus satisfy (1) by induction. We then have to check that $c_v(t)$ stays close to t/m for $t \in [0..Nm]$ (in the sense of (1)).

For the lower bound part, we use the fact that $c_v(t) \geq o_e(t)$, and that $o_e(t)$ stays close to t/m. For the upper bound part we use the following argument. Letting $d_v(t) = \max(c_{v_1}(t), \ldots, c_{v_h}(t))$, we can show that if $c_v(t_0) \geq d_v(t_0) + M$ at some time t_0, then there exists $t' \leq t_0$ such that $|o_e(t') - t'/m| \geq M/4$ or $|d_v(t') - t'/m| \geq M/4$ (the crucial point to establish this property, specific to the CU-process, is that in the regime where $c_v(t) > d_v(t)$, any step where $c_v(t)$ increases occurs when picking e). Since $o_e(t)$ and $d_v(t)$ stay close to t/m (for $d_v(t)$ we use induction on i, and the union bound), this property ensures that $c_v(t)$ is unlikely to exceed t/m. Given the lower bound part, $c_v(t)$ hence stays close to t/m (in the sense of (1)). Estimate (1) then guarantees that, in the N-uniform model, we have $|c_v^{(N)}/N - 1| \leq N^{-1/3}$ with probability exponentially close to 1. The same holds in the N-balanced model, by noting that the N-balanced model is the N-uniform model conditioned on the event that each edge is chosen N times, which occurs with a probability of order $N^{-m/2}$ (thus, any event that is almost sure with exponential rate in the N-uniform model is also almost sure with exponential rate in the N-balanced model).

The proof for $k \geq 3$ is analogous but requires some more ingredients. An additional difficulty is that, for $\lambda < \lambda_k$, a random edge e in $H^k_{n,\lambda n}$ may be in the giant component (if $\lambda \in (\frac{1}{k(k-1)}, \lambda_k)$). However, we rely on the fact that the peeling level of e is $O(1)$ w.h.p., and prove that for a vertex v of bounded level, we have $c_v^{(N)}/N = 1 + o(1)$ w.h.p. as $N \to \infty$, where the $o(1)$ term does not depend on the size of the giant component.

5.2 Proof of Theorem 3

The *excess* of a hypergraph H is $\mathrm{exc}(H) = |E| - |V|$.

Lemma 2. *Let $H = (V, E)$ be a k-uniform hypergraph. Then, for the N-balanced model, we have $\sum_{e\in E} R_e^{(N)} \geq \frac{1}{k}\mathrm{exc}(H)$.*

Proof. During the CU process, each time an edge is drawn, the counter of at least one of its extremities is increased by 1. Hence $\sum_{v\in V} c_v^{(N)} \geq N|E|$. Hence, with the notation $R_v^{(N)} := c_v^{(N)}/N - 1$, we have $\sum_{v\in V} R_v^{(N)} \geq \mathrm{exc}(H)$. Now, by Lemma 1, for each $v \in V$, there exists an edge e_v incident to v such that $c_{e_v}^{(N)} = c_v^{(N)}$ (if several incident edges have this property, an arbitrary one is chosen). Hence, $\sum_{v\in V} R_{e_v}^{(N)} \geq \mathrm{exc}(H)$. Note that, in this sum, every edge occurs at most k times (since it has k extremities), thus $\sum_{e\in E} R_e^{(N)} \geq \frac{1}{k}\mathrm{exc}(H)$. □

For $k = 2$ and $\lambda > 1/2$, it is known [32, Theorem 6] that there is an explicit constant $\tilde{f}(\lambda) > 0$ such that the excess of the giant component $G' = (V', E')$ of $G_{n,\lambda n}$ is concentrated around $\tilde{f}(\lambda)n$, with fluctuations of order $\sqrt{n}$. Thus, $\mathrm{exc}(G') \geq \frac{1}{2}\tilde{f}(\lambda)n$ w.h.p. as $n \to \infty$. Hence, by Lemma 2, w.h.p. as $n \to \infty$ (and for any $N \geq 1$), we have

$$\mathrm{err}_N(G_{n,\lambda n}) \geq \frac{1}{\lambda n} \sum_{e\in E'} R_e^{(N)} \geq \frac{1}{2\lambda n}\mathrm{exc}(G') \geq \frac{1}{4\lambda}\tilde{f}(\lambda) =: f(\lambda),$$

which implies Theorem 3.

6 Analysis for Some Non-peelable Hypergraphs

Analysing the asymptotic behaviour of the relative error of the CU-process on arbitrary hypergraphs seems to be a challenging task, even if we restrict ourselves to N-uniform and N-balanced models, as we do in this paper. Based on simulations, we expect that, for a fixed connected k-hypergraph $H = (V, E)$, and for $v \in V$, we have $c_v^{(N)}/N = C_v + o(1)$ w.h.p. as $N \to \infty$, for an explicit constant $C_v \in [1, \deg(v)]$. Since the number of increments at each step lies in $[1..k]$, constants C_v must verify $1 \leq \frac{1}{|E|}\sum_{v\in V} C_v \leq k$, where $\frac{1}{|E|}\sum_{v\in V} C_v$ can be seen as the average number of increments at a step. If H is peelable, then Theorem 2 implies that this concentration holds, with $C_v = 1$. We expect that, if no vertex of H is peelable, and if H is "sufficiently homogeneous", then the constants C_v should be all equal to the same constant $C > 1$, and thus the relative error $R_e^{(N)}$ of every edge is concentrated around $C - 1 > 0$ w.h.p. as $N \to \infty$. This, in particular, is supported by an experiment reported at the end of Sect. 4.2.

In this Section, we show that this is the case for a family of regular hypergraphs which are very sparse ($O(\sqrt{|V|})$ edges) but have a high order (an edge contains $O(\sqrt{|V|})$ vertices). The *dual* of a hypergraph H is the hypergraph H' where the roles of vertices and edges are interchanged: the vertices of H' are the edges of H, and the edges of H' are the vertices of H so that an edge of H' corresponding to a vertex v of H contains those vertices that correspond to edges incident to v in H.

Here we consider the hypergraph K_n' dual to the complete graph K_n. It is a $(n-1)$-uniform hypergraph with n edges and $\binom{n}{2}$ vertices, all of degree 2, therefore no vertex is peelable. For a fixed $n \geq 3$, we consider a CU-process on K_n', in the N-balanced model. Note that with regular Count-Min, the relative error of every edge is 1, since all vertices have degree 2 and $c_v^{(N)} = 2N$ for every vertex v of K_n'. We prove that with conservative updates, the relative error is reduced to a smaller constant $1/(n-1)$. The proof is omitted and can be found in [23].

Theorem 4. *For any fixed $n \geq 2$, in the N-uniform model (resp. in the N-balanced model), the counter of each vertex $v \in K_n'$ satisfies $c_v^{(N)}/N = n/(n-1) + o(1)$ w.h.p. as $N \to \infty$. Hence, the relative error $R_e^{(N)}$ of each edge e in K_n' satisfies $R_e^{(N)} = 1/(n-1) + o(1)$ w.h.p. as $N \to \infty$.*

7 Non-uniform Distributions

An interesting and natural question is whether the phase transition phenomenon holds for non-uniform distributions as well. This question is of practical importance, as in many practical situations keys are not distributed uniformly. In particular, Zipfian distributions often occur in various applications and are a common test case for Count-Min sketches [6,7,9,14,16]. We mention a recent learning-based variant of CountMin [25] (learning heavy hitters) and its study under a Zipfian distribution [1,15].

In Zipf's distributions, key probabilities in descending order are proportional to $1/i^\beta$, where i is the rank of the key and $\beta \geq 0$ is the *skewness* parameter. Note that for $\beta = 0$, Zipf's distribution reduces to the uniform one. It is therefore a natural question whether the phase transition occurs for Zipf's distributions with $\beta > 0$.

One may hypothesize that the answer to the question should be positive, as under Zipf's distribution, frequent keys tend to have no error, as it has been observed in earlier papers [5–7]. On the other hand, keys of the tail of the distribution have fairly similar frequencies, and therefore might show the same behavior as for the uniform case.

However, this hypothesis does not hold. Figure 3 shows the behavior of the average error for Zipf's distributions with $\beta \in \{0.2, 0.5, 0.7, 0.9\}$ vs. the uniform distribution ($\beta = 0$). The average error is defined here as the average error of all keys weighted by their frequencies[3], i.e. $\mathrm{err}_N(H) = \frac{1}{mN} \sum_{e \in E} o_e^{(N)} \frac{c_e^{(N)} - o_e^{(N)}}{o_e^{(N)}} = \frac{1}{mN} \sum_{e \in E} (c_e^{(N)} - o_e^{(N)})$. In other words, $\mathrm{err}_N(H)$ is the expected error of a randomly drawn key from the entire input stream of length mN (taking into account multiplicities).

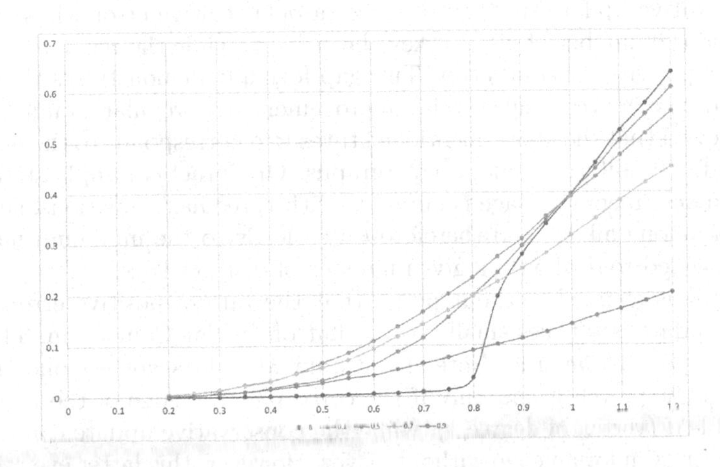

Fig. 3. Average error as a function of $\lambda = m/n$, for Zipf's distributions with $\beta \in \{0.0, 0.2, 0.5, 0.7, 0.9\}$. Plots obtained for $n = 1000$, $k = 3$, $N = 50,000$.

[3] This definition is natural for non-uniform distributions, as the error for a frequent key should have a larger contribution. Note that it is consistent with the definition of Sect. 4.1 in the N-balanced case, and in the N-uniform case it presents a negligible difference when N gets large.

We observe that the phase transition behavior disappears for $\beta > 0$. It turns out that even in the subcritical regime, frequent elements, while having no error themselves, heavily affect the error of certain rare elements, which raises the resulting average error. This phenomenon is analysed in more detail in our follow-up paper [23]. In the supercritical regime ($\lambda > 1$ in Fig. 3) the opposite happens: the uniform distribution shows the largest average error. This is because an increasingly large fraction of the keys (those in the core of the associated hypergraph) contribute to the error, while for skewed distributions, frequent keys tend to have no error, and thus the larger β (with frequent keys becoming more predominant) the smaller the average error. Note that this is in accordance with the observation of [7] that the estimates for the uniform distribution majorate the estimates of infrequent keys for skewed distributions.

8 Concluding Remarks

We presented an analysis of conservative update strategy for Count-Min sketch under the assumption of uniform distribution of keys in the input stream. Our results show that the behaviour of the sketch heavily depends on the properties of the underlying hash hypergraph. Assuming that hash functions are fully independent, the error produced by the sketch follows two different regimes depending on the density of the underlying hypergraph, that is the number of distinct keys relative to the size of the sketch. When this ratio is below the threshold, the conservative update strategy produces a $o(1)$ relative error when the input stream and the number of distinct keys both grow, while the regular Count-Min produces a positive constant error. This gap formally demonstrates that conservative update achieves a substantial improvement over regular Count-Min.

We showed that the above-mentioned threshold corresponds to the peelability threshold for k-uniform random hypergraphs. One practical implication of this is that the best memory usage is obtained with three hash functions, since λ_3 is maximum among all λ_k, and therefore $k = 3$ leads to the minimum number of counters needed to deal with a given number of distinct keys.

In [10] it is claimed, without proof, that the rate of positive errors of conservative update is k times smaller than that of regular Count-Min. This claim does not appear to be true. Note that Count-Min does not err on a key represented in the sketch if and only if the corresponding edge of the hypergraph includes a leaf (vertex of degree 1), while the conservative update can return an exact answer even for an edge without leaves. However, this latter event depends on the relative frequencies of keys and therefore on the specific distribution of keys and the input length. On the other hand, our experiments with uniformly distributed keys show that this event is relatively rare, and the rate of positive error for Count-Min and conservative update are essentially the same.

One important assumption of our analysis is the uniform distribution of keys in the input. We presented an experimental evidence that for skewed distributions, in particular for Zipf's distribution, the phase transition disappears when the skewness parameter grows. Therefore, the uniform distribution presents the smallest error in the subcritical regime. The situation is the opposite in the supercritical regime when the number of distinct keys is large compared to the number of counters: here the uniform distribution presents the largest average error. As mentioned earlier, for Zipf's distribution, frequent keys have essentially no error, whereas in the supercritical regime, low frequency keys have all similar overestimates. This reveals another type of phase transition in error approximation for Zipf's distribution, occurring between frequent and infrequent elements, having direct application to accurate detection of *heavy hitters* in streams. We refer to our follow-up work [23] for further insights regarding this issue.

Acknowledgments. We thank Djamal Belazzougui who first pointed out to us the conservative update strategy.

References

1. Aamand, A., Indyk, P., Vakilian, A.: (Learned) frequency estimation algorithms under Zipfian distribution. CoRR abs/1908.05198 (2019)
2. Almeida, P.S.: A case for partitioned Bloom filters. CoRR abs/2009.11789 (2020)
3. Behera, S., Gayen, S., Deogun, J.S., Vinodchandran, N.: Kmerestimate: a streaming algorithm for estimating k-mer counts with optimal space usage. In: Proceedings of the ACM International Conference on Bioinformatics, Computational Biology, and Health Informatics, pp. 438–447 (2018)
4. Behrisch, M., Coja-Oghlan, A., Kang, M.: Local limit theorems for the giant component of random hypergraphs. Comb. Probab. Comput. **23**(3), 331–366 (2014)
5. Ben Mazziane, Y., Alouf, S., Neglia, G.: Analyzing count min sketch with conservative updates. Comput. Netw. **217**, 109315 (2022)
6. Ben Mazziane, Y., Alouf, S., Neglia, G.: A formal analysis of the count-min sketch with conservative updates. CoRR abs/2203.14549 (2022)
7. Bianchi, G., Duffy, K., Leith, D.J., Shneer, V.: Modeling conservative updates in multi-hash approximate count sketches. In: Proceedings of the 24th International Teletraffic Congress, ITC 2012, Kraków, Poland, 4–7 September 2012, pp. 1–8. IEEE (2012)
8. Charikar, M., Chen, K., Farach-Colton, M.: Finding frequent items in data streams. Theoret. Comput. Sci. **312**(1), 3–15 (2004)
9. Chen, P., Wu, Y., Yang, T., Jiang, J., Liu, Z.: Precise error estimation for sketch-based flow measurement. In: Proceedings of the 21st ACM Internet Measurement Conference, pp. 113–121 (2021)
10. Cohen, S., Matias, Y.: Spectral bloom filters. In: Halevy, A.Y., Ives, Z.G., Doan, A. (eds.) Proceedings of the 2003 ACM SIGMOD International Conference on Management of Data, pp. 241–252 (2003)
11. Cormode, G.: Count-min sketch. In: Liu, L., Özsu, M.T. (eds.) Encyclopedia of Database Systems, 2nd edn. Springer, Cham (2018)
12. Cormode, G., Hadjieleftheriou, M.: Finding frequent items in data streams. Proc. VLDB Endowment **1**(2), 1530–1541 (2008)
13. Cormode, G., Muthukrishnan, S.: An improved data stream summary: the count-min sketch and its applications. J. Algorithms **55**(1), 58–75 (2005)
14. Cormode, G., Muthukrishnan, S.: Summarizing and mining skewed data streams. In: Proceedings of the 2005 SIAM International Conference on Data Mining, SDM 2005, Newport Beach, CA, USA, 21–23 April 2005, pp. 44–55 (2005)

15. Du, E., Wang, F., Mitzenmacher, M.: Putting the "learning" into learning-augmented algorithms for frequency estimation. In: Proceedings of the 38th International Conference on Machine Learning, ICML 2021, vol. 139, pp. 2860–2869 (2021)
16. Einziger, G., Friedman, R.: A formal analysis of conservative update based approximate counting. In: International Conference on Computing, Networking and Communications, ICNC 2015, pp. 255–259 (2015)
17. Erdos, P., Rényi, A., et al.: On the evolution of random graphs. Publ. Math. Inst. Hung. Acad. Sci **5**(1), 17–60 (1960)
18. Estan, C., Varghese, G.: New directions in traffic measurement and accounting. In: Mathis, M., Steenkiste, P., Balakrishnan, H., Paxson, V. (eds.) Proceedings of the ACM SIGCOMM 2002 Conference on Applications, Technologies, Architectures, and Protocols for Computer Communication, pp. 323–336. ACM (2002)
19. Fan, B., Andersen, D.G., Kaminsky, M., Mitzenmacher, M.D.: Cuckoo filter: Practically better than bloom. In: Proceedings of the 10th ACM International on Conference on Emerging Networking Experiments and Technologies, pp. 75–88 (2014)
20. Fan, L., Cao, P., Almeida, J., Broder, A.: Summary cache: a scalable wide-area web cache sharing protocol. IEEE/ACM Trans. Networking **8**(3), 281–293 (2000)
21. Frieze, A., Karoński, M.: Introduction to Random Graphs. Cambridge University Press, Cambridge (2015)
22. Fusy, É., Kucherov, G.: Phase transition in count approximation by Count-Min sketch with conservative updates. CoRR abs/2203.15496 (2022)
23. Fusy, É., Kucherov, G.: Count-min sketch with variable number of hash functions: an experimental study. CoRR abs/2302.05245 (2023)
24. Goodrich, M.T., Mitzenmacher, M.: Invertible bloom lookup tables. In: Proceedings of the 49th Annual Allerton Conference on Communication, Control, and Computing (Allerton), pp. 792–799. IEEE (2011)
25. Hsu, C., Indyk, P., Katabi, D., Vakilian, A.: Learning-based frequency estimation algorithms. In: Proceedings of the 7th International Conference on Learning Representations, ICLR 2019 (2019)
26. Karoński, M., Łuczak, T.: The phase transition in a random hypergraph. J. Comput. Appl. Math. **142**(1), 125–135 (2002)
27. Liu, H., Lin, Y., Han, J.: Methods for mining frequent items in data streams: an overview. Knowl. Inf. Syst. **26**(1), 1–30 (2011)
28. Majewski, B.S., Wormald, N.C., Havas, G., Czech, Z.J.: A family of perfect hashing methods. Comput. J. **39**(6), 547–554 (1996)
29. Mohamadi, H., Khan, H., Birol, I.: ntCard: a streaming algorithm for cardinality estimation in genomics data. Bioinformatics **33**(9), 1324–1330 (2017)
30. Molloy, M.: Cores in random hypergraphs and boolean formulas. Random Struct. Algorithms **27**(1), 124–135 (2005)
31. Pagh, R., Rodler, F.F.: Cuckoo hashing. J. Algorithms **51**(2), 122–144 (2004)
32. Pittel, B., Wormald, N.C.: Counting connected graphs inside-out. J. Comb. Theory, Series B **93**(2), 127–172 (2005)
33. Shibuya, Y., Kucherov, G.: Set-Min sketch: a probabilistic map for power-law distributions with application to k-mer annotation. bioRxiv (2020). https://www.biorxiv.org/content/10.1101/2020.11.14.382713v1
34. Walzer, S.: Random hypergraphs for hashing-based data structures. Ph.D. thesis, Technische Universität Ilmenau, Germany (2020)

Minimum-Link C-Oriented Paths Visiting a Sequence of Regions in the Plane

Kerem Geva[1], Matthew J. Katz[1(✉)], Joseph S. B. Mitchell[2], and Eli Packer[3]

[1] Ben Gurion University of the Negev, Beer Sheva, Israel
gevak@post.bgu.ac.il, matya@bgu.ac.il
[2] Stony Brook University, Stony Brook, NY, USA
joseph.mitchell@stonybrook.edu
[3] Track160, Tel Aviv-Yafo, Israel
eli.p@track160.com

Abstract. Let $E = \{e_1, \ldots, e_n\}$ be a set of C-oriented disjoint segments in the plane, where C is a given finite set of orientations that spans the plane, and let s and t be two points. We seek a minimum-link C-oriented tour of E, that is, a polygonal path π from s to t that visits the segments of E in order, such that, the orientations of its edges are in C and their number is minimum. We present an algorithm for computing such a tour in $O(|C|^2 \cdot n^2)$ time. This problem already captures most of the difficulties occurring in the study of the more general problem, in which E is a set of not-necessarily-disjoint C-oriented polygons.

1 Introduction

We consider the problem in which we are given a sequence of regions, $\mathcal{R} = (R_1, R_2, \ldots, R_n)$, where each R_i is a subset of an underlying geometric domain, and our goal is to compute a tour (a path or a cycle) within the domain that visits the regions $\mathcal{R}$ in the given order and is optimal in some prescribed sense. Optimality might be based on the Euclidean length of the tour, the number of turns in a polygonal tour (or, equivalently, the number of *links* (edges) in the tour), a weighted cost function, etc. There are also variants of the problem in which it is important to specify exactly what constraints there are on the ordered visitation of the regions, particularly if the regions are not disjoint. The problem arises naturally and is also motivated by applications in curve simplification (e.g., [6]), vehicle routing (e.g., the travelling salesperson problem (TSP); see [8]), search and exploration (e.g., [3]), computing structures on imprecise points [7], task sequencing in robotics (see [1,2]), etc.

In this paper we focus on the version of the problem in which the regions R_i are disjoint C-oriented line segments (with orientations/slopes from a finite set C) in the plane, the tour is required to be polygonal and C-oriented, and the optimality criterion is to minimize the number of links (equivalently, the number of turns, or vertices in the polygonal tour). We briefly mention generalizations (deferred to the full paper), including the case in which the regions R_i are more general than disjoint line segments.

M. Mavronicolas (Ed.): CIAC 2023, LNCS 13898, pp. 247–262, 2023.
https://doi.org/10.1007/978-3-031-30448-4_18

More formally, let C be a finite set of orientations, which can be thought of as points on a unit circle centered at the origin. We assume that (i) C spans the plane, i.e., for any two points p, q in the plane, there exists a two-link (directed) path from p to q (or a one-link path), such that the orientation of the edges in the path belong to C, and (ii) for any orientation $c_i \in C$, the orientation $\overline{c_i}$ is also in C, where $\overline{c_i}$ is the opposite orientation of c_i. The requirement for paths to be C-oriented arises in some settings (mechanical constraints) but also has advantages in lower/upper bounding of the turn angles, in comparison with polygonal paths having general links, which may form arbitrarily sharp turns.

We focus on the following problem: *Minimum-link C-oriented tour of a sequence of C-oriented segments*: Let $E = \{e_1, \ldots, e_n\}$ be a set of C-oriented disjoint segments, that is, if we think of $e \in E$ as a directed segment, by arbitrarily picking one of the two possible directions, then e's orientation belongs to C. Let s and t be two points that do not belong to any of the segments in E. A *tour* of E is a polygonal path π that begins at s and ends at t with the following property: There exists a sequence of points $p_1, \ldots, p_n$ on π, such that, p_i precedes p_{i+1}, for $1 \leq i \leq n-1$, and $p_i \in e_i$, for $1 \leq i \leq n$. A tour is C-oriented, if the orientation of each of its edges belongs to C. We wish to compute a C-oriented minimum-link tour of E, that is, a C-oriented tour consisting of a minimum number of links (i.e., edges).

Our main contribution is an efficient algorithm to compute a minimum-link C-oriented tour of a sequence of n disjoint C-oriented line segments, in time $O(|C|^2 \cdot n^2)$. (The algorithm becomes $O(n)$ in the special case of $|C| = 4$, e.g., axis-oriented paths).

Related Work

In the *touring polygons problem* (TPP), one seeks a tour that is shortest in Euclidean length that visits a sequence of polygons; such a tour is found in polynomial time if the polygons are convex and is NP-hard in general (and has an FPTAS) [3]. Minimization of the link length of a tour visiting a sequence of (possibly overlapping) disks is studied in [6], where the motivation for this "ordered stabbing" problem was curve and map simplification (see also [10]). In contrast with our problem specification, in [6] the path edges are allowed to be of arbitrary orientation, not required to be C-oriented. This assumption leads to particular efficiency, as one can use an extension of linear-time line stabbing methods (see Egyed and Wenger [4]) to execute a greedy algorithm efficiently. Computing a minimum-link C-oriented path from start to goal among obstacles has been studied as well, without requiring visitation of a sequence of regions; see [9,11].

2 Preliminaries

Notation. For any $1 \leq i \leq n$, let $l(e_i)$ be the number of links in a minimum-link path that begins at s and ends at a point on e_i. We only consider C-oriented paths to e_i that visit the segments $e_1, \ldots, e_i$, as defined above. We refer to the number of links in such a path as its *length*. We distinguish between paths to e_i both by their length and by the orientation of their last link. Let $I(e_i, c_j)$ ($I^+(e_i, c_j)$) be the set of maximal intervals on e_i formed by all paths of length $l(e_i)$ ($l(e_i) + 1$) from s to e_i, whose last link has orientation c_j. We set $I(e_i) = \bigcup_{c \in C} I(e_i, c)$ and $I^+(e_i) = \bigcup_{c \in C} I^+(e_i, c)$.

For an orientation $c_j \in C$, let c_{j+1} and c_{j-1} be the orientations in C that immediately succeed c_j and precede c_j in clockwise order, respectively. We denote by $\phi(c_j, c_k)$ the set of orientations in C between c_j and c_k (in clockwise order from c_j), not including c_j and c_k. Finally, we denote the ray emanating from p in orientation c_j by $Ray(p, c_j)$ and the line through p parallel to a segment of orientation c_j by $Line(p, c_j)$.

Let a be an interval on e_i that belongs to one of the sets $I(e_i)$ or $I^+(e_i)$. Then a has a length l_a (which is either $l(e_i)$ or $l(e_i) + 1$) and an orientation $c_a \in C$ associated with it. We denote the endpoints of a by a_1 and a_2, where a_1 is to the left of a_2, when approaching a through a path corresponding to a (i.e., a path starting at s and ending at a point in a, which is of length l_a and whose last link is of orientation c_a). Next, we use a to define two regions of the plane, namely, $PT(a)$ and $\psi(a, c_j)$.

Let $PT(a)$ denote the semi-slab consisting of all points that can be reached by extending the last link of a path corresponding to a. We refer to such a path as a path that *passes through* a and continues in the same orientation at which it reached a (i.e., c_a). Thus, the region $PT(a)$ is the semi-slab bounded by the rays $Ray(a_1, c_a), Ray(a_2, c_a)$ and the interval a (see, e.g., the red region in Fig. 1). Similarly, let $\psi(a, c_j)$ be the region of all points that can be reached by a path that passes through a and then, not necessarily immediately, turns and continues in orientation c_j. Thus, $\psi(a, c_j) = \bigcup_{q \in PT(a)} Ray(q, c_j)$, for example if $c_j = \overline{c_a}$, then $\psi(a, c_j)$ is the slab defined by the lines $Line(a_1, c_a)$ and $Line(a_2, c_a)$ (for additional examples see Fig. 8).

Finally, for an interval $b \in I^+(e_i)$, we set $\delta(b) = \{a \in I(e_i) | a \subseteq b\}$.

We now show that the sets $I(e_i)$ and $I^+(e_i)$ are sufficient, in the sense that there exists a minimum-link tour of E whose portion from s to e_i corresponds to an interval in $I(e_i) \cup I^+(e_i)$. Assume this is false, and let π be a minimum-link tour of E, such that its portion π_i from s to e_i does not correspond to an interval in $I(e_i) \cup I^+(e_i)$. Then, the length of π_i (denoted $|\pi_i|$) is at least $l(e_i) + 2$. Let p be the point on e_i where π_i ends, and denote the portion of π from p to t by π^i. Then $|\pi| \geq l(e_i) + 2 + |\pi^i|$, if π makes a turn at p, or $|\pi| = l(e_i) + 2 + |\pi^i| - 1$, otherwise. Consider any path π'_i from s to e_i that corresponds to an interval in $I(e_i)$ and let p' be the point on e_i where π'_i ends. Then, the tour obtained by π'_i, the edge $p'p$ and π^i is a tour of E of length at most $l(e_i) + 1 + |\pi^i| \leq |\pi|$. We have thus shown that

Claim 1. *There exists a minimum-link tour of E whose portion from s to e_i corresponds to an interval in $I(e_i) \cup I^+(e_i)$, for $1 \leq i \leq n$.*

Finally, since our assumptions on the set of orientations C imply that there exists a two-link path from p to q, for any pair of points p, q in the plane, we have

Claim 2. *$l(e_{i-1}) \leq l(e_i) \leq l(e_{i-1}) + 2$, for $1 \leq i \leq n$ (where $l(e_0) = 0$).*

3 The Main Algorithm

In this section, we present an algorithm for computing a minimum-link tour of E. The algorithm consists of two stages. In the first stage, it considers the segments of E, one at a time, beginning with e_1, and, at the current segment e_i, it computes the sets $I(e_i)$ and $I^+(e_i)$ from the sets $I(e_{i-1})$ and $I^+(e_{i-1})$, associated with the previous segment. In the second stage, it constructs a minimum-link tour of E, beginning from its last link, by consulting the sets $I(\cdot)$ and $I^+(\cdot)$ computed in the first stage.

We begin with several definitions that will assist us in the description of the algorithm. Given a set I of intervals on e_i, where each interval $a \in I$ is associated with some fixed length (link distance) $l_a = l$ and an orientation c_a, and $c_j \in C$, we define the sets of intervals *+0-intervals, +1-intervals, +2-intervals* on e_{i+1} with respect to I and c_j (the definition of the first set does not depend on c_j).

The **+0-intervals** on e_{i+1} consist of the intervals on e_{i+1} formed by passing through the intervals of I, without making any turns. It is constructed by computing the interval $b = PT(a) \cap e_{i+1}$, for each $a \in I$, and including it in the set, setting $l_b = l$ and $c_b = c_a$, if it is not empty.

The **+1-intervals** on e_{i+1} associated with orientation c_j consist of the intervals on e_{i+1} formed by passing through the intervals of I and then making a turn in orientation c_j. It is constructed by computing the interval $b = \psi(a, c_j) \cap e_{i+1}$, for each $a \in I$, and including it in the set, setting $l_b = l + 1$ and $c_b = c_j$.

The **+2-intervals** on e_{i+1} associated with orientation c_j consist of the intervals on e_{i+1} formed by passing through the intervals of I and then making two turns, where the first is in any orientation $c \neq \overline{c_a}$ and the second is in orientation c_j; see Lemma 3.

We construct it as follows. First, we check if there is an interval $a \in I$ such that $c_a \notin \{c_{j-1}, c_j, c_{j+1}\}$. If there is such an interval, we include the interval $b = e_{i+1}$, setting $l_b = l + 2$ and $c_b = c_j$, and stop; see Lemma 4. Otherwise, for each $a \in I$, we include the intervals $b^+ = \psi(a, \overline{c_a}_{+1}) \cap e_{i+1}$ and $b^- = \psi(a, \overline{c_a}_{-1}) \cap e_{i+1}$, provided that they are not empty, and set $l_{b^+} = l_{b^-} = l + 2$ and $c_{b^+} = c_{b^-} = c_j$; see paragraph following Lemma 4.

3.1 Stage I

We are now ready to describe the first stage of the algorithm. It is convenient to treat the points s and t as segments e_0 and e_{n+1}, respectively. We set $l(e_0) = 0$ and, for each $c_j \in C$, we insert the interval $a = e_0$, after setting $l_a = 0$ and $c_a = c_j$, into $I(e_0, c_j)$. Similarly, for each $c_j \in C$, we insert the interval $a = e_0$, after setting $l_a = 1$ and $c_a = c_j$, into $I^+(e_0, c_j)$.

We iterate over the segments $e_1, \ldots, e_{n+1}$, where in the i'th iteration, $1 \leq i \leq n + 1$, we compute $l(e_i)$ and the pair of sets $I(e_i)$ and $I^+(e_i)$. Assume we have already processed the segments $e_0, \ldots, e_i$, for some $0 \leq i \leq n$. We describe the next iteration, in which we compute $l(e_{i+1})$ and the sets $I(e_{i+1})$ and $I^+(e_{i+1})$.

For each $c_j \in C$, we compute the +0-intervals on e_{i+1} with respect to $I(e_i, c_j)$ and store them in $I(e_{i+1}, c_j)$. If at least one of the sets $I(e_{i+1}, c_j)$ is non-empty, we set $l(e_{i+1}) = l(e_i)$ (otherwise $l(e_{i+1}) > l(e_i)$). Next, for each $c_j \in C$, we compute the +0-intervals on e_{i+1} with respect to $I^+(e_i, c_j)$ and the +1-intervals on e_{i+1} with respect to $I(e_i) \backslash I(e_i, c_j)$ (and c_j). We store these intervals (if exist) either in $I^+(e_{i+1}, c_j)$, if $l(e_{i+1}) = l(e_i)$, or in $I(e_{i+1}, c_j)$, if $l(e_{i+1}) > l(e_i)$. If we performed the latter option, then we set $l(e_{i+1}) = l(e_i) + 1$. Finally, if we performed one of the two options, then we repeatedly merge overlapping intervals in the set (either $I^+(e_{i+1}, c_j)$ or $I(e_{i+1}, c_j)$), until there are no such intervals.

If $l(e_{i+1}) > l(e_i)$, then, for each $c_j \in C$, we compute the +2-intervals on e_{i+1} with respect to $I(e_i)$ and the +1-intervals on e_{i+1} with respect $I^+(e_i) \backslash I^+(e_i, c_j)$. We store these intervals (if exist) either in $I^+(e_{i+1}, c_j)$, if $l(e_{i+1}) = l(e_i) + 1$, or in $I(e_{i+1}, c_j)$, otherwise (i.e., we still have not fixed $l(e_{i+1})$). If we performed the latter option, then we set $l(e_{i+1}) = l(e_i) + 2$, and, as above, if we performed one of the two options, then we repeatedly merge overlapping intervals in the set, until there are no such intervals.

Finally, if $l(e_{i+1}) = l(e_i) + 2$, then, for each $c_j \in C$, we set $I^+(e_{i+1}, c_j) = e_{i+1}$; see Claim 5.

3.2 Stage II

In this stage we use the information collected in the first stage to construct a minimum-link tour π of E.

We construct π incrementally beginning at t and ending at s. That is, in the first iteration we add the portion of π from t to e_n, in the second iteration we add the portion from e_n to e_{n-1}, etc. Assume that we have already constructed the portion of π from t to e_i, where this portion ends at point p of interval a on e_i. In the full version of this paper [5], we describe how to compute the portion from e_i to e_{i-1}, which begins at the point p of interval a and ends at a point p' of interval b on e_{i-1} (where $b \in I(e_{i-1}) \cup I^+(e_{i-1})$) and consists of $l_a - l_b + 1$ links. Before continuing to the next iteration, we set $p = p'$ and $a = b$.

After adding the last portion, which ends at s, we remove all the redundant vertices from π, i.e., vertices at which π does not make a turn.

4 Analysis

In this section, we prove the correctness of our two-stage algorithm and bound its running time, via a sequence of lemmas and claims.

Lemma 1. *For any interval $a \in I(e_i)$ and for any $c_j \in C \backslash \{c_a\}$, there exists an interval $b \in I^+(e_i, c_j)$ such that $a \subseteq b$, for $1 \leq i \leq n$.*

Proof. Let $p \in a$, then there is a path π_i of length $l(e_i)$ that begins at s, ends at p, and whose last link is of orientation c_a. By making a turn at p in orientation c_j (without extending π_i), we obtain a path π_i' of length $l(e_i)+1$, whose last link is of orientation c_j. Therefore, there is an interval $b \in I^+(e_i, c_j)$ such that $p \in b$, and since (by construction) there are no overlapping intervals in $I^+(e_i, c_j)$, we conclude that $a \subseteq b$.

Lemma 2. *For any* $1 \leq i \leq n-1$ *and* $c_j \in C$*, if there is an interval* $a \in I(e_i, \overline{c_j}) \cup I^+(e_i, \overline{c_j})$ *such that* $PT(a) \cap e_{i+1} \neq \emptyset$*, then, for any interval* $b \in I(e_i, c_j) \cup I^+(e_i, c_j)$*, we have that* $PT(b) \cap e_{i+1} = \emptyset$*.*

Proof. If there exist intervals $a \in I(e_i, \overline{c_j}) \cup I^+(e_i, \overline{c_j})$ and $b \in I(e_i, c_j) \cup I^+(e_i, c_j)$, such that e_{i+1} intersects both $PT(a)$ and $PT(b)$, then e_{i+1} must intersect e_i (see Fig. 1)—contradiction.

The following claim bounds the number of intervals with associated length and orientation $l(e_i)+1$ and c_j, respectively, that are 'created' on e_{i+1}.

Claim 3. *At most* $\max\{|I(e_i, \overline{c_j})|, |I^+(e_i, c_j)|\} + 2$ *intervals with associated length and orientation* $l(e_i)+1$ *and* c_j*, respectively, are 'created' on* e_{i+1}*, during the execution of the algorithm.*

Proof. There are two ways to reach a point on e_{i+1} with a path of length $l(e_i)+1$ whose last link is of orientation c_j. The first is by passing through one of the intervals in $I(e_i) \backslash I(e_i, c_j)$ and then making a turn in orientation c_j. The second is by passing through one of the intervals in $I^+(e_i, c_j)$, without making any turn. That is, the intervals on e_{i+1} with associated length $l(e_i)+1$ and associated orientation c_j are determined by the intervals in $I^+(e_i, c_j) \cup (I(e_i) \backslash I(e_i, c_j))$.

Consider an interval $b \in I^+(e_i, c_j)$ (e.g., the blue interval in Fig. 2), and let $c \in C$ be the orientation of e_i when directed from b_1 to b_2. We divide $\delta(b) \backslash I(e_i, \overline{c_j})$ into four subsets as follows: $A = \{a \in \delta(b) \mid c_a \in \phi(c_j, c) \cup \{c\}\}$, $B = \{a \in \delta(b) \mid c_a \in \phi(c, \overline{c_j})\}$, $C = \{a \in \delta(b) \mid c_a \in \phi(\overline{c_j}, \overline{c})\}$, and $D = \{a \in \delta(b) \mid c_a \in \phi(\overline{c}, c_j) \cup \{\overline{c}\}\}$. We denote by $R_{b \cup A}$ the region of all points that can be reached by a path that passes through b, or passes through $a \in A$ and then makes a turn in orientation c_j (i.e., $R_{b \cup A} = PT(b) \cup \bigcup_{a \in A} \psi(a, c_j)$). We compute the boundary of $R_{b \cup A}$ from $PT(b)$, by adding the regions $\psi(a, c_j)$, one at a time, for each interval $a \in A$.

Let $\psi(a, c_j)$, for some $a \in A$, be the region that is added in the first step (see the red interval in Fig. 2). Since $(a_1, a_2) \subseteq (b_1, b_2)$ and $c_a \in \phi(c_j, c) \cup \{c\}$, $Ray(a_2, c_a)$ and $Ray(b_2, c_j)$ intersect at a point p_a (see Fig. 2). By passing through a and then turning before reaching $Ray(b_2, c_j)$ (i.e., at one of the points belonging to $PT(a) \cap PT(b)$), we cannot reach any point that is not already in $PT(b)$. However, by turning after crossing $Ray(b_2, c_j)$, we can reach points that are in the area bounded by $Ray(p_a, c_j)$ and $Ray(p_a, c_a)$ (the shaded area in Fig. 2). Thus, the region $R_{b \cup A}$ at the end of the first step, is bounded by $Ray(b_1, c_j)$, (b_1, b_2), (b_2, p_a) and $Ray(p_a, c_a)$, as can be seen in Fig. 2. Notice the semi-infinite convex 2-chain that we obtain at the end of the first step, namely,

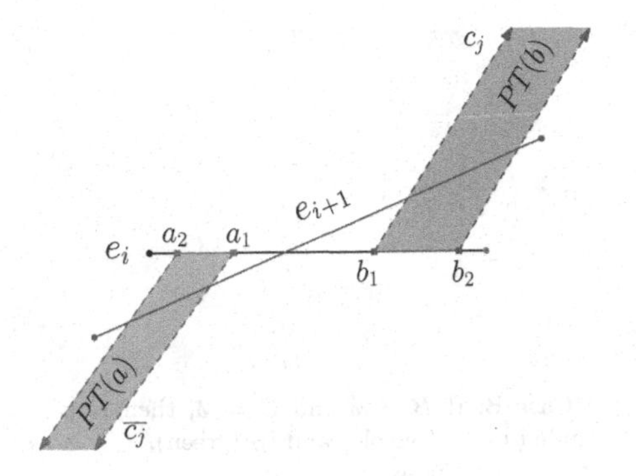

Fig. 1. If e_{i+1} intersects both $PT(a)$ (red) and $PT(b)$ (blue), then e_{i+1} must intersect e_i. (Color figure online)

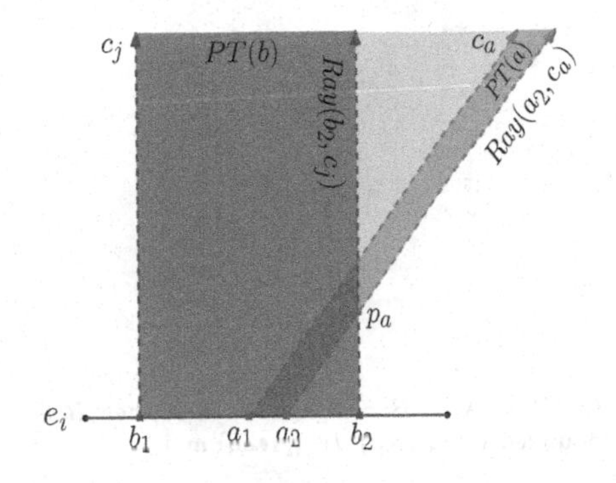

Fig. 2. The region $R_{b\cup A}$ at the end of the first step.

the chain consisting of (b_2, p_a) followed by $Ray(p_a, c_a)$. It is easy to see that the region $R_{b\cup A}$ at the end of the last step, is bounded by $Ray(b_1, c_j)$, (b_1, b_2), and a semi-infinite convex chain, denoted l_A, consisting of at most $|A| + 1$ edges (see red chain in Fig. 3c). Finally, if $A = \emptyset$, then $R_{b\cup A} = PT(b)$ and we set $l_A = Ray(b_2, c_j)$.

Next, we set $R_{b\cup D} = PT(b) \cup \bigcup_{a\in D} \psi(a, c_j)$, and compute the convex chain l_D, which, together with $Ray(b_2, c_j)$ and (b_1, b_2), defines the boundary of $R_{b\cup D}$. Once again, if $D = \emptyset$, we set $l_D = ray(b_1, c_j)$.

Finally, we compute in a similar manner the convex chains l_B, which defines (together with $Ray(b_1, c_j)$) the boundary of $R_{b\cup B} = PT(b) \cup \bigcup_{a\in B} \psi(a, c_j)$ (see purple chain in Fig. 3b), and l_C, which defines (together with $Ray(b_2, c_j)$) the boundary of $R_{b\cup C} = PT(b) \cup \bigcup_{a\in C} \psi(a, c_j)$.

We now set $R = R_{b\cup A} \cup R_{b\cup B} \cup R_{b\cup C} \cup R_{b\cup D}$, then R is the region of all points that can be reached by a path that passes through b, or passes through $a \in \delta(b)\backslash I(e_i, \overline{c_j})$ and then makes a turn in orientation c_j. Therefore, $R \cap e_{i+1}$ gives us the intervals on e_{i+1} with length $l(e_i) + 1$ and orientation c_j, which are created by passing through an interval in $\{b\} \cup \delta(b)\backslash I(e_i, \overline{c_j})$.

In order to find these intervals, we identify the boundary of R in each of the following four cases:

- **Case A:** $B = \emptyset$ and $C = \emptyset$ (as illustrated in Fig. 3a)
  In this case, $R = R_{b\cup A} \cup R_{b\cup D}$, since $R_{b\cup B} = R_{b\cup C} = PT(b)$ and $PT(b) \subseteq R_{b\cup A}, R_{b\cup D}$, and R's boundary is composed of l_A, l_D and b.
- **Case B:** $B \neq \emptyset$ and $C = \emptyset$ (as illustrated in Fig. 3b)
  In this case, the boundary of R is composed of l_B and l_D, since $R_{b\cup A} \subseteq R_{b\cup B}$.
- **Case C:** $B = \emptyset$ and $C \neq \emptyset$ (as illustrated in Fig. 3c)
  In this case, the boundary of R is composed of l_A and l_C, since $R_{b\cup D} \subseteq R_{b\cup C}$.
- **Case D:** $B \neq \emptyset$ and $C \neq \emptyset$ (as illustrated in Fig. 3d)
  in this case, $R = R_{b\cup B} \cup R_{b\cup C}$, and its boundary is the convex chain l that is obtained from the chains l_B and l_C, see Fig. 3d.

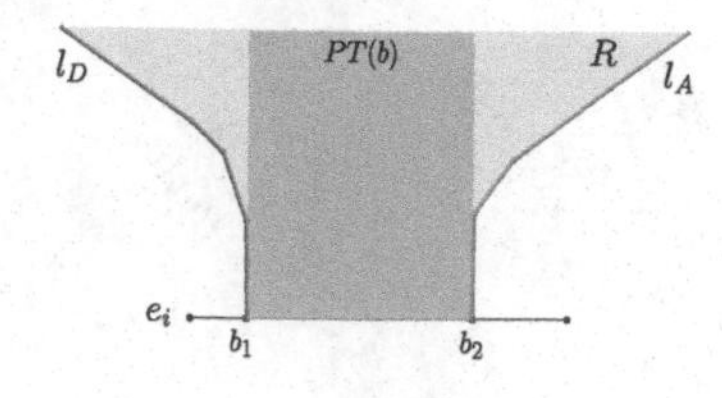

(a) Case A: If $B = \emptyset$ and $C = \emptyset$, then R is bounded by l_A (red), l_D (green) and b.

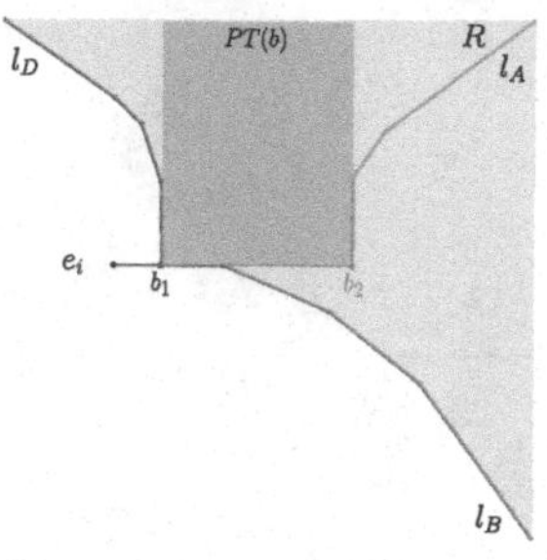

(b) Case B: If $B \neq \emptyset$ and $C = \emptyset$, then R is bounded by l_B (purple) and l_D (green).

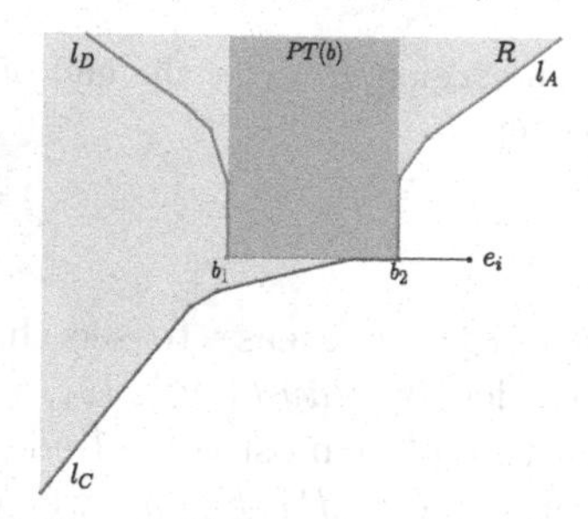

(c) Case C: If $B = \emptyset$ and $C \neq \emptyset$, then R is bounded by l_A (red) and l_C (dark blue).

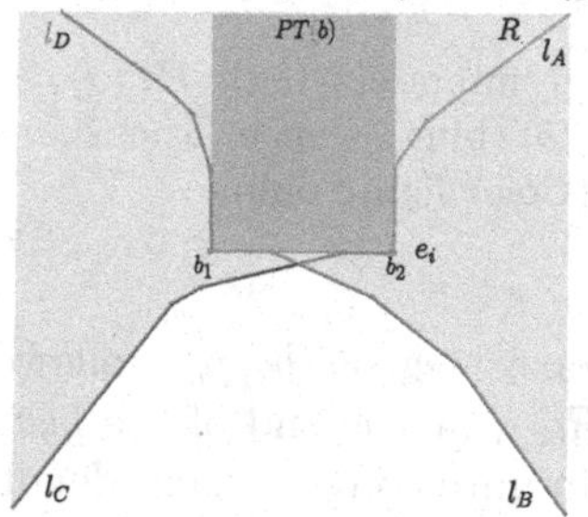

(d) Case D: If $B \neq \emptyset$ and $C \neq \emptyset$, then R is bounded by the chain l obtained from l_B (purple) and l_C (dark blue).

Fig. 3. The boundary of R.

We now examine how e_{i+1} can intersect R, in each of these cases. First, if e_{i+1} does not intersect the boundary of R, then either $R \cap e_{i+1} = e_{i+1}$ or $R \cap e_{i+1} = \emptyset$. In the former case, one interval is formed on e_{i+1}, which contains both its endpoints, and in the latter case, no interval is formed on e_{i+1}. Next, assume that e_{i+1} intersects the boundary of R. We distinguish between the case where there is an interval $h \in I(e_i, \overline{c_j})$ such that $PT(h) \cap e_{i+1} \neq \emptyset$, and the case where there is no such interval.

There is an interval $h \in I(e_i, \overline{c_j})$ such that $PT(h) \cap e_{i+1} \neq \emptyset$.
Then, by Lemma 2, $PT(b) \cap e_{i+1} = \emptyset$.

If Case A: Clearly, e_{i+1} cannot intersect both l_A and l_D, since this would imply $PT(b) \cap e_{i+1} \neq \emptyset$ (see Fig. 5a). Therefore, e_{i+1} intersects exactly one of these chains, either at a single point or at two points. If e_{i+1} intersects the chain at a single point q, then a single interval is formed on e_{i+1}, whose endpoints are q and the endpoint of e_{i+1} that lies in R (see the edge e^1_{i+1} in Fig. 4a). If e_{i+1} intersects the chain at two points, p and p', then two intervals are formed on e_{i+1}. The endpoints of these intervals are p and p' on one side and the corresponding endpoints of e_{i+1} on the other side (see the edge e^2_{i+1} in Fig. 4a).

If Case B: Unlike Case A, the fact that $PT(b) \cap e_{i+1} = \emptyset$ does not prevent e_{i+1} from intersecting both l_B and l_D. However, e_{i+1} can intersect these chains in at most two points (in total), and as in Case A at most two intervals are formed on e_{i+1}, where each of them contains an endpoint of e_{i+1} (see Fig. 5b).

If Case C: Since Cases B and C are symmetric, at most two intervals are formed on e_{i+1}, each of which contains an endpoint of e_{i+1}.

If Case D: If e_{i+1} intersects l at a single point q, then a single interval is formed on e_{i+1}, whose endpoints are q and the endpoint of e_{i+1} that lies in R. If e_{i+1} intersects l at two points p and p', then $R \cap e_{i+1}$ consist of all the points on e_{i+1}, except for those in the interior of (p, p'). Therefore, two intervals are formed on e_{i+1}, and their endpoints are p and p' on one side and the corresponding endpoints of e_{i+1} on the other side (see Fig. 4b)

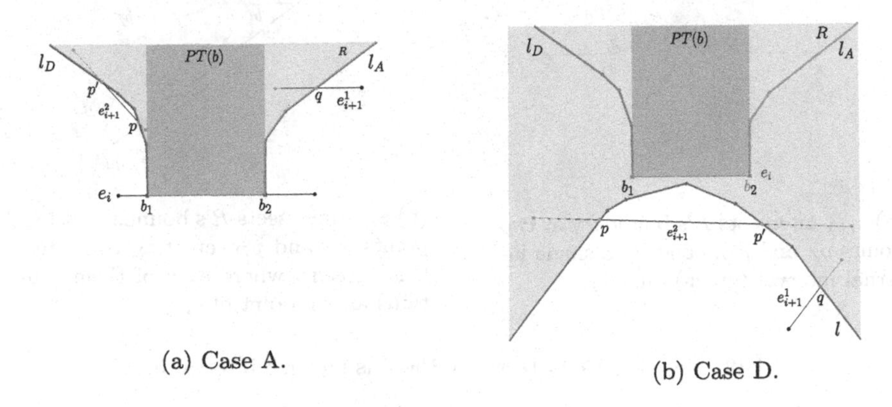

(a) Case A. (b) Case D.

Fig. 4. e_{i+1} intersects R's boundary either at a single point q or at two points p and p'.

We have shown that by passing through an interval in $\{b\} \cup \delta(b) \backslash I(e_i, \overline{c_j})$, at most two intervals (with associated length $l(e_i)+1$ and orientation c_j) are formed on e_{i+1}. Moreover, each of these intervals contains an endpoint of e_{i+1}. Therefore, the total number of such intervals that are formed on e_{i+1}, by passing through an interval in $\bigcup_{b \in I^+(e_i, c_j)} \{b\} \cup \delta(b) \backslash I(e_i, \overline{c_j})$ is at most two. (For each endpoint p of e_{i+1}, we retain only the longest interval with p as one of its endpoints.)

Finally, observe that by passing through an interval in $I(e_i, \overline{c_j})$ and turning backwards in orientation c_j, at most one interval is formed on e_{i+1}, which does not necessarily contain an endpoint of e_{i+1}.

We conclude that at most $|I(e_i, \overline{c_j})| + 2$ intervals (with associated length $l(e_i) + 1$ and orientation c_j) are formed on e_{i+1} during the execution of the algorithm (in the case that there is an interval $h \in I(e_i, \overline{c_j})$ such that $PT(h) \cap e_{i+1} \neq \emptyset$). We have used the equality $\bigcup_{b \in I^+(e_i, c_j)} \{b\} \cup \delta(b) = I^+(e_i, c_j) \cup I(e_i) \backslash I(e_i, c_j)$, which follows from Lemma 1.

We now proceed to the complementary case.

For Any Interval $h \in I(e_i, \overline{c_j})$, $PT(h) \cap e_{i+1} = \emptyset$. We defer the details of this case (which are similar to those of the previous case) to the full version of this paper [5]. These details lead to the conclusion that at most $|I^+(e_i, c_j)| + 2$ intervals (with associated length $l(e_i) + 1$ and orientation c_j) are formed on e_{i+1} during the execution of the algorithm in this case.

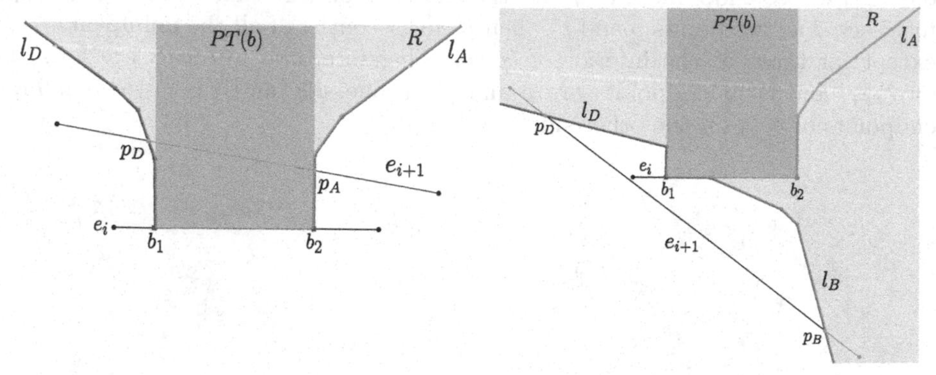

(a) e_{i+1} intersects R's boundary at two points p_A and p_D, creating a single internal interval (green) on e_{i+1}.

(b) e_{i+1} intersects R's boundary at two points p_B and p_D, creating two intervals (green), where each of them contains an endpoint of e_{i+1}.

Fig. 5. Cases A and B, where there is no such interval h.

Since only one of the two cases holds (i.e., either there is such an interval h or there is not), we conclude that at most $max\,\{|I(e_i, \overline{c_j})| + 2, |I^+(e_i, c_j)| + 2\}$ $= max\,\{|I(e_i, \overline{c_j})|, |I^+(e_i, c_j)|\} + 2$ intervals with associated length $l(e_i) + 1$ and orientation c_j are formed on e_{i+1} during the execution of the algorithm. This completes the proof of Claim 3.

Lemma 3. *For any interval $a \in I(e_i)$ and orientation $c_j \in C$, we do not need to compute the interval on e_{i+1} with associated length and orientation $l(e_i) + 2$ and c_j, respectively, which is formed by passing through a and then making two turns, where the first is in orientation $\overline{c_a}$.*

Proof. By Claim 2, $l(e_i) \leq l(e_{i+1}) \leq l(e_i) + 2$. So, the intervals on e_{i+1} of length $l(e_i) + 2$ are only relevant if $l(e_{i+1}) > l(e_i)$ (Claim 1). Assume therefore that $l(e_{i+1}) > l(e_i)$, and let $a \in I(e_i)$ (e.g., the red interval in Fig. 6). Let π be a tour of E that passes through a at a point p_i, makes a turn in orientation $\overline{c_a}$ at point p, and makes another turn in orientation c_j at point p', such that π_{i+1} (the portion of π from s to e_{i+1}) corresponds to an interval of length $l(e_i) + 2$.

We distinguish between two cases. If $pp' \cap e_i = \emptyset$ (i.e., the second turn is before π crosses e_i again), as shown in Fig. 6a, then pp' does not intersect e_{i+1}, since this would imply $l(e_i) = l(e_{i+1})$. Therefore, π reaches e_{i+1} only after the turn at p', and the tour π' which is obtained from π by deleting the link pp' (see Fig. 6b), is a tour of E of length $|\pi|-1$, hence π is not a minimum-link tour of E. Since our goal is to find a minimum-link tour of E, we do not need to compute the interval on e_{i+1} formed by paths such as π satisfying the condition above.

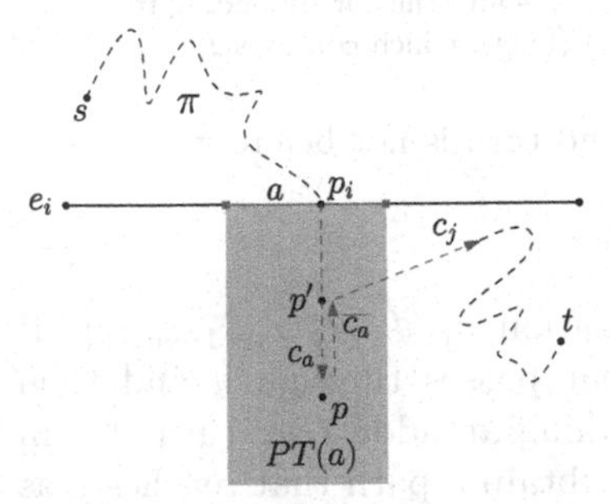

(a) A tour π that passes through $a \in I(e_i)$ (red), makes a turn in orientation $\overline{c_a}$, and another turn in orientation c_j.

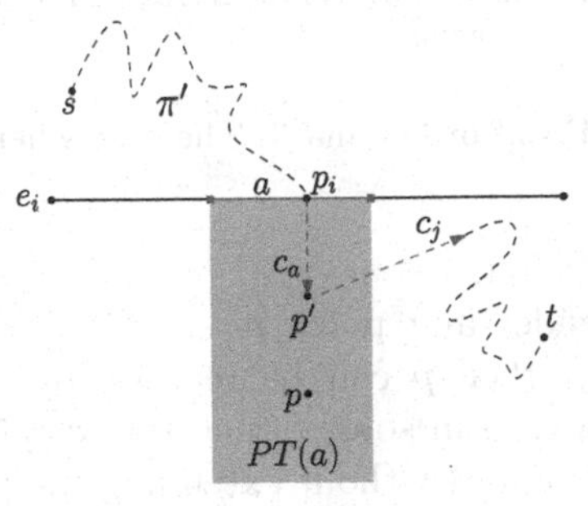

(b) The tour π' of length $|\pi|-1$, which is obtained by deleting the link pp' from π.

Fig. 6. Proof of Lemma 3. The case where the second turn is before π crosses e_i again.

If $pp' \cap e_i \neq \emptyset$ (i.e., the second turn is not before π crosses e_i again), let T denote the region of all points that can be reached by such paths, i.e., paths such as π satisfying the condition above (see the orange region in Fig. 7a). Then $T \cap e_{i+1}$ is the interval on e_{i+1} with associated length $l(e_i)+2$ and orientation c_j, formed by these paths. But, by Lemma 1, there exists $b \in I^+(e_i, \overline{c_a})$ such that $a \subseteq b$ (see the blue interval in Fig. 7b), and clearly $T \subseteq \psi(b, c_j)$, implying $T \cap e_{i+1} \subseteq \psi(b, c_j) \cap e_{i+1}$. The latter interval, i.e., $\psi(b, c_j) \cap e_{i+1}$ is computed by our algorithm, so we do not need to compute $T \cap e_{i+1}$.

Lemma 4. *For any interval $a \in I(e_i)$, any point $p \in \mathbb{R}^2$, and any orientation $c_j \notin \{c_a, c_{a+1}, c_{a-1}\}$, p can be reached by a path that passes through a and then makes a turn in some orientation $c \neq \overline{c_a}$ and another turn in orientation c_j.*

Proof. Consider any interval $a \in I(e_i)$. Recall that $\psi(a, \overline{c_{a+1}})$ ($\psi(a, \overline{c_{a-1}})$) denotes the region of all the points that can be reached by a path that passes through a and then makes a turn in orientation $\overline{c_{a+1}}$ ($\overline{c_{a-1}}$) (see Fig. 8). It is easy to see that $\psi(a, c) \subseteq \psi(a, \overline{c_{a+1}})$ for any $c \in \phi(\overline{c_a}, c_a)$ and $\psi(a, c) \subseteq \psi(a, \overline{c_{a-1}})$ for any $c \in \phi(c_a, \overline{c_a})$. Therefore, $\Delta_a = \psi(a, \overline{c_{a+1}}) \cup \psi(a, \overline{c_{a-1}})$ is the region of all the points that can be reached by a path that passes through a and then makes a turn in some orientation $c \neq \overline{c_a}$ (see Fig. 8c).

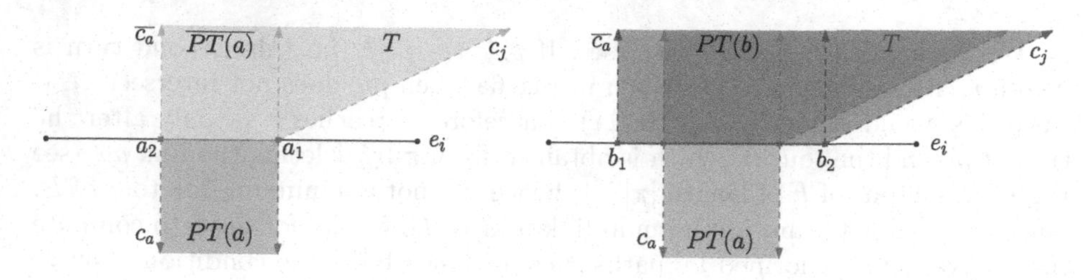

(a) An interval a (red) and the corresponding region T (orange).

(b) An interval b and the corresponding region $\psi(b, c_j)$ (blue), which contains T.

Fig. 7. Proof of Lemma 3. The case where the second turn is not before π crosses e_i again.

Consider any point $p \in \mathbb{R}^2$ and any orientation $c_j \notin \{c_a, c_{a+1}, c_{a-1}\}$. If $p \in \Delta_a$, then p can be reached by a path that passes through a and then makes a turn in some orientation $c \neq \overline{c_a}$. By making an additional turn at p in orientation c_j (without extending the path), we obtain a path that reaches p as required.

If $p \in \overline{\Delta_a} = \mathbb{R}^2 \backslash \Delta_a$, then $Ray(p, \overline{c_j}) \cap \Delta_a \neq \emptyset$, since $\overline{c_j} \notin \{\overline{c_a}_{-1}, \overline{c_a}, \overline{c_a}_{+1}\}$ (as shown in Fig. 9). Let p' be any point on $Ray(p, \overline{c_j}) \cap \Delta_a$, then p' can be reached by a path that passes through a and then makes a turn in some orientation $c \neq \overline{c_a}$, and by extending this path by adding the link $p'p$, we obtain a path that reaches p as required.

Consider the region $\Delta_a = \psi(a, \overline{c_a}_{+1}) \cup \psi(a, \overline{c_a}_{-1})$ defined in the proof of Lemma 4. Then, as mentioned in the proof of Lemma 4, Δ_a is the region of all the points that can be reached by a path that passes through a and then makes a turn in some orientation $c \neq \overline{c_a}$. In addition, we notice that by extending such a path by adding a link in orientation c_j, for $c_j \in \{c_a, c_{a+1}, c_{a-1}\}$, we cannot leave Δ_a (see Fig. 10), since for any point $q \in \Delta_a$, $Ray(q, c_j) \subseteq \Delta_a$.

The following claim bounds the number of intervals with associated length and orientation $l(e_i) + 2$ and c_j, respectively, that are 'created' on e_{i+1}.

Claim 4. *At most* $|I^+(e_i, \overline{c_j})| + 2$ *intervals with associated length and orientation* $l(e_i) + 2$ *and* c_j*, respectively, are 'created' on* e_{i+1}*, during the execution of the algorithm.*

Proof. The proof can be found in the full version of this paper [5]. Here, we only observe that there are two ways to reach a point on e_{i+1} with a path of length $l(e_i) + 2$ whose last link is of orientation c_j. The first is by passing through one of the intervals $a \in I(e_i)$ and then making two turns, where the first one is in orientation $c \neq \overline{c_a}$ and the second one is in orientation c_j (see Lemma 3). The second way is by passing through one of the intervals in $I^+(e_i) \backslash I^+(e_i, c_j)$, and then making a turn in orientation c_j. That is, the intervals on e_{i+1} with associated length $l(e_i) + 2$ and associated orientation c_j are determined by the intervals in $I(e_i) \cup (I^+(e_i) \backslash I^+(e_i, c_j))$.

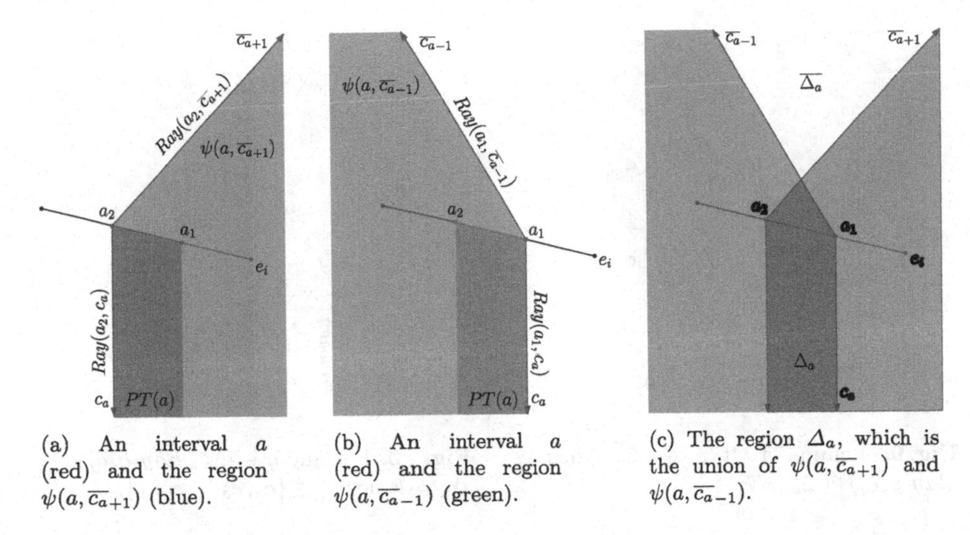

(a) An interval a (red) and the region $\psi(a, \overline{c_a}_{+1})$ (blue).

(b) An interval a (red) and the region $\psi(a, \overline{c_a}_{-1})$ (green).

(c) The region Δ_a, which is the union of $\psi(a, \overline{c_a}_{+1})$ and $\psi(a, \overline{c_a}_{-1})$.

Fig. 8. All the points that can be reached by a path that passes through a and then makes a turn in some orientation $c \neq \overline{c_a}$.

The following claim bounds the number of intervals with associated length and orientation $l(e_i) + 3$ and c_j, respectively, that are 'created' on e_{i+1}.

Claim 5. *For any $q \in e_{i+1}$ and for any $c_j \in C$, there exists a path of length $l(e_i) + 3$ from s to q, whose last link has orientation c_j, for $1 \leq i \leq n - 1$.*

Proof. Consider any path π from s to e_i that corresponds to an interval in $I(e_i)$, and let p be the point on e_i where π ends. Since C spans the plane, there exists a two-link path from p to q, and by making a turn at q in orientation c_j (without extending the path), we obtain a three-link path $\pi_{p,q}$ from p to q whose last link has orientation c_j. So, the path obtained by concatenating the paths π and $\pi_{p,q}$ is as desired.

Claim 6. *For any $0 \leq i \leq n+1$ and $c_j \in C$, $|I(e_i, c_j)| \leq 2i+1$ and $|I^+(e_i, c_j)| \leq 2i + 1$.*

Proof. The proof is by induction on i. For $i = 0$, the claim is clearly true; $|I(e_0, c_j)| = |I^+(e_0, c_j)| = 1$.

Assume now that the claim is true for i, $0 \leq i \leq n$, that is, for any $c_j \in C$, we have $|I(e_i, c_j)| \leq 2i + 1$ and $|I^+(e_i, c_j)| \leq 2i + 1$. We show below that it remains true for $i + 1$.

Recall that $l(e_i) \leq l(e_{i+1}) \leq l(e_i) + 2$ (Claim 2). We show that the claim remains true in each of the resulting three cases.

- **Case A:** $l(e_{i+1}) = l(e_i)$. In this case $I(e_{i+1}, c_j)$ stores the +0-intervals on e_{i+1} with respect to $I(e_i, c_j)$. Since, each interval $a \in I(e_i, c_j)$ 'creates' at most one +0-interval on e_{i+1}, we get that $|I(e_{i+1}, c_j)| \leq |I(e_i, c_j)| \leq 2i + 1$. Recall that $I^+(e_{i+1}, c_j)$ is the set of maximal intervals on e_{i+1} formed by all

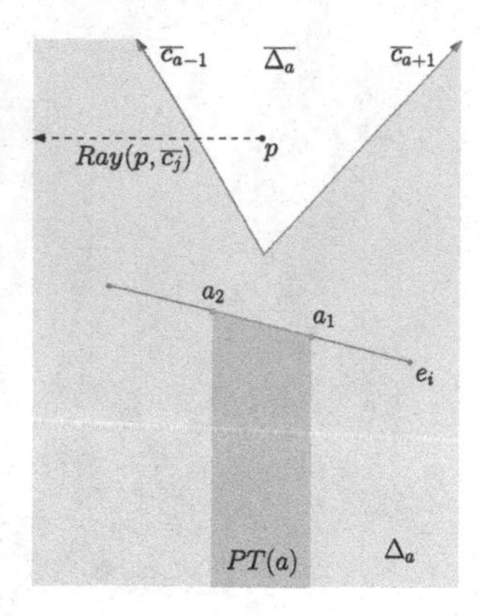

Fig. 9. Lemma 4. If $p \in \overline{\Delta_a}$, then $Ray(p, \overline{c_j}) \cap \Delta_a \neq \emptyset$.

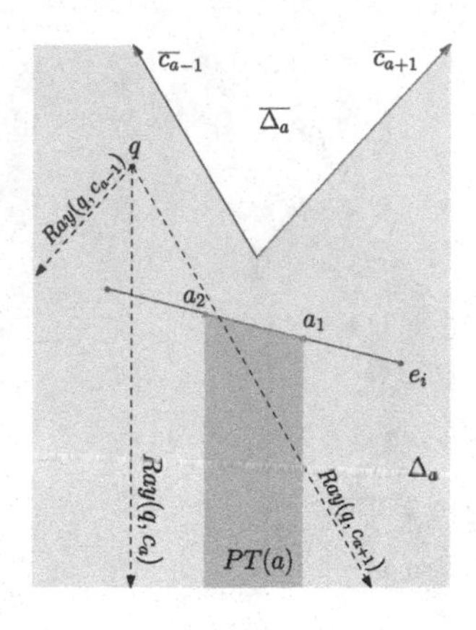

Fig. 10. For any $q \in \Delta_a$, $Ray(q, c_j) \cap \overline{\Delta_a} = \emptyset$, for $c_j \in \{c_a, c_{a+1}, c_{a-1}\}$.

paths of length $l(e_{i+1}) + 1 = l(e_i) + 1$, whose last link has orientation c_j. By Claim 3, $|I^+(e_{i+1}, c_j)| \leq \max\{|I(e_i, \overline{c_j})|, |I^+(e_i, c_j)|\} + 2$, and therefore $|I^+(e_{i+1}, c_j)| \leq \max\{2i+1, 2i+1\} + 2 = 2(i+1) + 1$.

- **Case B:** $l(e_{i+1}) = l(e_i)+1$. In this case, $I(e_{i+1}, c_j)$ is the set of maximal intervals on e_{i+1} formed by all paths of length $l(e_{i+1}) = l(e_i) + 1$, whose last link has orientation c_j. By Claim 3, $|I(e_{i+1}, c_j)| \leq \max\{|I(e_i, \overline{c_j})|, |I^+(e_i, c_j)|\}+2$, so, $|I(e_{i+1}, c_j)| \leq \max\{2i+1, 2i+1\} + 2 = 2(i+1) + 1$.
  Now, $I^+(e_{i+1}, c_j)$ is the set of maximal intervals on e_{i+1} formed by all paths of length $l(e_{i+1})+1 = l(e_i)+2$, whose last link has orientation c_j. By Claim 4, $|I^+(e_{i+1}, c_j)| \leq |I^+(e_i, \overline{c_j})| + 2$, and therefore $|I^+(e_{i+1}, c_j)| \leq 2i + 1 + 2 = 2(i+1) + 1$.
- **Case C:** $l(e_{i+1}) = l(e_i) + 2$. In this case, $I(e_{i+1}, c_j)$ is the set of maximal intervals on e_{i+1} formed by all paths of length $l(e_{i+1}) = l(e_i) + 2$, whose last link has orientation c_j. Thus, by Claim 4, $|I(e_{i+1}, c_j)| \leq |I^+(e_i, \overline{c_j})| + 2 \leq 2(i+1)+1$. Moreover, in this case, $I^+(e_{i+1}, c_j) = \{e_{i+1}\}$, so $|I^+(e_{i+1}, c_j)| = 1$.

Running Time. We bound the running time of each of the two stages of our algorithm. Consider the i'th iteration of the main loop of Stage I. We need $O(|I(e_i, c_j)| + |I^+(e_i, c_j)|)$ time to compute the +0-intervals on e_{i+1}, $O(|I(e_i) \backslash I(e_i, c_j)| + |I^+(e_i) \backslash I^+(e_i, c_j)|)$ time to compute the +1-intervals, and $O(|I(e_i)|)$ time to compute the +2-intervals. Since we perform this calculation for each $c_j \in C$, the running time of the i'th iteration is $O(|C| \cdot \{|I(e_i)| + |I^+(e_i)|\})$. By Claim 6 we conclude that $|I(e_i)| = O(|C| \cdot (2i+1))$ and $|I^+(e_i)| = O(|C| \cdot (2i+1))$, for $1 \leq i \leq n+1$. Therefore, the running time of Stage I is $\sum_{i=1}^{n+1} O(|C| \cdot |C| \cdot (2i+1)) = O(|C|^2 \cdot n^2)$.

In stage 2, we run the algorithm (described in the full version of this paper [5]) for each i from $n+1$ to 1. The running time of the algorithm is $O(|I(e_{i-1})| + |I^+(e_{i-1})|)$, and by Claim 6 we get $O(|C| \cdot i)$. Therefore, the running time of Stage II is $\sum_{i=1}^{n+1} O(|C| \cdot i) = O(|C| \cdot n^2)$.

Thus, the overall running time of the algorithm is $O(|C|^2 \cdot n^2)$, as summarized:

Theorem 1. *Given a set E of n disjoint C-oriented segments in the plane and points s and t that do not belong to any of the segments in E, one can compute a minimum-link C-oriented tour of E in $O(|C|^2 \cdot n^2)$ time.*

5 Extensions

In the case that $|C| = 4$ (e.g., axis-parallel paths and segments), the specialization of our analysis shows a constant upper bound on the number of intervals on each segment; this results in overall time $O(n)$. Also, our analysis only required that consecutive segments in E do not intersect each other; they can otherwise intersect. In ongoing and future work we consider more general polygonal regions, possibly overlapping arbitrarily. We also consider query versions of the problem in which we build data structures (shortest path maps) that allow link distance queries on subsequences of the input set of regions, between query points in the plane. Future work might examine problems in 3D.

Acknowledgements. M. Katz was partially supported by the US-Israel Binational Science Foundation (BSF project 2019715/NSF CCF-2008551). J. Mitchell was partially supported by the National Science Foundation (CCF-2007275) and the US-Israel Binational Science Foundation (BSF project 2016116).

References

1. Alatartsev, S., Mersheeva, V., Augustine, M., Ortmeier, F.: On optimizing a sequence of robotic tasks. In: 2013 IEEE/RSJ International Conference on Intelligent Robots and Systems, pp. 217–223. IEEE (2013)
2. Alatartsev, S., Stellmacher, S., Ortmeier, F.: Robotic task sequencing problem: a survey. J. Intell. Robot. Syst. **80**, 279–298 (2015)
3. Dror, M., Efrat, A., Lubiw, A., Mitchell, J.S.B.: Touring a sequence of polygons. In: 2003 Proceedings of the 35th Annual ACM Symposium on Theory of Computing, pp. 473–482. ACM (2003)
4. Egyed, P., Wenger, R.: Ordered stabbing of pairwise disjoint convex sets in linear time. Discret. Appl. Math. **31**(2), 133–140 (1991)
5. Geva, K., Katz, M.J., Mitchell, J.S.B., Packer, E.: Minimum-link C-oriented paths visiting a sequence of regions in the plane. CoRR, abs/2302.06776 (2023)
6. Guibas, L.J., Hershberger, J., Mitchell, J.S.B., Snoeyink, J.: Approximating polygons and subdivisions with minimum link paths. Int. J. Comput. Geometry Appl. **3**(4), 383–415 (1993)
7. Löffler, M.: Existence and computation of tours through imprecise points. Int. J. Comput. Geomet. Appl. **21**(1), 1–24 (2011)
8. Mitchell, J.S.B.: Shortest paths and networks. In: Goodman, J.E., Tóth, C., O'Rourke, J. (eds.) Handbook of Discrete and Computational Geometry, 3rd edn., chap. 31, pp. 811–848. Chapman & Hall/CRC, Boca Raton (2017)
9. Mitchell, J.S.B., Polishchuk, V., Sysikaski, M.: Minimum-link paths revisited. Comput. Geomet. **47**(6), 651–667 (2014)

10. Neyer, G.: Line simplification with restricted orientations. In: Dehne, F., Sack, J.-R., Gupta, A., Tamassia, R. (eds.) WADS 1999. LNCS, vol. 1663, pp. 13–24. Springer, Heidelberg (1999). https://doi.org/10.1007/3-540-48447-7_2
11. Speckmann, B., Verbeek, K.: Homotopic C-oriented routing with few links and thick edges. Comput. Geom. **67**, 11–28 (2018)

Grouped Domination Parameterized by Vertex Cover, Twin Cover, and Beyond

Tesshu Hanaka[1], Hirotaka Ono[2](✉), Yota Otachi[2], and Saeki Uda[2]

[1] Kyushu University, Fukuoka, Japan
hanaka@inf.kyushu-u.ac.jp
[2] Nagoya University, Nagoya, Japan
{ono,otachi}@nagoya-u.jp, uda.saeki.z4@s.mail.nagoya-u.ac.jp

Abstract. A dominating set S of graph G is called an *r-grouped dominating set* if S can be partitioned into $S_1, S_2, \ldots, S_k$ such that the size of each unit S_i is r and the subgraph of G induced by S_i is connected. The concept of r-grouped dominating sets generalizes several well-studied variants of dominating sets with requirements for connected component sizes, such as the ordinary dominating sets ($r = 1$), paired dominating sets ($r = 2$), and connected dominating sets (r is arbitrary and $k = 1$). In this paper, we investigate the computational complexity of r-Grouped Dominating Set, which is the problem of deciding whether a given graph has an r-grouped dominating set with at most k units. For general r, r-Grouped Dominating Set is hard to solve in various senses because the hardness of the connected dominating set is inherited. We thus focus on the case in which r is a constant or a parameter, but we see that r-Grouped Dominating Set for every fixed $r > 0$ is still hard to solve. From the observations about the hardness, we consider the parameterized complexity concerning well-studied graph structural parameters. We first see that r-Grouped Dominating Set is fixed-parameter tractable for r and treewidth, which is derived from the fact that the condition of r-grouped domination for a constant r can be represented as monadic second-order logic (MSO_2). This fixed-parameter tractability is good news, but the running time is not practical. We then design an $O^*(\min\{(2\tau(r+1))^\tau, (2\tau)^{2\tau}\})$-time algorithm for general $r \geq 2$, where τ is the twin cover number, which is a parameter between vertex cover number and clique-width. For paired dominating set and trio dominating set, i.e., $r \in \{2, 3\}$, we can speed up the algorithm, whose running time becomes $O^*((r+1)^\tau)$. We further argue the relationship between FPT results and graph parameters, which draws the parameterized complexity landscape of r-Grouped Dominating Set.

Keywords: Dominating Set · Paired Dominating Set · Parameterized Complexity · Graph Structural Parameters

---

Partially supported by JSPS KAKENHI Grant Numbers JP17H01698, JP17K19960, JP18H04091, JP20H05793, JP20H05967, JP21K11752, JP21H05852, JP21K17707, JP21K19765, and JP22H00513.

M. Mavronicolas (Ed.): CIAC 2023, LNCS 13898, pp. 263–277, 2023.
https://doi.org/10.1007/978-3-031-30448-4_19

1 Introduction

1.1 Definition and Motivation

Given an undirected graph $G = (V, E)$, a vertex set $S \subseteq V$ is called a *dominating set* if every vertex in V is either in S or adjacent to a vertex in S. The dominating set problem is the problem of finding a dominating set with the minimum cardinality. Since the definition of dominating set, i.e., covering all the vertices via edges, is natural, many practical and theoretical problems are modeled as dominating set problems with additional requirements; many variants of dominating set are considered and investigated. Such variants somewhat generalize or extend the ordinary dominating set based on theoretical or applicational motivations. In this paper, we focus on variants that require the dominating set to satisfy specific connectivity and size constraints. One example considering connectivity is the connected dominating set. A dominating set is called a *connected dominating set* if the subgraph induced by a dominating set is connected. Another example is the paired dominating set. A paired dominating set is a dominating set of a graph such that the subgraph induced by it admits a perfect matching.

This paper introduces the r-grouped dominating set, which generalizes the connected dominating set, the paired dominating set, and some other variants. A dominating set S is called an *r-grouped dominating set* if S can be partitioned into $\{S_1, S_2, \dots, S_k\}$ such that each S_i is a set of r vertices and $G[S_i]$ is connected. We call each S_i a *unit*. The r-grouped dominating set generalizes both the connecting dominating set and the paired dominating set in the following sense: a connecting dominating set with r vertices is equivalent to an r-grouped dominating set of one unit, and a paired dominating set with k pairs is equivalent to a 2-grouped dominating set with k units.

This paper investigates the parameterized complexity of deciding whether a given graph has an r-grouped dominating set with k units. The parameters that we focus on are so-called graph structural parameters, such as vertex cover number and twin-cover number. The results obtained in this paper are summarized in Our Contribution (Sect. 1.3).

1.2 Related Work

An enormous number of papers study the dominating set problem, including the ones strongly related to the r-grouped dominating set.

The dominating set problem is one of the most important graph optimization problems. Due to its NP-hardness, its tractability is finely studied from several aspects, such as approximation, solvable graph classes, fast exact exponential-time solvability, and parameterized complexity. Concerning the parameterized complexity, the dominating set problem is W[2]-complete for solution size k; it is unlikely to be fixed-parameter tractable [12]. On the other hand, since the dominating set can be expressed in MSO_1, it is FPT when parametrized by clique-width or treewidth (see, e.g., [23]).

The connected dominating set is a well-studied variant of dominating set. This problem arises in communication and computer networks such as mobile ad hoc networks. It is also W[2]-hard when parameterized by the solution size [12]. Furthermore, the connected dominating set also can be expressed in MSO_1; it is FPT when parametrized by clique-width and treewidth as in the ordinary dominating set problem. Furthermore, single exponential-time algorithms for connected dominating set parameterized by treewidth can be obtained by the Cut & Count technique [13] or the rank-based approach [2].

The notion of the paired dominating set is introduced in [21,22] by Haynes and Slater as a model of dominating sets with pairwise backup. It is NP-hard on split graphs, bipartite graphs [8], graphs of maximum degree 3 [6], and planar graphs of maximum degree 5 [29], whereas it can be solved in polynomial time on strongly-chordal graphs [7], distance-hereditary graphs [26], and AT-free graphs [29]. There are several graph classes (e.g., strongly orderable graphs [28]) where the paired dominating set problem is tractable, whereas the ordinary dominating set problem remains NP-hard. For other results about the paired dominating set, see a survey [14].

1.3 Our Contributions

This paper provides a unified view of the parameterized complexity of dominating set problem variants with connectivity and size constraints.

As mentioned above, an r-grouped dominating set of G with 1 unit is equivalent to a connected dominating set with size r, which implies that some hardness results of r-Grouped Dominating Set for general r are inherited directly from Connected Dominating Set. From these, we mainly consider the case where r is a constant or a parameter.

Unfortunately, r-Grouped Dominating Set for $r = 1, 2$ is also hard to solve again because 1-Grouped Dominating Set and 2-Grouped Dominating Set are respectively the ordinary dominating set problem and the paired dominating set problem. Thus, it is worth considering whether a larger but constant r enlarges, restricts, or leaves unchanged the graph classes for which similar hardness results hold. A way to classify or characterize graphs of certain classes is to focus on graph-structural parameters. By observing that the condition of r-grouped dominating set can be represented as monadic second-order logic (MSO_2), we can see that r-Grouped Dominating Set is fixed-parameter tractable for r and treewidth. Recall that the condition of the connected dominating set can be represented as monadic second-order logic (MSO_1), which implies that there might exist a gap between $r = 1$ and 2, or between $k = 1$ and $k > 1$. Although this FPT result is good news, the running time is not practical. From these, we focus on less generalized graph structural parameters, vertex cover number ν or twin cover number τ as a parameter, and design single exponential fixed-parameter algorithms for r-Grouped Dominating Set.

Our algorithm is based on dynamic programming on nested partitions of a vertex cover, and its running time is $O^*(\min\{(2\nu(r+1))^\nu, (2\nu)^{2\nu}\})$ for general $r \geq 2$. For paired dominating set and trio dominating set, i.e., $r \in \{2, 3\}$, we can tailor the algorithm to run in $O^*((r+1)^\nu)$ time by observing that the nested partitions of a vertex cover degenerate in some sense.

We then turn our attention to a more general parameter, the twin cover number. We show that, given a twin cover, r-GROUPED DOMINATING SET admits an optimal solution in which twin-edges do not contribute to the connectivity of r-units. This observation implies that these edges can be removed from the graph, and thus we can focus on the resultant graph of bounded vertex cover number. Hence, we can conclude that our algorithms still work when the parameter ν in the running time is replaced with twin cover number τ.

We further argue the relationship between FPT results and graph parameters. The perspective is summarized in Fig. 1, which draws the parameterized complexity landscape of r-GROUPED DOMINATING SET. We omit the proofs of some of the results, which can be found in the full version of this paper [20].

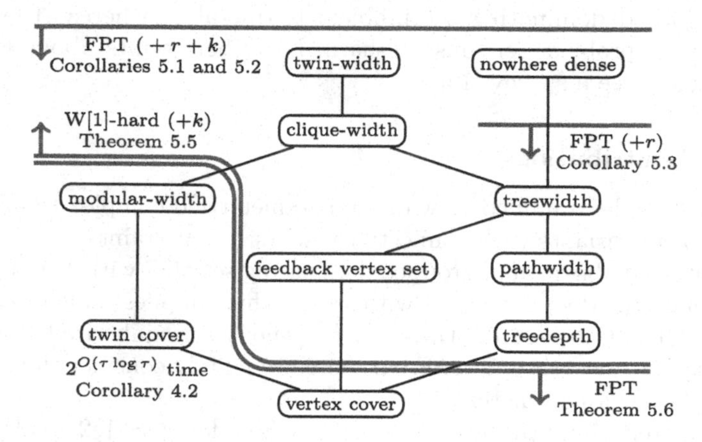

Fig. 1. The complexity of r-GROUPED DOMINATING SET with respect to structural graph parameters. An edge between two parameters indicates that there is a function in the one above that lower-bounds the one below (e.g., treewidth $\leq$ pathwidth).

2 Preliminaries

Let $G=(V,E)$ be an undirected graph. For a vertex subset $V' \subseteq V$, the subgraph induced by V' is denoted by $G[V']$. Let us denote by $N(v)$ and $N[v]$ the open neighborhood and the closed neighborhood of v, respectively. The degree of a vertex v is defined by $d(v)=|N(v)|$, and Δ denotes the maximum degree of G.

A vertex set S is a *vertex cover* of G if for every edge $\{u,v\} \in E$, at least one of u, v is in S. The *vertex cover number* ν of G is defined by the size of a minimum vertex cover of G. A minimum vertex cover of G can be found in $O^*(1.2738^\nu)$ time [5].[1]

[1] The O^* notation suppresses the polynomial factors of the input size.

Two vertices u and v are (*true*) *twins* if $N[u] = N[v]$. An edge $\{u, v\} \in E$ is a *twin edge* if u and v are true twins. A vertex set S is a *twin cover* if for every edge $\{u, v\} \in E$, either $\{u, v\}$ is a twin edge, or at least one of u, v is in S. The size τ of a minimum twin cover of G is called the *twin cover number* of G. A minimum twin cover of G can be found in the same running time as vertex cover, i.e., $O^*(1.2738^\tau)$ [16].

2.1 r-Grouped Dominating Set

An *r-grouped dominating set with k units* in G is a family $\mathcal{D} = \{D_1, \dots, D_k\}$ of subsets of V such that D_i's are mutually disjoint, $|D_i| = r$, $G[D_i]$ is connected for $1 \le i \le k$, and $\bigcup_{D \in \mathcal{D}} D$ is a dominating set of G. For simplicity, let $\bigcup \mathcal{D}$ denote $\bigcup_{D \in \mathcal{D}} D$. We say that $\mathcal{D}$ is a minimum r-grouped dominating set if it is an r-grouped dominating set with the minimum number of units.

> r-Grouped Dominating Set
> **Input:** A graph G and positive integers r and k.
> **Question:** Is there an r-grouped dominating set with at most k units in G?

3 Basic Results

In this section, we see some preliminary results without proof (see the full version for the proofs). We first see that r-Grouped Dominating Set is W[2]-hard but XP when parameterized by $k+r$. Furthermore, it is NP-hard even on planar bipartite graphs of maximum degree 3.

Theorem 3.1. *For every fixed $r \ge 1$, r-Grouped Dominating Set is W[2]-hard when parameterized by k even on split graphs or bipartite graphs.*

Theorem 3.2. *For every fixed $k \ge 1$, r-Grouped Dominating Set is W[2]-hard when parameterized by r even on split graphs or bipartite graphs.*

Theorems 3.1 and 3.2 imply that r-Grouped Dominating Set is unlikely to be FPT when parameterized by k or r. On the other hand, we can show that the problem is XP when parameterized by $k + r$.

Theorem 3.3. r-Grouped Dominating Set *can be solved in $O^*(\Delta^{O(kr^2)})$ time.*

Corollary 3.4. r-Grouped Dominating Set *belongs to XP when parameterized by $k + r$.*

Tripathi et al. [29] showed that Paired Dominating Set (equivalently, 2-Grouped Dominating Set) is NP-complete for planar graphs with maximum degree 5. We obtain a more potent and broader result, that is, r-Grouped Dominating Set is NP-hard even on planar bipartite graphs of maximum degree 3 for every fixed $r \ge 1$.

Theorem 3.5. *For every fixed $r \geq 1$, r-*GROUPED DOMINATING SET *is NP-complete on planar bipartite graphs of maximum degree 3.*

Corollary 3.6. *For every fixed $r \geq 1$, r-*GROUPED DOMINATING SET *cannot be solved in $2^{o(n+m)}$ time on bipartite graphs unless ETH fails.*

4 Fast Algorithms Parameterized by Vertex Cover Number and by Twin Cover Number

In this section, we present FPT algorithms for r-GROUPED DOMINATING SET parameterized by vertex cover number ν. Our algorithm is based on dynamic programming on nested partitions of a vertex cover, and its running time is $O^*((2\nu(r+1))^\nu)$ for general $r \geq 2$. For the cases of $r \in \{2,3\}$, we can tailor the algorithm to run in $O^*((r+1)^\nu)$ time by focusing on the fact that the nested partitions of a vertex cover degenerate in some sense.

We then turn our attention to a more general parameter twin cover number. We show that, given a twin cover, r-GROUPED DOMINATING SET admits an optimal solution in which twin-edges do not contribute to the connectivity of r-units. This implies that these edges can be removed from the graph, and thus we can focus on the resultant graph of bounded vertex cover number. Hence, we can conclude that our algorithms still work when the parameter ν in the running time is replaced with twin cover number τ.

Theorem 4.1. *For graphs of twin cover number τ, r-*GROUPED DOMINATING SET *can be solved in $O^*((2\tau(r+1))^\tau)$ time. For the cases of $r \in \{2,3\}$, it can be solved in $O^*((r+1)^\tau)$ time.*

With a simple observation, Theorem 4.1 implies that r-GROUPED DOMINATING SET parameterized solely by τ is fixed-parameter tractable.

Corollary 4.2. *For graphs of twin cover number τ, r-*GROUPED DOMINATING SET *can be solved in $O^*((2\tau)^{2\tau})$ time.*

Proof. If $r < 2\tau - 1$, then the problem can be solved in $O^*((2\tau)^{2\tau})$ time by Theorem 4.1. Assume that $r \geq 2\tau - 1$. Let C be a connected component of the input graph. If $|V(C)| < r$, then we have a trivial no-instance. Otherwise, we construct a connected dominating set D of C with size exactly r, which works as a unit dominating C. We initialize D with a non-empty twin cover of size at most τ. Note that such a set can be found in $O^*(1.2738^\tau)$ time: if C is a complete graph, then we pick an arbitrary vertex $v \in V(C)$ and set $D = \{v\}$; otherwise, just find a minimum twin cover. Since C is connected, D is a dominating set of C. If $C[D]$ is not connected, we update D with a new element v adjacent to at least two connected components of $C[D]$. Since $|D| \leq \tau$ at the beginning, we can repeat this update at most $\tau - 1$ times, and after that $C[D]$ becomes connected and $|D| \leq 2\tau - 1 \leq r$. We finally add $r - |D|$ vertices arbitrarily and obtain a desired set. □

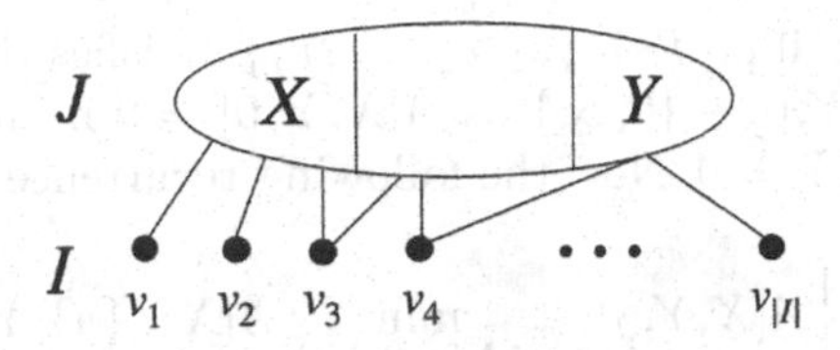

Fig. 2. Partitioning a vertex cover into three parts.

In the next subsection, we first present an algorithm for 2-GROUPED DOMINATING SET parameterized by vertex cover number, which gives a basic scheme of our dynamic programming based algorithms. We then see how we extend the idea to 3-GROUPED DOMINATING SET. As explained above, these algorithms are based on dynamic programming (DP), and they compute certain function values on partitions of a vertex cover. Unfortunately, it is not obvious how to extend the strategy to general r. Instead, we consider nested partitions of a vertex cover for DP tables, which makes the running time a little slower though. In the last subsection, we see how a vertex cover can be replaced with a twin cover in the same running time in terms of order.

4.1 Algorithms Parameterized by Vertex Cover Number

Algorithm for 2-Grouped Dominating Set. We first present an algorithm for the simplest case $r = 2$, i.e., the paired dominating set. Let $G = (V, E)$ be a graph and J be a vertex cover of G. Then, $I = V \setminus J$ is an independent set. The basic scheme of our algorithm follows the algorithm for the dominating set problem by Liedloff [25], which focuses on a partition of a given vertex cover J. For a minimum dominating set D, the vertex cover J is partitioned into three parts: $J \cap D$; $(J \setminus D) \cap N(J \cap D)$, that is, the vertices in $J \setminus D$ that are dominated by $J \cap D$; and $J \setminus N[J \cap D]$, that is, the remaining vertices. Note that the remaining vertices in $J \setminus N[J \cap D]$ are dominated by $I \cap D$. Once $J \cap D$ is fixed, a minimum $I \cap D$ is found by solving the set cover problem that reflects the condition that $J \setminus N[J \cap D]$ must be dominated by $I \cap D$. The algorithm computes a minimum dominating set by solving set cover problems defined by all candidates of $J \cap D$.

To adjust the algorithm to 2-GROUPED DOMINATING SET, we need to handle the condition that a dominating set contains a perfect matching.

For each subset $J_D \subseteq J$, we find a subset $I_D \subseteq I$ (if any exists) of the minimum size such that $J_D \cup I_D$ can form a 2-grouped dominating set. Let X and Y be disjoint subsets of J, and let $I = \{v_1, v_2, \dots, v_{|I|}\}$ (see Fig. 2). For $j = 0, \dots, |I|$, we define an auxiliary table $A[X, Y, j]$ as the minimum size of $I' \subseteq \{v_1, v_2, \dots, v_j\}$ that satisfies the following conditions.

1. $Y \subseteq N(I')$,
2. $I' \cup X$ has a partition $\mathcal{D}^{(2)} = \{D_1^{(2)}, D_2^{(2)}, \dots, D_p^{(2)}\}$ with $p \leq k$ such that for all $i = 1, \dots, p$, $|D_i^{(2)}| = 2$ and $G[D_i^{(2)}]$ is connected.

We set $A[X,Y,j] = \infty$ if no $I' \subseteq \{v_1, v_2, \dots, v_j\}$ satisfies the conditions. We can easily compute $A[X,Y,0] \in \{0,\infty\}$ as $A[X,Y,0] = 0$ if and only if $G[X]$ has a perfect matching and $Y = \emptyset$. Now the following recurrence formula computes A:

$$A[X,Y,j+1] = \min\left\{A[X,Y,j], \min_{u\in N(v_{j+1})\cap X} A[X \setminus \{u\}, Y \setminus N(v_{j+1}), j] + 1\right\}.$$

The recurrence finds the best way under the condition that we can use vertices from $v_1, v_2, \dots, v_j, v_{j+1}$ in a dominating set: not using v_{j+1}, or pairing v_{j+1} with $u \in N(v_{j+1}) \cap X$. We can compute all entries of A in $O^*(3^{|J|})$ time in a DP manner as there are only $3^{|J|}$ ways for choosing disjoint subsets X and Y of J.

Now we compute the minimum number of units in a 2-grouped dominating set of G (if any exists) by looking up some appropriate table entries of A. Let $\mathcal{D}$ be a 2-grouped dominating set of G with $J_D = J \cap \bigcup \mathcal{D}$ and $I_D = I \cap \bigcup \mathcal{D}$. Since $\bigcup \mathcal{D}$ is a dominating set with no isolated vertex in $G[\bigcup \mathcal{D}]$, J_D dominates all vertices in I. Let $J_Y = J \setminus N[J_D]$. Then the definition of A implies that $A[J_D, J_Y, |I|] = |I_D|$. Conversely, if $X \subseteq J$ dominates I, $Y = J \setminus N[X]$, and $A[X,Y,|I|] \neq \infty$, then there is a 2-grouped dominating set with $(|X| + A[X,Y,|I|])/2$ units. Therefore, the minimum number of units in a 2-grouped dominating set of G is $\min\{(|X| + A[X, J \setminus N[X], |I|])/2 \mid X \subseteq J \text{ and } I \subseteq N(X)\}$, which can be computed in $O^*(2^{|J|})$ time given the table A. Thus the total running time is $O^*(3^{|J|})$.

Algorithm for 3-Grouped Dominating Set. Next, we consider the case $r = 3$, i.e., the trio dominating set. Let $G = (V, E)$ be a graph, J be a vertex cover of G, and $I = V \setminus J$. The basic idea is the same as the case $r = 2$ except that we partition the vertex cover into four parts in the DP, and thus the recurrence formula for A is different. In the DP, the vertex cover J is partitioned into four parts depending on the partial solution corresponding to each table entry.

For each subset $J_D \subseteq J$, we find a subset $I_D \subseteq I$ (if any exists) of the minimum size such that $J_D \cup I_D$ can form a 3-grouped dominating set. Let X, F, and Y be disjoint subsets of J, and let $I = \{v_1, v_2, \dots, v_{|I|}\}$. Intuitively, the set F represents partial units that will later be completed to full units. For $j = 0, \dots, |I|$, we define $A[X,F,Y,j]$ as the minimum size of $I' \subseteq \{v_1, v_2, \dots, v_j\}$ that satisfies the following conditions:

1. $Y \subseteq N(I')$,
2. I' can be partitioned into two parts I'_2, I'_3 satisfying the following conditions:
   - $I'_2 \cup F$ has a partition $\mathcal{D}^{(2)} = \{D^{(2)}_1, D^{(2)}_2, \dots, D^{(2)}_p\}$ with $p \leq k$ such that for all $i = 1, \dots, p$, $|D^{(2)}_i| = 2$ and $G[D^{(2)}_i]$ is connected.
   - $I'_3 \cup X$ has a partition $\mathcal{D}^{(3)} = \{D^{(3)}_1, D^{(3)}_2, \dots, D^{(3)}_q\}$ with $q \leq k$ such that for all $i = 1, \dots, q$, $|D^{(3)}_i| = 3$ and $G[D^{(3)}_i]$ is connected.

We set $A[X,F,Y,j] = \infty$ if no $I' \subseteq \{v_1, v_2, \dots, v_j\}$ satisfies the conditions. We can easily compute $A[X,F,Y,0] \in \{0,\infty\}$ as $A[X,F,Y,0] = 0$ if and only if $F = Y = \emptyset$ and $G[X]$ admits a partition into connected graphs of 3-vertices.

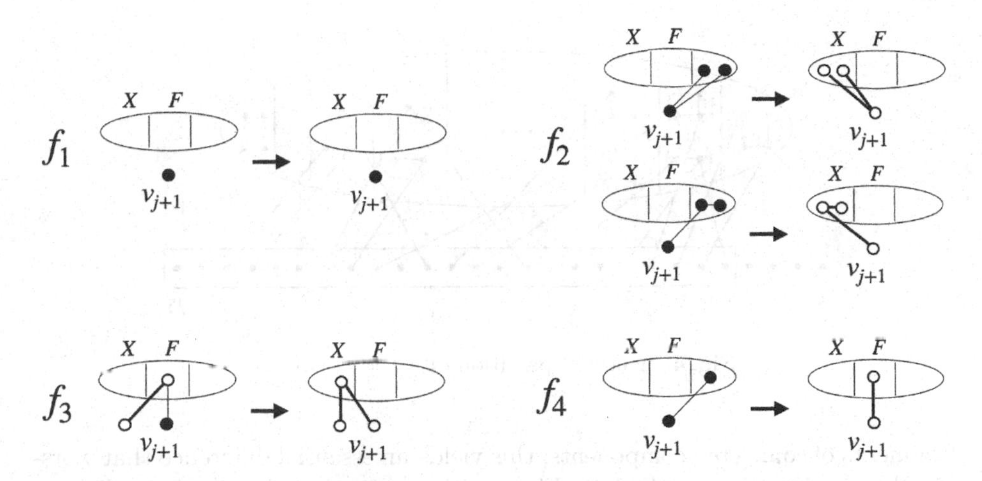

Fig. 3. How v_{j+1} is used. (The white vertices belong to a dominating set.)

The last condition can be checked in $O(2^{|J|} \cdot |J|^3)$ time for all $X \subseteq J$ by recursively considering all possible ways for removing three vertices from X; that is, $A[X, \emptyset, \emptyset, 0] = \min_{\{x,y,z\} \in \binom{X}{3}} A[X \setminus \{x, y, z\}, \emptyset, \emptyset, 0]$ if $|X| \geq 3$. The following recurrence formula holds: $A[X, F, Y, j+1] = \min\{f_1, f_2, f_3, f_4\}$, where

$$f_1 = A[X, F, Y, j],$$

$$f_2 = \min_{\alpha, \beta \in X, |E(G[\{\alpha,\beta,v_{j+1}\}])| \geq 2} A[X \setminus \{\alpha, \beta\}, F, Y \setminus \{N(\{\alpha, \beta, v_{j+1}\})\}, j] + 1,$$

$$f_3 = \min_{\alpha \in X \cap N(v_{j+1})} A[X \setminus \{\alpha\}, F \cup \{\alpha\}, Y \setminus N(v_{j+1}), j] + 1,$$

$$f_4 = \min_{\beta \in F \cap N(v_{j+1})} A[X, F \setminus \{\beta\}, Y \setminus N(\{\beta, v_{j+1}\}), j] + 1.$$

The four options f_1, f_2, f_3, and f_4 assume different ways of the role of v_{j+1} and compute the optimal value under the assumptions (see Fig. 3). Concretely, f_1 is the case when v_{j+1} does not belong to the solution, and f_2 is the case when v_{j+1} belongs to the solution together with two vertices in J in a connected way. In f_3, v_{j+1} forms a triple in the solution with a vertex in F and a vertex in I_j. In f_4, v_{j+1} currently forms a pair in J and will form a triple with a vertex in $I \setminus I_j$. We can compute all entries of A in $O^*(4^{|J|})$ time.

Similarly to the previous case of $r = 2$, we can compute the minimum number of units in a 3-grouped dominating set as $\min\{(|X| + A[X, \emptyset, J \setminus N[X], |I|])/3 \mid X \subseteq J \text{ and } I \subseteq N(X)\}$. Given the table A, this can be done in $O^*(2^{|J|})$ time. Thus the total running time of the algorithm is $O^*(4^{|J|})$.

Algorithm for r-Grouped Dominating Set. We now present our algorithm for general $r \geq 4$. Let $G = (V, E)$ be a graph, J be a vertex cover of G, and $I = V \setminus J$. Our algorithm is still based on a similar framework to the previous cases, though connected components of general r can be built up from smaller

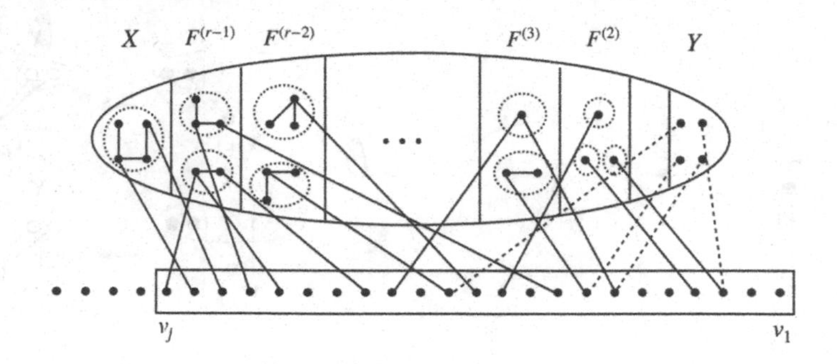

Fig. 4. A nested partition of a vertex cover.

fragments of connected components; this yields an essential difference that worsens the running time. In the DP, J is partitioned into $r+1$ parts depending on the partial solution corresponding to each table entry, and then some of the parts in the partition are further partitioned into smaller subsets. In other words, each table entry corresponds to a nested partition of the vertex cover.

As in the previous algorithms, for each subset $J_D \subseteq J$, we find a subset $I_D \subseteq I$ (if any exists) of the minimum size such that $J_D \cup I_D$ can form an r-grouped dominating set. Let X, $F^{(r-1)}, \ldots, F^{(3)}, F^{(2)}, Y$ be disjoint subsets of J, and let $I = \{v_1, v_2, \ldots, v_{|I|}\}$. For $i = 2, \ldots, r-1$, let $\mathcal{F}^{(i)}$ be a partition of $F^{(i)}$, where $\mathcal{F}^{(i)} = \{F_1^{(i)}, F_2^{(i)}, \ldots, F_{|\mathcal{F}^{(i)}|}^{(i)}\}$. The number of such nested partitions $(X, \mathcal{F}^{(r-1)}, \ldots, \mathcal{F}^{(2)}, Y)$ is at most $(r+1)^{|J|}|J|^{|J|}$. For $j = 0, \ldots, |I|$, we define $A[X, \mathcal{F}^{(r-1)}, \ldots, \mathcal{F}^{(2)}, Y, j]$ as the minimum size of $I' \subseteq \{v_1, v_2, \ldots, v_j\}$ that satisfies the following conditions:

1. $Y \subseteq N(I')$,
2. I' can be partitioned into $r-1$ parts $I'_2, I'_3, \ldots, I'_r$ satisfying the following conditions:
   - for $i = 2, \ldots, r-1$, $I'_i \cup F^{(i)}$ has a partition $\mathcal{D}^{(i)} = \{D_1^{(i)}, D_2^{(i)}, \ldots, D_{|\mathcal{F}^{(i)}|}^{(i)}\}$ such that for all $p = 1, \ldots, |\mathcal{F}^{(i)}|$, $D_p^{(i)}$ includes at least one vertex of I' and is a superset of $F_p^{(i)}$, and $|D_p^{(i)}| = i$ and $G[D_p^{(i)}]$ is connected.
   - $I'_r \cup X$ has a partition $\mathcal{D}^{(r)} = \{D_1^{(r)}, D_2^{(r)}, \ldots, D_q^{(r)}\}$ such that for all $i = 1, \ldots, q$, $|D_i^{(r)}| = r$ and $G[D_i^{(r)}]$ is connected.

We set $A[X, \mathcal{F}^{(r-1)}, \ldots, \mathcal{F}^{(2)}, Y, j] = \infty$ if no $I' \subseteq \{v_1, v_2, \ldots, v_j\}$ satisfies the conditions. We can compute $A[X, \mathcal{F}^{(r-1)}, \ldots, \mathcal{F}^{(2)}, Y, 0]$, which is 0 or ∞, as it is 0 if and only if $F^{(r-1)} = \cdots = F^{(2)} = Y = \emptyset$ and $G[X]$ admits a partition into connected graphs of r vertices. The last condition can be checked in $O(|J|^{|J|})$ time for all $X \subseteq J$ by checking all possible partitions of J.

Assume that all entries of A with $j \leq c$ for some c are computed. Since the degree of v_{c+1} is at most $|J|$, the number of possible ways of how v_{c+1} extends a partial solution is at most $2^{|J|}$. Thus from each table entry of A with $j = c$, we obtain at most $2^{|J|}$ candidates of the table entries with $j = c+1$. Thus, we can compute all entries of A in $O^*(2^{|J|}(r+1)^{|J|}|J|^{|J|})$ time.

Given A, we can compute the minimum number of units in an r-grouped dominating set as $\min\{(|X| + A[X, \emptyset, \ldots, \emptyset, J \setminus N[X], |I|])/r \mid X \subseteq J \text{ and } I \subseteq N(X)\}$. Again this takes only $O^*(2^{|J|})$ time. Thus the total running time of the algorithm is $O^*(2^{|J|}(r+1)^{|J|}|J|^{|J|}) = O^*((2\nu(r+1))^{\nu})$.

4.2 Algorithms Parameterized by Twin Cover Number

In this subsection, we show that the algorithms presented above still work when the parameter ν in the running time is replaced with twin cover number τ. To show this, we prove the following lemma. It says that twin-edges do not contribute to the connectivity of units for some minimum r-grouped dominating sets and can be removed from the graph. As a result, the vertex cover number can be replaced with the twin cover number.

Lemma 4.3. *Let G be a graph and K be a twin cover of G. If G has an r-grouped dominating set, then there exists a minimum r-grouped dominating set such that every unit has at least one vertex in K.*

Proof. Let $G = (V, E)$ be a graph, and K be a twin cover of G. Suppose that a minimum r-grouped dominating set $\mathcal{D}$ exists and one of its units $D = \{v_1, v_2, \ldots, v_r\}$ has no vertex in K. Since K is a twin cover, $N[v_1] = \cdots = N[v_r]$ holds. Let $K_D = K \cap N(v_1)$. Then, there is at least one vertex x in K_D such that $x \notin \bigcup \mathcal{D}$. Suppose to the contrary that there is no such x, and thus all vertices in K_D belong to $\bigcup \mathcal{D}$. This implies that no vertex is dominated only by D and that D itself is dominated by some vertices in K_D. Thus, $\mathcal{D} \setminus \{D\}$ is an r-grouped dominating set. This contradicts the minimality of $\mathcal{D}$. Let $D' = D \setminus \{v_1\} \cup \{x\}$, then $\mathcal{D}' = \mathcal{D} \setminus \{D\} \cup \{D'\}$ is also a minimum r-grouped dominating set of G (see Fig. 5). By repeating this process, we can obtain a minimum r-grouped dominating set such that every unit has at least one vertex in K. □

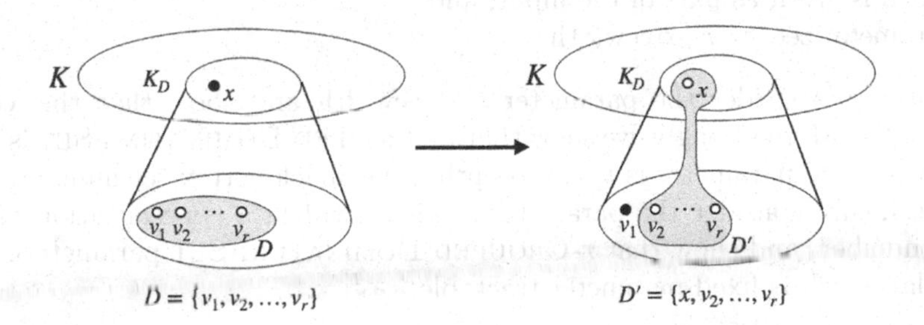

Fig. 5. An example for exchange. White vertices belong to a dominating set.

5 Beyond Vertex Cover and Twin Cover

In this section, we further explore the parameterized complexity of r-GROUPED DOMINATING SET with respect to structural graph parameters that generalize vertex cover number and twin cover number. We show that if we do not try to optimize the running time of algorithms, then we can use known algorithmic meta-theorems that automatically give fixed-parameter algorithms parameterized by certain graph parameters.

For the sake of brevity, we define only the parameters for which we need their definitions. For example, we do not need the definition of treewidth for applying the meta-theorem described below. On the other hand, to contrast the results here with the ones in the previous sections, it is important to see the picture of the relationships between the parameters. See Fig. 1 for the hierarchy of the graph parameters we deal with.

Roughly speaking, the algorithmic meta-theorems we use here say that if a problem can be expressed in a certain logic (e.g., FO, MSO_1, or MSO_2), then the problem is fixed-parameter tractable parameterized by a certain graph parameter (e.g., twin-width, treewidth, or clique-width). Such theorems are extremely powerful and used widely for designing fixed-parameter algorithms [23]. On the other hand, the generality of the meta-theorems unfortunately comes with very high dependency on the parameters [15]. When our target parameter is vertex cover number, the situation is slightly better, but still a double-exponential $2^{2^{\Omega(\nu)}}$ lower bound of the parameter dependency is known under ETH [24]. This implies that our "slightly superexponential" $2^{O(\tau \log \tau)}$ algorithm in Sect. 4 cannot be obtained by applications of known meta-theorems.

In the rest of this section, we first introduce FO, MSO_1, and MSO_2 on graphs. We then observe that the problem can be expressed in FO when r and k are part of the parameter and in MSO_2 when r is part of the parameter. These observations combined with known meta-theorems immediately imply that r-Grouped Dominating Set is fixed-parameter tractable when

- parameterized by $r + k$ on nowhere dense graph classes;
- parameterized by $r+k+$twin-width if a contraction sequence of the minimum width is given as part of the input; and
- parameterized by $r+$treewidth.

We then consider the parameter $k+$treewidth and show that this case is intractable. More strongly, we show that r-Grouped Dominating Set is W[1]-hard when the parameter is $k+$ treedepth + feedback vertex set number.

We finally consider the parameter modular-width, a generalization of twin cover number, and show that r-Grouped Dominating Set parameterized by modular-width is fixed-parameter tractable.

5.1 Results Based on Algorithmic Meta-theorems and Related Results

The *first-order logic* on graphs (FO) allows variables representing vertices of the graph under consideration. The atomic formulas are the equality $x = y$ of variables and the adjacency $E(x, y)$ meaning that $\{x, y\} \in E$. The FO formulas are defined recursively from atomic formulas with the usual Boolean connectives ($\neg$, $\wedge$, $\vee$, $\Rightarrow$, $\Leftrightarrow$), and quantification of variables ($\forall$, $\exists$). We also use the existential quantifier with a dot ($\dot{\exists}$) to quantify distinct objects. For example, $\dot{\exists} a, b \colon \phi$ means $\exists a, b \colon (a \neq b) \wedge \phi$. We write $G \models \phi$ if G satisfies (or *models*) ϕ. Given a graph G and an FO formula ϕ, FO MODEL CHECKING asks whether $G \models \phi$.

It is straightforward to express the property of having an r-grouped dominating set of k units with an FO formula whose length depends only on $r + k$:

$$\phi_{r,k} = \dot{\exists} v_1, v_2, \ldots, v_{rk} \colon \\ \mathsf{dominating}(v_1, \ldots, v_{rk}) \wedge \bigwedge_{0 \le i \le k-1} \mathsf{connected}(v_{ir+1}, \ldots, v_{ir+r}),$$

where $\mathsf{dominating}(\cdots)$ is a subformula expressing that the rk vertices form a dominating set and $\mathsf{connected}(\cdots)$ is the one expressing that the r vertices induce a connected subgraph. This implies that r-GROUPED DOMINATING SET parameterized by $r+k$ is fixed-parameter tractable on graph classes on which FO MODEL CHECKING parameterized by the formula length $|\phi|$ is fixed-parameter tractable. Such graph classes include nowhere dense graph classes [18] and graphs of bounded twin-width (given with so called contraction sequences) [3].

Corollary 5.1. r-GROUPED DOMINATING SET *parameterized by* $r + k$ *is fixed-parameter tractable on nowhere dense graph classes.*

Corollary 5.2. r-GROUPED DOMINATING SET *parameterized by* $r + k+$ twin-width *is fixed-parameter tractable if a contraction sequence of the minimum width is given as part of the input.*

The *monadic second-order logic* on graphs (MSO_1) is an extension of FO that additionally allows variables representing vertex sets and the inclusion predicate $X(x)$ meaning that $x \in X$. MSO_2 is a further extension of MSO_1 that also allows edge variables, edge-set variables, and an atomic formula $I(e, x)$ representing the edge-vertex incidence relation. Given a graph G and an MSO_1 (MSO_2, resp.) formula $\phi(X)$ with a free set variable X, MSO_1 (MSO_2, resp.) OPTIMIZATION asks to find a minimum set S such that $G \models \phi(S)$.

It is not difficult to express the property of a vertex set being the union of r-units of a r-grouped dominating set with an MSO_2 formula whose length depending only on r:[2]

$$\psi_r(X) = \mathsf{dominating}(X) \wedge \\ (\exists F \subseteq E \colon \mathsf{span}(F, X) \wedge (\forall C \subseteq X \colon \mathsf{cc}(F, C) \Rightarrow \mathsf{size}_r(C))),$$

where $\mathsf{dominating}(X)$ is a subformula expressing that X is a dominating set, $\mathsf{span}(F, X)$ is the one expressing that X is the set of all endpoints of the edges in F, $\mathsf{cc}(F, C)$ expresses that C is the vertex set of a connected component of the subgraph induced by F, and $\mathsf{size}_r(C)$ means that C contains exactly r elements. Since MSO_2 OPTIMIZATION parameterized by treewidth is fixed-parameter tractable [1,4,9], we have the following result.

[2] Note that there is no equivalent MSO_1 formula of length depending only on r. This is because $G \models \psi_2(V)$ expresses the property of having a perfect matching, for which an MSO_1 formula does not exist (see e.g., [10]).

Corollary 5.3. *r-*Grouped Dominating Set *parameterized by $r+$* treewidth *is fixed-parameter tractable.*

Now the natural question regarding treewidth and r-Grouped Dominating Set would be the complexity parameterized by $k+$ treewidth. Unfortunately, this case is W[1]-hard even if treewidth is replaced with a possibly much larger parameter pathwidth + feedback vertex set number and the graphs are restricted to be planar. Furthermore, if the planarity is not required, we can replace pathwidth in the parameter with treedepth.

Theorem 5.4. *r-*Grouped Dominating Set *parameterized by $k+$* pathwidth + feedback vertex set number *is W[1]-hard on planar graphs.*

It is known that on general (not necessarily planar) graphs, Equitable Connected Partition parameterized by $k+$treedepth + feedback vertex set number is W[1]-hard [17]. Since adding pendants to all vertices increases treedepth by at most 1 (see e.g., [27]), the same reduction shows the following hardness.

Theorem 5.5. *r-*Grouped Dominating Set *parameterized by $k+$*treedepth + feedback vertex set number *is W[1]-hard.*

As a parameter between twin-cover number and clique-width, modular-width is well studied. It is known that the modular-width of a graph and a recursive partition certificating it can be computed in linear time [11,19]. We can show the following theorem.

Theorem 5.6. *r-*Grouped Dominating Set *parameterized by modular-width is fixed-parameter tractable.*

References

1. Arnborg, S., Lagergren, J., Seese, D.: Easy problems for tree-decomposable graphs. J. Algorithms **12**(2), 308–340 (1991)
2. Bodlaender, H.L., Cygan, M., Kratsch, S., Nederlof, J.: Deterministic single exponential time algorithms for connectivity problems parameterized by treewidth. Inf. Comput. **243**, 86–111 (2015)
3. Bonnet, É., Kim, E.J., Thomassé, S., Watrigant, R.: Twin-width I: tractable FO model checking. J. ACM **69**(1), 1–46 (2022)
4. Borie, R.B., Parker, R.G., Tovey, C.A.: Automatic generation of linear-time algorithms from predicate calculus descriptions of problems on recursively constructed graph families. Algorithmica **7**(5&6), 555–581 (1992)
5. Chen, J., Kanj, I.A., Xia, G.: Improved upper bounds for vertex cover. Theor. Comput. Sci. **411**(40–42), 3736–3756 (2010)
6. Chen, L., Lu, C., Zeng, Z.: Hardness results and approximation algorithms for (weighted) paired-domination in graphs. Theor. Comput. Sci. **410**(47), 5063–5071 (2009)
7. Chen, L., Lu, C., Zeng, Z.: A linear-time algorithm for paired-domination problem in strongly chordal graphs. Inf. Process. Lett. **110**(1), 20–23 (2009)
8. Chen, L., Lu, C., Zeng, Z.: Labelling algorithms for paired-domination problems in block and interval graphs. J. Comb. Optim. **19**(4), 457–470 (2010)

9. Courcelle, B.: The monadic second-order logic of graphs. I. recognizable sets of finite graphs. Inf. Comput. **85**(1), 12–75 (1990)
10. Courcelle, B., Engelfriet, J.: Graph Structure and Monadic Second-Order Logic - A Language-Theoretic Approach. Cambridge University Press, Cambridge (2012)
11. Cournier, A., Habib, M.: A new linear algorithm for modular decomposition. In: Tison, S. (ed.) CAAP 1994. LNCS, vol. 787, pp. 68–84. Springer, Heidelberg (1994). https://doi.org/10.1007/BFb0017474
12. Cygan, M., et al.: Parameterized Algorithms. Springer, Cham (2015). https://doi.org/10.1007/978-3-319-21275-3
13. Cygan, M., Nederlof, J., Pilipczuk, M., Pilipczuk, M., Van Rooij, J.M.M., Wojtaszczyk, J.O.: Solving connectivity problems parameterized by treewidth in single exponential time. ACM Trans. Algorithms **18**(2) (2022)
14. Desormeaux, W.J., Haynes, T.W., Henning, M.A.: Paired domination in graphs. In: Haynes, T.W., Hedetniemi, S.T., Henning, M.A. (eds.) Topics in Domination in Graphs. DM, vol. 64, pp. 31–77. Springer, Cham (2020). https://doi.org/10.1007/978-3-030-51117-3_3
15. Frick, M., Grohe, M.: The complexity of first-order and monadic second-order logic revisited. Ann. Pure Appl. Log. **130**(1–3), 3–31 (2004)
16. Ganian, R.: Improving vertex cover as a graph parameter. Discret. Math. Theor. Comput. Sci. **17**(2), 77–100 (2015)
17. Gima, T., Otachi, Y.: Extended MSO model checking via small vertex integrity. In: 33rd International Symposium on Algorithms and Computation (ISAAC 2022). Leibniz International Proceedings in Informatics (LIPIcs), vol. 248, pp. 20:1–20:15. Schloss Dagstuhl - Leibniz-Zentrum für Informatik, Dagstuhl, Germany (2022)
18. Grohe, M., Kreutzer, S., Siebertz, S.: Deciding first-order properties of nowhere dense graphs. J. ACM **64**(3), 1–32 (2017)
19. Habib, M., Paul, C.: A survey of the algorithmic aspects of modular decomposition. Comput. Sci. Rev. **4**(1), 41–59 (2010)
20. Hanaka, T., Ono, H., Otachi, Y., Uda, S.: Grouped domination parameterized by vertex cover, twin cover, and beyond. arXiv preprint arXiv:2302.06983 (2023)
21. Haynes, T.W., Slater, P.J.: Paired-domination and the paired-domatic number. Congressus Numerantium 65–72 (1995)
22. Haynes, T.W., Slater, P.J.: Paired-domination in graphs. Networks **32**(3), 199–206 (1998)
23. Kreutzer, S.: Algorithmic meta-theorems. In: Esparza, J., Michaux, C., Steinhorn, C. (eds.) Finite and Algorithmic Model Theory, London Mathematical Society Lecture Note Series, vol. 379, pp. 177–270. Cambridge University Press (2011)
24. Lampis, M.: Algorithmic meta-theorems for restrictions of treewidth. Algorithmica **64**(1), 19–37 (2012)
25. Liedloff, M.: Finding a dominating set on bipartite graphs. Inf. Process. Lett. **107**(5), 154–157 (2008)
26. Lin, C.C., Ku, K.C., Hsu, C.H.: Paired-domination problem on distance-hereditary graphs. Algorithmica **82**(10), 2809–2840 (2020)
27. Nešetřil, J., Ossona de Mendez, P.: Sparsity. AC, vol. 28. Springer, Heidelberg (2012). https://doi.org/10.1007/978-3-642-27875-4
28. Pradhan, D., Panda, B.: Computing a minimum paired-dominating set in strongly orderable graphs. Discret. Appl. Math. **253**, 37–50 (2019)
29. Tripathi, V., Kloks, T., Pandey, A., Paul, K., Wang, H.-L.: Complexity of paired domination in AT-free and planar graphs. In: Balachandran, N., Inkulu, R. (eds.) CALDAM 2022. LNCS, vol. 13179, pp. 65–77. Springer, Cham (2022). https://doi.org/10.1007/978-3-030-95018-7_6

Broadcasting in Split Graphs

Hovhannes A. Harutyunyan and Narek Hovhannisyan(✉)

Department of Computer Science and Software Engineering, Concordia University, Montreal, QC H3G 1M8, Canada
haruty@cs.concordia.ca, narek.hovhannisyan@concordia.ca

Abstract. Broadcasting is an information dissemination primitive where a message is passed from one node (called originator) to all other nodes in the network. With the increasing interest in interconnection networks, an extensive amount of research was dedicated to broadcasting. Two main research goals of this area are finding inexpensive network structures that maintain efficient broadcasting and finding the broadcast time for well-known and widely used network topologies.

In the scope of this paper, we will mainly focus on determining the broadcast time and the optimal broadcasting scheme for graphs. Determination of the broadcast time of a node x in an arbitrary network G is known to be NP-hard. Polynomial time solutions are known only for a few classes of networks. There also exist various heuristic and approximation algorithms for different network topologies. In this paper, we will consider networks that can be represented as split graphs. We will present a polynomial time 2-approximation algorithm for the broadcast time problem and will introduce an algorithm for generating optimal or near-optimal broadcast schemes in split graphs.

Keywords: Interconnection networks · Information dissemination · Broadcasting · Approximation algorithms

1 Introduction

Broadcasting is one of the most important information dissemination processes in an interconnected network. Over the last four decades, a large number of research work has been published concerning broadcasting in networks under different models. These models can have different numbers of originators, numbers of receivers at each time unit, distances of each call, numbers of destinations, and other characteristics of the network such as the knowledge of the neighborhood available to each node. In the context of this paper, we are going to focus on the classical model of broadcasting. The network is modeled as an undirected connected graph $G = (V, E)$, where $V(G)$ and $E(G)$ denote the vertex set and the edge set of G, respectively. The classical model follows the below-mentioned basic assumptions.

M. Mavronicolas (Ed.): CIAC 2023, LNCS 13898, pp. 278–292, 2023.
https://doi.org/10.1007/978-3-031-30448-4_20

(1) The broadcasting process is split into discrete time units.
(2) The only vertex that has the message at the first time unit is called *originator.*
(3) In each time unit, an informed vertex (*sender*) can *call* at most one of its uninformed neighbors (*receiver*).
(4) During each unit, all calls are performed in parallel.
(5) The process halts as soon as all the vertices in the graph are informed.

We can represent each call in this process as an ordered pair of two vertices (u, v), where u is the sender and v is the receiver. The *broadcast scheme* is the order of calls made by each vertex during a broadcasting process and can be represented as a sequence $(C_1, C_2, ..., C_t)$, where C_i is the set of calls performed in time unit i. An informed vertex v is *idle* in time unit t if v does not make any call in time t. Given that every vertex, other than the originator, can be informed by exactly one vertex, the broadcast scheme forms a directed spanning tree (*broadcast tree*) rooted at the originator. We are also free to omit the direction of each call in the broadcast tree.

Definition 1. *The* **broadcast time** *of a vertex v in a given graph G is the minimum number of time units required to broadcast in G if v is the originator and is denoted by $b(v, G)$. The broadcast time of a given graph G, is the maximum broadcast time from any originator in G, formally $b(G) = max_{v \in V(G)}\{b(v, G)\}$.*

A broadcast scheme for an originator v that uses $b(v, G)$ time units is called an optimal broadcast scheme. Obviously, by the assumption (3), the number of informed vertices after each time unit can at most be doubled. Meaning, in general, the number of informed vertices after time unit i is upper bounded by 2^i. Therefore, it is easy to see that $b(v, G) \geq \lceil \log n \rceil$, where n is the number of vertices in G, which implies that $b(G) \geq \lceil \log n \rceil$.

The general *broadcast time* decision problem is formally defined as follows. Given a graph $G = (V, E)$ with a specified set of vertices $V_0 \subseteq V$ and a positive integer k, is there a sequence $V_0, E_1, V_1, E_2, V_2, \ldots, E_k, V_k$ where $V_i \subseteq V$, $E_i \subseteq E (1 \leq i \leq k)$, for every $(u, v) \in E_i$, $u \in V_{i-1}$, $v \in V_i$, $v \notin V_{i-1}$, and $V_k = V$. Here k is the total broadcast time, V_i is the set of informed vertices at round i, and E_i is the set of edges used at round i. It is obvious that when $| V_0 |= 1$ then this problem becomes our broadcast problem of determining $b(v, G)$ for an arbitrary vertex v in an arbitrary graph G.

Generally, the broadcast time decision problem in an arbitrary graph is NP-complete [12,29]. Moreover, the broadcast time problem was proved to be NP-complete even for some specific graph classes, such as 3-regular planar graphs [24]. There is a very limited number of graph families, for which an exact algorithm with polynomial time complexity is known for the broadcast time problem. Exact linear time algorithms are available for the broadcast time problem in trees [27,29], in connected graphs with only one cycle (unicyclic graphs) [17,18], in necklace graphs (chain of rings) [15], in fully connected trees [13], and in Harary-like graphs [5,6]. For a more detailed introduction to broadcasting, we refer the reader to [11,16,19,20].

A split graph is a graph in which the vertices can be partitioned into a clique and an independent set [23]. Split graphs were first studied in [9] and independently introduced in [31]. An edge of a graph is a chord of a cycle if it joins two nodes of the cycle but is not itself in the cycle. A graph is chordal if and only if every cycle of length greater than three has a chord. Split graphs are a subclass of chordal graphs and are exactly those chordal graphs whose complement is also chordal [14]. Therefore, all problems which are polynomial-time solvable for chordal graphs are also solvable for split graphs [12,14]. Chordal graphs play a central role in techniques for exploiting sparsity in large semidefinite optimization problems [33] and in related convex optimization problems involving sparse positive semidefinite matrices. Chordal graph properties are also fundamental to several classical results in combinatorial optimization, linear algebra, statistics, signal processing, machine learning, and nonlinear optimization [32,34].

In [3], authors showed that in the limit as n goes to infinity, the fraction of n-vertex chordal graphs that are split approaches one. Less formally, they showed that almost all chordal graphs are split graphs, thus making split graphs an important area of research. Moreover, split graphs are widely used as an interconnection network topology. Networks that have an important group of tightly coupled nodes (or in other words the *core* of the network), and a number of independent nodes that are only connected to the network core. In terms of social networks, split graphs correspond to the variety of interpersonal and intergroup relations [1]. The interaction between the cliques (socially strong and trusty groups) and the independent sets (fragmented and non-connected groups of people) is naturally represented as a split graph. Different optimization problems were studied in split graphs due to their many important characteristics [2,4,7,8,25]. Given all of the above, it is interesting to study the problem of broadcasting on split graphs.

[21] is one of the first works in literature that study the minimum broadcast time problem in split graphs. In the paper, the authors prove that it is NP-complete to decide whether the broadcast time from multiple originators in split graphs (and in chordal graphs) is equal to 2. In the same paper, the authors also show that the minimum broadcast time problem from a single originator in chordal graphs is NP-complete. Later, in [30], the minimum broadcast and multicast time problems are considered for a subclass of split graphs where the degree of all vertices in the independent set is one. The authors introduce a polynomial time exact algorithm for that subclass of split graphs.

The rest of this paper is organized as follows. In Sect. 2 we introduce a polynomial time approximation algorithm for the broadcast time problem in split graphs that guarantees a 2 approximation ratio. Further, in Sect. 3, we analyze some important aspects of an optimal broadcast scheme and present an exact algorithm scheme. Lastly, Sect. 4 concludes the paper.

2 An Approximation Algorithm for Broadcasting in Split Graphs

Consider a split graph $G = (V, E)$ such that the vertex set can be partitioned into a clique K on n vertices and an independent set I on m vertices. Note that split graphs may have more than one partitioning into a clique and an independent set. Clearly, we can assume that the clique K is a maximal (and also maximum) clique, otherwise, we could change the partitioning by adding some vertices from I to K to make it maximal. Given a connected split graph G, a clique K, an independent set I, and an arbitrary originator $v \in V(G)$, we are going two consider two cases: $v \in K$ or $v \in I$.

Let t be a positive integer. A *t-star-matching* of graph G is a collection of mutually vertex disjoint subgraphs $K_{1,i}$ (stars) of G with $1 \leq i \leq t$ [22]. A *perfect t-star-matching* of a graph G is a t-star-matching that covers every vertex of the graph G. A *proper t-star-matching* of a split graph G is a perfect t-star-matching such that any vertex $v \in K$ is a center of a star.

First, let us consider a broadcasting problem with multiple originators. We are given a split graph $G = (V, E)$ with a partitioning of a maximum clique K and an independent set I, and a set of informed vertices (originators) S such that $S = K$. The goal of the problem is to find the minimum broadcast time $b(K, G)$ needed to inform all the vertices in the independent set. It is easy to see that in this case of multiple originators, any broadcast scheme forms a directed spanning forest (instead of a tree), where each tree in the forest is rooted at one of the originators. Moreover, as broadcasting in split graphs has only one level, the broadcast scheme will form a set of stars, where the center of each star is one of the originators. In other words, a broadcasting scheme with broadcast time b induces a proper b-star-matching for the given split graph G. Hence, if we find the minimum positive integer t, for which there exists a proper t-star-matching, we can claim that $b(K, G) = t$.

2.1 Finding a Proper Star-Matching with Minimum Maxdegree

In order to find the minimum positive integer t, for which there exists a proper t-star-matching, we will consider the following decision problem.

Problem 1. Proper star-matching problem
Instance: (G, K, I, t), where $G = (V, E)$ is a split graph, K is a clique in G, I is an independent set in G, and t is a natural number.
Output: "Yes" if G contains a proper t-star-matching for; "No" otherwise.

Next, we will reduce an instance of the above-defined problem to an instance of a maximum flow problem [10,28].

Problem 2. Maximum flow
Instance: (G, s, q, μ), where $G = (V, E)$ is a flow graph, s is the source vertex, q is the sink vertex, and μ is a natural number.
Output: "Yes" if the maximum flow for graph G from s to q is greater than or equal to μ; "No" otherwise.

The reduction consists of the steps described in Algorithm 1, and an example of the reduction is presented in Fig. 1.

Algorithm 1. Reduction Algorithm

Input A split graph G, a clique K, an independent set I
Output An instance of the maximum flow problem

1: **procedure** MAXFLOWREDUCTION
2: Create a copy G' of the graph G
3: Remove all edges between a pair of vertices u and v, where $u, v \in K$
4: Assign direction from u to v to each edge (u, v), where $u \in K$, $v \in I$
5: Assign capacity 1 to each edge (u, v), where $u \in K$, $v \in I$
6: Add a source vertex s and connect it by an outgoing edge to every vertex in K. Assign a capacity t to those edges
7: Add a sink vertex q and connect it by an ingoing edge to every vertex in I. Assign a capacity 1 to those edges
8: **return** $(G', s, q, |I|)$
9: **end procedure**

Proposition 1. *There exists a proper t-star-matching for the graph G if and only if the maximum flow in the graph G' is equal to m, where m is the cardinality of the independent set I.*

Proof. It is easy to see that when there exists a proper t-star-matching, the same edges can be used in graph G' to achieve flow with value m. And similarly, if the maximum flow is equal to m, the paths used to construct the flow can be used to find a proper star-matching for the graph G. Since each vertex in K has an incoming edge capacity of t, it will result in a proper t-star-matching for the graph G. □

Thus, we can find the minimum integer t for which there exists a proper t-star-matching. The procedure *MaxFlowReduction* (Algorithm 1) can be implemented with complexity $\mathcal{O}(n) + \mathcal{O}(m) = \mathcal{O}(|V|)$. The time to find the maximum flow in a graph highly depends on the selected algorithm. One of the best algorithms for maximum flow has a complexity of $\mathcal{O}(|E||V|)$ [26]. Thus, considering the binary search time, the overall time complexity of this algorithm would be $\mathcal{O}(|E||V| \log |I|)$.

2.2 Broadcasting from a Vertex in the Clique

Algorithm 2 shows the steps of the broadcasting algorithm from an originator in the clique. Clearly, after the algorithm halts, all vertices of the graph G are informed as the clique is informed in step 1 (line 2), and the independent set is informed in step 3 (line 4).

Let $b_{Alg}(v, G)$ be the broadcast time generated by Algorithm 2. It is easy to see that it has two components: the broadcast time of the clique K and

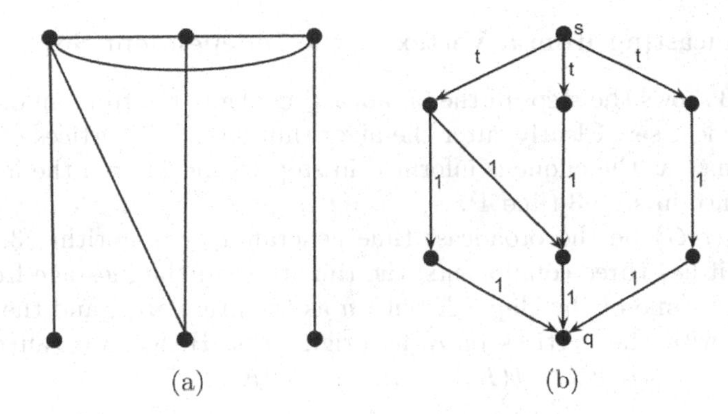

Fig. 1. Example of a reduction in Algorithm 1. Subfigure (a) shows the graph G before the reduction and subfigure (b) shows the graph G' after the reduction.

Algorithm 2. Approximation Algorithm

Input A split graph G, a clique K, an independent set I, and an originator $v \in K$

Output A broadcast scheme with time $b_{Alg}(v, G)$ for the graph G and the originator v

1: **procedure** BroadcastingFromClique
2: Broadcast in the clique K starting from the vertex v
3: Find a proper t-star-matching M with minimum t
4: Use the edges of M to broadcast from K to I
5: **end procedure**

---

the broadcast time in G with the vertices of K as originators. Hence, we can claim that $b_{Alg}(v, G) = b(K_n) + b(K, G) = \lceil \log n \rceil + t$, where $n = |K|$. Recall that for any graph G, $b(v, G) \geq \lceil \log |V(G)| \rceil$. As in our case $n \leq |V(G)|$, then $b(v, G) \geq \lceil \log n \rceil$. Moreover, it is trivial that the broadcast time from a single originator in the clique cannot be less than the broadcast time where all the vertices in the clique are originators. Hence, $b(v, G) \geq b(K, G) = t$. From the two lower bounds of the minimum broadcast time described above, it follows that:

$$b(v, G) \geq \frac{\lceil \log n \rceil}{2} + \frac{t}{2} \tag{1}$$

Thus,

$$\frac{b_{Alg}(v, G)}{b(v, G)} \leq \frac{\lceil \log n \rceil + t}{\frac{\lceil \log n \rceil + t}{2}} = 2 \tag{2}$$

We can see that the broadcast time generated by Algorithm 2 is guaranteed to be at most twice as big as the optimal broadcast time.

As steps 2 and 4 of Algorithm 2 have known broadcast times, they have $\mathcal{O}(1)$ time complexity. Thus, the proposed algorithm has a complexity of $\mathcal{O}(|E||V| \log |I|)$, making it a polynomial time 2-approximation algorithm for the broadcast time problem in split graphs from an originator in the clique.

2.3 Broadcasting from a Vertex in the Independent Set

Algorithm 3 shows the steps of the broadcasting algorithm from an originator in the independent set. Clearly, after the algorithm halts, all vertices of the graph G are informed as the clique is informed in step 1 (line 2), and the independent set is informed in step 3 (line 4).

Let $b_{Alg}(v, G)$ be the broadcast time generated by Algorithm 3. It is easy to see that it has three components: the time to send the message from v to u, the broadcast time of the clique K with u as the originator, and the broadcast time in G' with the vertices of K as originators. Hence, we can claim that $b_{Alg}(v, G) = 1 + b(u, K_n) + b(K, G')$ where $n = |K|$.

$$b_{Alg}(v, G) = 1 + b(u, K_n) + b(K, G') = 1 + \lceil \log n \rceil + t \tag{3}$$

Algorithm 3. Approximation Algorithm

Input A split graph G, a clique K, an independent set I, and an originator $v \in I$
Output A broadcast scheme with time $b_{Alg}(v, G)$ for the graph G and the originator v

1: **procedure** BROADCASTINGFROMINDEPENDENTSET
2: Pass the message to an arbitrary neighbor $u \in K$ of v
3: Complete broadcasting in the clique K
4: From a graph G' from G by removing the vertex v
5: Find a proper t-star-matching M with minimum t for the split graph G'
6: Use the edges of M to broadcast from K to I
7: **end procedure**

Recall that for any graph G, $b(v, G) \geq \lceil \log |V(G)| \rceil$. In our case $n = |K|$ and $m = |I|$, then $b(v, G) \geq \lceil \log(n + m) \rceil \geq \lceil \log n \rceil$. Moreover, it is trivial that the broadcast time from a single originator in the independent set is greater by at least one than the broadcast time where all the vertices in the clique are originators and the rest of the independent set should be informed. Hence, $b(v, G) \geq 1 + b(K, G') = 1 + t$. From the two lower bounds of the minimum broadcast time described above, it follows that:

$$b(v, G) \geq \frac{\lceil \log n \rceil + t + 1}{2} \tag{4}$$

Thus,

$$\frac{b_{Alg}(v, G)}{b(v, G)} \leq \frac{\lceil \log n \rceil + t + 1}{\frac{\lceil \log n \rceil + t + 1}{2}} = 2 \tag{5}$$

Similar to Algorithm 2, Algorithm 3 has a complexity of $\mathcal{O}(|E||V| \log |I|)$, making it a polynomial time 2-approximation algorithm for the broadcast time problem in split graphs from an originator in the independent set.

2.4 Tightness of Approximation

In this section, we will prove that the approximation ratio of 2 is tight for the approximation algorithms introduced above (Algorithm 2 and 3). For that, we will construct an infinite subfamily of split graphs for which 2 is a tight approximation ratio when the originator is in the clique or in the independent set.

Claim. For every positive integer $0 < \epsilon < 1$, there exists a split graph instance G and an originator v for which $b_{Alg}(v, G) > (2 - \epsilon) \cdot b(v, G)$, where $b_{Alg}(v, G)$ is the broadcast time returned by Algorithm 2 or Algorithm 3 (depending on the originator) and $b(v, G)$ is the minimum broadcast time.

Proof. Let G be a split graph with a clique partition K and an independent set partition I, such that $n = |K| = 2^t + 1$ and $m = |I| = t$ for some positive integer t. Assume vertex $v_1 \in K$ is adjacent to every vertex $w_i \in I$, $1 \leq i \leq m$, and there exist no more edges between the clique and the independent set (Fig. 2). We will analyze the broadcast time in 2 cases: v_1 is the originator in the clique and w_1 is the originator in the independent set.

Consider the following broadcast scheme S_1 for originator v_1.

(1) Place a call from v_1 to v_2 in the first time unit.
(2) In the next t time units
   (a) v_1 informs its neighbors in I,
   (b) vertices in $K \setminus \{v_1\}$ finish broadcasting in the clique with v_2 as the originator.

Clearly, the broadcast scheme S_1 will have a broadcast time of $b(S_1) = t+1 = \lceil \log n \rceil = b(v_1, G)$, which is the trivial lower bound. Whereas, Algorithm 2 will return a broadcast time $b_{Alg}(v, G) = \lceil \log n \rceil + m = 2t + 1$. Hence, after simple calculations, we can show that for every $0 < \epsilon < 1$, $b_{Alg}(v, G) > (2 - \epsilon) \cdot b(v, G)$ when $t > \dfrac{1 - \epsilon}{\epsilon}$.

Next, consider the broadcast scheme S_2 for originator w_1.

(1) Place a call from w_1 to v_1 in the first time unit.
(2) Place a call from v_1 to v_2 in the second time unit.
(3) In the next t time units
   (a) v_1 informs its uninformed neighbors in I,
   (b) vertices in $K \setminus \{v_1\}$ finish broadcasting in the clique with v_2 as the originator.

The broadcast scheme S_2 will have a broadcast time of $b(S_1) = t + 2 = 1 + \lceil \log n \rceil = b(w_1, G)$, which is again the trivial lower bound. Whereas, Algorithm 3 will return a broadcast time $b_{Alg}(v, G) = \lceil \log n \rceil + (m-1) + 1 = 2t + 1$. Hence, after simple calculations, we can show that for every $0 < \epsilon < 1$, $b_{Alg}(v, G) > (2 - \epsilon) \cdot b(v, G)$ when $t > \dfrac{3 - 2\epsilon}{\epsilon}$. □

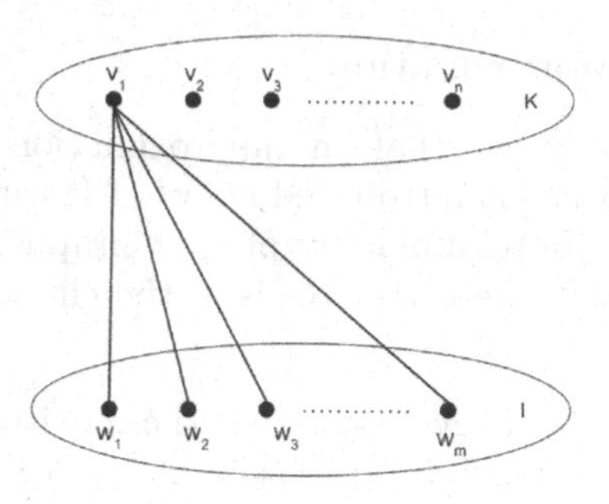

Fig. 2. Example of a split graph with tight approximation ratio.

3 Analysis of an Optimal Broadcast Scheme

In this section, we will be focusing on the decision version of the broadcast time problem.

Problem 3. Broadcast time problem
Instance: (G, v, t), where $G = (V, E)$ is a graph, $v \in V$ is the originator, and t is a natural number.
Output: "Yes" if $b(v, G) \leq t$; "No" otherwise.

Lemma 1. *Let G be a split graph, K be a clique in G, I be an independent set in G, and q be a vertex in K. There exists an optimal broadcast scheme S for the graph G starting from the originator q, such that every vertex $u \in I$ has no uninformed neighbors after the time unit when u gets informed.*

Proof. In other words, Lemma 1 claims that there exists an optimal broadcast scheme where no calls are placed from a vertex in the independent set to a vertex in the clique. Let OPT be an optimal broadcast scheme. If OPT does not contain any calls placed from the independent set to the clique then the lemma is proved. Assume in scheme OPT, a vertex $u \in I$ is informed by $v \in K$ in time unit i and in time units $i \leq t_1 \leq t_2 ... \leq t_k$ u informs its uninformed neighbors $w_1, w_2, ..., w_k \in K$, respectively. Then we construct a new broadcast scheme S, where in time unit i, v informs w_1 as they are both in the clique and in the rest of the broadcasting process v continues as in OPT. Whereas, in time unit t_j, vertex w_j passes the message to w_{j+1}, where $1 \leq j \leq k-1$ and, in time unit t_k, vertex u is informed by its neighbor w_p. This transformation is visualized in Fig. 3, where subfigure (a) shows the order of calls in scheme OPT and subfigure (b) shows the order of calls in scheme S. This shuffle will not affect the broadcast time, as after time unit t_k we will have the same set of vertices informed. Moreover, at each time unit $t_1, .., t_k$, we will have the same number of informed vertices, and informed vertices will be able to proceed following the broadcast scheme OPT. The same transformation can be applied to any other

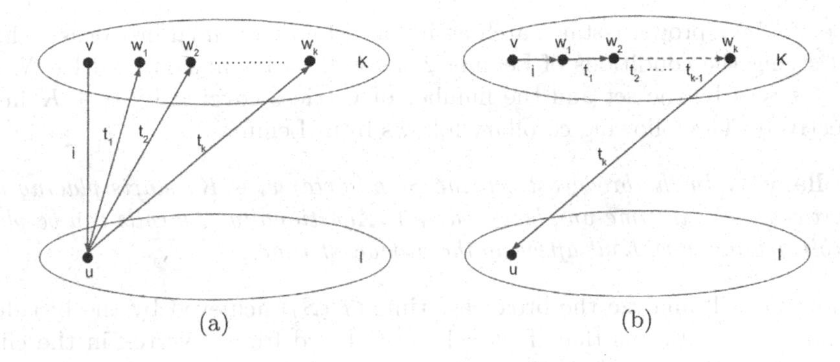

Fig. 3. Example of a transformation discussed in Lemma 1.

situation where a call is placed from the independent set to the clique. Hence, the created broadcast scheme S will have a broadcast time equal to that of OPT, thus becoming an optimal broadcast scheme that satisfies the conditions of the lemma. □

From Lemma 1 it is clear that any calls placed from a vertex $v \in K$ to its neighbors in K should be performed before calls towards its neighbors in I.

Lemma 2. *Let G be a split graph, K be a clique in G, I be an independent set in G, and q be a vertex in K. There exists an optimal broadcast scheme S satisfying the conditions of Lemma 1, such that no vertex $v \in K$ informs a vertex $u \in K$ after a vertex $w \in I$.*

Proof. In other words, vertices in the clique start placing calls to their neighbors in the independent set only when they have already informed vertices in the clique. Let $((u, v), (u, w))$ be a pair of calls (pair of directed edges), such that $u, w \in K$ and $v \in I$. We say that $((u, v), (u, w))$ is a *bad pair of calls* in a broadcast scheme S, if the call (u, v) was placed before the call (u, w). Let S be an optimal broadcast scheme satisfying the conditions of Lemma 1. It is clear that if S contains no bad pair of calls then it satisfies Lemma 2. Otherwise, if the scheme S contains some bad pairs of calls, we will do the following.

(1) Select an arbitrary bad pair of calls $((u, v), (u, w))$, where (u, v) was placed in time unit t_1, and (u, w) was placed in time unit $t_2 > t_1$.
(2) Swap the order of the call in the pair: use the edge (u, w) in time unit t_1 and the edge (u, v) in time unit t_2.
(3) Leave the rest of the calls unchanged (even if it causes idling).

After the above modification, the scheme S will finish broadcasting in the same amount of time because by Lemma 1, the vertex v does not place any calls after being informed. Moreover, the above transformation will reduce the number of bad pairs of calls in scheme S by at least 1. Hence, after a finite number of steps, scheme S will have optimal broadcast time containing no bad pairs of calls, thus, proving the lemma. □

Let M be a proper t-star-matching induced by an optimal broadcast scheme S satisfying the conditions of Lemma 2 (as well as Lemma 1), and let N_i and $d_i, 1 \leq i \leq n$ be the set and the number of vertices covered by $v_i \in K$ in M, respectively. The following corollary follows from Lemma 2.

Corollary 1. *In the broadcast scheme S, a vertex $v_i \in K$, starts placing calls to vertices in N_i in time unit $b(S) - d_i + 1$. Additionally, the calls can be placed in arbitrary order without affecting the broadcast time.*

Now we will analyze the broadcast time ($B(S)$) achieved by the broadcast scheme S. It is obvious that if no calls are placed from a vertex in the clique to the independent set then, theoretically, by the time the broadcasting is over $2^{B(S)}$ vertices could be informed in the clique (informed vertices are doubled every time unit). Now assume that a vertex u spends the last p time units of the broadcasting process placing calls to the vertices of the independent set I. Since these p time units could be used to inform $2^p - 1$ vertices in the clique, then the overall number of vertices that will, theoretically, be informed in the clique would be at most $2^{B(S)} - 2^p + 1$. According to Corollary 1, each vertex $v_i \in K$ spends d_i time unit broadcasting to I, and hence, after $B(S)$ time units, the number of informed vertices in the clique will be upper bounded by the following number.

$$2^{B(S)} - 2^{d_1} + 1 - 2^{d_2} + 1 - ... - 2^{d_n} + 1 = 2^{B(S)} - \sum_{i=1}^{n} 2^{d_i} + n \tag{6}$$

As the scheme S is a valid broadcast scheme and informs all the vertices in the graph within $B(S)$ time units, the following inequality is obvious.

$$2^{B(S)} - \sum_{i=1}^{n} 2^{d_i} + n \geq n \tag{7}$$

$$2^{B(S)} \geq \sum_{i=1}^{n} 2^{d_i} \tag{8}$$

For a given proper t-star-matching M, let *cost* of M, denoted by $C(M)$, be $C(M) = \sum_{i=1}^{n} 2^{d_i}$, where d_i is the degree of vertex $v_i \in K$ in the subgraph induced by M. Let M^* be a proper t-star-matching that minimizes the cost $C(M^*)$. We define the following decision problem for the star-matching with minimum cost.

Problem 4. Minimum cost star-matching
Instance: (G, c), where $G = (V, E)$ is a split graph, and c is a natural number.
Output: "Yes" if there exists a proper star-matching M^* of G such that $C(M^*) \leq c$; "No" otherwise.

The following corollary obviously follows from the Inequality 8.

Corollary 2. *The broadcast time of a split graph G from an originator $v \in K$ is lower bounded by $b(v, G) \geq \log C(M^*)$. Moreover, there exists an optimal broadcast scheme S, such that the star-matching induced by S is exactly equal to M^*.*

Hence, the problem of finding the minimum broadcast time of a given split graph can be solved if the minimum cost star-matching problem is solved. Based on everything discussed previously, the following exact algorithm for the minimum broadcast time problem could be derived.

Algorithm 4. Exact Algorithm

Input A split graph G, a clique K, an independent set I, an originator $v \in K$, and a natural number t

Output "Yes" if $b(v, G) \leq t$; "No" otherwise.

1: **procedure** SplitGraphBroadcast
2: Find a proper t-star-matching M with minimum cost $C(M)$
3: **if** $2^t \leq C(M)$ **then**
4: **return** "No"
5: **end if**
6: $d \leftarrow []$
7: $d[i] \leftarrow$ degree of $v_i \in K$ in M
8: Sort the degree array d in a non-increasing fashion
9: Broadcast in the clique prioritizing the vertices based on their order in d
10: **if** a vertex $v_i \in K$ is not informed in time unit $t - d_i + 1$ **then**
11: **return** "No"
12: **else**
13: Use the edges of v_i in M in arbitrary order starting from time unit $t - d_i + 1$
14: **end if**
15: **return** "Yes"
16: **end procedure**

Algorithm 4 follows the claims made previously in this section. The only exception is the conditional operation (if) on line 10. The statement checks if all the vertices that need to start placing calls towards the independent set are informed before their determined time unit. If this is not the case then the broadcasting cannot terminate within the given amount of time.

To sum up, Algorithm 4 gives a strategy that can generate an exact algorithm if the proper star-matching with minimum cost can be found. However, for any selected matching strategy the algorithm will produce a heuristic, the performance of which will only depend on the cost of the selected matching. For instance, if we used the same proper star-matching with min maxdegree that was used in the approximation algorithm introduced earlier, it would generate a valid heuristic. Clearly, it is possible to have many idle vertices in the approximation algorithm that we discussed earlier, depending on the structure of the graph instance. Thus, the heuristic will, obviously, generate broadcast times not worse than those of the 2-approximation algorithm, as it will have the same behavior without any vertex idling. However, we were unable to provide a better approximation ratio for that algorithm.

3.1 Split Graphs Achieving the Lower Bound of the Broadcast Time

As we mentioned earlier in this paper, the broadcast time is lower bounded by $\lceil \log |V| \rceil$. So for a split graph G, with clique partition $|K| = n$ and independent set partition $|I| = m$, the broadcast time is lower bounded by $\lceil \log(n + m) \rceil$.

Assume that for a given graph G and an originator $v \in K$, $b(v, G) = \lceil \log(n + m) \rceil$. Let M be a proper t-star-matching induced by an optimal broadcast scheme for G, and let $d_i, 1 \leq i \leq n$ be the number of vertices covered by $v_i \in K$ in M. For any vertex v_i, $d(v_i)$ denotes the number of vertices in I adjacent to v_i, or more formally, $d(v_i) = N(v_i) \cap I$, where $N(v_i)$ is the open neighborhood of v_i. Note that $0 \leq d_i \leq d(v_i)$ for any i. From the Inequality 8, we can claim that $b(v, G) = \lceil \log(n + m) \rceil$ if

$$2^{\lceil \log(n+m) \rceil} \geq \sum_{i=1}^{n} 2^{d(v_i)} \text{ or } 2^{\lceil \log(n+m) \rceil} \geq \sum_{i=1}^{n} 2^{d_i} \tag{9}$$

Thus, any split graph satisfying the above-mentioned inequality will have a broadcast time equal to the lower bound of $\lceil \log(n+m) \rceil$. For instance, let $n = 2^m$ and let each vertex in the clique have at most 1 neighbor in the independent set ($d(v_i) \leq 2$, for any $1 \leq i \leq n$). In that case, we can see that $2^{\lceil \log(n+m) \rceil} = 2^{m+1}$ and $\sum_{i=1}^{n} 2^{d(v_i)} \leq 2 \cdot n = 2^{m+1}$. Hence, Inequality 9 holds for these graphs, and we can claim that they have broadcast time of $\lceil \log(n + m) \rceil$.

4 Conclusion and Future Work

In this paper, we presented approximation algorithms for the minimum broadcast time problem in split graphs that guarantee an approximation ratio of 2, both for an originator in the clique or in the independent set. Moreover, we showed that our calculations for the approximation ratio of these algorithms are tight.

We also proved several important characteristics of an optimal broadcast scheme, which helped to design a scheme for generating optimal broadcasting algorithms or heuristics for split graphs. For a large infinite subfamily of split graphs, we proved that the minimum broadcast time is equal to the lower bound of $\lceil \log n \rceil$. For the future, it is interesting to try calculating an approximation ratio for the heuristic that is generated by the introduced scheme, when using the proper star-matching with minimum maxdegree. Additionally, we defined a new problem for finding a minimum cost star-matching of a given split graph. We showed that solving the minimum cost star-matching problem would finally close the minimum broadcast time problem in split graphs. This would be the first non-tree-like graph class with a polynomial-time algorithm for the minimum broadcast time problem.

References

1. Belik, I.: The analysis of split graphs in social networks based on the k-cardinality assignment problem. Int. J. Netw. Sci. **1**(1), 53–62 (2016)
2. Belmonte, R., Kim, E.J., Lampis, M., Mitsou, V., Otachi, Y., Sikora, F.: Token sliding on split graphs. Theory Comput. Syst. **65**(4), 662–686 (2021)
3. Bender, E.A., Richmond, L.B., Wormald, N.C.: Almost all chordal graphs split. J. Aust. Math. Soc. **38**(2), 214–221 (1985)
4. Bertossi, A.A.: Dominating sets for split and bipartite graphs. Inf. Process. Lett. **19**(1), 37–40 (1984)
5. Bhabak, P., Harutyunyan, H.A., Kropf, P.: Efficient broadcasting algorithm in Harary-like networks. In: 46th International Conference on Parallel Processing Workshops (ICPPW 2017), pp. 162–170. IEEE (2017)
6. Bhabak, P., Harutyunyan, H.A., Tanna, S.: Broadcasting in Harary-like graphs. In: 17th International Conference on Computational Science and Engineering (CSE 2014), pp. 1269–1276. IEEE (2014)
7. Bodlaender, H.L., Kloks, T., Tan, R.B., van Leeuwen, J.: λ-coloring of graphs. In: Reichel, H., Tison, S. (eds.) STACS 2000. LNCS, vol. 1770, pp. 395–406. Springer, Heidelberg (2000). https://doi.org/10.1007/3-540-46541-3_33
8. Collins, K.L., Trenk, A.N.: Finding balance: split graphs and related classes. arXiv preprint arXiv:1706.03092 (2017)
9. Földes, S., Hammer, P.L.: Split graphs. In: Proceedings of the Eighth Southeastern Conference on Combinatorics, Graph Theory and Computing (SECCGTC 1977), vol. XIX, pp. 311–315. Winnipeg: Utilitas Math (1977)
10. Ford, L.R., Fulkerson, D.R.: Flows in networks. In: Flows in Networks. Princeton University Press (2015)
11. Fraigniaud, P., Lazard, E.: Methods and problems of communication in usual networks. Discret. Appl. Math. **53**(1–3), 79–133 (1994)
12. Garey, M.R., Johnson, D.S.: Computers and Intractability, vol. 174. Freeman, San Francisco (1979)
13. Gholami, S., Harutyunyan, H.A., Maraachlian, E.: Optimal broadcasting in fully connected trees. J. Interconnect. Netw. **23**(01), 2150037:1–2150037:20 (2023)
14. Golumbic, M.C.: Algorithmic Graph Theory and Perfect Graphs. Elsevier, Amsterdam (2004)
15. Harutyunyan, H.A., Hovhannisyan, N., Maraachlian, E.: Broadcasting in chains of rings. In: Submitted to International Workshop on Combinatorial Algorithms (IWOCA 2023) (2023)
16. Harutyunyan, H.A., Liestman, A.L., Peters, J.G., D., R.: Broadcasting and gossiping. In: Handbook of Graph Theory, pp. 1477–1494. Chapman and Hall (2013)
17. Harutyunyan, H., Maraachlian, E.: Linear algorithm for broadcasting in unicyclic graphs. In: Lin, G. (ed.) COCOON 2007. LNCS, vol. 4598, pp. 372–382. Springer, Heidelberg (2007). https://doi.org/10.1007/978-3-540-73545-8_37
18. Harutyunyan, H.A., Maraachlian, E.: On broadcasting in unicyclic graphs. J. Comb. Optim. **16**(3), 307–322 (2008)
19. Hedetniemi, S.M., Hedetniemi, S.T., Liestman, A.L.: A survey of gossiping and broadcasting in communication networks. Networks **18**(4), 319–349 (1988)
20. Hromkovič, J., Klasing, R., Monien, B., Peine, R.: Dissemination of information in interconnection networks (broadcasting & gossiping). In: Du, D.Z., Hsu, D.F. (eds.) Combinatorial Network Theory. Applied Optimization, vol. 1, pp. 125–212. Springer, Boston (1996). https://doi.org/10.1007/978-1-4757-2491-2_5

21. Jansen, K., Müller, H.: The minimum broadcast time problem for several processor networks. Theoret. Comput. Sci. **147**(1–2), 69–85 (1995)
22. Lin, W., Lam, P.C.B.: Star matching and distance two labelling. Taiwan. J. Math. **13**(1), 211–224 (2009)
23. Merris, R.: Split graphs. Eur. J. Comb. **24**(4), 413–430 (2003)
24. Middendorf, M.: Minimum broadcast time is np-complete for 3-regular planar graphs and deadline 2. Inf. Process. Lett. **46**(6), 281–287 (1993)
25. Müller, H.: Hamiltonian circuits in chordal bipartite graphs. Discret. Math. **156**(1–3), 291–298 (1996)
26. Orlin, J.B.: Max flows in $o(nm)$ time, or better. In: Proceedings of the Forty-Fifth Annual ACM Symposium on Theory of Computing (STOC 2013), pp. 765–774 (2013)
27. Proskurowski, A.: Minimum broadcast trees. IEEE Trans. Comput. **30**(05), 363–366 (1981)
28. Schrijver, A.: On the history of the transportation and maximum flow problems. Math. Program. **91**(3), 437–445 (2002)
29. Slater, P.J., Cockayne, E.J., Hedetniemi, S.T.: Information dissemination in trees. SIAM J. Comput. **10**(4), 692–701 (1981)
30. Tamura, H., Tasaki, F., Sengoku, M., Shinoda, S.: Scheduling problems for a class of parallel distributed systems. In: IEEE International Symposium on Circuits and Systems (ISCAS 2005), pp. 176–179. IEEE (2005)
31. Tyshkevich, R.I., Chernyak, A.A.: Canonical partition of a graph defined by the degrees of its vertices. Isv. Akad. Nauk BSSR Ser. Fiz.-Mat. Nauk (in Russian) **5**, 14–26 (1979)
32. Vandenberghe, L., Andersen, M.S.: Chordal graphs and semidefinite optimization. Found. Trends® Optim. **1**(4), 241–433 (2015)
33. Zhang, R.Y., Lavaei, J.: Sparse semidefinite programs with near-linear time complexity. In: 2018 IEEE Conference on Decision and Control (CDC 2018), pp. 1624–1631. IEEE (2018)
34. Zheng, Y.: Chordal sparsity in control and optimization of large-scale systems. Ph.D. thesis, University of Oxford (2019)

Partitioning Subclasses of Chordal Graphs with Few Deletions

Satyabrata Jana[1], Souvik Saha[1(✉)], Abhishek Sahu[3], Saket Saurabh[1,2], and Shaily Verma[1]

[1] The Institute of Mathematical Sciences, HBNI, Chennai, India
{souviks,saket,shailyverma}@imsc.res.in
[2] University of Bergen, Bergen, Norway
[3] National Institute of Science, Education and Research, An OCC of Homi Bhabha National Institute, Bhubaneswar, India

Abstract. In the (VERTEX) k-WAY CUT problem, input is an undirected graph G, an integer s, and the goal is to find a subset S of edges (vertices) of size at most s, such that $G - S$ has at least k connected components. Downey et al. [Electr. Notes Theor. Comput. Sci. 2003] showed that k-WAY CUT is W[1]-hard parameterized by k. However, Kawarabayashi and Thorup [FOCS 2011] showed that the problem is fixed-parameter tractable (FPT) in general graphs with respect to the parameter s and provided a $\mathcal{O}(s^{s^{\mathcal{O}(s)}} n^2)$ time algorithm, where n denotes the number of vertices in G. The best-known algorithm for this problem runs in time $s^{\mathcal{O}(s)} n^{\mathcal{O}(1)}$ given by Lokshtanov et al. [ACM Tran. of Algo. 2021]. On the other hand, VERTEX k-WAY CUT is W[1]-hard with respect to either of the parameters, k or s or $k + s$. These algorithmic results motivate us to look at the problems on special classes of graphs.

In this paper, we consider the (VERTEX) k-WAY CUT problem on subclasses of chordal graphs and obtain the following results.

- We first give a sub-exponential FPT algorithm for k-WAY CUT running in time $2^{\mathcal{O}(\sqrt{s} \log s)} n^{\mathcal{O}(1)}$ on chordal graphs.
- It is "known" that VERTEX k-WAY CUT is W[1]-hard on chordal graphs, in fact on split graphs, parameterized by $k + s$. We complement this hardness result by designing polynomial-time algorithms for VERTEX k-WAY CUT on interval graphs, circular-arc graphs and permutation graphs.

Keywords: chordal graphs · FPT · interval graphs · circular-arc graphs · permutation graphs

1 Introduction

Graph partitioning problems have been extensively studied because of their applications in VLSI design, parallel supercomputing, image processing, and clustering [1]. In this paper, we consider one of the classical graph partitioning problems, namely, the (VERTEX) k-WAY CUT problem. In this problem the

M. Mavronicolas (Ed.): CIAC 2023, LNCS 13898, pp. 293–307, 2023.
https://doi.org/10.1007/978-3-031-30448-4_21

objective is to partition the graph into k components by deleting as few (vertices) edges as possible. Formally, the problems we study are defined as follows.

k-WAY CUT

Input: A graph $G = (V, E)$ and two integers s and k.
Parameter: s
Question: Does there exist a set $S \subseteq E$ of size at most s, such that $G - S$ has at least k connected components?

VERTEX k-WAY CUT

Input: A graph $G = (V, E)$ and two integers s and k.
Parameter: s
Question: Does there exist a set $S \subseteq V$ of size at most s, such that $G - S$ has at least k connected components?

These problems are decision versions of natural generalization of the GLOBAL MIN CUT problem, which seeks to delete a set of edges of minimum cardinality such that the graph gets partitioned into two parts ($k = 2$). In other words, the graph becomes disconnected. We first give a brief account of the history of known results on the problem to set the context of our study.

Algorithmic History of the Problem. There is a rich algorithmic study of (VERTEX) k-WAY CUT problem. In 1996, Goldschmidt and Hochbaum [6] showed that the k-WAY CUT problem is NP-hard for arbitrary k, but polynomial-time solvable when k is fixed and gave a $\mathcal{O}(n^{(1/2-o(1))k^2})$ time algorithm, where n is the number of vertices in the graph. Later, Karger and Stein [10] gave an edge contraction based randomized algorithm with running time $\tilde{\mathcal{O}}(n^{(2k-1)})$. The notation $\tilde{\mathcal{O}}$ hides the poly-logarithimic factor in the running time. Recently, Li [13] obtained an improved randomized algorithm with running time $\tilde{\mathcal{O}}(n^{(1.981+o(1))k})$. To date, the best known deterministic exact algorithm is given by Chekuri et al. [2] which runs in $\mathcal{O}(mn^{(2k-3)})$ time.

In terms of approximation algorithms, several approximation algorithms are known for the k-WAY CUT problem with approximation factor $(2 - o(1))$, that run in time polynomial in n and k [17]. Recently, Manurangsi [15] proved that the approximation factor cannot be improved to $(2 - \epsilon)$ for every $\epsilon > 0$, assuming small set expansion hypothesis. Lately, this problem has received significant attention from the perspective of parameterized approximation as well. Gupta et al. [8] gave the first FPT approximation algorithm for the problem with approximation factor 1.9997 which runs in time $2^{\mathcal{O}(k^6)}n^{\mathcal{O}(1)}$. The same set of authors [9] also gave an $(1 + \epsilon)$-approximation algorithm with running time $(k/\epsilon)^{\mathcal{O}(k)}n^{k+\mathcal{O}(1)}$, and an approximation algorithm with a factor 1.81 running in time $2^{\mathcal{O}(k^2)}n^{\mathcal{O}(1)}$. Later, Kawarabayashi and Lin [11] gave a $(5/3 + \epsilon)$-approximation algorithm for the problem with running time $2^{\mathcal{O}(k^2 \log k)}n^{\mathcal{O}(1)}$. Recently, Lokshtanov et al. [14] designed $(1 + \epsilon)$-approximation algorithm for every $\epsilon > 0$, running in time $(k/\epsilon)^{\mathcal{O}(k)}n^{\mathcal{O}(1)}$ improving upon the previous result.

Table 1. Complexity of the problems for different parameterizations

Problems	Parameter(s)		
	k	s	$k+s$
VERTEX k-WAY CUT	W[1]-hard [5]	W[1]-hard [16]	W[1]-hard [16]
k-WAY CUT	W[1]-hard [5]	FPT [12]	FPT [4]

From the parameterized perspective, Downey et al. [5] proved that the k-WAY CUT and VERTEX k-WAY CUT problems are W[1]-hard when parameterized by k. On the other hand, when parameterized by the cut size s, it is known that finding a VERTEX k-WAY CUT of size s is also W[1]-hard [16]; however finding a k-WAY CUT of size s is FPT [12]. Kawarabayashi and Thorup [12] gave a $\mathcal{O}(s^{s^{\mathcal{O}(s)}} \cdot n^2)$ time FPT algorithm for the k-WAY CUT problem. Recently, Lokshtanov et al. [4] designed a faster algorithm with running time $s^{\mathcal{O}(s)}n^{\mathcal{O}(1)}$. These tractable and intractable results (see Table 1) are a starting point of our work. That is, we address the following question: *What is the complexity of* (VERTEX) k-WAY CUT *problem on well-known graph classes?*

Our Results. In this paper we obtain a a sub-exponential-FPT algorithm for k-WAY CUT running in time $2^{\mathcal{O}(\sqrt{s}\log s)}n^{\mathcal{O}(1)}$ on chordal graphs (Sect. 3) and polynomial-time algorithms for VERTEX k-WAY CUT on interval graphs, circular-arc graphs, and permutation graphs (Sect. 4).

2 Preliminaries

All graphs considered in this paper are finite, simple, and undirected. We use the standard notation and terminology that can be found in the book of graph theory [18]. We use $[n]$ to denote the set of first n positive integers $\{1, 2, 3, \ldots, n\}$. For a graph G, we denote the set of vertices of the graph by $V(G)$ and the set of edges of the graph by $E(G)$. We denote $|V(G)|$ and $|E(G)|$ by n and m respectively, where the graph is clear from context. We abbreviate an edge (u, v) as uv sometimes. For a set $S \subseteq V(G)$, the subgraph of G induced by S is denoted by $G[S]$ and it is defined as the subgraph of G with vertex set S and edge set $\{(u, v) \in E(G) : u, v \in S\}$ and the subgraph obtained after deleting S (and the edges incident to the vertices in S) is denoted by $G - S$. For $v \in V(G)$, we will use $G - v$ to denote $G - \{v\}$ for ease of notation. All vertices adjacent to a vertex v are called neighbours of v and the set of all such vertices is called the open neighbourhood of v, denoted by $N_G(v)$. For a set of vertices $S \subseteq V(G)$, we define $N_G(S) = (\cup_{v \in S} N(v) \backslash S)$. We define the closed neighbourhood of a vertex v in the graph G to be $N_G[v] := N_G(v) \cup \{v\}$ and closed neighbourhood of a set of vertices $S \subseteq V(G)$ to be $N_G[S] := N_G(S) \cup S$. We drop the subscript G when the graph is clear from the context. For $C \subseteq V(G)$, if $G[C]$ is connected and $N(C) = \emptyset$, then we say that $G[C]$ is a connected component of G. For both the problems k-WAY CUT and VERTEX k-WAY CUT, in the given instance, we assume that $k > 1$, otherwise the input itself is an optimal solution with zero cut size. A partition of G in to k components is a partition of $V(G)$ into k sets $V_1, \ldots, V_k$ such that each $G[V_i]$ is a connected. We say a partition is *non-trivial* when $k > 1$.

Definition 1. *A tree-decomposition of a connected graph G is a pair (T, β), where T is a tree and and $\beta\colon V(T) \to V(G)$ such that*

- $\bigcup_{x \in V(T)} \beta(x) = V(G)$*, we call $\beta(x)$ as the bag of x,*
- *for every edge $(u, v) \in E(G)$, there exists $x \in V(T)$ such that $\{u, v\} \subseteq \beta(x)$, and*
- *for every vertex $v \in V(G)$, the subgraph of T induced by the set $\beta^{-1}(v) := \{x\colon v \in \beta(x)\}$ is connected.*

Chordal Graphs: A graph G is a *chordal graph* if every cycle in G of length at least 4 has a *chord* i.e., an edge joining two non-consecutive vertices of the cycle. A *clique-tree* of G is a tree-decomposition of G where every bag is a maximal clique. We further insist that every bag of the clique-tree is distinct. There are several ways to obtain a clique-tree decomposition of G; one way is by using perfect elimination ordering (PEO) of G [3]. The following lemma shows that the class of chordal graphs is exactly the class of graphs that have a clique-tree.

Lemma 1 ([7]). *A connected graph G is a chordal graph if and only if G has a clique-tree.*

Let $\mathscr{F}$ be a non-empty family of sets. A graph G is called an *intersection graph* for $\mathscr{F}$ if there is a one-to-one correspondence between $\mathscr{F}$ and G where two sets in $\mathscr{F}$ have nonempty intersection if and only if their corresponding vertices in G are adjacent. We call $\mathscr{F}$ an *intersection model* of G and we use $G(\mathscr{F})$ to denote the intersection graph for $\mathscr{F}$. If $\mathscr{F}$ is a family of intervals on a real line, then $G(\mathscr{F})$ is called an *interval graph* for $\mathscr{F}$. A *proper interval graph* is an interval graph that has an intersection model in which no interval properly contains another. If $\mathscr{F}$ is a family of arcs on a circle in the plane, then $G(\mathscr{F})$ is called an *circular-arc graph* for $\mathscr{F}$. If $\mathscr{F}$ is a family of line segments in the plane whose endpoints lie on two parallel lines, then the intersection graph of $\mathscr{F}$ is called the *permutation graph* for $\mathscr{F}$.

3 Sub-exponential FPT Algorithm on Chordal Graphs

Chordal graphs belong to the class of perfect graphs that contains several other graph classes such as split graphs, interval graphs, threshold graphs, and block graphs. A graph G is a *chordal graph* if every cycle in G of length at least 4 has a *chord* i.e., an edge joining two non-consecutive vertices of the cycle. Chordal graphs are also characterized as the intersection graph of sub-trees of a tree. Every chordal graph has a tree-decomposition where every bag induces a clique. In this section, we obtain a sub-exponential FPT algorithm for the k-Way Cut problem in chordal graphs parameterized by s, the number of cut edges. We first give a characterization of the k-Way Cut on a clique in Lemma 3. Later, we use this characterization to design our algorithm.

Lemma 2. *Let $\mathbb{K}$ be a clique and s be an integer. Then we can not partition the clique into more than one component by deleting s edges if one of the following conditions holds.*

(i) $|\mathbb{K}| > (s+1)$,
(ii) $|\mathbb{K}| > (2\sqrt{s}+1)$, *and size of every component in the partition is at most* $\sqrt{s}$.

Proof. (i) If $|\mathbb{K}| > (s+1)$, the size of min-cut of $\mathbb{K}$ is at least $s+1$ and hence we cannot partition $\mathbb{K}$ by deleting s edges. (ii) In the second condition, the size of every component in the partition is at most $\sqrt{s}$ and hence every vertex v in any component must be disconnected from at least $2\sqrt{s}+2-\sqrt{s} = \sqrt{s}+2$ vertices that are in other components. Thus the total number of edges that needs to be deleted is at least $(2\sqrt{s}+2)(\sqrt{s}+2)/2 > s$. Hence the clique can not be partitioned by deleting s edges. □

Lemma 3. *Let $\mathbb{K}$ be a clique and s be an integer such that $(2\sqrt{s}+1) < |\mathbb{K}| < (s+2)$, then any non-trivial partition of $\mathbb{K}$ obtained by deleting at most s edges, has a component of size at least $(|\mathbb{K}| - \sqrt{s})$.*

Proof. Let $\mathbb{K}$ be a clique such that $(2\sqrt{s}+1) < |\mathbb{K}| < (s+2)$ and we have to partition the clique into k components by deleting at most s edges. Let γ be the size of the largest component in the partition.

$$|E(\mathbb{K})| = |E(\text{Largest component})| + |\text{E(other components)}| + |\text{cut edges}|$$
$$\implies \binom{|\mathbb{K}|}{2} \leq \binom{\gamma}{2} + \binom{|\mathbb{K}|-\gamma}{2} + |\text{cut edges}|$$
$$\implies \binom{|\mathbb{K}|}{2} \leq \binom{\gamma}{2} + \binom{|\mathbb{K}|-\gamma}{2} + s$$
$$\implies |\mathbb{K}|(|\mathbb{K}|-1) \leq \gamma(\gamma-1) + (|\mathbb{K}|-\gamma)(|\mathbb{K}|-\gamma-1) + 2s$$
$$\implies 0 \leq \gamma^2 - \gamma|\mathbb{K}| + s$$

Therefore, either $\gamma \leq \frac{|\mathbb{K}|-\sqrt{|\mathbb{K}|^2-4s}}{2}$, or $\gamma \geq \frac{|\mathbb{K}|+\sqrt{|\mathbb{K}|^2-4s}}{2}$ holds. If the first inequality holds, then it implies $\gamma \leq \frac{|\mathbb{K}|-\sqrt{|\mathbb{K}|^2}+\sqrt{4s}}{2}$ (by using the inequality $\sqrt{a}-\sqrt{b} \leq \sqrt{a-b}$ *for* $0 < b \leq a$). It follows that $\gamma \leq \sqrt{s}$. However, Lemma 2 implies that if $\gamma \leq \sqrt{s}$ and $|\mathbb{K}| > 2\sqrt{s}+1$, then there is no non-trivial partition of $\mathbb{K}$. Thus in this case, $\mathbb{K}$ has no non-trivial partition. If the second inequality holds, then $\gamma \geq \frac{|\mathbb{K}|+\sqrt{|\mathbb{K}|^2-4s}}{2}$, which implies that $\gamma \geq (|\mathbb{K}|-\sqrt{s})$. Hence any non-trivial partition of $\mathbb{K}$, obtained by deleting at most s edges, has a component of size at least $(|\mathbb{K}| - \sqrt{s})$. □

Lemma 4. *There are $2^{\mathcal{O}(\sqrt{s}\log s)}$ many possible choices for any non-trivial partition of a clique $\mathbb{K}$ obtained by deleting at most s edges.*

Proof. We have the following three cases depending on the size of $\mathbb{K}$.

Case 1. $|\mathbb{K}| \geq (s+2)$.

In this case, no non-trivial partition exists by Lemma 2.

Case 2. $|\mathbb{K}| \leq (2\sqrt{s}+1)$.

In this case, there are $k^{2\sqrt{s}+1}$ ways of partitioning the clique into k components. Since $k \leq (s+1)$, $k^{2\sqrt{s}+1} \leq 2^{\mathcal{O}(\sqrt{s}\log s)}$.

Case 3. $(2\sqrt{s}+1) < |\mathbb{K}| < (s+2)$.
From Lemma 3, in a partition of $\mathbb{K}$ into k components, there exists a component with at least $(|\mathbb{K}| - \sqrt{s})$ many vertices. So, we guess $(|\mathbb{K}| - \sqrt{s})$ many vertices in a component. Now, the rest $\sqrt{s}$ vertices are partitioned into k components. The total number of choices for such a partition of $\mathbb{K}$ is bounded by $\binom{|\mathbb{K}|}{|\mathbb{K}|-\sqrt{s}} \cdot k^{\sqrt{s}} \cdot k$. Since both k and $|\mathbb{K}|$ are bounded by $(s+1)$, we have $|\mathbb{K}|^{\sqrt{s}} \cdot k^{\sqrt{s}} \cdot k \leq 2^{\mathcal{O}(\sqrt{s}\log s)}$.
□

Now we prove the following theorem.

Theorem 1. *k*-WAY CUT *problem on a chordal graph with n vertices can be solved in time $2^{\mathcal{O}(\sqrt{s}\log s)}n^{\mathcal{O}(1)}$.*

To prove Theorem 1, we design a dynamic-programming algorithm for the k-WAY CUT problem on chordal graphs, which exploits its clique-tree decomposition. Let G be a chordal graph and $\mathcal{T} = (T, \{K_t\}_{t\in V(t)})$ be its clique-tree decomposition.

Let T be a clique-tree of G rooted at some node r. For a node t of T, K_t is the set of vertices contained in t and let V_t be the set of all vertices of the sub-tree of T rooted at t. The parent node of t is denoted by $\mathsf{parent}(t)$. We follow a bottom-up dynamic-programming approach on T to design our algorithm.

For a set of vertices U, we use $\mathsf{P}(U)$ to denote a partition $\{A_1, A_2, \ldots, A_k\}$ of U where each A_i is a set in the partition. Given the partitions of two sets $U_1, U_2 \subseteq V(G)$, say $\mathsf{P}(U_1) = \{A_1, A_2, \ldots, A_k\}$ and $\mathsf{P}(U_2) = \{B_1, B_2, \ldots, B_k\}$, we call these partitions *mutually compatible*, if for each vertex u in $U_1 \cap U_2$, $u \in A_i$ if and only if $u \in B_i$ for some $i \in [k]$. We denote the mutually compatible operation by $\perp$. For any node t, a partition $\mathsf{P}(K_t)$ and an integer w where $0 \leq w \leq (k-1)$, a feasible solution for $(t, \mathsf{P}(K_t), w)$ is a k-way cut in $G[V_t]$ with the following properties: ($\mathsf{P}(V_t)$ is the partition induced on V_t by the above k-way cut).

- $\mathsf{P}(K_t) \perp \mathsf{P}(V_t)$,
- Exactly w components in $\mathsf{P}(V_t)$ contain no vertex from K_t, that is, these w components are completely contained inside $G[V_t \backslash K_t]$.

Next, we define the dynamic-programming table whose entry is denoted by $M[t; \mathsf{P}(K_t), w]$ for a node t and integer w, $0 \leq w \leq k$. The entry $M[t; \mathsf{P}(K_t), w]$ stores the size of the smallest such feasible solution. From Lemma 4, the number of sub-problems (or number of entries that we have to compute) for each node in the tree is bounded by $2^{\mathcal{O}(\sqrt{s}\log s)}$ as each node is a clique. Below we give a recurrence relation to compute $M[t; \mathsf{P}(K_t), w]$ for each tuple $(t, \mathsf{P}(K_t), w)$. The case where t is a leaf, corresponds to the base case of the recurrence, whereas the values of $M[t; ., .]$ for a non-leaf node t depends on the value of $M[t', .]$ for each child t' of node t (which have already been computed). By applying the formula in a bottom-up manner on T, we compute $M(r; \mathtt{P}(K_r), k-1)$ for the root node r. Note that the value of $M(r; \mathtt{P}(K_r), k-1)$ is exactly the size of an optimal solution for our problem, because in any optimal solution there are exactly $k-1$ components that are completely contained in $G - K_r$. Here without loss of generality, we can assume that K_r contains exactly one vertex of G. For a partition $\mathtt{P}(U)$ of U, we define $\mathtt{CUT}(\mathtt{P}(U))$ as the set of edges whose endpoints belong to different sets in the partition. Now, we describe the recursive formulas to compute the value of $M[t; ., .]$, for each node t.

Leaf Node. Let t be a leaf node. Then for each partition $\mathtt{P}(K_t)$, we define

$$M[t; \mathtt{P}(K_t), w] = \begin{cases} |\mathtt{CUT}(\mathtt{P}(K_t))| & \text{if } w = 0, \\ +\infty & \text{otherwise.} \end{cases}$$

Non-leaf Node. Let t be a non-leaf node. Assume that the node t has ℓ children $t_1, \ldots, t_\ell$. For a pair of distinct vertices u, v in K_t, let $\mathtt{Child\_Pair}(t; u, v)$ denote the number of children of t containing both the vertices u and v. For a partition $\mathtt{P}(K_t)$, let $\mathtt{Child}(\mathtt{P}(K_t))$ denote the sum of the number of occurrences (with repetitions) of the edges from $\mathtt{CUT}(\mathtt{P}(K_t))$ in all the children nodes of t, that is, $\mathtt{Child}(\mathtt{P}(K_t)) = \sum_{(u,v) \in \mathtt{CUT}(\mathtt{P}(K_t))} \mathtt{Child\_Pair}(t; u, v)$. Let $\psi(\mathtt{P}(K_t))$ denote the number of sets in $\mathtt{P}(K_t)$ that have no common vertex with the parent node of t. Therefore, the recurrence relation for computing $M(t; ., .)$ for t is as follows:

$$M[t; \mathtt{P}(K_t), w] = |\mathtt{CUT}(\mathtt{P}(K_t))| - \mathtt{Child}(\mathtt{P}(K_t)) + \min_{\substack{\forall (\mathtt{P}(K_{t_i}), w_i): \\ \mathtt{P}(K_{t_i}) \perp \mathtt{P}(K_t) \\ w = \sum_i (w_i + \psi(\mathtt{P}(K_{t_i})))}} \sum_{i=1}^{\ell} M[t_i; \mathtt{P}(K_{t_i}), w_i].$$

Next, we prove the correctness of the above recurrence relation.

Correctness. Let R denote the value of the right side expression above. To prove the recurrence relation, first we show $M[t; \mathtt{P}(K_t), w] \leq R$ and then $M[t; \mathtt{P}(K_t), w] \geq R$. Let t be a node in T having ℓ children $t_1, t_2, \ldots, t_\ell$. Any set of ℓ compatible partitions, one for each child of t together with $\mathtt{P}(K_t)$ leads to a feasible solution for $(t, \mathtt{P}(K_t), w)$ if $w = \sum_i (w_i + \psi(\mathtt{P}(K_{t_i})))$. Now for each child node t_i of t and for any pair of vertices u, v in K_t, if the vertices u and v are in different sets in each of the partitions $\mathtt{P}(K_t)$ and $\mathtt{P}(K_{t_i})$, then the (to be deleted) edge (u, v) is counted twice, once in $\mathtt{CUT}(\mathtt{P}(K_t))$ and once in $M[t_i; \mathtt{P}(K_{t_i}), w_i]$. Now if the edge (u, v) is present in c many children of t, then in the entry $\sum_{i=1}^{\ell} M[t_i; \mathtt{P}(K_{t_i}), w_i]$ this edge gets counted c times. To avoid over-counting of the edge (u, v) in $M[t_i; ., .]$, we must consider the edge (u, v) exactly once and for this purpose we use $\mathtt{Child}(\mathtt{P}(K_t))$ in the recurrence relation. Considering this over counting, the set of edges corresponding to $M[t_1; \mathtt{P}(K_{t_1}), w_1], M[t_2; \mathtt{P}(K_{t_2}), w_2], \ldots, M[t_\ell; \mathtt{P}(K_{t_\ell}), w_\ell]$ with size $\sum_{i=1}^{\ell} M[t_i; \mathtt{P}(K_{t_i}), w_i] - \mathtt{Child}(\mathtt{P}(K_t))$, together with the edges corresponding to $\mathtt{CUT}(\mathtt{P}(K_t))$ gives us a feasible solution for $(t, \mathtt{P}(K_t), w)$. Hence, $M[t; \mathtt{P}(K_t), w] \leq |\mathtt{CUT}(\mathtt{P}(K_t))| - \mathtt{Child}(\mathtt{P}(K_t)) + \sum_{i=1}^{\ell} M[t_i; \mathtt{P}(K_{t_i}), w_i]$, where $\mathtt{P}(K_t) \perp \mathtt{P}(K_{t_i})$ for each $i \in [\ell]$ and $w = \sum_i (w_i + \psi(\mathtt{P}(K_{t_i})))$.

Next, we show that $M[t; \mathtt{P}(K_t), w] > R$. Let Y be a set of cut edges corresponding to the entry $M[t; \mathtt{P}(K_t), w]$. Let $Y' \subseteq Y$ be the set of edges that are not present in K_t. So $Y \backslash Y'$ determines the partition in K_t. Let $Y' = Y_1 \cup \ldots \cup Y_\ell$, where each Y_i is the set of edges for $G[V(t_i)]$. Let $X_1 \cup \ldots \cup X_\ell \subseteq (Y \backslash Y')$, where $X_i = (Y \backslash Y') \cap E(K(t_i))$. Now it is easy to see that $Y_i \cup X_i$ is a feasible solution for $(t_i, \mathtt{P}(K_{t_i}), w_i)$, where $\mathtt{P}(K_t) \perp \mathtt{P}(K_{t_i})$ for each $i \in [\ell]$ and $w = \sum_i (w_i + \psi(\mathtt{P}(K_{t_i})))$. Since $Y \backslash Y'$ determines the partition only in K_t, $|Y \backslash Y'| = |\mathtt{CUT}(\mathtt{P}(K_t))|$. Thus, we get $M[t; \mathtt{P}(K_t), w] - |\mathtt{CUT}(\mathtt{P}(K_t))| + \mathtt{Child}(\mathtt{P}(K_t)) \geq \sum_{i=1}^{\ell} M[t_i; \mathtt{P}(K_{t_i}), w_i]$. Hence the correctness of the recurrence relation follows.

Time Complexity. There are $\mathcal{O}(n)$ many nodes in the clique tree of the given graph G. The number of entries $M[.;.,.]$ for any node can be upper bounded by $k 2^{\mathcal{O}(\sqrt{s} \log s)}$ (from Lemma 4). To compute one such entry, we look at the entries with the compatible partitions in the children nodes. Now, we describe how we compute $M[t; \mathtt{P}(K_t), w]$ in a node for a fixed partition $\mathtt{P}(K_t)$ and a fixed integer $w \leq k$. We apply an incremental procedure to find this. Consider an ordering $t_1 \prec t_2 \prec \ldots \prec t_\ell$ of child nodes of t. In the dynamic-programming, we store the entries $M[t_i; \mathtt{P}(K_{t_i}), w_i]$ for each $\mathtt{P}(K_{t_i}) \perp \mathtt{P}(K_t)$ and $w_i \leq k$. For each t_i, we compute the entries $D_i(z)$ for $0 \leq z \leq k$, where $D_i(z) = \min_z \{M[t_i; \mathtt{P}(K_{t_i}), w^*] : \mathtt{P}(K_{t_i}) \perp \mathtt{P}(K_t),\ z = w^* + \psi(\mathtt{P}(K_{t_i}), w^* \leq k\}$. Next we create a set of entries for D, defined by $D(1, 2, \ldots, i; z) = \min_{z = z_1 + z_2} \{D(1, 2, \ldots, i-1; z_1) + D_i(z_2)\}$, for $i \in [\ell]$. $D(1; z) = D_1(z), \forall z$ (the base case). It takes $\mathcal{O}(\ell k^3)$ time to compute all the entries of the table D. Now using the entries of the table D, we compute $M[t; \mathtt{P}(K_t), w]$, i.e. $M[t; \mathtt{P}(K_t), z] = |\mathtt{CUT}(\mathtt{P}(K_t))| - \mathtt{Child}(\mathtt{P}(K_t)) + D(1, 2, \ldots, \ell; z)$.

Since there are $2^{\mathcal{O}(\sqrt{s} \log s)}$ many partitions of each node t, computing all DP table entries at each node takes $2^{\mathcal{O}(\sqrt{s} \log s)} \mathcal{O}(\ell k^3)$ time. Because $\ell, k \leq n$, and there are $\mathcal{O}(n)$ many nodes in the clique tree, the total running time is upper-bounded by $2^{\mathcal{O}(\sqrt{s} \log s)} n^{\mathcal{O}(1)}$.

4 Polynomial Time Algorithmic Results

In this section, we obtain polynomial-time algorithms for the optimization version of the VERTEX k-WAY CUT on interval graphs, circular-arc graphs, and permutation graphs.

4.1 Interval Graphs

Here, we design a dynamic-programming algorithm for the optimization version of the VERTEX k-WAY CUT on interval graphs. Let G be an interval graph with vertex set $V(G) = \{v_1, v_2, \ldots, v_n\}$. Since G is an interval graph, there exists a corresponding geometric intersection representation of G, where each vertex $v_i \in V(G)$ is associated with an interval $I_i = (\ell(I_i), r(I_i))$ in the real line, where $\ell(I_i)$ and $r(I_i)$ denote left and right endpoints, respectively in I_i. Two vertices v_i and v_j are adjacent in G if and only if their corresponding intervals I_i and I_j intersect with each other. Without loss of generality we can assume that along with the graph, we are also given the corresponding underlying intervals on the real line. We use $\mathcal{I}$ to denote the set $\{I_i \colon v_i \in V\}$ of intervals and P to denote the set of all endpoints of these intervals, i.e., $P = \cup_{I \in \mathcal{I}}\{\ell(I), r(I)\}$. In the remaining section, we use v_i and I_i interchangeably. For a pair of points a and b on the real line with $a \leq b$ (we say $a \leq b$ when x-coordinate of a is not greater than x-coordinate of b), we define $I_{a,b}$ to denote the intervals which are properly contained in $[a, b]$, formally $I_{a,b} = \{I \in \mathcal{I} \colon a \leq \ell(I) \leq r(I) \leq b\}$. Let $I_{\geqslant b}$ be the set of intervals whose left endpoints are greater than b and $I_{<b}$ be the set of intervals whose left endpoint is strictly less than b, formally $I_{\geqslant b} = \{I \in \mathcal{I} \colon \ell(I) \geq b\}$ and $I_{<b} = \{I \in \mathcal{I} \colon \ell(I) < b\}$.

We now define a table for dynamic-programming algorithm. For every tuple (i, x, y), where $1 \leq i \leq k$ and $x, y \in P$ with $x < y$, any cut where $G[I_{x,y}]$ is the i-th component with respect to the cut in $G[I_{<y}]$ is a feasible cut for the tuple (i, x, y) and $T[i; x, y]$ stores the minimum size among all such feasible cuts for the tuple (i, x, y). Notice that any two connected components do not intersect. Hence we can order the components from left to right. In particular, for a pair of components C_j and $C_{j'}$, we say $C_j \prec C_{j'}$ if for any pair of intervals $I \in C_j$ and $I' \in C_{j'}$ the condition $r(I) < \ell(I')$ holds. In the base case, we compute the values for $T[1; x, y]$ for each possible pair x, y in P where $x < y$. $T[1; x, y]$ stores the number of intervals in $G[I_{<y}]$ that have either left endpoint strictly less than x or right endpoint strictly greater than y, formally $T[1; x, y] = |I_{<y}| - |I_{x,y}|$.

In the next lemma, we give a recursive formula for computing the values $T[i; x, y]$ for $i > 1$.

Lemma 5. *For every integer i and every pair of points x, y in P where $2 \leq i \leq k$ and $x < y$, the following holds:*

$$T[i; x, y] = \min_{\substack{x', y' \in P \\ x' < y' < x}} \{T[i-1; x', y'] + |I_{<y} \cap I_{\geqslant y'}| - |I_{x,y}|\}.$$

Proof. We prove the recurrence relation by showing inequalities in both directions. In one direction, let $(C_1, C_2, \ldots, C_i)$ be a feasible cut corresponding to the entry $T[i; x, y]$. Here $C_i = G[I_{x,y}]$. Let x' and y' be the left endpoint and right endpoint of the component C_{i-1}, so $C_{i-1} \subseteq G[I_{x',y'}]$. Clearly, $x' < y' < x < y$. Now the intervals of the set $(I_{<y} \cap I_{\geqslant y'}) \backslash I_{x,y}$ are part of cut vertices corresponding to the entry $T[i; x, y]$. Here we can get a set of $(i-1)$ components $C_1, C_2, \ldots, C_{i-1}$ in the graph $G[I_{<y'}]$ with $C_{i-1} = G[I_{x',y'}]$ and cut of size at

most $T[i; x, y] - (|I_{<y} \cap I_{>y'}| - |I_{x,y}|)$. Therefore, by the definition of $T[i; x, y]$, $T[i-1; x', y'] \leq T[i; x, y] - (|I_{<y} \cap I_{>y'}| - |I_{x,y}|)$.

In the other direction, let $(C'_1, C'_2, \ldots, C'_{i-1})$ be a feasible cut corresponding to the entry $T[i-1; x', y']$, where $x' < y' < x < y$ and $C_{i-1} = G[I_{x',y'}]$. Now the component induced by $I_{x,y}$ together with $C'_1, C'_2, \ldots, C'_{i-1}$ produces a feasible cut for $T[i; x, y]$. Therefore, the cut corresponding to $T[i-1; x', y']$ together with $(I_{<y} \cap I_{\geqslant y'}) \backslash I_{x,y}$ gives a cut with the components $C'_1, \ldots, C'_{i-1}, C'_i = G[I_{x,y}]$. Hence, $T[i-1; x', y'] + |I_{<y} \cap I_{\geqslant y'}| - |I_{x,y}| \geq T[i; x, y]$. This completes the proof of the lemma. □

With the insight of Lemma 5, we can now state the following theorem.

Theorem 2. Vertex k-Way Cut *in interval graphs with* n *vertices can be solved in* $\mathcal{O}(kn^4)$ *time.*

Proof. Let G be a given graph with $\mathcal{I}$ as an interval representation where P denotes the set of endpoints of all the intervals. In the pre-processing step, we do the following: (i) for every point $p \in P$, we construct $I_{<p}$ and $I_{\geqslant p}$, (ii) for every pair of points p, q in P, we compute $|I_{p,q}|$ and $|I_{<p} \cap I_{\geqslant q}|$. It will take $\mathcal{O}(n^2)$ time to perform both these pre-processing steps. Now in the recurrence formula, to obtain $T[i; x, y]$, we use the already computed values $T[i; x', y']$ for each possible pair $x', y' \in P$ with $x' < y' < x < y$. Computing any entry takes $\mathcal{O}(n^2)$ time. Since i ranges from 1 to k, we can compute all the values $T[i; x, y]$ in $\mathcal{O}(kn^4)$ time. Notice that the entry $T[k; ., .]$ with minimum value gives us the size of a minimum vertex k-way cut in G. Hence, the theorem holds. □

4.2 Proper Interval Graphs

In this subsection, we design a dynamic-programming algorithm for the optimization version of the Vertex k-Way Cut on proper interval graphs. In proper interval graphs, each vertex is associated with an interval in the real line such that no interval is completely contained in another interval. We use the notations $\mathcal{I}$, I_i, $\ell(I_i)$, $r(I_i)$ and P with the same definitions as used in the previous subsection. Let $\mathcal{I}$ be the set of all intervals with ordering $I_1 < I_2 < \ldots < I_n$ according to their left endpoints. Observe that for proper interval graphs, the ordering of intervals with respect to their left endpoints is same as with respect to their right endpoints. More explicitly, for any two intervals I_i and I_j where $\ell(I_i) < \ell(I_j)$, $r(I_i)$ must be less than $r(I_j)$. Let $\mathcal{I}_i = \{I_1, I_2, \ldots, I_i\}$ and $G[\mathcal{I}_i]$ denotes the subgraph of G induced by $\mathcal{I}_i$. Also for an interval I_i, I_i^ℓ denotes the interval in $\mathcal{I}$ which has leftmost left endpoint among all the intervals containing $\ell(I_i)$, formally, $I_i^\ell = I_c$, where $c = \min\{j; I_j \in \mathcal{I}, \ell(I_j) < \ell(i) < r(I_j)\}$.

We now define a table for dynamic-programming algorithm. For every pair (i, t), where $1 \leq i \leq n$ and $1 \leq t \leq k$, we define two entries. $T[\in;\ i,\ t]$ and $T[\notin;\ i,\ t]$. For every tuple $(\in, i, t)$, any cut where the interval I_i lies in one of the t components with respect to the cut in $G[\mathcal{I}_i]$ is a feasible cut for the tuple $(\in, i, t)$ and $T[\in;\ i,\ t]$ stores the minimum size among all such feasible cuts for the tuple $(\in, i, t)$. For every tuple $(\notin, i, t)$, any cut where the interval I_i does not lie in any of the t components with respect to the cut in $G[\mathcal{I}_i]$ is a feasible cut for the tuple $(\notin, i, t)$ and $T[\notin;\ i,\ t]$ stores the minimum size among all such feasible cuts for the tuple $(\notin, i, t)$. Similar to interval graphs, here also we order the components from left to right. In particular, for a pair of components C_j and $C_{j'}$, we say $C_j \prec C_{j'}$ if for any pair of intervals $I \in C_j$ and $I' \in C_{j'}$ the condition $r(I) < \ell(I')$ holds.

In the base case, the values $T[\in;\ i,\ 1] = 0$ and $T[\notin;\ i,\ 1] = 1$, for $i \in [n]$.

In the next two lemmas, we give recursive formulas for computing the values $T[\in;\ i,\ t]$ and $T[\notin;\ i,\ t]$, for $i \in [n]$, $1 < t \leq k$.

Lemma 6. *For every t and i where $2 \leq t \leq k$ and $1 \leq i < n$, the following holds:*

$$T[\notin;\ i+1,\ t] = 1 + \min\{T[\in;\ i, t],\ T[\notin;\ i,\ t]\}.$$

Proof. We prove the given recurrence by showing inequalities in both directions. In one direction, let $(C_1, C_2, \ldots, C_t)$ be a feasible cut corresponding to the entry $T[\notin;\ i+1,\ t]$. We distinguish the following two cases. Case 1: If $I_i \in C_t$, then $(C_1, C_2, \ldots, C_t)$ is a feasible cut corresponding to the entry $T[\in;\ i,\ t]$. Case 2: If $I_i \notin C_t$ then $(C_1, C_2, \ldots, C_t)$ is a feasible cut corresponding to the entry $T[\notin;\ i,\ t]$. In both these cases, the cut size is one less than a cut corresponding to $T[\notin;\ i+1,\ t]$. Therefore, $T[\notin;\ i+1,\ t] - 1 \geq \min\{T[\in;\ i,\ t],\ T[\notin;\ i,\ t]\}$.

In the other direction, let $(C'_1, C'_2, \ldots, C'_t)$ be a feasible cut respecting the tuple $(\in, i, t)$, where X_1 is the corresponding set of cut vertices. Now $(C'_1, C'_2, \ldots, C'_t)$ is also a feasible cut for $T[\notin;\ i+1,\ t]$ with $X_1 \cup \{I_{i+1}\}$ considered as the set of cut vertices. Similarly, let $(C''_1, C''_2, \ldots, C''_t)$ be a feasible cut corresponding to the entry $T[\notin;\ i,\ t]$, where X_2 is a set of cut vertices. Now $(C''_1, C''_2, \ldots, C''_t)$ is also a feasible cut corresponding to the entry $T[\notin;\ i+1,\ t]$ where $X_2 \cup \{I_{i+1}\}$ is a set of cut vertices. Thus, $T[\notin;\ i+1,\ t] \leq 1 + \min\{T[\in;\ i,\ t],\ T[\notin;\ i,\ t]\}$. Hence the lemma holds. □

Lemma 7. *Let d_i be the number of intervals passing through $\ell(I_i)$ and i' be the index corresponding to the interval I_i^{ℓ}. Then for every $2 \leq t \leq k$ the following holds:*

$$T[\in;\ i+1,\ t] = \min\{T[\in;\ i,\ t],\ T[\notin;\ i',\ t-1] + d_{i+1} - 1\}.$$

Proof. We prove the recurrence relation by showing inequalities in both directions. In one direction, let $(C_1, C_2, \ldots, C_t)$ be a feasible cut corresponding to the entry $T[\in;\ i+1,\ t]$. We distinguish the following two cases. If $I_i \in C_t$ then $(C_1, C_2, \ldots, (C_t \backslash \{I_{i+1}\}))$ is a feasible cut corresponding to the entry $T[\in;\ i,\ t]$. If $I_i \notin C_t$, then $(C_1, C_2, \ldots, C_{t-1})$ is a feasible cut corresponding to the entry $T[\notin;\ i',\ t-1]$, but in this case the cut size decreases by $d_{i+1} - 1$. So $T[\in;\ i+1,\ t] \geq \min\{T[\in;\ i,\ t],\ T[\notin;\ i',\ t-1] + d_{i+1} - 1\}$. In the other direction, let $(C'_1, C'_2, \ldots, C'_t)$ be a feasible cut corresponding to the entry $T[\in;\ i,\ t]$, where X_1 is the set of cut vertices. Now $(C'_1, C'_2, \ldots, C'_t \cup \{I_{i+1}\})$ is also a feasible cut corresponding to the entry $T[\in;\ i+1,\ t]$ with the same cut X_1. Similarly, let

$(C_1'', C_2'', \ldots, C_{t-1}'')$ be a feasible cut corresponding to the entry $T[\notin;\ i',\ t-1]$, where X_2 is the set of cut vertices. Let Z denote the set of intervals containing $\ell(I_{i+1})$ except I_{i+1}. Now $(C_1'', C_2'', \ldots, C_{t-1}'', I_{i+1})$ is also a feasible cut corresponding to the entry $T[\in;\ i+1,\ t]$ with $X_2 \cup Z$ as a set of cut vertices. Since $|Z| = d_{i+1}$, then $T[\in;\ i+1,\ t] \leq \min\{T[\in;\ i,\ t],\ T[\notin;\ i',\ t-1] + d_{i+1} - 1\}$. □

With the insight of Lemma 6 and Lemma 7, we can now state the following theorem.

Theorem 3. Vertex k-Way Cut *in proper interval graph with n vertices can be solved in $\mathcal{O}(kn)$ time assuming that the interval model is given..*

Proof. Let G be a given proper interval graph with corresponding set $\mathcal{I}$ of n intervals. Let P denote the set of all endpoints of these intervals. Here we assume that we are given the set of intervals with the ordering based on left endpoints as an input. In the pre-processing step, we do the following: compute I_i^ℓ and d_i, for each interval $I_i \in \mathcal{I}$. It will take $\mathcal{O}(n)$ time to perform all the pre-processing steps. Now in the recurrence formula, to obtain $T[\notin;\ i+1,\ t]$ and $T[\in;\ i+1,\ t]$, we use $\mathcal{O}(1)$ many computations. So computing any entry takes $\mathcal{O}(1)$ time. Since i ranges from 1 to up to n, and $t \leq k$, we can compute all the entries of the table in $\mathcal{O}(kn)$ time. Notice that the entry $T[.; n, k]$ with minimum value gives us the size of a minimum vertex k-way cut in G. Hence, the theorem holds. □

4.3 Circular-Arc Graphs

A graph G is said to be a circular-arc graph if there exists a corresponding geometric intersection representation $\mathcal{A}(G)$ of G, where each vertex $v \in G$ is associated with an arc on a fixed circle. Two vertices u and v are adjacent in G if and only if the corresponding arcs intersect each other. It is easy to observe that this graph class contains interval graphs.

Here we design a polynomial-time algorithm for the optimization version of Vertex k-Way Cut problem on circular-arc graphs. Let S be an optimal solution of Vertex k-Way Cut problem on G and C be a component in $G \setminus S$. Assume I is the circular-arc representation of C in $\mathcal{A}(G)$ and $I_1 \in I$ be the arc that has the last endpoint, say u, in the clockwise direction in the circular-arc representation of $G \setminus S$. Let I' be the set of arcs in $\mathcal{A}(G)$ that intersect u, excluding I_1. Since S is a k-way cut it must contain all the vertices corresponding to the arcs in I'. Now assume we cut the circle corresponding to the circular-arc representation of $G \setminus S$ at u and convert the circular-arc to a real line to get an instance of Vertex k-Way Cut problem on interval graphs. We claim that $S \setminus I'$ is an optimal solution to the Vertex k-Way Cut problem on the interval graph instance that we construct.

Claim. $S \setminus I'$ is a solution to the Vertex k-Way Cut problem on the interval graph instance $G \setminus I'$.

Proof. Let S' be an optimal solution on the VERTEX k-WAY CUT problem on the interval graph induced by $G\backslash I'$. If $|S'| = |S\backslash I'|$, we are done. Else, $|S'| < |S\backslash I'|$ then $S\backslash I'$ is not an optimal solution to the VERTEX k-WAY CUT problem on the interval graph instance $G\backslash I'$. Observe that $G\backslash(S' \cup I')$ has at least k components, and $|S' \cup I'| = |S'| + |I'| < |S| + |I'| = |S \cup I'|$. Thus $S' \cup I'$ is an optimal solution to VERTEX k-WAY CUT problem on G with size strictly smaller than S which is a contradiction to our assumption that S is an optimal solution. □

Now given an instance G for VERTEX k-WAY CUT problem on circular-arc graphs we convert it to an instance of interval graph for all the $2n$ endpoints and run the algorithm for VERTEX k-WAY CUT problem, designed in Sect. 4.1, on each of those interval graphs and store the corresponding S', I'. As a solution, we return the set $S' \cup I'$ that has minimum size. Since algorithm for interval graph runs in $\mathcal{O}(kn^4)$ time (Theorem 2); so we have the following theorem.

Theorem 4. VERTEX k-WAY CUT *in circular-arc graphs with* n *vertices can be solved in* $\mathcal{O}(kn^5)$ *time.*

4.4 Permutation Graphs

This subsection presents a dynamic-programming algorithm for the optimization version of the VERTEX k-WAY CUT problem on permutation graphs. Let G be a permutation graph with vertex set $V(G)$ and edge set $E(G)$. There exists a corresponding geometric intersection representation for a permutation graph G similar to interval graphs, where each vertex v in G is associated with a line segment $S(v)$ with endpoints $x(v)$ and $y(v)$ being on two parallel lines X and Y, respectively. Without loss of generality, we can assume that both the lines X and Y are horizontal. Two vertices u and v are adjacent in G if and only if the segments $S(u)$ and $S(v)$ intersect with each other. Assume that along with the graph, we have the set of corresponding line segments as an input. Here, we use $\mathcal{S}$ to denote the segments $\{S(v) \colon v \in V\}$. Let P_X and P_Y denote the set of all endpoints of S on the lines X and Y, respectively. Let $P = P_X \cup P_Y$.

For a pair of vertices u and v, we write $x(u) < x(v)$ (similarly, $y(u) < y(v)$) to indicate that $x(v)$ is to the right of $x(u)$ (similarly, $y(v)$ is to the right of $y(u)$). If both $x(u) < x(v)$ and $y(u) < y(v)$ hold, then we say $S(u) \prec S(v)$. In the rest of this subsection, we interchangeably use v and $S(v)$. For a pair of points α and β where $\alpha \in X, \beta \in Y$, we denote the set of segments in $\mathcal{S}$ whose one endpoint lies either to the left of α or to the left of β by S^{α}_{β}. We use $G[\alpha, \beta]$ to denote the subgraph induced by S^{α}_{β} in G. Additionally, for any set of four points, $\alpha_1, \alpha_2 \in X$ and $\beta_1, \beta_2 \in Y$ such that $\alpha_1 < \alpha_2$ and $\beta_1 < \beta_2$, we define $S^{\alpha_1,\alpha_2}_{\beta_1,\beta_2} = \{S(v) \colon \alpha_1 \leq x(v) \leq \alpha_2,\ \beta_1 \leq y(v) \leq \beta_2\}$. We use $G[(\alpha_1, \alpha_2), (\beta_1, \beta_2)]$ to denote the subgraph of G induced by the segments $S^{\alpha_1,\alpha_2}_{\beta_1,\beta_2}$.

We now define a table for our dynamic-programming algorithm. For every tuple (i, p, q, r, s), where $p, q \in P_X$ with $p < q$ and $r, s \in P_Y$ with $r < s$, any cut where $G[(p, q), (r, s)]$ is the i-th component with respect to the cut in $G[q, s]$ is a feasible cut for the tuple (i, p, q, r, s) and $T[i; p, q, r, s]$ stores the minimum size among all such feasible cut for the tuple (i, p, q, r, s). Notice that any two connected components do not intersect. Hence we can order the components from left to right. In particular, for a pair of components C_j and $C_{j'}$, we say $C_j \prec C_{j'}$ if for any pair of line segments $u \in C_j$ and $v \in C_{j'}$, $S(u) \prec S(v)$.

For the base case, the value $T[1; p, q, r, s]$ is the number of segments in $G[q, s]$ whose one endpoint lies either strictly to the left of p or r, or strictly to the right of q or s, formally $T[1; p, q, r, s] = |S^q_s| - |S^{p,q}_{r,s}|$. In the next lemma, we give a recursive formula for computing the values $T[i; p, q, r, s]$, for $i > 1$.

Lemma 8. *For every i, $2 < i < k$ and any set of four points p, q, r, s, where $p, q \in P_X$ with $p < q$ and $r, s \in P_Y$ with $r < s$, the following holds:*

$$T[i; p, q, r, s] = \min_{\substack{p', q' \in P_X \ \& \ r', s' \in P_Y \\ p' < q' < p, \ r' < s' < r}} \{T[i-1; p', q', r', s'] + |S^q_s| - |S^{q'}_{s'}| - |S^{p,q}_{r,s}|\}.$$

Proof. We prove the recurrence by showing inequalities in both directions. In one direction, let $(C_1, C_2, \ldots, C_i)$ be a feasible cut corresponding to the entry $T[i; p, q, r, s]$. Here $C_i = G[(p, q), (r, s)]$. Let p', q', r', s' be four points such that $C_{i-1} = G[(p', q'), (r', s')]$, $p', q' \in P_X$ and $r', s' \in P_Y$. Clearly, $p' < q' < p$ and $r' < s' < r$ hold. Now, the segments of the set $S^q_s \backslash (S^{q'}_{s'} \cup S^{p,q}_{r,s})$ are cut vertices corresponding to the entry $T[i; p, q, r, s]$. Here we get a set of $(i-1)$ components $C_1, C_2, \ldots, C_{i-1}$ in the graph $G[q', s']$ with $C_{i-1} \subseteq G[(p', q'), (r', s')]$ and cut size at most $T[i; p, q, r, s] - (|S^q_s| - |S^{q'}_{s'}| - |S^{p,q}_{r,s}|)$. Therefore, $T[i-1; p', q', r', s'] \leq T[i; p, q, r, s] - (|S^q_s| - |S^{q'}_{s'}| - |S^{p,q}_{r,s}|)$.

In the other direction, let $(C'_1, C'_2, \ldots, C'_{i-1})$ be a feasible cut corresponding to the entry $T[i-1; p', q', r', s']$, where $p' < q' < p$, $r' < s' < r$ and $C_{i-1} = G[(p', q'), (r', s')]$. The component induced by the subgraph $G[(p, q), (r, s)]$ together with $C'_1, C'_2, \ldots, C'_{i-1}$ produces a feasible cut for $T[i; p, q, r, s]$. Now the cut corresponding to the entry $T[i-1; p', q', r', s']$ together with $(|S^q_s| - |S^{q'}_{s'}| - |S^{p,q}_{r,s}|)$ gives a cut that yields the set of components $C'_1, C'_2, \ldots, C'_{i-1}$, $C'_i = G[(p, q), (r, s)]$. Hence, $T[i-1; p', q', r', s'] + |S^q_s| - |S^{q'}_{s'}| - |S^{p,q}_{r,s}| \geq T[i; p, q, r, s]$. This completes the proof of the lemma. □

With the insight of Lemma 8, we can now state the following theorem.

Theorem 5. Vertex *k*-Way Cut *in permutation graph with n vertices can be solved in $\mathcal{O}(kn^8)$ time.*

Proof. Let G be a given graph with a set $\mathcal{S}$ of n line segments. Recall that we use P_X and P_Y to denote the set of all endpoints of line segments in X and Y, respectively. In the pre-processing step, we do the following: (i) we construct S^α_β, for every pair of points $\alpha \in P_X$ and $\beta \in P_Y$. (ii) we compute $|S^{\alpha_1,\alpha_2}_{\beta_1,\beta_2}|$ for each possible set of four points $\alpha_1, \alpha_2 \in P_X$ and $\beta_1, \beta_2 \in P_Y$. It takes $\mathcal{O}(n^5)$ time to perform all these pre-processing steps. Now in the recurrence formula, to obtain $T[i; p, q, r, s]$, we use the already computed values, where $p', q' \in P_X$ and $r', s' \in P_Y$ with $p' < q' < p$ and $r' < s' < r$. Computing any entry takes $\mathcal{O}(n^4)$

time. Since i ranges from 1 to k, we can compute all the values $T[i; p, q, r, s]$ in $\mathcal{O}(kn^8)$ time. Notice that the entry $T[k; ., .]$ with minimum value gives us the size of a minimum vertex k-way cut in G. Hence, the theorem holds. □

References

1. Augeri, C.J., Ali, H.H.: New graph-based algorithms for partitioning VLSI circuits. In: 2004 IEEE International Symposium on Circuits and Systems (IEEE Cat. No. 04CH37512), vol. 4, pp. IV-IV. IEEE (2004)
2. Chekuri, C., Quanrud, K., Chao, X.: LP relaxation and tree packing for minimum k-cut. SIAM J. Discret. Math. **34**(2), 1334–1353 (2020)
3. Corneil, D.G.: The complexity of generalized clique packing. Discrete Appl. Math. **12**(3), 233–239 (1985)
4. Cygan, M., et al.: Randomized contractions meet lean decompositions. ACM Trans. Algorithms, **17**(1), 6:1–6:30 (2021)
5. Downey, R.G., Estivill-Castro, V., Fellows, M., Prieto, E., Rosamund, F.A.: Cutting up is hard to do: the parameterised complexity of k-cut and related problems. Electron. Notes Theor. Comput. Sci. **78**, 209–222 (2003)
6. Goldschmidt, O., Hochbaum, D.S.: A polynomial algorithm for the k-cut problem for fixed k. Math. Oper. Res. **19**(1), 24–37 (1994)
7. Golumbic, M.C.: Algorithmic Graph Theory and Perfect Graphs. Elsevier, Amsterdam (2004)
8. Gupta, A., Lee, E., Li, J.: An FPT algorithm beating 2-approximation for k-cut. In: Proceedings of the Twenty-Ninth Annual ACM-SIAM Symposium on Discrete Algorithms, pp. 2821–2837. SIAM (2018)
9. Gupta, A., Lee, E., Li, J.: Faster exact and approximate algorithms for k-cut. In: 2018 IEEE 59th Annual Symposium on Foundations of Computer Science (FOCS), pp. 113–123. IEEE (2018)
10. Karger, D.R., Stein, C.: A new approach to the minimum cut problem. J. ACM (JACM) **43**(4), 601–640 (1996)
11. Kawarabayashi, K.I., Lin, B.: A nearly 5/3-approximation FPT algorithm for min-k-cut. In: Proceedings of the Fourteenth Annual ACM-SIAM Symposium on Discrete Algorithms, pp. 990–999. SIAM (2020)
12. Kawarabayashi, K.I., Thorup, M.: The minimum k-way cut of bounded size is fixed-parameter tractable. In: 2011 IEEE 52nd Annual Symposium on Foundations of Computer Science, pp. 160–169. IEEE (2011)
13. Li, J.: Faster minimum k-cut of a simple graph. In: 2019 IEEE 60th Annual Symposium on Foundations of Computer Science (FOCS), pp. 1056–1077. IEEE (2019)
14. Lokshtanov, D., Saurabh, S., Surianarayanan, V.: A parameterized approximation scheme for min k-cut. In: 2020 IEEE 61st Annual Symposium on Foundations of Computer Science (FOCS), pp. 798–809. IEEE (2020)
15. Manurangsi, P.: Inapproximability of maximum edge biclique, maximum balanced biclique and minimum k-cut from the small set expansion hypothesis. In: 44th International Colloquium on Automata, Languages, and Programming (ICALP 2017). Schloss Dagstuhl-Leibniz-Zentrum fuer Informatik (2017)
16. Marx, D.: Parameterized graph separation problems. Theor. Comput. Sci. **351**(3), 394–406 (2006)
17. Saran, H., Vazirani, V.V.: Finding k cuts within twice the optimal. SIAM J. Comput. **24**(1), 101–108 (1995)
18. West, D.B., et al.: Introduction to Graph Theory, vol. 2. Prentice Hall, Upper Saddle River (2001)

Complete Decomposition of Symmetric Tensors in Linear Time and Polylogarithmic Precision

Pascal Koiran and Subhayan Saha(✉)

Univ Lyon, EnsL, UCBL, CNRS, LIP, LYON Cedex 07, 69342 Lyon, France
{pascal.koiran,subhayan.saha}@ens-lyon.fr

Abstract. We study symmetric tensor decompositions, i.e., decompositions of the form $T = \sum_{i=1}^{r} u_i^{\otimes 3}$ where T is a symmetric tensor of order 3 and $u_i \in \mathbb{C}^n$. In order to obtain efficient decomposition algorithms, it is necessary to require additional properties from the u_i. In this paper we assume that the u_i are linearly independent. This implies $r \leq n$, i.e., the decomposition of T is *undercomplete*. We will moreover assume that $r = n$ (we plan to extend this work to the case $r < n$ in a forthcoming paper.) We give a randomized algorithm for the following problem: given T, an accuracy parameter ϵ, and an upper bound B on the *condition number* of the tensor, output vectors u_i' such that $||u_i - u_i'|| \leq \epsilon$ (up to permutation and multiplication by phases) with high probability. The main novel features of our algorithm are:

- We provide the first algorithm for this problem that works in the computation model of finite arithmetic and requires only polylogarithmic (in n, B and $\frac{1}{\epsilon}$) many bits of precision.
- Moreover, this is also the first algorithm that runs in linear time in the size of the input tensor. It requires $O(n^3)$ arithmetic operations for all accuracy parameters $\epsilon = \frac{1}{\text{poly}(n)}$.

In order to obtain these results, we rely on a mix of techniques from algorithm design and algorithm analysis. The algorithm is a modified version of Jennrich's algorithm for symmetric tensors. In terms of algorithm design, our main contribution lies in replacing the usual appeal to resolution of a linear system of equations [5,12] by a matrix trace-based method. The analysis of the algorithm depends on the following components:

1. We use the fast and numerically stable diagonalisation algorithm from [1]. We provide better guarantees for the approximate solution returned by the diagonalisation algorithm when the input matrix is diagonalisable.
2. We show strong anti-concentration bounds for certain families of polynomials when the randomness is sampled uniformly from a discrete grid.

Keywords: Tensor Decomposition · Finite precision arithmetic · Jennrich's Algorithm · Computational Linear Algebra

M. Mavronicolas (Ed.): CIAC 2023, LNCS 13898, pp. 308–322, 2023.
https://doi.org/10.1007/978-3-031-30448-4_22

1 Introduction

Tensor decompositions have generated significant interest in recent years due to their applications in different fields such as signal processing, computer vision, chemometrics, neuroscience and others (see [14] for a comprehensive survey on the applications and available software for this problem). In fact, a number of learning algorithms for certain models have been developed through the fundamental machinery of tensor decompositions. Numerous algorithms have been devised for solving the tensor decomposition problem with different assumptions on the input tensor and different efficiency and accuracy bounds [5,8,9,16].

In this paper, we study the algorithmic problem of *approximately* decomposing an arbitrary symmetric order-3 tensor $T \in \mathbb{C}^n \otimes \mathbb{C}^n \otimes \mathbb{C}^n$ uniquely (up to permutation and scaling) into a sum of rank-one tensors. To do this efficiently, we need to impose certain restrictions on the "independence" of the rank-one components. More formally, we assume that the rank-one components are of the form $u_i \otimes u_i \otimes u_i$ where the u_i's are linearly independent. We will explore these restrictions in more detail in Sect. 1.1.

While this problem is well-studied when the underlying model of computation is exact real arithmetic (even when the input tensor has some noise), not much work has been done in the setting where the underlying model of computation is finite precision arithmetic (see Sect. 1.3 for a presentation of this model). The key difficulties lie in the fact that every arithmetic operation in this model is done approximately and the stored numbers can also have some adversarial error (even the input). An iterative algorithm is called numerically stable if it can be implemented using polylogarithmically many bits in finite precision arithmetic [1,20]. The central contribution of this paper is a rigorous analysis of a numerically stable algorithm that runs in linear time in the input size. This algorithm is inspired by Jennrich's algorithm and a high level presentation of the algorithm appears as Algorithm 1 in Sect. 1.6.

The presentation of this paper has been severely shortened to accommodate the page restrictions. For a full version of this paper, refer to [13].

1.1 Symmetric Tensor Decomposition

Let $T \in \mathbb{C}^n \otimes \mathbb{C}^n \otimes \mathbb{C}^n$ be a symmetric tensor of order 3. We recall that such an object can be viewed as a 3-dimensional array $(T_{ijk})_{1 \le i,j,k \le n}$ that is invariant under all 6 permutations of the indices i, j, k. This is therefore a 3-dimensional generalization of the notion of symmetric matrix. In this paper, we study symmetric tensor decompositions, i.e., decompositions of the form

$$T = \sum_{i=1}^{r} u_i \otimes u_i \otimes u_i \tag{1}$$

where $u_i \in \mathbb{C}^n$. The smallest possible value of r is the symmetric tensor rank of T and it is NP-hard to compute already for order-3 tensors. This was shown by Shitov [19], and a similar NP-hardness result for ordinary tensors was obtained much earlier by Håstad [10]. In this paper, we impose an additional linear independence condition on the u_i. Under this assumption, such a decomposition is unique if it exists, up to a permutation of the u_i's and scaling by cube roots of unity [9,15]. There is a traditional distinction between *undercomplete* decompositions, where $r \le n$ in (1), and *overcomplete* decompositions, where $r > n$. In this paper we consider only undercomplete decompositions because of the linear independence condition on the u_i. Moreover, we will impose the additional condition that r is exactly equal to n, i.e., we focus on *complete decompositions*. We say that a tensor is *diagonalisable* if it satisfies these two conditions. The results of the present paper will be extended to general undercomplete decomposition in a forthcoming work by reduction to the complete case.

1.2 Approximate Tensor Decomposition

As explained above, an order-3 symmetric tensor $T \in \mathbb{C}^n \otimes \mathbb{C}^n \otimes \mathbb{C}^n$ is called diagonalisable if there exist linearly independent vectors $u_i \in \mathbb{C}^n$ such that $T = \sum_{i=1}^n u_i^{\otimes 3}$. The objective of the ϵ-approximation problem for tensor decomposition is to find linearly independent vectors $u'_1, ..., u'_n$ such that there exists a permutation $\pi \in S_n$ where $||\omega_i u_{\pi(i)} - u'_i|| \le \epsilon$ with ω_i being a cube root of unity. Here ϵ is the desired accuracy parameter given as input. Hence the problem is essentially that of approximating the vectors u_i appearing in the decomposition of T.

1.3 Model of Computation

We are chiefly interested in the finite precision model of arithmetic. Some algorithms are also presented in exact real arithmetic as an intermediate step toward their derivation in the finite precision model. For the latter model, we use like [1] the standard floating point axioms from [11].

1.4 Our Results

Recall that an order-3 tensor $T \in (\mathbb{C}^n)^{\otimes 3}$ is called diagonalisable if there exist linearly independent vectors $u_1, ..., u_n \in \mathbb{C}^n$ such that T can be decomposed as in (1).

Definition 1 (Condition number of a diagonalisable symmetric tensor). *Let T be a diagonalisable symmetric tensor over $\mathbb{C}$ such that $T = \sum_{i=1}^n u_i^{\otimes 3}$. Let $U \in M_n(\mathbb{C})$ be the matrix with rows $u_1, \dots, u_n$. We define the tensor decomposition condition number of T as: $\kappa(T) = ||U||_F^2 + ||U^{-1}||_F^2$.*

It can be shown that $\kappa(T)$ is well defined: for a diagonalisable tensor the condition number is independent of the choice of U. (Refer to Lemma 6.3 in [13] for a formal proof.) Note that when U is close to a singular matrix, the corresponding tensor is poorly conditioned, i.e., has a large condition number. This is not surprising since our goal is to find a decomposition where the vectors u_i are linearly independent.

Our main result is a randomized polynomial time algorithm in the finite precision model which on input a diagonalisable tensor, an estimate B for the condition number of the tensor and an accuracy parameter ϵ, returns a forward approximate solution to the tensor decomposition problem (following the definition in Sect. 1.2).

In the following, we denote by $T_{MM}(n)$ the number of arithmetic operations required to multiply two $n \times n$ matrices in a numerically stable manner. If ω denotes the exponent of matrix multiplication, it is known that $T_{MM}(n) = O(n^{\omega+\eta})$ for all $\eta > 0$ (see Sect. 2 in [13] for details).

Theorem 1 (Main Theorem). *There is an algorithm which, given a diagonalisable tensor T, a desired accuracy parameter ϵ and some estimate $B \geq \kappa(T)$, outputs an ϵ-approximate solution to the tensor decomposition problem for T in*

$$O(n^3 + T_{MM}(n) \log^2 \frac{nB}{\epsilon})$$

arithmetic operations on a floating point machine with

$$O(\log^4(\frac{nB}{\epsilon}) \log n)$$

bits of precision, with probability at least $\left(1 - \frac{1}{n} - \frac{12}{n^2}\right)\left(1 - \frac{1}{\sqrt{2}n} - \frac{1}{n}\right)$.

A simplified version of this algorithm, designed for the exact arithmetic model, is presented in Sect. 1.6 (Refer to Sect. 6 in [13] for the version compatible with the finite-precision arithmetic model.) The following are the important conclusions from the above theorem:

- The number of bits of precision required for this algorithm is polylogarithmic in n, B and $\frac{1}{\epsilon}$.
- The running time as measured by the number of arithmetic operations is $O(n^3)$ for all $\epsilon = \frac{1}{\text{poly(n)}}$, i.e., it is linear in the size of the input tensor. This requires the use of fast matrix multiplication. With standard matrix multiplication, the running time is quasilinear instead of linear (i.e., it is multiplied by a polylogarithmic factor). The bit complexity of the algorithm is also quasilinear.
- The algorithm can provide inverse exponential accuracy, i.e., it still runs in polynomial time even when the desired accuracy parameter is $\epsilon = \frac{1}{\exp{(n)}}$.

In order to obtain this result we combine techniques from algorithm design and algorithm analysis; the main ideas are outlined in Sect. 1.6 and Sect. 1.4.3 in [13]. To the best of our knowledge, this is the first tensor decomposition algorithm shown to work in polylogarithmic precision. Moreover, this algorithm is also the first to run in a linear number of arithmetic operations (i.e., prior to this work no linear time algorithm was known, *even in the exact arithmetic model*).

1.5 Related Work and Discussion

Our algorithm can be viewed as an optimized version of Jennrich's algorithm [9,16,17]. This algorithm, also referred to in the literature as the "simultaneous diagonalisation algorithm," was one of the first to give provable guarantees for tensor decomposition. In fact, if the input tensor satisfies certain genericity conditions this algorithm returns the unique decomposition (up to permutation and scaling) almost surely. It was shown in [5] that this algorithm runs in polynomial time in the exact arithmetic computational model, i.e., when the model has the underlying assumption that all the steps of the algorithm can be performed exactly. Moreover, it is shown in the same paper that the algorithm is robust to noise in the input.

One may also drop the genericity condition and attempt to decompose an arbitrary low-rank tensor given as input. For symmetric tensors with constant rank, such an algorithm can be found in [4]. This algorithm was recently extended to slightly superconstant rank in [18]. Still other algorithms for symmetric tensor decomposition can be found in the algebraic literature, see e.g. [3,6]. These two papers do not provide any complexity analysis for their algorithms.

Condition Numbers: There is no universally accepted definition of a "condition number" in numerical analysis, but a common one, used in [2], is as follows. Suppose we wish to compute a map $f : X \to Y$. The condition number of f at an input x is a measure of the variation of the image $f(x)$ when x is perturbed by a small amount. This requires the choice of appropriate distances on the spaces X and Y. The condition number is therefore a quantitative measure of the continuity of f at x. In particular, it is independent of the choice of an algorithm for computing f. In finite arithmetic, we cannot hope to approximate $f(x)$ with a low precision algorithm at an input x with a high condition number since we do not even assume that the input is stored exactly. Moreover, designing algorithms that work in low precision at well-conditioned inputs is often a challenging task. For the purpose of this paper we work with the somewhat ad-hoc choice of $\kappa(T)$ as our condition number because this parameter controls the numerical precision needed for our main algorithm, as shown by Theorem 1[1]. In particular, we have found it more convenient to work with $\kappa(T)$ than with a quantity such as $||U||.||U^{-1}||$, commonly used as a condition number in numerical linear algebra. The results presented in this paper are in stark contrast with those of Beltrán et al. [2]. That paper analyzes a class of tensor decomposition algorithms related to Jennrich's algorithm. Their conclusion is that all these "pencil-based algorithms" are numerically unstable. Indeed, the pencil-based algorithms of [2] are all deterministic. Beltrán et al. conclude their paper with the following sentence: "We hope that these observations may (re)invigorate the search for numerically stable algorithms for computing CPDs."[2] The algorithm presented in this paper answers their call, at least for the case of complete decomposition of symmetric

[1] $\kappa(T)$ also appears in the sublinear term for the arithmetic complexity of the algorithm.

[2] CPD stands for *Canonical Polyadic Decomposition, i.e., decomposition as a sum of rank-1 tensors.*

tensors. A precise comparison of our results with the numerical instability result of [2] is delicate because we do not work in the same setting. In particular, they work with ordinary instead of symmetric tensors; they do not work with the same condition number; and their result is obtained for undercomplete rather than complete decompositions. We believe that the main reason why we obtain a positive result is due to yet another difference, namely, the use of randomization in step (i) of our algorithm. In the setting of [2] one would have to take two *fixed* linear combinations $T^{(a)}, T^{(b)}$ of the slices. Essentially, they show that for every fixed choice of a pair of linear combinations, there are input tensors for which this choice is bad; whereas we show that for every (well conditioned) input T, most choices of a and b are good. We believe that our techniques can also be applied to decomposition of ordinary tensors. In this paper we have chosen to focus on symmetric tensors because this setting is somewhat simpler technically.

1.6 The Algorithm

Before giving a high-level presentation of our algorithm, we introduce a few notations. A symmetric tensor $T \in \mathbb{C}^n \otimes \mathbb{C}^n \otimes \mathbb{C}^n$ can be cut into n slices $T_1, \ldots, T_n$ where $T_k = (T_{ijk})_{1 \le i,j \le n}$. Each slice is a symmetric matrix of size n. In the algorithm below we also make use of a "change of basis" operation, which applies a linear map of the form $A \otimes A \otimes A$ to a tensor. Here, $A \in M_n(\mathbb{C})$ and we apply A to the 3 components of the input tensor. In particular, for rank-1 symmetric tensors we have $(A \otimes A \otimes A).(u \otimes u \otimes u) = (A^T u)^{\otimes 3}$. We give more details on this operation at the beginning of Sect. 2. The algorithm proceeds as follows.

Algorithm 1: Algorithm for tensor decomposition in exact arithmetic

---

1. Pick vectors $a = (a_1, ..., a_n)$ and $b = (b_1, ..., b_n)$ at random from a finite set and compute two random linear combinations $T^{(a)} = \sum_{i=1}^n a_i T_i$ and $T^{(b)} = \sum_{i=1}^n b_i T_i$ of the slices of T.
2. Diagonalise $(T^{(a)})^{-1} T^{(b)} = V D V^{-1}$. Let $v_1, ..., v_n$ be the columns of V.
3. Let $u_1, ..., u_n$ be the rows of V^{-1}.
4. Let $T' = (V \otimes V \otimes V).T$. Let $T'_1, ..., T'_n$ be the slices of T'. Define $\alpha_i = \mathrm{Tr}(T'_i)$. We will refer to the computation of $Tr(T'_i)$ as the trace of slices after a change of basis (TSCB).
5. Output $(\alpha_1)^{\frac{1}{3}} u_1, ..., (\alpha_n)^{\frac{1}{3}} u_n$.

The above algorithm is a modified version of Jennrich's algorithm for symmetric tensors. In terms of algorithm design, our main contribution lies in step (iv). Previous versions of Jennrich's algorithm have appealed instead to the resolution of a linear system of equations: see e.g. [5,17] for the case of ordinary tensors. In the symmetric case, the algebraic algorithm in [12] for decomposition of a polynomial as a sum of powers of linear forms also appeals to the resolution of a linear system for essentially the same purpose. Our trace-based version of step (iv) is more efficient, and this is crucial for the derivation of the complexity bounds in Theorem 1. Steps (i) and (iv) are indeed the most expensive: they are responsible for the $O(n^3)$ term in the arithmetic complexity of the algorithm.

Step (ii) (especially the diagonalisation step) is the one that contributes most significantly to the bound on the number of bits of precision.

2 Slices After a Change of Basis

Given tensors $T, T' \in \mathbb{C}^{n\times n\times n}$, we say that there is a change of basis $A \in \mathrm{M}_n(\mathbb{C})$ that takes T to T' if $T' = (A\otimes A\otimes A).T$. This notation was already introduced in Sect. 1.6. In the present section we give a fast and numerically stable algorithm for computing the trace of the slices after a change of basis. More formally, given a tensor T and a matrix V, it computes $Tr(S_1), ..., Tr(S_n)$ where $S_1, ..., S_n$ are the slices of the tensor $S = (V\otimes V\otimes V).T$ with small error in $O(n^3)$ many arithmetic operations. Written in standard basis notation, the equality $T' = (A\otimes A\otimes A).T$ corresponds to the fact that for all $i_1, i_2, i_3 \in [n]$,

$$T'_{i_1 i_2 i_3} = \sum_{j_1, j_2, j_3 \in [n]} A_{j_1 i_1} A_{j_2 i_2} A_{j_3 i_3} T_{j_1 j_2 j_3}. \tag{2}$$

Note that if $T = u^{\otimes 3}$ for some vector $u \in \mathbb{C}^n$, then $(A \otimes A \otimes A).T = (A^T u)^{\otimes 3}$.

Norms: We denote by $||x||$ the ℓ^2 (Hermitian) norm of a vector $x \in \mathbb{C}^n$. For $A \in M_n(\mathbb{C})$, we denote by $||A||$ its operator norm and by $||A||_F$ its Frobenius norm: $||A||_F^2 = \sum_{i,j=1}^n |A_{ij}|^2$. We always have $||A|| \le ||A||_F$. For a given invertible matrix V, we define $\kappa_F(V) = ||V||_F^2 + ||V^{-1}||_F^2$.

Definition 2 (Tensor Norm). *Given a tensor $T \in (\mathbb{C}^n)^{\otimes 3}$, we define the Frobenius norm $||T||_F$ of T as $||T||_F = \sqrt{\sum_{i,j,k=1}^n |T_{i,j,k}|^2}$.*

Then if $T_1, ..., T_n$ are the slices of T, we also have that $\sum_{i=1}^n ||T_i||_F^2 = ||T||_F^2$.

Algorithm 2: Trace of the slices after a change of basis (TSCB)

Input: An order-3 symmetric tensor $T \in \mathbb{C}^{n\times n\times n}$, a matrix $V = (v_{i,j}) \in \mathbb{C}^{n\times n}$.

Let $T_1, ..., T_n$ be the slices of T.

1 Compute $W = V^T V$ on a floating point machine.

2 Compute $x_{m,k} = (WT_m)_{k,k}$ on a floating point machine for all $m, k \in [n]$.

3 Compute $x_m = \sum_{k=1}^n x_{m,k}$ on a floating point machine for all $m \in [n]$.

4 Compute $\tilde{s}_i = \sum_{m=1}^n v_{m,i} x_m$ on a floating point machine for all $i \in [n]$.

Output $\tilde{s}_1, ..., \tilde{s}_n$

The following is the main theorem of this section.

Theorem 2. *Let us assume that a tensor $T \in (\mathbb{C}^n)^{\otimes 3}$ and a matrix $V \in M_n(\mathbb{C})$ are given as input to Algorithm 2. Set $S = (V\otimes V\otimes V).T$ following the definition in (2) and let $S_1, ..., S_n$ be the slices of S. Then the algorithm returns $\tilde{s}_1, ..., \tilde{s}_n$ such that $|\tilde{s}_i - Tr(S_i)| \le \mu_{CB}(n) \cdot \boldsymbol{u} \cdot ||V||_F^3 ||T||_F$ where $\mu_{CB}(n) \le 14n^{\frac{3}{2}}$. It performs $T_{CB}(n) = O(n^3)$ operations on a machine with precision $\boldsymbol{u} < \frac{1}{10n}$.*

Proof. Let $S' \in \mathbb{C}^{n \times n \times n}$ be such that $S' = (V \otimes V \otimes V).T$. Let $S'_1, ..., S'_n$ be the slices of S'. We first claim that $\sum_{m=1}^{n} v_{m,i}\Big(\sum_{k=1}^{n}(V^T V T_m)_{k,k}\Big) = Tr(S'_i)$. We know that $S'_i = V^T D_i V$ where $D_i = \sum_{m=1}^{n} v_{m,i} T_m$. Using the cyclic property and the linearity of the trace operator, we get that

$$\begin{aligned} Tr(S'_i) = Tr(V^T D_i V) = Tr(V^T V D_i) = Tr(V^T V(\sum_{m=1}^{n} v_{m,i} T_m)) \\ = \sum_{m=1}^{n} v_{m,i} Tr(V^T V T_m) = \sum_{m=1}^{n} v_{m,i}\Big(\sum_{k=1}^{n}(V^T V T_m)_{k,k}\Big). \end{aligned} \tag{3}$$

From this, we conclude that if Algorithm 2 is run in exact arithmetic, it computes exactly the trace of the slices S'_i of S'. A proof of the bound on the number of arithmetic operations required and the correctness analysis of the algorithm in finite precision arithmetic can be found in Sect. 3 in [13].

3 Diagonalisation Algorithm for Diagonalisable Matrices

In their recent breakthrough result, a numerically stable algorithm for matrix diagonalisation was given in [1] that runs in nearly matrix multiplication time in the finite precision arithmetic model. We address two related issues in this section:

(i) Strengthening the conditioning guarantee from [1] on the similarity V that approximately diagonalises the input matrix A.
(ii) Relaxing the assumption $||A|| \leq 1$ on the input.

Our contribution regarding (i) appears in Theorem 3 and comes at the expense of additional assumptions on A: this matrix must be diagonalisable with distinct eigenvalues. The bounds in that theorem are expressed as a function of the condition number of the eigenproblem (5), already defined in [1], and of the Frobenius eigenvector condition number (4). Regarding (ii), we need to slightly modify the algorithm from [1] in order to scale the input matrix. The main result of this section is Theorem 3 where we combine (i) and (ii). In particular, the (routine) error analysis due to the scaling of A is worked out in the proof of Theorem 3. All the omitted proofs of this section and the corresponding definitions can be found in Sect. 4 in [13]. Note that the diagonalization algorithm of [1] is responsible for the number of bits of precision needed in our main result (Theorem 1).

Definition 3 (Eigenpair and eigenproblem). *[1] An eigenpair of a matrix $A \in \mathbb{C}^{n \times n}$ is a tuple $(\lambda, v) \in \mathbb{C} \times \mathbb{C}^n$ such that $Av = \lambda v$ and $||v||_2 = 1$. The eigenproblem is the problem of finding a maximal set of linearly independent eigenpairs (λ_i, v_i) of a given matrix A. Note that an eigenvalue may appear more than once if it has geometric multiplicity greater than one. In the case when A is diagonalizable, the solution consists of exactly n eigenpairs, and if A has distinct eigenvalues, then the solution is unique, up to the phases of v_i.*

Definition 4 (δ-forward approximation for the eigenproblem). *Let (λ_i, v_i) be true eigenpairs for a diagonalizable matrix A. Given an accuracy parameter δ, the problem is to find pairs (λ'_i, v'_i) such that $||v_i - v'_i|| \leq \delta$ and $|\lambda_i - \lambda'_i| \leq \delta$ i.e., to find a solution close to the exact solution.*

Condition numbers. If A is diagonalizable, we define following [1] its *eigenvector condition number*: $\kappa_V(A) = \inf_V ||V|| \cdot ||V^{-1}||$, where the infimum is over all invertible V such that $V^{-1}AV$ is diagonal. Its minimum eigenvalue gap is defined as $\text{gap}(A) := \min_{i \neq j} |\lambda_i(A) - \lambda_j(A)|$, where λ_i are the eigenvalues of A (with multiplicity). Instead of the eigenvector condition number, it is sometimes more convenient to work instead with the *Frobenius eigenvector condition number*

$$\kappa_V^F(A) = \inf_V (||V||_F^2 + ||V^{-1}||_F^2) = \inf_V \kappa_F(V), \tag{4}$$

where the infimum is taken over the same set of invertible matrices. We always have $\kappa_V^F(A) \geq 2\kappa_V(A)$. Following [1], we define the condition number of the eigenproblem to be:

$$\kappa_{\text{eig}}(A) := \frac{\kappa_V(A)}{\text{gap}(A)} \in [0, \infty]. \tag{5}$$

Lemma 1. *Suppose that A has n distinct eigenvalues $\lambda_1, \ldots, \lambda_n$, with $v_1, \ldots, v_n$ the corresponding eigenvectors. Let W be the matrix with columns $v_1, \ldots, v_n$; let $u_1, \ldots, u_n$ be the left eigenvectors of A, i.e., the rows of W^{-1}. Then $\kappa_V^F(A) = 2\sum_{i=1}^n ||u_i|| \cdot ||v_i||$, and the infimum in (4) is reached for the matrix V obtained from W by multiplication of each column by $\sqrt{||u_i||/||v_i||}$.*

The following lemma states that if a matrix A' is "close" to a matrix A with distinct eigenvalues that is "well-conditioned", then A' is also "well-conditioned" with respect to the Frobenius condition number.

Lemma 2. *Let $A, A' \in M_n(\mathbb{C})$ be such that A has n distinct eigenvalues and $||A - A'|| \leq \delta$ where $\delta < \frac{1}{8\kappa_{eig}(A)}$. Then $\kappa_V^F(A') \leq 6n\kappa_V(A) \leq 3n\kappa_V^F(A)$.*

Lemma 3. *Let $A \in M_n(\mathbb{C})$ be a diagonalisable matrix with distinct eigenvalues and let $A = VDV^{-1}$ such that for all $i \in [n]$, for each column v_i of V, $\Big| ||v_i|| - 1 \Big| \leq \delta$. Then $\kappa_F(V) \leq n(1+\delta)^2 + \frac{(\kappa_V^F(A))^2}{4(1-\delta)^2}$.*

The next theorem is the main result of this section. It relies in particular on the three above lemmas. It states that there exists a numerically stable algorithm for computing the δ-forward approximation for the eigenproblem of a diagonalisable matrix A that runs in nearly matrix multiplication time. From an algorithm design point of view, this contains a simple scaling operation on top of the forward approximation algorithm from [1] in order to handle matrices with norm > 1 and the corresponding error analysis can be found in the proof of this theorem in Sect. 5 in [13]. The following is the most important consequence of this theorem: Let $w_1, ..., w_n$ be the forward approximate eigenvectors returned by this algorithm on some diagonalisable matrix A. We show that the matrix with columns w_i has nice additional conditioning guarantees. It is especially useful for small values of δ. In comparison, the bounds on $\kappa(W)$ that can obtained using the corresponding bounds from [1] are the following: If $\kappa_{\text{eig}}(A) \leq K_{\text{eig}}$ and $K_{\text{norm}} > \max\{||A||_F, 1\}$, then $\kappa(W) < \frac{Cn^{3.5}K_{\text{eig}}K_{\text{norm}}}{\delta}$ for some appropriate constant C.

Theorem 3. *Given a diagonalisable matrix $A \in M_n(\mathbb{C})$, a desired accuracy parameter $\delta \in (0, \frac{1}{2})$ and estimates $K_{norm} > \max\{||A||_F, 1\}$ and $K_{eig} > \kappa_{eig}(A)$ as input, there is an algorithm EIG-FWD that outputs vectors $w_1, ..., w_n \in \mathbb{C}^n$ such that the following properties are satisfied with probability at least $1 - \frac{1}{n} - \frac{12}{n^2}$:*

- *If $v_1^{(0)}, ..., v_n^{(0)}$ are the true normalized eigenvectors of A, then we have $||v_i^{(0)} - w_i|| < \delta$ up to multiplication by phases.*
- *Let W be the matrix with columns $w_1, ..., w_n$. Then*

$$\kappa(W) \leq \frac{\kappa_F(W)}{2} \leq \frac{1}{2}(\frac{9n}{4} + 81n^4(\kappa_V^F(A))^2).$$

The algorithm requires $O(T_{MM}(n) \log^2 \frac{nK_{eig}K_{norm}}{\delta})$ arithmetic operations on a floating point machine with $O(\log^4(\frac{nK_{eig}K_{norm}}{\delta}) \log n)$ bits of precision.

4 Probability Analysis of Condition Numbers and Gap

The central theme of this section is to deduce anti-concentration inequalities about certain families of polynomials arising in the analysis of the Algorithm 1 in Sect. 1.6 in finite-precision arithmetic. There are two families of polynomials: (i) quadratic polynomials arising from the analysis of the eigenvalue gap of a matrix and (ii) linear polynomials arising from the analysis of the condition number of a matrix. We focus on (i) in this section, discussing briefly why these polynomials are important and prove the respective anti-concentration inequalities for these polynomials. Compared to [5], an interesting novelty of these inequalities is that the underlying distribution for the random variables is discrete and that they are applicable to polynomials from $\mathbb{R}^n$ to $\mathbb{C}$. We first study some polynomial norms and then prove these results. We define the norm of a complex-valued polynomial following the definition used by Forbes and Shpilka [7] for real-valued polynomials to construct so-called "robust hitting sets".

Definition 5 *(Norm of a complex-valued polynomial) For an n-variate polynomial $f(x) \in \mathbb{C}[\boldsymbol{x}]$, we denote $||f||_2 := (\int_{[-1,1]^n} |f(\boldsymbol{x})|^2 d\mu(x))^{\frac{1}{2}}$ where $\mu(x)$ is the uniform probability measure on $[-1,1]^n$. We also denote $||f||_\infty = \max_{v\in[-1,1]^n} |f(v)|$.*

The following theorem states that if the l_2 norm of a polynomial is not too small, then on inputs picked uniformly and independently at random from $[-1,1)^n$, the value of the polynomial is not too small with high probability. We follow the presentation of this theorem from [7] and extend it to the case of complex-valued polynomials.

Theorem 4 (Carbery-Wright for complex-valued polynomials). *There exists an absolute constant C_{CW} such that if $f : \mathbb{R}^n \to \mathbb{C}$ is a polynomial of degree at most d, then for $\alpha > 0$, it holds that*

$$Pr_{v \in_U [-1,1)^n}[|f(v)| \geq \alpha] \geq 1 - 2C_{CW}d\Big(\frac{\alpha}{||f||_2}\Big)^{\frac{1}{d}}.$$

Discussion for the choice of the polynomial associated with gap: Let $T \in \mathbb{C}^n \otimes \mathbb{C}^n \otimes \mathbb{C}^n$ be a diagonalisable tensor with bounded *condition number* i.e. $T = (U \otimes U \otimes U).(\sum_{i=1}^{n} e_i^{\otimes 3})$ where e_is are the standard basis vectors for $\mathbb{C}^n$ and $U \in M_n(\mathbb{C})$ is an invertible matrix such that $\kappa(T) = \kappa_F(U) \leq B$. Let $a, b \in \mathbb{C}^n$ and let $T^{(a)} = \sum_{i=1}^{n} a_i T_i$ and $T^{(b)} = \sum_{i=1}^{n} b_i T_i$ where $T_1, ..., T_n$ are the slices of T (refer to Definition 1 for a definition of the condition number of a tensor and Sect. 1.6 for a definition of the slices). Notice from Algorithm 1, at step (ii) of the algorithm we want to diagonalise the matrix $(T^{(a)})^{-1}T^{(b)}$. Our algorithm for complete tensor decomposition in finite-precision arithmetic requires a numerically stable *forward*-approximation algorithm for the eigenproblem on input $D := (T^{(a)})^{-1}T^{(b)}$ (refer to Sect. 3 for a definition). From Theorem 3 in the same section, this would require computing an upper bound on the *eigenvector condition number* of D. If we can show that the eigenvalue gap of D is greater than some parameter K, then using the fact that $\kappa_V(D) \leq \frac{\kappa_F(U)}{2} \leq \frac{B}{2}$, we can show that $\kappa_{\text{eig}}(D) \leq \frac{B}{2k}$.

If $u_1, ..., u_n$ are the rows of U, then the eigenvalues of $(T^{(a)})^{-1}T^{(b)}$ are $\frac{\langle b, u_k \rangle}{\langle a, u_k \rangle}$ for all $k \in [n]$. Following the definition of eigenvalue gap in Sect. 3,

$$gap((T^{(a)})^{-1}T^{(b)}) = \min_{k \neq l \in [n]} \left| \frac{\langle b, u_k \rangle}{\langle a, u_k \rangle} - \frac{\langle b, u_l \rangle}{\langle a, u_l \rangle} \right| = \min_{k \neq l \in [n]} \left| \frac{\langle b, u_k \rangle \langle a, u_l \rangle - \langle b, u_l \rangle \langle a, u_k \rangle}{\langle a, u_k \rangle \langle a, u_l \rangle} \right|.$$

Since by Cauchy-Schwarz inequality, we can already show that $|\langle a, u_k \rangle \langle a, u_l \rangle| \leq \frac{nB}{2}$ for all $k \neq l \in [n]$ when $a \in [-1, 1]^n$, we just need to show that the numerator $|\langle b, u_k \rangle \langle a, u_l \rangle - \langle b, u_l \rangle \langle a, u_k \rangle|$ is bounded below. So we choose the polynomials $P^{kl}(\mathbf{x}, \mathbf{y}) = \sum_{i,j \in [n]} p_{ij}^{kl} x_i y_j$ to be the quadratic polynomial defined for all $k, l \in [n]$ by its coefficients $p_{ij}^{kl} = u_{ik}u_{jl} - u_{il}u_{jk}$. Notice that $|P^{kl}(a, b)| = |\langle b, u_k \rangle \langle a, u_l \rangle - \langle b, u_l \rangle \langle a, u_k \rangle|$.

The goal of this section is to show that for most choices of $a, b \in [-1, 1]^n$, $|P^{kl}(a, b)| > K$ for some parameter K with high probability. Firstly, we show that such a result is true when a, b are picked uniformly and independently at random from $[-1, 1]^n$. Then we will show that it is true even when a, b are picked uniformly at random from a discrete grid $G_\eta \subset [-1, 1]^n$.

Applying the Carbery-Wright Theorem to P^{kl}: First we give a lower bound for the l_2 norm of the polynomial.

Lemma 4. *Let $U = (u_{ij}) \in GL_n(\mathbb{C})$ be such that $\kappa_F(U) \leq B$. Then, for all $k, l \in [n]$, $\sum_{i,j \in [n]} |u_{ik}u_{jl} - u_{il}u_{jk}|^2 \geq \frac{2}{B^2}$.*

Proof. We construct a submatrix $U_2 \in M_{n,2}(\mathbb{C})$ with the k-th and l-th columns of U. Let $k = 1$ and $l = 2$ without loss of generality. Since $\kappa_F(U) \leq B$, it follows that for all $y \in \mathbb{C}^n$, $||Uy|| \geq \epsilon||y||$ where $\epsilon = \frac{1}{\sqrt{B}}$. Then for all $y \in \mathbb{C}^2$, we have $||U_2y|| \geq \epsilon||y||$. This implies that $||U_2y||^2 \geq \epsilon^2||y||^2$ and consequently, $y^*U_2^*U_2y \geq \epsilon^2 y^*y$. The minimum singular value $\sigma_{\min}$ of U_2 is defined as $\sigma_{\min}^2 = \min_{y\in\mathbb{C}^n, y\neq 0} \frac{y^*U_2^*U_2y}{y^*y}$. Therefore, $\sigma_{\min}^2(U_2) \geq \epsilon^2$. Since $U_2^*U_2$ is a Hermitian matrix, $\sigma_{\min}^2(U_2) = \lambda_{\min}(U_2^*U_2)$ where $\lambda_{\min}^2$ refers to the smallest eigenvalue. This gives us that $\lambda_{\min}(U_2^*U_2) \geq \epsilon^2$. Let $a = (a_1, ..., a_n)$ and $b = (b_1, ..., b_n)$ be the columns of U_2. Then $U_2^*U_2 = \begin{pmatrix} ||a||^2 & a^*b \\ b^*a & ||b||^2 \end{pmatrix}$. Also, $\det(U_2^*U_2) \geq \lambda_{\min}^2(U_2^*U_2)$, i.e.,$||a||^2||b||^2 - |a^*b|^2 \geq \lambda_{\min}^2(U_2^*U_2) \geq \epsilon^4$. Now from the complex form of Lagrange's identity, we know that $||a||^2||b||^2 - |a^*b|^2 = \frac{1}{2}\sum_{i,j=1}^n |a_ib_j - a_jb_i|^2$. As a result, $\sum_{i,j=1}^n |a_ib_j - a_jb_i|^2 \geq 2\epsilon^4$. Choosing $\epsilon = \frac{1}{\sqrt{B}}$, we finally conclude that for all $k, l \in [n]$, $\sum_{i,j\in[n]} |u_{ik}u_{jl} - u_{il}u_{jk}|^2 \geq \frac{2}{B^2}$.

Theorem 5. *Let $U = (u_{ij}) \in GL_n(\mathbb{C})$ be such that $\kappa_F(U) \leq B$. Let $P^{kl}(\boldsymbol{x}, \boldsymbol{y}) = \sum_{i,j\in[n]} p_{ij}x_iy_j$ where $p_{ij} = u_{ik}u_{jl} - u_{il}u_{jk}$. Then*

$$Pr_{\boldsymbol{v}\in_U[-1,1)^n}[|f(\boldsymbol{v})| \geq \frac{\sqrt{2}\alpha}{3B}] \geq 1 - 4C_{CW}\alpha^{\frac{1}{2}}.$$

Proof. Applying Theorem 4 to P^{kl} with $d = 2$ shows that

$$\Pr_{\mathbf{v}\in_U[-1,1)^n}[|f(\mathbf{v})| \geq \alpha||P^{kl}||_2] \geq 1 - 4C_{CW}\alpha^{\frac{1}{2}}. \tag{6}$$

Now we claim that $||P^{kl}||_2 \geq \frac{\sqrt{2}}{3B}$. Recall that $||P^{kl}||_2^2 = \int_{[-1,1]^{2n}} |P^{kl}(\mathbf{x}, \mathbf{y})|^2 d\mu(\mathbf{x}, \mathbf{y})$ where $\mu(\mathbf{x}, \mathbf{y})$ is the uniform probability distribution on $[-1, 1]^{2n}$. Let us define $p_{ij}^{(r)}$ and $p_{ij}^{(i)}$ as the real and imaginary parts respectively of p_{ij}. We can estimate $|P^{kl}||_2^2$ as follows:

$$\int_{[-1,1]^{2n}} |\sum_{i,j\in[n]} p_{ij}x_iy_j|^2 d\mu(\mathbf{x}, \mathbf{y}) = \int_{[-1,1]^{2n}} |(\sum_{i,j\in[n]} p_{ij}^{(r)}x_iy_j) + \iota(\sum_{i,j\in[n]} p_{ij}^{(i)}x_iy_j)|^2 d\mu(\mathbf{x}, \mathbf{y})$$

$$= (\sum_{i,j,k,l\in[n]} p_{ij}^{(r)}p_{kl}^{(r)}(\int_{[-1,1]^{2n}} x_iy_jx_ky_l d\mu(\mathbf{xy}))) + (\sum_{i,j,k,l\in[n]} p_{ij}^{(i)}p_{kl}^{(i)}(\int_{[-1,1]^{2n}} x_iy_jx_ky_l d\mu(\mathbf{xy}))$$

$$= \sum_{i,j,k,l\in[n]} (p_{ij}^{(r)}p_{kl}^{(r)} + p_{ij}^{(i)}p_{kl}^{(i)})(\int_{[-1,1]^n} x_ix_k d\mu(\mathbf{x}))(\int_{[-1,1]^n} y_jy_l d\mu(\mathbf{y})).$$

When $i \neq k$, $\int_{[-1,1]^n} x_ix_k d\mu(\mathbf{x}) = (\int_{[-1,1]} x_i d\mu(x_i))^2 = 0$. Similarly $\int_{[-1,1]^n} y_jy_l d\mu(\mathbf{y}) = 0$ for $j \neq l$. This gives us that

$$||P^{kl}||_2^2 = \sum_{i,j\in[n]} \left((p_{ij}^{(r)})^2 + (p_{ij}^{(i)})^2\right)(\int_{[-1,1]^n} x_i^2 d\mu(\mathbf{x}))(\int_{[-1,1]^n} y_j^2 d\mu(\mathbf{y})).$$

Since $\int_{[-1,1]^n} x_i^2 d\mu(\mathbf{x}) = \frac{1}{2}\int_{-1}^1 x_i^2 dx_i = \int_{[-1,1]^n} y_j^2 d\mu(\mathbf{y}) = \frac{1}{2}\int_{-1}^1 y_j^2 dy_j = \frac{1}{3}$, we get that $||P^{kl}||_2^2 = \frac{1}{9}\sum_{i,j\in[n]} |p_{ij}|^2$. Now, from Lemma 4, it follows that $||P^{kl}||_2^2 \geq \frac{2}{9B^2}$. Using this in (6), we can conclude the desired result.

Extending the result when the randomness is obtained uniformly from a discrete grid: Our next goal is to show a similar probabilistic result for both families of polynomials (linear and quadratic), but replacing the previous continuous distribution over $[-1,1)^n$ by a distribution where the inputs are chosen uniformly and independently at random from a discrete grid. To formalise this distribution, we describe another equivalent random process of picking an element at random from $[-1,1)^n$ and rounding it to the nearest point on the grid. We use the presentation from [7].

Definition 6 (Rounding function). *Given $\eta \in (0,1)$ such that $\frac{1}{\eta}$ is an integer, for any point $(\boldsymbol{a},\boldsymbol{b}) \in [-1,1]^{2n}$, we define $g_\eta(\boldsymbol{a},\boldsymbol{b})$ to be the point $(\boldsymbol{a}',\boldsymbol{b}')$ such that the i-th element, $(\boldsymbol{a}',\boldsymbol{b}')_i = m_i\eta$, where $m_i\eta \le (\boldsymbol{a},\boldsymbol{b})_i < (m_i+1)\eta$.*

We also define $G_\eta = \{-1, -1+\eta, -1+2\eta, ..., 1-2\eta, 1-\eta\}^{2n}$. Note here that for any point $(a,b) \in [-1,1)^{2n}$, $g_\eta(a,b) \in G_\eta$. Also, note that the process of picking (a,b) uniformly and independently at random from $[-1,1)^{2n}$ and then using the rounding function g_η on (a,b) is equivalent to the process of picking an element uniformly and independently at random from G_η.

Theorem 6 (Multivariate Markov's Theorem). *Let $f : \mathbb{R}^n \to \mathbb{R}$ be a homogeneous polynomial of degree r, that for every $\boldsymbol{v} \in [-1,1]^n$ satisfies $|f(\boldsymbol{v})| \le 1$. Then, for every $||\boldsymbol{v}|| \le 1$, it holds that $||\nabla(f)(\boldsymbol{v})|| \le 2r^2$.*

Theorem 7. *Let $f : \mathbb{R}^{2n} \to \mathbb{C}$ be a homogeneous polynomial of degree at most d. Let $\eta > 0$ be such that $\frac{1}{\eta}$ is an integer. Let $\boldsymbol{a},\boldsymbol{b} \in [-1,1)^{2n}$ and $(\boldsymbol{a}',\boldsymbol{b}') = g_\eta(\boldsymbol{a},\boldsymbol{b})$. Then $|f(\boldsymbol{a},\boldsymbol{b}) - f(\boldsymbol{a}',\boldsymbol{b}')| \le 4\eta\sqrt{n}||f||_\infty d^2$.*

Proof. We write $f = \Re(f) + \iota\Im(f)$ where $\Re(f), \Im(f) : \mathbb{R}^n \to \mathbb{R}$. By the mean value theorem, there exists a point $(\mathbf{a}_0,\mathbf{b}_0)$ on the line segment connecting $(\mathbf{a},\mathbf{b})$ and $(\mathbf{a}',\mathbf{b}')$, such that $|\Re(f)(\mathbf{a},\mathbf{b}) - \Re(f)(\mathbf{a}',\mathbf{b}')| = ||(\mathbf{a},\mathbf{b}) - (\mathbf{a}',\mathbf{b}')|| \cdot |(\Re(f))'(\mathbf{a}_0,\mathbf{b}_0)|$ where $(\Re(f))'(\mathbf{a}_0,\mathbf{b}_0)$ is the derivative of $\Re(f)$ in the direction $(\mathbf{a},\mathbf{b}) - (\mathbf{a}',\mathbf{b}')$ evaluated at $\mathbf{a}_0,\mathbf{b}_0$. From Theorem 6, it follows that $|(\Re(f))'(\mathbf{a}_0,\mathbf{b}_0)| \le 2||\Re(f)||_\infty d^2$. Similarly, we also get that $|(\Im(f))'(\mathbf{a}_0,\mathbf{b}_0)| \le 2||\Im(f)||_\infty d^2$. This finally gives us that

$$\begin{aligned}
|f(\mathbf{a},\mathbf{b}) - f(\mathbf{a}',\mathbf{b}')| &= |\left(\Re(f)(\mathbf{a},\mathbf{b}) - \Re(f)(\mathbf{a}',\mathbf{b}')\right) + \iota\left(\Im(f)(\mathbf{a},\mathbf{b}) - \Im(f)(\mathbf{a}',\mathbf{b}')\right)| \\
&= \sqrt{\left(\Re(f)(\mathbf{a},\mathbf{b}) - \Re(f)(\mathbf{a}',\mathbf{b}')\right)^2 + \left(\Im(f)(\mathbf{a},\mathbf{b}) - \Im(f)(\mathbf{a}',\mathbf{b}')\right)^2} \\
&\le ||(\mathbf{a},\mathbf{b}) - (\mathbf{a}',\mathbf{b}')|| \cdot \sqrt{4||\Re(f)||_\infty^2 d^4 + 4||\Im(f)||_\infty^2 d^4} \le 4\eta\sqrt{n}||f||_\infty d^2.
\end{aligned}$$

The last inequality follows from the fact that $||\Re(f)||_\infty, ||\Im(f)||_\infty \le ||f||_\infty$.

Theorem 8. *Let $U = (u_{ij}) \in GL_n(\mathbb{C})$ be such that $\kappa_F(U) \le B$. Let $P^{kl}(\boldsymbol{x},\boldsymbol{y}) = \sum_{i,j\in[n]} p_{ij}x_iy_j$ where $p_{ij} = u_{ik}u_{jl} - u_{il}u_{jk}$. Let C_{CW} be the absolute constant guaranteed by Theorem 4. Then*

$$Pr_{(\boldsymbol{a},\boldsymbol{b})\in_U G_\eta}[|P^{kl}(\boldsymbol{a},\boldsymbol{b})| \ge \frac{\sqrt{2}\alpha}{3B} - 16\eta n^{\frac{3}{2}} B] \ge 1 - 4C_{CW}\alpha^{\frac{1}{2}}.$$

Corollary 1. *Let* $U = (u_{ij}) \in GL_n(\mathbb{C})$ *be such that* $\kappa_F(U) \leq B$. *Let* $P^{kl}(\boldsymbol{x}, \boldsymbol{y}) = \sum_{i,j \in [n]} p_{ij} x_i y_j$ *where* $p_{ij} = u_{ik}u_{jl} - u_{il}u_{jk}$. *Let* C_{CW} *be the absolute constant guaranteed by Theorem 4. Then*

$$Pr_{(\boldsymbol{a},\boldsymbol{b}) \in_U G_\eta}[|P^{kl}(\boldsymbol{a}, \boldsymbol{b})| \geq k] \geq 1 - 4C_{CW}\Big(\frac{3B(k + 16\eta B n^{\frac{3}{2}})}{\sqrt{2}}\Big)^{\frac{1}{2}}).$$

References

1. Banks, J., Garza-Vargas, J., Kulkarni, A., Srivastava, N.: Pseudospectral shattering, the sign function, and diagonalization in nearly matrix multiplication time. In: IEEE 61st Annual Symposium on Foundations of Computer Science (FOCS) (2020). https://doi.org/10.1109/FOCS46700.2020.00056
2. Beltrán, C., Breiding, P., Vannieuwenhoven, N.: Pencil-based algorithms for tensor rank decomposition are not stable. SIAM J. Matrix Anal. Appl. **40**(2), 739–773 (2019). https://doi.org/10.1137/18M1200531
3. Bernardi, A., Gimigliano, A., Idà, M.: Computing symmetric rank for symmetric tensors. J. Symb. Comput. **46**(1), 34–53 (2011). https://doi.org/10.1016/j.jsc.2010.08.001
4. Bhargava, V., Saraf, S., Volkovich, I.: Reconstruction algorithms for low-rank tensors and depth-3 multilinear circuits. In: Proceedings of the 53rd Annual ACM SIGACT Symposium on Theory of Computing, pp. 809–822 (2021). https://doi.org/10.1145/3406325.3451096
5. Bhaskara, A., Charikar, M., Moitra, A., Vijayaraghavan, A.: Smoothed analysis of tensor decompositions. In: Proceedings of the Forty-Sixth Annual ACM Symposium on Theory of Computing (STOC), pp. 594–603 (2014). https://doi.org/10.1145/2591796.2591881
6. Brachat, J., Comon, P., Mourrain, B., Tsigaridas, E.P.: Symmetric tensor decomposition. In: 17th European Signal Processing Conference (EUSIPCO). IEEE (2009). https://ieeexplore.ieee.org/document/7077748/
7. Forbes, M.A., Shpilka, A.: A PSPACE construction of a hitting set for the closure of small algebraic circuits. In: Proceedings of the 50th Annual ACM SIGACT Symposium on Theory of Computing (STOC), pp. 1180–1192 (2018). https://doi.org/10.1145/3188745.3188792
8. Goyal, N., Vempala, S., Xiao, Y.: Fourier PCA and robust tensor decomposition. In: Proceedings of the Forty-Sixth Annual ACM Symposium on Theory of Computing (STOC), pp. 584–593 (2014). https://doi.org/10.1145/2591796.2591875
9. Harshman, R.: Foundations of the PARAFAC procedure: models and conditions for an "explanatory" multi-mode factor analysis. UCLA Working Papers in Phonetics (1970). https://www.psychology.uwo.ca/faculty/harshman/wpppfac0.pdf
10. Håstad, J.: Tensor rank is NP-complete. In: Ausiello, G., Dezani-Ciancaglini, M., Della Rocca, S.R. (eds.) ICALP 1989. LNCS, vol. 372, pp. 451–460. Springer, Heidelberg (1989). https://doi.org/10.1007/BFb0035776
11. Higham, N.J.: Accuracy and Stability of Numerical Algorithms. Society for Industrial and Applied Mathematics, 2nd edn. (2002). https://doi.org/10.1137/1.9780898718027
12. Kayal, N.: Efficient algorithms for some special cases of the polynomial equivalence problem. In: Proceedings of the Twenty-Second Annual ACM-SIAM Symposium on Discrete Algorithms (SODA), pp. 1409–1421 (2011). https://doi.org/10.1137/1.9781611973082.108

13. Koiran, P., Saha, S.: Complete decomposition of symmetric tensors in linear time and polylogarithmic precision (2022). https://doi.org/10.48550/ARXIV.2211.07407
14. Kolda, T.G., Bader, B.W.: Tensor decompositions and applications. SIAM Rev. **51**(3), 455–500 (2009). https://doi.org/10.1137/07070111X
15. Kruskal, J.B.: Three-way arrays: rank and uniqueness of trilinear decompositions, with application to arithmetic complexity and statistics. Linear Algebra Appl. **18**(2), 95–138 (1977). https://doi.org/10.1016/0024-3795(77)90069-6
16. Leurgans, S.E., Ross, R.T., Abel, R.B.: A decomposition for three-way arrays. SIAM J. Matrix Anal. Appl. **14**(4), 1064–1083 (1993). https://doi.org/10.1137/0614071
17. Moitra, A.: Algorithmic Aspects of Machine Learning. Cambridge University Press, Cambridge (2018) https://books.google.fr/books?id=ruVqDwAAQBAJ
18. Peleg, S., Shpilka, A., Volk, B.L.: Tensor reconstruction beyond constant rank (2022). https://doi.org/10.48550/ARXIV.2209.04177
19. Shitov, Y.: How hard is the tensor rank? (2016). https://doi.org/10.48550/ARXIV.1611.01559
20. Smale, S.: Complexity theory and numerical analysis. Acta Numer **6**, 523–551 (1997). https://doi.org/10.1017/S0962492900002774

Improved Deterministic Leader Election in Diameter-Two Networks

Manish Kumar[(✉)], Anisur Rahaman Molla, and Sumathi Sivasubramaniam

Indian Statistical Institute, Kolkata, India
manishsky27@gmail.com, anisurpm@gmail.com, sumathivel89@gmail.com

Abstract. In this paper, we investigate the leader election problem in diameter-two networks. Recently, Chatterjee et al. [DC 2020] studied the leader election in diameter-two networks. They presented a $O(\log n)$-round deterministic implicit leader election algorithm which incurs optimal $O(n \log n)$ messages, but a drawback of their algorithm is that it requires knowledge of n. An important question– whether it is possible to remove the assumption on the knowledge of n was left open in their paper. Another interesting open question raised in their paper is whether *explicit* leader election can be solved in $\tilde{O}(n)$ messages deterministically. In this paper, we give an affirmative answer to them. Further, we solve the *broadcast problem*, another fundamental problem in distributed computing, deterministically in diameter-two networks with $\tilde{O}(n)$ messages and $\tilde{O}(1)$ rounds without the knowledge of n. In fact, we address all the open questions raised by Chatterjee et al. for the deterministic leader election problem in diameter-two networks. In particular, our results are:

1. We present a deterministic *explicit* leader election algorithm which takes $O(\log \Delta)$ rounds and $O(n \log \Delta)$ messages, where n in the number of nodes and Δ is the maximum degree of the network. The algorithm works without the knowledge of n. The message bound is tight due to the matching lower bound, showed Chatterjee et al. [DC 2020].
2. We show that *broadcast* can be solved deterministically in $O(\log \Delta)$ rounds using $O(n \log \Delta)$ messages. More precisely, a broadcast tree can be computed with the same complexities and the depth of the tree is $O(\log \Delta)$. This also doesn't require the knowledge of n.

To the best of our knowledge, this is the first $\tilde{O}(n)$ deterministic result for the explicit leader election in the diameter-two networks, that too without the knowledge of n.

Keywords: Distributed Algorithm · Leader Election · Message Complexity · Diameter-two graphs

1 Introduction

In the four decades since its inception, leader election has remained a well explored and fundamental problem in distributed networks [20,21,23]. The basic premise of leader

The work of A.R. Molla and S. Sivasubramaniam were supported, in part, by ISI DCSW/TAC Project (file no. G5446 and G5719).

M. Mavronicolas (Ed.): CIAC 2023, LNCS 13898, pp. 323–335, 2023.
https://doi.org/10.1007/978-3-031-30448-4_23

election is simple: given a group of n nodes, a unique node is elected as a leader (where n denotes the number of nodes in the network). Depending on the nodes knowledge of the leader, there are two popular versions. In the first version (known as the *implicit* leader election), the non-leader nodes are not required to know the leader's identity; it is enough for them to know that they are not the leader. The *implicit* leader election is quite well studied in literature [2,16–18,22]. In the other version (known as *explicit* leader election), the non-leader nodes are required to learn the leader's identity. The implicit version of the leader election is the generalized version of the (explicit) leader election. Clearly, there is a lower bound of $\Omega(n)$ for message complexity in the explicit version of the problem. In this paper, we study the explicit version of the problem. In particular, we show an improvement on the existing deterministic solution for the implicit leader election algorithm presented in [4] and provide an algorithm for turning the implicit leader election explicit without any additional overhead on messages.

Leader election has been studied extensively with respect to both message and round complexity in various graph structures like rings [21,30], complete graphs [1,3,7,11,13,14,28], diameter-two networks [4] etc., as well as in general graphs [5,6,18,22,24][1]. Earlier works were primarily focused on providing deterministic solutions. However, eventually, randomized algorithms were explored to reduce mainly the message complexity (see [3,6,17,18] and the references there in). Kutten et al. gave the fundamental lower bound for leader election in general graphs with $\Omega(m)$ message complexity and $\Omega(D)$ round complexity [17], where m is the number of edges and D is the diameter of the graph. This bound is applicable for all graphs with diameters greater than two, whether the algorithm is deterministic or randomized. For the clique, recently a tight message lower bound of $\Omega(n \log n)$ is established by Kutten et al. [19] for the deterministic algorithms under simultaneous wake-up of the nodes. The same lower bound was shown earlier by Afek and Gafni (1991) [1], but assumes adversarial wake-up. Table 1 presents an overview of the results (deterministic). Recently, diameter-two networks were explored, and the message complexity was settled by providing a deterministic algorithm with $O(n \log n)$ message complexity [4].

Our work is closely related to the work by Chatterjee et al. [4]. In their work, the authors studied leader election (the *implicit* version) in diameter-two networks. They presented a deterministic algorithm with $O(n \log n)$ message complexity and $O(\log n)$ round complexity. Crucially, their algorithm requires prior knowledge on the size of the network, n. In comparison to this, our algorithm elects a leader *explicitly* without prior knowledge of n. Our algorithm uses $O(n \log \Delta)$ messages and finishes in $O(\log \Delta)$ rounds, where Δ is the maximum degree of the graph (see, Table 2). In addition to this, we show how to leverage the edges used during the leader election protocol to create a broadcast tree for the diameter-two graphs with a message and round complexity of $O(n \log \Delta)$ and $O(\log \Delta)$, respectively. Computing a broadcast tree efficiently is another fundamental problem in distributed computing. A broadcast tree can be used as a subroutine to many distributed algorithms which look for message efficiency. Finding a deterministic $\tilde{O}(n)$-message and $\tilde{O}(1)$-round broadcast algorithm in diameter-two networks was also left open in [4]. We have addressed it.

[1] We interchangeably use the word "graph" and "network" throughout the paper.

Table 1. Best known deterministic leader election results on networks with different diameters. Since $\Delta = \Omega(\sqrt{n})$ in diameter-two graphs, $\log \Delta = O(\log n)$, see the Remark 1 below. So our upper bound doesn't violate the message lower bound in [4]. * Attaining $O(1)$ time requires $\Omega(n^{1+\Omega(1)})$ messages in cliques, whereas achieving $O(n \log n)$ messages requires $\Omega(\log n)$ rounds; see [1]. ** $\Omega(1)$ is a trivial lower bound.

DETERMINISTIC (EXPLICIT) LEADER ELECTION RESULTS

Paper	Message Complexity	Round Complexity	Graph of Diameter
Afek-Gafni [1]	$O(n \log n)$	$O(\log n)$ *	$D = 1$
Kutten et al. [19]	$\Omega(n \log n)$	$\Omega(1)$	$D = 1$
This paper	$O(n \log \Delta)$	$O(\log \Delta)$	$D = 2$
Chatterjee et al. [4]	$\Omega(n \log n)$	$\Omega(1)$ **	$D = 2$
Kutten et al. [17]	$O(m \log n)$	$O(D \log n)$	$D \geq 3$
Kutten et al. [17]	$\Omega(m)$	$\Omega(D)$	$D \geq 3$

Paper Organization: In the rest of this Sect. 1, we state our results. In Sect. 2, we present our model and definitions. We briefly introduce various related works in Sect. 3. We present our algorithms for deterministic leader election and broadcast tree formation in Sect. 4. And finally, we conclude in Sect. 5 with several open problems.

1.1 Our Results

Our work focuses on the deterministic leader election in diameter-two networks without the knowledge of number of nodes. Apart from this, by leveraging the leader election protocol, we show that *broadcast* can be solved deterministically, matching the complexity of the leader election algorithm. Specifically, we have the following results.

1. We present a deterministic *explicit* leader election algorithm which takes $O(\log \Delta)$ rounds and $O(n \log \Delta)$ messages, where n in the number of nodes and Δ is the maximum degree of the network. The algorithm works without the knowledge of n. The message bound is tight due to the matching lower bound, showed by Chatterjee et al. in [4].
2. We show that *broadcast* can be solved deterministically in $O(\log \Delta)$ rounds using $O(n \log \Delta)$ messages. More precisely, we show that a broadcast tree, of depth at most $O(\log \Delta)$ can be computed with the same complexities.

2 Model and Definition

Our model is similar to the one in [4]. We consider the distributed network to be an undirected graph $G = (V, E)$ of n nodes and diameter $D = 2$. Each node has a unique ID of size $O(\log n)$ bits. The model is a *clean network model* in the sense that the nodes are unaware of their neighbors' IDs initially, also known as KT_0 model [25]. The network is synchronous. The nodes communicate via passing messages in a synchronous round. We limit each message to be of size at most $O(\log n)$ bits as in the CONGEST communication model in distributed networks [25]. In each round, nodes may send messages, receive messages and perform some local computation. The round complexity of an algorithm is the total number of rounds of communication taken by the algorithm before termination. The message complexity is the total number of messages exchanged in the network throughout the execution of the algorithm. Throughout this paper, we assume that all nodes are awake initially and simultaneously start executing the algorithm.

We will now formally define the implicit and explicit version of leader election in our model.

Definition 1 (Implicit Leader Election). *Consider an n-node distributed network. Let each node maintain a state variable that can be set to a value in $\{\perp, NONELECTED, ELECTED\}$, where $\perp$ denotes the 'undecided' state. Initially, all nodes set their state to $\perp$. In the implicit version of leader election, it requires that exactly one node has its state variable set to $ELECTED$ and all other nodes are in state $NONELECTED$. The unique node whose state is $ELECTED$ is the leader.*

Definition 2 (Explicit Leader Election). *Consider an n-node distributed network. Let each node maintain a state variable that can be set to a value in $\{\perp, NONELECTED, ELECTED\}$, where $\perp$ denotes the 'undecided' state. Initially, all nodes set their state to $\perp$. In the explicit version of leader election, it requires that exactly one node has its state variable set to $ELECTED$ and all other nodes are in state $NONELECTED$. Further, the $NONELECTED$ nodes must know the identity of the node, whose state is $ELECTED$, the leader.*

3 Related Work

In 1977, the leader election problem was introduced by Le Lann in the ring network [21]. Since then the problem has been studied extensively in different settings. The leader election problem has been explored in both implicit and explicit versions over the years [6,9,15,18,22,24] for a variety of models and settings, and for various graph topologies such as cliques, cycles, mesh, etc., (see [8–10,16,24,26,27,29] and the references therein for more details). In general, the implicit leader election suffices for most networks.

Both deterministic and randomized solutions exist for leader election. For the randomized case, for complete graphs, Kutten et al. [18] showed that $\tilde{\Theta}(\sqrt{n})$ is a tight message complexity bound for randomized (implicit) leader election. For any graph with diameter greater than 2, the authors in [18] showed that $\Omega(D)$ is a lower bound for the

Table 2. Comparison of the current paper to the state-of-the-art.

DETERMINISTIC LEADER ELECTION IN DIAMETER-TWO GRAPHS				
Paper	Message Complexity	Round Complexity	Type	Knowledge of n
Chatterjee et al. [4]	$O(n \log n)$	$O(\log n)$	Implicit	YES
This paper	$O(n \log \Delta)$	$O(\log \Delta)$	Explicit	NO

number of rounds for leader election using a randomized algorithm (they also showed a lower bound for the message complexity, $\Omega(m)$). Recently, Chatterjee et al., [4] showed a lower bound of $\Omega(n)$ for the message complexity of randomized leader election in diameter-two graphs.

In the deterministic case, it is known that $\Theta(n \log n)$ is a tight bound on the message complexity for complete graphs [1,19]. This tight bound also carries over to the general case as seen from [1,12,14]. In our work, we restrict our model to graphs of diameter-two. For diameter-two graphs, Chatterjee and colleagues provide a $O(\log n)$ round algorithm that uses $O(n \log n)$ messages. However, their algorithm requires knowledge of n, our algorithm provides an algorithm that requires no prior knowledge of n and runs in $O(\log n)$ rounds with $O(n \log n)$ message complexity.

4 Deterministic Leader Election in Diameter-Two Networks

We present a deterministic (explicit) leader election algorithm for diameter-two networks with n nodes in which the value of n is unknown to nodes in the network. In this section, we answer several questions raised in [4]. Specifically, we address the following: (i) Can explicit leader election be performed in $\tilde{O}(n)$ messages in diameter-two graphs deterministically? (ii) Given the leader election algorithm, can broadcast can be solved deterministically in diameter-two graphs with $\tilde{O}(n)$ message complexity and $O(\text{polylog}\, n)$ rounds if n is known, and crucially (iii) "Removing the assumption of the knowledge of n (or showing that it is not possible) for deterministic, implicit leader election algorithms with $\tilde{O}(n)$ message complexity and running in $\tilde{O}(1)$ rounds is open as well". In this section, we solve the explicit leader election with $\tilde{O}(n)$ message complexity, along with that our algorithm solves the explicit leader election without the knowledge of n; thus addressing the questions (i) and (iii). We further present a solution for the question (ii) that too without the knowledge of n.

4.1 Algorithm

Our algorithm is inspired from the work done by Chatterjee et al. [4]. They presented an algorithm for implicit leader election that ran in $O(\log n)$ rounds with $O(n \log n)$ message complexity (with the knowledge of n). Our Algorithm 1, achieves somewhat better result without the knowledge of n and also elects the leader explicitly.

As mentioned earlier (in Sect. 2), each node has a unique ID. For any node $v \in V$, let's denote the degree of v by d_v and the ID of v by ID_v. The *priority* $\mathcal{P}_v$, of node v, is a combination of the degree and ID of the node v such that $\mathcal{P}_v = \langle d_v, ID_v \rangle$. The leader is elected based on the priority, which is decided by the degree of the node. In the case of a tie, the higher ID gets the higher priority. Essentially, the node with the highest priority becomes the leader.

Our algorithm runs in two phases of $O(\log d_v)$ rounds each. In the first $O(\log d_v)$ rounds, we eliminate as many invalid candidates as possible. In the second phase, all candidates except the actual leader are also eliminated, culminating in the election of a unique leader.

Detailed Description of the Algorithm:

Initially, every node is a "candidate" and has an "active" status. Each node v numbers its neighbors from 1 to d_v arbitrarily, denoted by $w_{v,1}, w_{v,2}, \ldots, w_{v,d_v}$. For the first $i = 1$ to $\log d_v$ rounds, if v is active, then node v sends a message containing $\mathcal{P}_v$ to its neighbors $w_{v,2^{i-1}}, \cdots, w_{v,\min\{d_v,2^i-1\}}$. If v encounters a priority higher than its own from its neighbors (either because a neighbor has a higher priority or has heard of a node with higher priority) then v becomes "inactive" and "non-candidate". That is, v does not send any further messages to its neighbors containing v's priority. Although, v may send higher priority message based on the received message's priority (explained later). Let L_v denotes the ID of the current highest priority node known to v. At the beginning of the execution, L_v is simply $\mathcal{P}_v$. If at the end of the first $\log d_v$ rounds, $L_v = \mathcal{P}_v$ then v declares itself leader temporarily. Further, v waits for $\log d_v$ rounds. If at the end of $\log d_v$ rounds v is still the candidate node (v has not heard from a node about the higher priority) then v becomes the leader.

There are two major phases to the algorithm. For the first $\log d_v$ rounds, we eliminate as many invalid candidates (the node which has encountered higher priority node) as possible, as follows. N_v contains the ID of the neighbor that informed v about the current highest priority. As mentioned before, L_v contains the current highest priority known to v. Let χ_v denote the (possibly empty) set of v's neighbors from whom v has received messages in a round during this phase, and $\mathcal{P}(\chi_v)$ be the set of $\mathcal{P}$s sent to v by the members of χ_v such that $\mathcal{P}_u$ be the highest $\mathcal{P}$ in $\mathcal{P}(\chi_v)$. If $\mathcal{P}_u$ is higher than that of L_v then v stores the highest priority seen so far in L_v. Further, v informs N_v about $L_v = \mathcal{P}_u$, i.e., the highest $\mathcal{P}$ it has seen so far. This particular step exploits the neighborhood intersection property to ensure that information about higher priority nodes is disseminated quickly. Then v updates N_v. Finally, v tells every member of χ_v about L_v, i.e., the highest $\mathcal{P}$ it has seen so far. If $L_v \neq P_v$ then v becomes "inactive" and "non-candidate". Notice that an "inactive" and "non-candidate" node v only disseminates the information of higher priorities it hears, to N_v.

At the end of the first $\log d_v$ rounds, we begin the final phase of the election. If v is still the candidate node then v waits for $\log(d_v)$ rounds. Furthermore, if v does not receive any higher priority message then v declares itself as the leader and informs its neighbors. Then each neighbor of v, say u, informs their neighbors about the election of v via set of Ψ_u nodes. On the other hand, if there exists a node whose priority is higher than the priority of v then v gets to know about the leader and informs all the nodes to whom v has communicated (so far) about the leader's ID (that is the set Ψ_v)

and exits. Hence, All the nodes elect the same leader whose priority is the highest. Our claim is that given certain properties of the degree (see Lemma 1) we can guarantee that the second phase of waiting for $\log \Delta$ rounds eliminates all but a unique candidate, which then becomes leader.

Now, we would discuss some important lemmas and the correctness of the algorithm. Finally, we conclude the result in Theorem 1.

Lemma 1. *Let v be a node whose degree, d_v, is the highest among its neighbors and Δ is the maximum degree of the graph. There does not exist any diameter-two graph with n nodes ($n > 4$) such that $\Delta > d_v^2$.*

Proof. For a node v, all nodes are at most 2 hop distance away from v, since the diameter of the graph is 2. Node v has degree d_v and its neighbors have degree at most d_v, by assumption. This gives an upper bound on n, that is, $n \leq d_v(d_v - 1) + 1$, because each of the d_v neighbors can have at most other $d_v - 1$ neighbors (excluding v) each, and by the distance assumption there are no other nodes in the graph. Also, Δ can be at most $n - 1$. Therefore, $\Delta < n < d_v^2 + 1$. Consequently, $d_v^2 > \Delta$. Hence, the lemma. □

Remark 1. It is clear that there does not exist any diameter-two network whose nodes are neither connected to v nor its neighbor. Therefore, $d_v^2 \geq n$. This implies $d_v \geq \sqrt{n}$. Hence, $\Delta \geq \sqrt{n}$.

Lemma 2. *Algorithm 1 solves the leader election in $O(\log \Delta)$ rounds, where Δ is the maximum degree of the graph.*

Proof. A candidate node v becomes the leader if its priority is the highest among its neighbors (Line 22). From Lemma 1, we know that $\Delta < d_v^2$. Therefore, the node v with degree d_v waits for $\log d_v$ rounds, in that time, the node with degree Δ inform about its priority to v (if any) and v becomes inactive. Otherwise, v consider d_v as Δ and inform all its neighbors about its election. v's neighbor further conveys the message to all other nodes in $\log \Delta$ rounds. Therefore, the round complexity of the algorithm is $O(\log \Delta)$. □

For the message complexity analysis we adapt a couple of results from [4], since our algorithm (Algorithm 1) uses the similar approach to keep a node active. In particular, we use the Lemma 11 and Lemma 12 from [4], which used ID of the nodes to take a decision on the "active" or "inactive" nodes whereas our algorithm uses priority (which depends on degree and ID). Hence, the results also applies to our algorithm. The following two lemmas are adapted from Lemma 11 and Lemma 12 in [4].

Lemma 3 ([4]). *At the end of the round i, there are at most $\frac{n}{2^i}$ "active" nodes.*

Algorithm 1. DETERMINISTIC-LEADER-ELECTION: CODE FOR A NODE v

Input: A two diameter connected anonymous network. Each node possess unique ID.
Output: Leader Election.

1: v becomes a "candidate" and "active".
2: Let $\mathcal{P}_v = \langle d_v, ID_v \rangle$ be the priority of v. Priority is determined by degree, the node with the higher degree (d_v) has higher priority. The node's ID is used to break any ties.
3: $L_v \leftarrow \mathcal{P}_v$ ▷ L_v is the current highest priority known to v.
4: $N_v \leftarrow \mathcal{P}_v$ ▷ N_v is the neighbor which informed about L_v.
5: v creates an arbitrary assignment of its neighbors based on its degree (from 1 to d_v) which are called $w_{v,1}, w_{v,2}, \cdots, w_{v,d_v}$ respectively.
6: **for** rounds $i = 1$ to $\log d_v$ **do**
7: **if** v is active **then**
8: v sends a "probe" message containing its priority $\mathcal{P}$ to its neighbors $w_{v,2^{i-1}}, \cdots, w_{v,\min\{d_v, 2^i-1\}}$.
9: **end if**
10: Let χ_v be the possibly empty subset of $v's$ neighbors from which v received messages in this round.
11: Let $\Psi_v = \bigcup_1^i \chi_v$.
12: Let $\mathcal{P}(\chi_v)$ be the set of $\mathcal{P}$s sent to v by the members of χ_v.
13: Let $\mathcal{P}_u$ be the highest $\mathcal{P}$ in $\mathcal{P}(\chi_v)$.
14: **if** $\mathcal{P}_u > L_v$ **then**
15: $L_v \leftarrow \mathcal{P}_u$
16: v tells N_v about $L_v = \mathcal{P}_u$, i.e., the highest $\mathcal{P}$ it has seen so far.
17: $N_v \leftarrow x$. ▷ v remembers neighbor who told v about L_v.
18: v becomes "inactive" and "non-candidate".
19: **end if**
20: v tells every member of χ_v about L_v, i.e., the highest $\mathcal{P}$ it has seen so far.
21: **end for**
22: **if** $L_v = \mathcal{P}_v$ **then**
23: v waits for $\log(d_v)$ rounds. If at the end of $\log(d_v)$ rounds, $L_v = \mathcal{P}_v$ then v declares itself as leader and inform all the neighbors as well as exit the protocol.
24: **end if**
25: **if** v knows about the leader and v is not the leader **then**
26: Let Φ_v be the set of neighbors of v to whom v sent the messages before knowing about the leader.
27: Let $\Psi_v = \Psi_v \bigcup \Phi_v$.
28: v informs Ψ_v about the leader's ID and exit.
29: **end if**
30: All the nodes elect the same leader whose priority is the highest.

Proof. Consider a node v that is active at the end of round i. This implies that the if-clause of Line 14 of Algorithm 1 has not so far been satisfied for v, which in turn implies that $\mathcal{P}_v > \mathcal{P}_{w_{v,j}}$ for $1 \leq j \leq 2^i - 1$, therefore none of $w_{v,1}, w_{v,2}, \ldots, w_{v,2^i-1}$ is active after round i. Thus, for every active node at the end of round i, there are at least $2^i - 1$ inactive nodes. We call this set of inactive nodes, together with v itself, the "kingdom" of v after round i i.e.,

$$KINGDOM_i(v) \stackrel{\text{def}}{=} \{v\} \cup w_{v,1}, w_{v,2}, \ldots, w_{v,2^i-1} \text{ and } |KINGDOM_i(v)| = 2^i.$$

If we can show that these kingdoms are disjoint for two different active nodes, then we are done.
Proof by contradiction. Suppose not. Suppose there are two active nodes u and v such that

$$u \neq v \text{ and } KINGDOM_i(u) \cap KINGDOM_i(v) = \phi$$

(after some round i, $1 \leq i \leq \log n$). Let x be such that $x \in KINGDOM_i(u) \cap KINGDOM_i(v)$. Since an active node obviously cannot belong to the kingdom of another active node, this x equals neither u nor v, and therefore,

$$x \in \left\{w_{v,1}, w_{v,2}, \ldots, w_{v,2^i-1}\right\} \cap \left\{w_{u,1}, w_{u,2}, \ldots, w_{u,2^i-1}\right\},$$

that is, both u and v have sent their respective probe-messages to x. Then it is straightforward to see that x would not allow u and v to be active at the same time. Case-by-case analysis can be found in [4]. □

Lemma 4 ([4]). *In round i, Algorithm 1 transmits at most* $3n$ *messages in the for loop (from Line 6 to Line 21).*

Proof. In round i, each active node sends exactly $2^i - 1$ probe messages, and each probe-message generates at most two responses (corresponding to Lines16 and 20 of Algorithm 1). Thus, in round i, each active node contributes to, directly or indirectly, at most $3 \cdot (2^i - 1)$ messages. The result immediately follows from Lemma 3. □

Lemma 5. *The message complexity of the Algorithm 1 is* $O(n \log \Delta)$.

Proof. Each round transmits at most $3n$ messages (Lemma 4) and the execution of the Algorithm 1 (from Line 6 to Line 21) takes place in $O(\log \Delta)$ rounds (Lemma 2). Further, leader informs about its election via Ψ edges which are $O(n \log \Delta)$. Therefore, the total number of message transmitted throughout the execution are: $3n \cdot O(\log \Delta) + O(n \log \Delta) = O(n \log \Delta)$.

Correctness of the Algorithm: In this, we show that all the nodes agree on a leader and the leader is unique. First, we show that all the nodes agree on a leader. If a node v is still a candidate node at the end of the first phase, then it must have both i) explored all its neighbors and ii) never encountered a priority higher than its own. Thus, it can declare itself leader after waiting $\log d_v$ rounds. Note that a waiting period of $\log d_v$ is enough because from Lemma 1 we know that $\Delta < d_v^2$. This guarantees that the highest degree is made leader.

Now, we show that the known leader is unique. If not, then suppose there exist two nodes u and v such that u agrees on a leader l_1 and v agree on a leader l_2. From algorithm 1, l_1 should have the highest priority in its neighbors and similarly, l_2 should have the highest priority in its neighbors. Since it is a diameter two graph, therefore, there should be at least one node common among l_1 and l_2. Therefore, both the node can't have the highest priority among their neighbors, which is a contradiction. Therefore, we can say all the nodes agree on the unique leader.

From the above discussion, we conclude the following result.

Theorem 1. *There exists a deterministic (explicit) leader election algorithm for n-node anonymous networks with diameter two that sends $O(n \log \Delta)$ messages and terminates in $O(\log \Delta)$ rounds, where Δ is the maximum degree of the network.*

Remark 2. The implicit deterministic leader election algorithm presented in [4] can be converted to an explicit leader election algorithm in the same way as done in Algorithm 1.

4.2 Broadcast Tree Formation

In Algorithm 1, the nodes agree on the leader explicitly. In this section, we exploit the edges used during the leader election algorithm (Algorithm 1) and create a broadcast tree of height $O(\log \Delta)$ (Algorithm 2). This also allows to reduce the message complexity. The process is simple. The leader, say ℓ, initiates the flooding process by broadcasting its ID to its neighbors, forming the root of the tree T. All of its neighbors become a part of T. At any point in the algorithm, the leaves of T do the following. Let v be a leaf in T in some round. In that round, v sends its own ID to the nodes in Ψ_v (used in Algorithm 1). Non tree nodes which receive an ID v earlier become a part of T with v as its parent. If a non-tree node receives multiple messages, then it chooses the higher ID as its parent. The algorithm ends when all nodes have become a part of T. Note that since only the leaves send out messages in each round and each node (except the root node, i.e., leader node) possess only one parent, we avoid the creation of cycles.

Let us now show some important lemmas which support the correctness of the algorithm. In particular, Lemma 7 shows Algorithm 2 forms a tree of height $O(\log \Delta)$. The round complexity and message complexity of the Algorithm 2 is shown by Lemma 6 and Lemma 8, respectively. Finally, we conclude with message and round complexity as well as height of the tree in Theorem 2.

Lemma 6. *In $O(\log \Delta)$ rounds, all nodes are guaranteed to be part of the tree T.*

Proof. This is guaranteed from the use of leader election algorithm. Consider the graph G' constructed as follows. Let ℓ's neighbors be its neighbors in G. For every other node $v \neq \ell$ its neighbors are ψ_v. Clearly, from Algorithm 1, G' is connected (as every node learns of ℓ) and of diameter $O(\log \Delta)$. Let level i denote all nodes that are at most i hops away from ℓ in G'. We claim that in i rounds, all nodes in level i would become a part of the tree T. By using induction, this is clearly true for is $i = 1$. Assuming it's true for i, nodes of $i + 1$ would become part of the tree next as they are in the Ψ_v of at least one node in level i and thus would get an invite. And since the number of levels can be at most $O(\log \Delta)$, all nodes become part of T in $O(\log \Delta)$ rounds. □

Lemma 7. *Algorithm 2 forms a tree of height $O(\log \Delta)$.*

Algorithm 2. BROADCAST-TREE-FORMATION

Input: A diameter-2 connected network graph G in which each node possess unique ID.
Output: Tree Structure T.

1: First run Algorithm 1 to elect the leader ℓ. Each node also keeps track of its Ψ_v (created during the course of the algorithm).
2: ℓ becomes root of T. ℓ then broadcasts its ID as an invite to all its neighbors. And its neighbors become its children in T.
3: **while** there are nodes outside of T **do** ▷ Takes $O(\log \Delta)$ rounds.
4: Each node $v \in T$ broadcasts its ID to the nodes in Ψ_v.
5: **if** node $u \notin T$ receives IDs from nodes in tree T **then**
6: u accept invitation based on the highest priority node, say v, and becomes v's child in T.
7: **end if**
8: **end while**

Proof. Since in each iteration of the while loop, the height of the tree is extended by at most 1 (that is by attaching children to the leaves of T). And since the algorithm ensures that all nodes have become a part of T in $O(\log \Delta)$ iterations of the while loop, the height of T can not be more than $O(\log \Delta)$. Notice that since each node accepts only one invite, there can be no creation of a cycle. □

Remark 3. The diameter of the graph created by Algorithm 2 is $O(\log \Delta)$.

Lemma 8. *Algorithm 2 takes $O(n \log \Delta)$ messages.*

Proof. In Algorithm 1, for every node v communication takes place via Ψ_v edges in $O(n \log \Delta)$ messages (Theorem 1). In Algorithm 2 (from Line 2 to Line 8) communication also takes place via same edges (Ψ_v) for two times. Therefore, message complexity remain unchanged to $O(n \log \Delta)$. □

Thus, from the above discussion, we conclude the following result.

Theorem 2. *There exists an algorithm which solve the broadcast problem in $O(n \log \Delta)$ messages and $O(\log \Delta)$ rounds which generate a tree of height $O(\log \Delta)$.*

5 Conclusion and Future Work

We studied the leader election problem in diameter-two networks. We settled all the questions raised by Chatterjee et al. [4] w.r.t. deterministic setting. Various open problems come to light due to our work. These are as follows:

1. We presented an $O(\log \Delta)$-round and $O(n \log \Delta)$-message complexity algorithm for the explicit leader election. An interesting question is to reduce the round complexity to $O(1)$ while keeping the message complexity $O(n \log n)$?
2. Tree formed by broadcast has height $O(\log \Delta)$. An interesting question rises whether this is optimal when the message and round complexity remain unchanged or constant height is possible.

3. Is it possible to have a randomized algorithm (with high probability) with message complexity $O(n \log n)$ and constant round complexity without the knowledge of n?
4. With or without the knowledge of n, what would be the complexity and lower bound (in deterministic setting) in the LOCAL model where nodes can communicate with arbitrary message size in a round?

References

1. Afek, Y., Gafni, E.: Time and message bounds for election in synchronous and asynchronous complete networks. SIAM J. Comput. **20**(2), 376–394 (1991)
2. Attiya, H., Welch, J.L.: Distributed Computing - Fundamentals, Simulations, and Advanced Topics. Wiley series on parallel and distributed computing, 2nd edn. Wiley, Hoboken (2004)
3. Augustine, J., Molla, A.R., Pandurangan, G.: Sublinear message bounds for randomized agreement. In: PODC, pp. 315–324. ACM (2018)
4. Chatterjee, S., Pandurangan, G., Robinson, P.: The complexity of leader election in diameter-two networks. Distributed Comput. **33**(2), 189–205 (2020). https://doi.org/10.1007/s00446-019-00354-2
5. Gallager, R.G., Humblet, P.A., Spira, P.M.: A distributed algorithm for minimum-weight spanning trees. ACM Trans. Program. Lang. Syst. **5**(1), 66–77 (1983)
6. Gilbert, S., Robinson, P., Sourav, S.: Leader election in well-connected graphs. In: Proceedings of the 2018 ACM Symposium on Principles of Distributed Computing, PODC 2018, Egham, United Kingdom, 23–27 July 2018
7. Humblet, P.: Selecting a leader in a clique in O(N log N) messages (1984)
8. Kapron, B.M., Kempe, D., King, V., Saia, J., Sanwalani, V.: Fast asynchronous byzantine agreement and leader election with full information. In: SODA, pp. 1038–1047. SIAM (2008)
9. Khan, M., Kuhn, F., Malkhi, D., Pandurangan, G., Talwar, K.: Efficient distributed approximation algorithms via probabilistic tree embeddings. Distrib. Comput. **25**(3), 189–205 (2012)
10. King, V., Saia, J., Sanwalani, V., Vee, E.: Scalable leader election. In: SODA, pp. 990–999. ACM Press (2006)
11. Korach, E., Kutten, S., Moran, S.: A modular technique for the design of efficient distributed leader finding algorithms. ACM Trans. Program. Lang. Syst. **12**(1), 84–101 (1990)
12. Korach, E., Moran, S., Zaks, S.: Tight lower and upper bounds for some distributed algorithms for a complete network of processors. In: Proceedings of the Third Annual ACM Symposium on Principles of Distributed Computing, Vancouver, B.C., Canada, 27–29 August 1984, pp. 199–207. ACM (1984)
13. Korach, E., Moran, S., Zaks, S.: The optimality of distributive constructions of minimum weight and degree restricted spanning trees in a complete network of processors. SIAM J. Comput. **16**(2), 231–236 (1987)
14. Korach, E., Moran, S., Zaks, S.: Optimal lower bounds for some distributed algorithms for a complete network of processors. Theor. Comput. Sci. **64**(1), 125–132 (1989)
15. Kowalski, D.R., Mosteiro, M.A.: Time and communication complexity of leader election in anonymous networks. In: ICDCS 2021, pp. 449–460. IEEE (2021)
16. Kumar, M., Molla, A.R.: Brief announcement: on the message complexity of fault-tolerant computation: leader election and agreement. In: PODC 2021, pp. 259–262. ACM (2021)
17. Kutten, S., Pandurangan, G., Peleg, D., Robinson, P., Trehan, A.: On the complexity of universal leader election. J. ACM **62**(1), 7:1–7:27 (2015)

18. Kutten, S., Pandurangan, G., Peleg, D., Robinson, P., Trehan, A.: Sublinear bounds for randomized leader election. Theor. Comput. Sci. **561**, 134–143 (2015)
19. Kutten, S., Robinson, P., Tan, M.M., Zhu, X.: Improved tradeoffs for leader election (2023). https://doi.org/10.48550/arXiv.2301.08235
20. Lamport, L., Shostak, R.E., Pease, M.C.: The byzantine generals problem. ACM Trans. Program. Lang. Syst. **4**(3), 382–401 (1982)
21. Lann, G.L.: Distributed systems - towards a formal approach. In: Information Processing, Proceedings of the 7th IFIP Congress 1977, pp. 155–160. North-Holland (1977)
22. Lynch, N.A.: Distributed Algorithms. Morgan Kaufmann, Burlington (1996)
23. Pease, M.C., Shostak, R.E., Lamport, L.: Reaching agreement in the presence of faults. J. ACM **27**(2), 228–234 (1980)
24. Peleg, D.: Time-optimal leader election in general networks. J. Parallel Distrib. Comput. **8**(1), 96–99 (1990)
25. Peleg, D.: Distributed Computing: A Locality-Sensitive Approach (2000)
26. Refai, M., Sharieh, A.A.A., Alshammari, F.: Leader election algorithm in 2D torus networks with the presence of one link failure. Int. Arab J. Inf. Technol. **7**(2), 105–114 (2010)
27. Santoro, N.: Design and Analysis of Distributed Algorithms. Wiley series on parallel and distributed computing, Wiley, Hoboken (2007)
28. Singh, G.: Efficient distributed algorithms for leader election in complete networks. In: ICDCS, pp. 472–479. IEEE Computer Society (1991)
29. Tel, G.: Introduction to Distributed Algorithms (1994)
30. Yifrach, A., Mansour, Y.: Fair leader election for rational agents in asynchronous rings and networks. In: Newport, C., Keidar, I. (eds.) Proceedings of the 2018 ACM Symposium on Principles of Distributed Computing, PODC 2018, Egham, United Kingdom, 23–27 July 2018, pp. 217–226. ACM (2018)

Fast Cauchy Sum Algorithms for Polynomial Zeros and Matrix Eigenvalues

Victor Y. Pan[1,2,3(✉)], Soo Go[2], Qi Luan[3], and Liang Zhao[1,2]

[1] Department of Computer Science, Lehman College of CUNY, Bronx, New York, USA
{victor.pan,Liang.Zhao1}@lehman.cuny.edu
[2] Program in Computer Science, The Graduate Center of the City University of New York, New York, USA
sgo@gradcenter.cuny.edu
[3] Program in Mathematics, The Graduate Center of the City University of New York, New York, USA
qi_luan@yahoo.com
http://comet.lehman.cuny.edu/vpan

Abstract. Given a black box oracle that evaluates a univariate polynomial $p(x)$ of a degree d, we seek its zeros, aka the roots of the equation $p(x) = 0$. At FOCS 2016, Louis and Vempala approximated within $1/2^b$ an absolutely largest zero of such a real-rooted polynomial at the cost of the evaluation of Newton's ratio $\frac{p(x)}{p'(x)}$ at $O(b\log(d))$ points x and then extended this algorithm to approximation of an absolutely largest eigenvalue of a symmetric matrix at a record Boolean cost. By applying **distinct approach and techniques** we obtain *much more general results at the same computational cost.* Our use of Cauchy integrals and randomization is non-trivial and pioneering in this field. Somewhat surprisingly, the Boolean complexity of the accelerated versions of our algorithms in [25,26] reached below the known lower bounds on the Boolean complexity of polynomial root-finding.

Keywords: Symbolic-numeric computing · Polynomial roots · Computer algebra

1 Introduction

1.1 The Classical Problem of Univariate Polynomial Root-Finding

Can be stated as follows: given complex coefficients of a polynomial

$$p = p(x) := \sum_{i=0}^{d} p_i x^i := p_d \prod_{j=1}^{d} (x - x_j),\ p_d \neq 0, \quad (1)$$

approximate, within a fixed tolerance ϵ to output errors, all d complex zeros x_j of $p(x)$, aka roots[1] of the equation $p(x) = 0$, or approximate just the roots that

[1] Hereafter we frequently refer to them just as *roots* or *zeros*.

M. Mavronicolas (Ed.): CIAC 2023, LNCS 13898, pp. 336–352, 2023.
https://doi.org/10.1007/978-3-031-30448-4_24

lie in a fixed region (e.g., disc or square) of the complex plane. According to [17], univariate polynomial root-findings had been the central problem of Mathematics and Computational Mathematics for about 4,000 years, since Sumerian times and well into the 19th century. It is still a popular research subject with various applications to scientific computing [15,16]. Its intensive study in the 1980s and 1990s has culminated at STOC 1995 with a solution in nearly optimal Boolean time (see [18,19,21], and the bibliography therein).

1.2 Black Box Polynomial Root-Finding: The Problem

The cited root-finders operate with the coefficients of p and do not apply to the important class of *black box polynomials* – given with an oracle (black box) for their evaluation. One can evaluate p at $d_+ > d$ points and then interpolate to it, albeit at the price of destroying sparsity and blowing up precision and Boolean cost of the computations. These problems are paramount for root-finders operating with coefficients of p but disappear for black box polynomial root-finders.

Moreover, one must use at least $2d$ arithmetic operations *(ops)* to evaluate a general polynomial p of (1) given with its $d+1$ coefficients (see [32, Section "Pan's method"] or [12]), whereas $O(\log(d))$ ops are sufficient to evaluate $p(x)$ for a large and important class of dth degree polynomials, including sparse polynomials, more generally, the sums of shifted monomials such as $p := \alpha(x-a)^d + \beta(x-b)^d+\gamma(x-c)^d$ for six constants a, b, c, α, β, and γ, and the Mandelbrot polynomial $p(x) = p_k(x)$, where $p_0 := 1$, $p_1(x) := x$, $p_{i+1}(x) := xp_i^2(x) + 1$, $i = 0, 1, \ldots, k$, and $d = 2^k - 1$. Interpolation to such polynomials can *dramatically slow down fast polynomial root-finders* in a region that contains a small number of roots.

Instead of the above observations, Louis and Vempala in [3] were motivated by the following one: black box polynomial root-finders, unlike those operating with the coefficients, can be readily extended to the highly important classical problem of approximation of eigenvalues of a matrix as the roots of its characteristic polynomial $p(x)$. Evaluation of its coefficients blows up precision and Boolean cost of the computations, but black box root-finders avoid this hurdle.

1.3 Black Box Polynomial Root-Finding: The State of the Art

In [14] Louis and Vempala proposed novel high order Newton's iterations, which approximate within $1/2^b$ an absolutely largest root of a black box polynomial of a degree d at NR cost of the evaluation of Newton's ratio $\mathrm{NR}(x) := \frac{p(x)}{p'(x)}$ at $O(b\log(d))$ points. The root-finder performs with $O(\log(d)$-bit precision and at overall Boolean time $O(b\log(d)\mu(\log(d))$ provided that it can access the values of $p(x)$ supplied by a black box oracle where

$$\mu(s) = O(s\log(s)\log(\log(s))) \tag{2}$$

bounds the Boolean cost of multiplication of two integers modulo 2^s [12].

By applying their algorithm to the characteristic polynomial $p(x) = \det(xI - M)$ of a $d \times d$ symmetric matrix M and combining it with Storjohann's Las Vegas randomized algorithm of [31] for computing the determinant of an integer matrix, Louis and Vempala approximated, within a fixed tolerance ϵ to output errors, an absolutely largest eigenvalue of M at a record expected Boolean complexity $O(d^\omega \log^5(d) b^2)$. Here ω denotes an exponent of *feasible or unfeasible* matrix multiplication,[2] $b := \log(||M||_F/\epsilon)$, $||\cdot||_F$ denotes the Frobenius matrix norm, and $\mu(s) = O(s \log(s) \log(\log(s)))$ (cf. 2).

Extension of [14] to approximation of all d roots of a black box polynomial p and all d eigenvalues of a $d \times d$ matrix or all $m \le d$ roots and eigenvalues in a fixed region of interest on the complex plane (such as a disc or a square) has remained a natural research challenge with no progress since [14].

1.4 Our Results

Like [14], we focus on estimating NR cost of our polynomial root-finder; it dominates the cost of other ops involved (see Observation 15 and Theorem 18). Our approach and techniques, however, are *novel, have nothing to do with those of* [14], and enable us to solve much more general root-finding and eigenvalue problems within the same NR and Boolean cost bounds.

Namely, for a black box polynomial p of a degree d, we approximate all its m roots lying in a fixed square on the complex plane[3] at NR cost in $O(m^3 b \log(d))$ (cf. Corollary 2 and Observation 15). For a constant m this matches the cost bounds of the root-finder of [14], relaxing its restrictions on the input and output.

For any pair of an integer $\gamma \ge 1$ and a real $v \ge 1$ such that $m \gg v\gamma$, we decrease our NR cost bounds for the approximation of roots and eigenvalues by a factor of $\frac{m}{v\gamma}$ – by applying Las Vegas randomization; in that case we allow output errors with a probability at most $1/\gamma^v$ and verify correctness of the output at a dominated cost (cf. Corollary 2, Theorem 18, and Remark 3).

We readily extend to our algorithms the bound of $O(\log(d))$ bits of [14] on the computational precision provided that we can access the values of a polynomial $p(x)$ approximated by a black box oracle within a relative error $1/d^{O(1)}$ for x reasonably isolated from the zeros of $p(x)$. Then, as in [14], we immediately extend our root-finders to yield Las Vegas expected Boolean time $\tilde{O}(m^2 b^2 d^\omega)$ (for b and ω of the previous subsection and for $\tilde{O}(w)$ denoting $w \log^{O(1)}(w)$, that is, w up to polylogarithmic factors) for approximation of all m eigenvalues of a $d \times d$ symmetric or unsymmetric matrix M that lie in a fixed disc on the complex plane isolated from the external eigenvalues. Then again, for a constant m we match [14] but relax its restrictions on the input and the output.

Our techniques and auxiliary results can be of independent interest. In particular, we devise fast *exclusion/inclusion (e/i) tests*, which decide whether a fixed disc on the complex plane contains any zero of a black box polynomial

[2] The current records are about 2.7734 for feasible exponent of MM [4,10], unbeaten since 1982, and about 2.37 for unfeasible one [2].

[3] We count roots with their *multiplicities* and can readily extend our study to various other convex domains on the complex plane such as a disc or a polygon.

$p(x)$. This computational problem is fundamental for polynomial root-finding and has long been studied, but its known solution algorithms operate with the coefficients of $p(x)$. Our randomized acceleration of root-finding and application of Cauchy integrals to e/i test are nontrivial and pioneering in the field.

1.5 Classical Subdivision

Polynomial root-finders, traced back to [5, 33], seek all m roots lying in a fixed *suspect* square on the complex plane. A subdivision iteration divides every suspect square into four congruent sub-squares and to each of them applies e/i test: we either discard the square if the test certifies that it contains no roots or call it also *suspect* and process it in the next iteration otherwise (see Fig. 1).

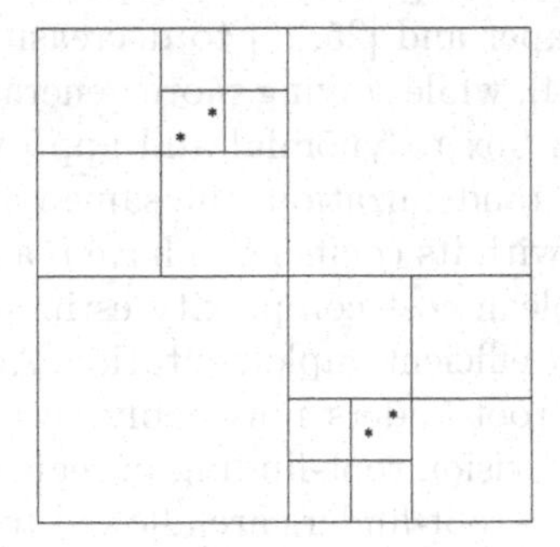

Fig. 1. Four roots are marked by asterisks; sub-squares containing them are suspect; the other sub-squares are discarded.

Observation 1. *At every subdivision iteration (i) at most 4m suspect squares are processed (since any root lies in 1, 2, or 4 suspect squares); (ii) the centers of these suspect squares approximate all m roots within the half-diameter of the squares, (iii) which decreases by twice at every iteration.*

Corollary 2. *Subdivision iterations applied to a square with m roots and side length* Δ *approximate these roots within* $\epsilon = \Delta/2^b$ *by applying e/i test to at most* $4mk$ *suspect squares for* $k \le \lceil \log_2(\frac{\Delta}{\epsilon}\sqrt{2}) \rceil = \lceil b + 0.5 \rceil$.

1.6 Soft Exclusion/Inclusion Tests

To complete classical subdivision root-finding we devise *soft e/i tests* for a black box polynomial. They can certify that the minimal disc $D(c, \rho) := \{x : |x - c| \le \rho\}$ covering a fixed square S contains no roots, and then we report *exclusion* and discard S, or that a little larger concentric disc $D(c, \sigma\rho)$ for a fixed $\sigma > 1$ contains a root or roots, and then we report *σ-soft inclusion*, call the square S suspect, and subdivide it. Both criteria for exclusion and σ-soft inclusion can hold simultaneously, but we stop as soon as we verify any of them. The estimate $4mk$ of Corollary 2 increases to at most $\alpha(\sigma)mk$ for $\alpha(\sigma)$ exceeding 4 but bounded by a constant for a constant σ.

1.7 The Known e/i Tests

The known e/i tests operate with the coefficients of an input polynomial, e.g., they are stated in terms of the coefficients of the auxiliary polynomials obtained from $p(x)$ by means of map (4) and recursive root-squaring (Pellet's theorem), [3], involve the higher order derivatives $p^{(i)}(x)$, $i = 1, 2, \dots, d$ [5,29], or Newton's identities and fast root-squaring algorithms based on convolution [20].

1.8 New Progress

New progress in the design and analysis of subdivision root-finders for a black box polynomial, based in particular on novel e/i tests, began in 2018 in arXiv preprint [24]. It contained 139 pages in August 2022, was partly covered in [6–8,13,22,23], and is extended in this paper and [25,26] to decreasing the complexity bounds of the pioneering paper [14], while solving more general computational problems of root-finding for a black box polynomial and approximation of matrix eigenvalues. Furthermore, with randomization, the same new algorithms applied to a general polynomial given with its coefficients have reached and then even significantly decreased the Boolean cost complexity estimates of [18,21].[4] Moreover, the algorithms allow highly efficient implementation. According to extensive tests of 2020 in [6,7], the new root-finders noticeably accelerated the previous best implementation [9] of subdivision root-finding in regions containing only a small number of roots, even where root-finders are allowed to operate with coefficients. In the implementation of 2022 of more advanced variants of these root-finders, which are still significantly slower than those of [25,26], acceleration became dramatic, and even for approximation of all d zeros of a polynomial p given with its coefficients, the new root-finders became at least competitive with user's choice package MPSolve [8], which was dramatically faster than [9] for that task.[5]

1.9 Cauchy-Based e/i Tests: Outline and an Extension

The power sums of the roots in a complex domain $\mathcal{D}$ are given by Cauchy integrals over its boundary contour $\mathcal{C}$ (cf. [1]):

$$s_h = s_h(p, \mathcal{D}) := \sum_{x_j \in \mathcal{D}} x_j^h = \frac{1}{2\pi\sqrt{-1}} \int_{\mathcal{C}} \frac{p'(x)}{p(x)} \, x^h \, dx, \; h = 0, 1, \dots . \qquad (3)$$

[4] Up to small poly-logarithmic factors these estimates reach lower bound for approximation of even a single zero of $p(x)$, but as we specify in Sect. 7, the algorithms of [25,26] greatly accelerate those of [18,21] for approximation of all $m = o(d)$ zeros of p that lie in a disc reasonably well isolated from the $d - m$ external zeros.

[5] The complexity of a subdivision root-finder is proportional to the number of roots in a region, while MPSolve is about as fast and slow for all roots as for their fixed subset. MPSolve implements Ehrlich-Aberth's root-finding iterations, which empirically converge to all d roots very fast right from the start but with no formal support and so far only under an initialization that operates with the coefficients of p (see [28]).

In particular s_0 is the number $\#(\mathcal{D})$ of roots in $\mathcal{D}$, and we can devise certified e/i tests by approximating s_0 or even just $\Re(s_0)$ within less than 1/2 and rounding the approximation to the nearest integer. We only deal with discs, where $\mathcal{D} = D(c, \rho)$, and furthermore reduce our study to the case of the unit disc $D(0,1)$ based on the univalent map:

$$x \mapsto \frac{x-c}{\rho},\ D(c,\rho) \mapsto D(0,1),\ \text{and}\ p(x) \mapsto t(x) = p\Big(\frac{x-c}{\rho}\Big). \tag{4}$$

In this case we approximate the integral s_0 with a finite sum $s_{0,q}$ at the qth roots of unity at NR cost q. Then $|s_{0,q} - s_0| < 1/2$ already for $q = O(\log(d))$ if no roots lie on or near the boundary circle $C(0,1)$. If $|s_{0,q}| > 1/2$, however, then $s_0 \geq 1$ unless some roots lie near the circle $C(0,1)$, and in both cases we certify soft inclusion. Otherwise, exclusion seems to be likely, but is not certified yet because boundary roots can corrupt approximation of s_0 by $s_{0,q}$. Next we outline further computations required to complete our e/i test.

Assume that the unit disc $D(0,1)$ contains at most m roots and apply e/i tests to $2m + 1$ concentric discs, whose boundary circles are pairwise isolated enough, so that any root can affect e/i test just for a single disc, and hence m roots can affect at most m out of all $2m + 1$ tests. Then majority vote certifies that the innermost disc contains no roots or that the outermost disc contains a root. The ratio $\sigma > 1$ of the radii of these two discs bounds the softness of the resulting test, and we readily keep σ below 1.2, say.

NR cost of $2m + 1$ applications of an e/i test is $2m + 1$ times the cost of a single test, but in case of larger m we decrease the overall NR cost by applying our alternative randomized e/i test w times for $w \ll m$. Namely, we first apply our soft e/i test to a disc whose center is fixed (say, at the origin) and whose radius ρ is sampled at random in a fixed range. Then the output of the test is correct if no roots lie near the boundary circle, and this occurs with a probability $1 - \gamma > 0$ for γ inversely proportional to the NR cost q of our e/i test. We fix an integer $v \geq 1$, reapply our test for v *independent identically distributed (iid)* values ρ of the radius and certify inclusion unless $|s_{0,q}| < 1/2$ in all of the v outputs. In the latter case we claim exclusion with an error probability $P \leq \gamma^v$; γ^v decreases exponentially in v, while NR cost increases by a factor of v.

1.10 Organization of the Paper

We devote the next two sections to background. Sections 4 and 5 cover our deterministic and randomized e/i tests, respectively. In Sect. 6 we extend the estimates of [14] for the precision of computing and hence for the Boolean cost of approximation of matrix eigenvalues to our more general case. In Sect. 7 we outline further progress in [25,26].

2 Definitions and Basic Properties

- *NR cost* (*NR complexity*) of a root-finder is the number of points x at which it evaluates *Newton's ratio* $\mathrm{NR}(x)$ and *Newton's inverse ratio* $\mathrm{NIR}(x)$.

- Define square, disc, circle (circumference), and annulus on the complex plane: $S(c,\rho) := \{x : |\Re(c-x)| \le \rho, |\Im(c-x)| \le \rho\}$, $D(c,\rho) := \{x : |x-c| \le \rho\}$, $C(c,\rho) := \{x : |x-c| = \rho\}$, $A(c,\rho,\rho') := \{x : \rho \le |x-c| \le \rho'\}$.
- $\Delta(\mathbb{R})$, $X(\mathbb{R})$, and $\#(\mathbb{R})$ are the diameter, root set, and index (root set's cardinality) of a region $\mathbb{R}$ on the complex plane, respectively.
- A disc $D(c,\rho)$, a circle $C(c,\rho)$, or a square $S(c,\rho)$ is *θ-isolated* for $\theta > 1$ if $X(D(c,\rho)) = X(D(c,\theta\rho))$, $X(C(c,\rho)) = X(A(c,\rho/\theta,\rho/\theta))$, or $X(S(c,\rho)) = X(S(c,\theta\rho))$, respectively.
- The largest upper bound on such a value θ is said to be the *isolation* of the disc $D(c,\rho)$, the circle $C(c,\rho)$, or the square $S(c,\rho)$, respectively, and is denoted $i(D(c,\rho))$, $i(C(c,\rho))$, and $i(S(c,\rho))$, respectively.
- $r_1(c,t) = |y_1 - c|, \dots, r_d(c,t) = |y_d - c|$ in non-increasing order are the *d root radii*, that is, the distances from a complex center c to the roots $y_1 \dots, y_d$ of a dth degree polynomial $t(x)$. $r_j(c) := r_j(c,p)$, $r_j := r_j(0)$ for $j = 1, \dots, d$.
- Differentiate factorization (1) of $p(x)$ to express $\text{NIR}(x)$ as follows:

$$\text{NIR}(x) := \frac{p'(x)}{p(x)} = \sum_{j=1}^{d} \frac{1}{x - x_j}. \tag{5}$$

Definition 3. *For $p(x)$ of (1), $\sigma > 1$, and integers ℓ and m such that $1 \le \ell \le m \le d$ and $\#(D(0,1)) \le m$, a σ-**soft** ℓ-**test**, or just ℓ-**test** for short (1-test being e/i test), either outputs 1 and stops if it detects that $r_{d-\ell+1} \le \sigma$, that is, $\#(D(0,\sigma)) \ge \ell$, or outputs 0 and stops if it detects that $r_{d-\ell+1} > 1$, that is,*[6] *$\#(D(0,1)) < \ell$. ℓ-test$_{c,\rho}$, aka ℓ-test for the disc $D(c,\rho)$, is an ℓ-test applied to the polynomial $t(y)$ of (4).*

Observation 4. *For a complex c and $\rho > 0$ it holds that (i) $i(S(c,\rho))/\sqrt{2} \le i(D(c,\rho)) \le i(S(c,\rho\sqrt{2}))/\sqrt{2}$, (ii) $i(C(c,\rho)) \le i(D(c,\rho))$ and if $i(D(c,\rho)) \ge \theta^2 \ge 1$, then $i(C(c,\rho\theta)) \ge \theta$.*

3 The Power Sums of the Roots and Cauchy Sums

3.1 The Power Sums of the Roots and Cauchy Sums in the Unit Disc

For a positive integer q define Cauchy sums by means of discretization of Cauchy integral (3) in the case where $\mathcal{D}$ is the unit disc $D(0,1)$:

$$s_{h,q} := \frac{1}{q} \sum_{g=0}^{q-1} \zeta^{(h+1)g} \frac{p'(\zeta^g)}{p(\zeta^g)}, \text{ for } h = 0, 1, \dots, q-1 \text{ and } \zeta := \exp\left(\frac{2\pi\sqrt{-1}}{q}\right) \tag{6}$$

denoting a primitive q-th root of unity.

[6] Both bounds $r_{d-\ell+1} \le \sigma$ and $r_{d-\ell+1} > 1$ can hold simultaneously, but as soon as an ℓ-test verifies any of them, it stops without checking if the other bound also holds.

Remark 1. We call the values $s_{h,q}$ *Cauchy sums* by following [24]. Schönhage used these values in [30] for $h > 0$ to deflate a factor of p, but we only use them in the case of $h = 0$, ignored by Schönhage, and only for e/i tests, not considered in [30]; we know of no application of the Cauchy sums to root-finding between [30] and [24]. As in [24] we certify inclusion if $p(\zeta^g) = 0$ for any g, $0 \le g < q$.

Theorem 5 [6,24]. *For a polynomial $p(x)$ of (1) and an integer $q \ge 1$ let $\prod_{j=1}^{d}(x_j^q - 1) \ne 0$. Then $s_{h,q} = \sum_{j=1}^{d} \frac{x_j^h}{1-x_j^q}$ for $s_{h,q}$ of (6) and $h = 0, 1, \ldots, q-1$.*

3.2 Approximation Errors and Root-Counting in the Unit Disc

Corollary 6 [6,24,30].[7] *Let q be a positive integer and let the unit circle $C(0,1)$ be θ-isolated for $\theta > 1$. Then $|s_{h,q} - s_h| \le \frac{d\theta^h}{\theta^q - 1}$* for $h = 0, 1, \ldots, q-1$.

Algorithm 7. Root-counting in the unit disc at NR cost $q = O(\log_\theta(d))$.

INPUT: a black box polynomial p of a degree d and $\theta > 1$.
INITIALIZATION: Compute the integer $q = \lfloor \log_\theta(4d+2) \rfloor > \log_\theta(2d+1)$.
COMPUTATIONS: Compute Cauchy sum $s_{0,q}$ and output an integer $\bar{s}_0$ closest to it,[8] *which is also closest to its real part $\Re(s_{0,q})$.*

Observation 8. *(i) Algorithm 7 runs at NR cost $q = \lfloor \log_\theta(4d+2) \rfloor$. (ii) It outputs $\bar{s}_0 = \#(D(0,1))$ if the circle $C(0,1)$ is θ-isolated for $\theta > 1$. (iii) If the algorithm outputs $\bar{s}_0 > 0$, then $\#(D(0,\theta)) > 0$.*

Proof. Corollary 6 immediately implies claim (ii) but also implies that $\#(D(0,1)) > 0$ if $\bar{s}_0 > 0$ unless $\#(A(0,1/\theta,\theta)) > 0$. In both cases $\#(D(0,\theta)) > 0$.

3.3 Extension to Any Disc

Map (4) reduces the computation of Cauchy sum $s_{h,q}$ in any disc $D(c,\rho)$ to the case of the unit disc $D(0,1)$ as follows.

Definition 9. *For a disc $D(c,\rho)$, polynomials $p(x)$ and $t(y) = p(\frac{y-c}{\rho})$, a positive integer q, and $\zeta = \zeta_q := \exp(\frac{2\pi\sqrt{-1}}{q})$, define Cauchy sums*

$$s_{h,q}(p,c,\rho) := s_{h,q}(t,0,1) := \frac{\rho^{h+1}}{q} \sum_{g=0}^{q-1} \zeta^{(h+1)g} \frac{p'(c+\rho\zeta^g)}{p(c+\rho\zeta^g)} \text{ for } h = 0, 1, \ldots, q-1.$$

Observation 10. *Given a polynomial $p(x)$, a complex c, a positive ρ, a positive integer q, and the qth roots of unity, evaluation of $s_{0,q}(p,c,\rho)$ can be reduced to the evaluation of $NIR(\zeta^g)$ for $g = 0, 1, \ldots, q-1$ at NR cost q and performing $q+1$ divisions, $2q-1$ multiplications, and $2q-1$ additions.*

[7] [30] proved this corollary directly; [24] and then [6] deduced it from Theorem 5.

[8] Given the coefficients of $p(x)$ one can fix $q := 2^k$ for $k = \lceil \log_2 \lfloor \log_\theta(4d+2) \rfloor \rceil$ and then evaluate $p(x)$ and $p'(x)$ at all qth roots of unity by applying FFT.

Observation 11. *(4) maps the roots x_j of $p(x)$ into the roots $y_j = \frac{x_j - c}{\rho}$ of $t(y)$, for $j = 1, \ldots, d$, and preserves the index $\#(D(c, \rho))$ and the isolation $i(D(c, \rho))$.*

Refer to Algorithm 7 applied to the polynomial $t(y)$ of (4) as *Algorithm* $7_{c,\rho}$ and also as *Algorithm* 7 *applied to the disc* $D(c, \rho)$ *and the circle* $C(c, \rho)$; recall Observation 11 and extend Observation 8 as follows.

Observation 12. *(i) Algorithm $7_{c,\rho}$ runs at NR cost $q = \lfloor \log_\theta(4d + 2) \rfloor$. (ii) It outputs $\bar{s}_0 = \#(D(c, \rho))$ if the circle $C(c, \rho)$ is θ-isolated for $\theta > 1$. (iii) If the algorithm outputs $\bar{s}_0 > 0$, then $\#(D(c, \theta\rho)) > 0$.*

Remark 2. Algorithm 7 reduces computation of $s_{0,q}$ to evaluation of $\mathrm{NIR}(\zeta^g)$ for $g = 0, 1, \ldots, q - 1$. If $v := \max_{g=0}^{q-1} |\mathrm{NIR}(\zeta^g)|$ is small for a reasonably large q, then $\#(D(0, 1)) = 0$ for a large class of polynomials $p(x)$. Indeed, $\mathrm{NIR}(x) = \sum_{j=1}^{d} \frac{1}{x - x_j}$ (see (5)), and this sum vanishes only on an algebraic variety of a smaller dimension in the space of the zeros $x_1, \ldots, x_d$ of $p(x)$. If v is small, then all the qth roots of unity lie near that variety, which strongly restricts the input class of polynomials p where q is large. Extensive experiments in [6,8] have been performed for both synthetic and real world inputs and q of order $\log^2(d)$. In these experiments the unit disc $D(0, 1)$ contained no roots, with no single exception, unless v exceeded some fixed reasonably small upper bound.[9] This empirically supported a very fast heuristic e/i test, even though it fails on specially concocted polynomials $p(x) = x^{q+1} - (q+1)x + w$ of degree q where $|w|$ is small. Clearly, the extensive tests of [6,8] have never encountered such inputs; our next e/i test is a little slower but never fails on any input.

4 Deterministic ℓ-Test Under No Isolation Assumption

Algorithm $7_{c,\rho}$ is a reliable root-counter unless some roots lie on or near the circle $C(c, \rho)$. To counter their adverse impact, apply the algorithm to $2m + 1$ concentric discs $D(c, \rho_i)$, $i = 0, 1, \ldots, 2m$, such that the $2m + 1$ concentric circles $C_i = C(c, \rho_i)$, $i = 0, 1, \ldots, 2m$, are sufficiently well isolated pairwise, allowing a single root to corrupt Algorithm $7_{c,\rho_i}$ for only a single i. Now majority vote certifies that at least ℓ roots lie in the outermost disc of the family or less than ℓ roots lie in its innermost disc. This defines a σ-soft ℓ-test for $\sigma > 1$ equal to the ratio of the radii of the latter two discs. Next we specify this recipe. Without loss of generality *(wlog)* let $c = 0$, $\rho = 1$, and $C(c, \rho) = C(0, 1)$.

Algorithm 13. Deterministic *ℓ-test.*

> *INPUT: A black box polynomial $p(x)$ of a degree d and two positive integers ℓ and $m \geq \ell$ such that the unit disc $D(0, 1)$ contains at most m roots.*

---

[9] Actually, [6,8] tested the assumption that $|s_{h,q}|$ were small for $h = 0, 1, 2$, but this follows if v is small because $v \geq |s_{h,q}|$ for $h = 0, 1, \ldots, q - 1$.

INITIALIZATION: Fix

$$\sigma > 1,\ \theta = \sigma^{\frac{1}{4m+1}},\ q = \lfloor \log_\theta(4d+2) \rfloor,\ \rho_i = \frac{1}{\theta^{2i+1}},\ i = 0, 1, \dots, 2m. \quad (7)$$

OUTPUT: 0 if $\#(D(0, 1/\sigma)) < \ell$ or 1 if $\#(D(0, 1)) \geq \ell$.
COMPUTATIONS: Apply Algorithm $7_{0,\rho_i}$ for q of (7) and $i = 0, 1, \dots, 2m$. Output 1 if among all $2m + 1$ output integers $\bar{s}_0$ at least $m + 1$ integers are at least ℓ. Otherwise output 0. (We can output 1 already on a single output $\bar{s}_0 > 0$ if $\ell = 1$; we can output 0 already on a single output $\bar{s}_0 < m$ if $\ell = m$.)

Theorem 14. *Algorithm 13 is a σ-soft ℓ-test for $\sigma > 1$ of our choice (see (7)).*

Proof. Corollary 6 implies that for every i the output $\bar{s}_0$ of Algorithm $7_{0,\rho_i}$ is equal to $\#(D(0, \rho_i))$ unless there is a root in the open annulus $A_i := A(0, \rho_i/\theta, \theta\rho_i)$.

The $2m+1$ open annuli A_i, $i = 0, 1, \dots, 2m$, are disjoint. Hence a single root cannot lie in two annuli A_i, and so Algorithm $7_{0,\rho_i}$ can fail only for a single i among $i = 0, \dots, 2m$. Therefore, at least $m + 1$ outputs of the algorithm are correct.

To complete the proof, combine this property with the relationships $\rho_{2m} = 1/\theta^{4m+1}$ (cf. (7)), $\rho_0 = 1/\theta < 1$, and $\#(D(0, \rho_{i+1})) \leq \#(D(0, \rho_i))$, which hold because $D(0, \rho_{i+1}) \subseteq (D(0, \rho_i)$ for $i = 0, 1, \dots, 2m - 1$.

Observation 15 *[Cf. Observation 10]. Algorithm 13 runs at NR cost $\mathcal{A} = (2m + 1)q$ for $q = \lfloor \log_\theta(4d + 2) \rfloor = \lfloor (4m + 1) \log_\sigma(4d + 2) \rfloor$, which dominates the cost of the remaining ops involved, even in the cases of the Mandelbrot and sparse input polynomials, and which stays in $O(m^2 \log(d))$ if $\sigma - 1$ exceeds a positive constant.*

5 Randomized Root-Counting and ℓ-Tests

5.1 Solution with a Crude Bound on Error Probability

Next accelerate ℓ-test$_{c,\rho}$ in case of larger integers m by applying randomization and using fewer calls for Algorithm $7_{c,\rho}$ at the price of only certifying the output with a high probability *(whp)* rather than deterministically. We fix the center c of a circle $C(c, \rho)$ but choose its radius ρ at random in a small range. Then whp $i(C(c, \rho))$ is sufficiently well separated from 1; this excludes the adverse impact of boundary roots, while a soft ℓ-test accepts a small variation of the radius ρ. Wlog let $c = 0$.

Algorithm 16. Basic randomized root-counter.

INPUT: $\gamma \geq 1$ and a d-th degree black box polynomial $p(x)$ having at most m roots in the disc $D(0, \sqrt{2})$.
INITIALIZATION: Sample a random value w in the range $[0.2, 0.4]$ under the uniform probability distribution in that range and output $\rho = 2^w$.

COMPUTATIONS: By applying Algorithm $7_{0,\psi}$ compute the 0-th Cauchy sum $s_{0,q}$ in the disc $D(0,\rho)$ for $p(x)$ and $q = \lfloor 10m\gamma \log_2(4d+2)\rfloor$. Output an integer $\bar{s}_0$ closest to $s_{0,q}$.

Theorem 17. *Algorithm 16, running at NR cost $q = \lfloor 10m\gamma \log_2(4d+2)\rfloor$ for any fixed $\gamma \geq 1$, outputs $\bar{s}_0 = \#(D(0,\rho))$ with a probability at least $1 - 1/\gamma$.*

Proof. By assumption, the annulus $A(0,1,\sqrt{2})$ contains at most m roots. Hence at most m root radii $r_j = 2^{e_j}$, for $j = 1,\ldots,m' \leq m$, lie in the range $[1,\sqrt{2}]$, that is, $0 \leq e_j \leq 0.5$ for at most $m' \leq m$ integers j. Fix m' intervals, centered at e_j, for $j = 1,\ldots,m'$, each of a length at most $1/(5m'\gamma)$. Then the overall length of these intervals is at most $1/(5\gamma)$. Let $\mathbb{U}$ denote the union of these intervals.

Sample a random u in the range $[0.2, 0.4]$ under the uniform probability distribution in that range and notice that Probability $(u \in \mathbb{U}) \leq 1/\gamma$.

Hence with a probability at least $1 - 1/\gamma$ the circle $C(0,w)$ is θ-isolated for $\theta = 2^{\frac{1}{10m\gamma}}$, in which case Algorithm 16 outputs $\bar{s}_0 = s_0 = \#(D(0,w))$ by virtue of claim (ii) of Observation 12.

5.2 Refining the Bound on Error Probability

Next we reapply Algorithm 16 v times, increasing its NR cost v times, but this will decrease the error probability dramatically – below $1/\gamma^v$. First readily reduce the root radius approximation problem to the decision problem of ℓ-test. Namely, *narrow Algorithm* 16 *to an ℓ-test* for a fixed ℓ in the range $1 \leq \ell \leq m$, so that with a probability at least $1 - 1/\gamma$ this ℓ-test outputs 0 if $\#(D(0,1)) < \ell$ and outputs 1 if $\#(D(0,2^{0.5})) \geq \ell$. Moreover, the output 1 is certified by virtue of claim (iii) of Observation 12 if $\ell = 1$; similarly the output 0 is certified if $\ell = m$.

Next extend the ℓ-test by means of applying Algorithm 16 for v iid random variables w in the range $0.2 \leq w \leq 0.4$. Call this ℓ-test **Algorithm 16v, ℓ.**

Specify its output g for $\ell = 1$ as follows: let $g = 1$ if Algorithm $16v, 1$ outputs 1 at least once in its v applications; otherwise let $g = 0$. Likewise, let $g = 0$ if Algorithm $16v, m$ outputs 0 at least once in its v applications; otherwise let $g = 1$. Recall Theorem 17 and then readily verify the following theorem.

Theorem 18. *(i) For two integers $1 \leq \ell \leq m$ and $v \geq 1$ and a real $\gamma \geq 1$, Algorithm 16v, ℓ runs at NR cost $\lfloor 10m\gamma \log_2(4d+2)\rfloor v$, which dominates the cost of the other ops involved, even in the cases of the Mandelbrot and sparse input polynomials (cf. Observation 10). (ii) The output 1 of Algorithm 16v, 1 and the output 0 of Algorithm 16v, m are certified. (iii) The output 0 of Algorithm 16v, 1 and the output 1 of Algorithm 16v, m are correct with a probability at least $1 - 1/\gamma^v$.*

The theorem bounds the cost of ℓ-tests for $\ell = 1$ (e/i tests) and $\ell = m$, but [27] extends it to ℓ-tests for any integer ℓ in the range $1 < \ell < m$.

Theorem 19 [27]. *For an integer ℓ such that $1 < \ell < m \leq d$ and real $v \geq 1$ and $\gamma > 1$, Algorithms 16v, ℓ runs at NR cost $\lfloor 10m\gamma \log_2(4d+2) \rfloor v$, and its output is correct with a probability at least $1 - (4/\gamma)^{v/2}$.*

Remark 3. Root-finding with error detection and correction, see more in [27]. A randomized root-finder can lose some roots, albeit with a low probability, but we can detect such a loss at the end of root-finding process, simply by observing that among the m roots in an input disc only $m - w$ *tame* roots have been closely approximated,[10] while $w > 0$ *wild* roots remain at large. Then we can recursively apply the same or another root-finder until we approximate all m roots.[11]

6 Precision of Computing in Our Root-Finders

High order Newton's iterations in [14] involve high order derivatives of $p(x)$, approximated in [14] with high order divided differences, but we only involve $\mathrm{NIR}(x) = \frac{p'(x)}{p(x)}$ and approximate it with the first order divided difference:

$$\mathrm{NIR}_\delta(x) := \frac{1}{\delta} - \frac{p(x-\delta)}{\delta\, p(x)} = \frac{p(x) - p(x-\delta)}{p(x)\delta} = \frac{p'(y)}{p(x)} \approx \mathrm{NIR}(x) \text{ for } \delta \approx 0. \quad (8)$$

Here y lies in the line segment $[x - \delta, x]$ by virtue of Taylor-Lagrange's formula, and so $y \approx x$ for $\delta \approx 0$. Next we estimate precision of computing that supports approximation of $\mathrm{NIR}(x)$ required in our root-finders. We only need to consider our e/i tests for θ-isolated discs $D(c, \rho)$ with $\theta - 1 > 0$ of order $\frac{1}{m}$. Map (4) reduces such a test for $p(x)$ to the disc $D(0, 1)$ for the polynomial $t(x) = p(\frac{x-c}{\rho})$, and we only need to certify that the output errors of our e/i tests in Algorithms 13 and 16 are less than $\frac{1}{2}$. We obtain such a certification by performing our algorithms with a precision of $O(\log(d))$ bits for NR cost q of order $m \log(d)$ provided that a black box oracle evaluates for us the ratio $\frac{t(x-\delta)}{t(x)}$ within $\frac{1}{8\delta}$ for any fixed positive $\delta = O(1/d^{O(1)})$.

Let us specify this certification. Our e/i tests amount to computing the sum of the q values $\rho x \frac{p'(c+\rho x)}{p(c+\rho x)} = x \frac{t'(x)}{t(x)}$, for $x = \zeta^g$ and $g = 0, 1, \ldots, q-1$, and dividing the sum by q. By slightly abusing notation write $\mathrm{NIR}(x) := \frac{t'(x)}{t(x)}$ rather than $\mathrm{NIR}(x) = \frac{p'(x)}{p(x)}$ and notice that it is sufficient to compute $\zeta^g \mathrm{NIR}(\zeta^g)$ within, say, $\frac{3}{8}$ for every q. In that case the sum of the q error bounds is at most $\frac{3q}{8}$ and decreases below $\frac{1}{2}$ in division by q if the overall rounding error of those

[10] We call a complex point c a tame root for a fixed error tolerance ϵ if it is covered by an isolated disc $D(c, \epsilon)$. Given such a disc $D(c, \rho)$, we can readily compute $\#(D(c, \rho))$ by applying Corollary 6.

[11] This recipe detects output errors of any root-finder at the very end of computations. In the case of subdivision root-finders we can detect the loss of a root earlier – whenever we notice that at a subdivision step the indices of all suspect squares sum to less than m.

summation and division is less than $\frac{1}{8}$, and we readily ensure such a bound for division by q (cf. [11, Lemma 3.4]).

To bound the precision of computing $\zeta^g \mathrm{NIR}(\zeta^g)$, first recall Eq. (5) and represent $x\mathrm{NIR}(x)$ for $|x| = 1$ within $\frac{1}{8}$ by using a precision of $O(\log(d))$ bits.

Theorem 20. *Write $C := C(c, \rho)$ and assume that $|x| = 1$ and the circle C is θ-isolated for $\theta > 1$. Then $|x\mathrm{NIR}(x)| \le \frac{d}{1-1/\theta} = \frac{\theta d}{\theta - 1}$.*

Proof. Equation (5) implies that $|\mathrm{NIR}(x)| = |\sum_{j=1}^{d} \frac{1}{|x-x_j|}| \le \sum_{j=1}^{d} \frac{1}{|x-x_j|}$ where $|x - x_j| \ge (1 - 1/\theta)\rho$ for $|x_j| < 1$, while $|x - x_j| \ge (\theta - 1)\rho$ for $|x_j| > 1$ since $i(C) \ge \theta$. Combine these bounds, write $m := \#(D(c, \rho))$, and obtain $|x\mathrm{NIR}(x)| \le \frac{m}{1-1/\theta} + \frac{d-m}{\theta-1} \le \frac{d}{1-1/\theta}$ for $|x| = 1$.

Corollary 21. *One can represent $x\mathrm{NIR}(x)$ within $\frac{1}{8}$ for $|x| = 1$ by using a precision of $3 + \lceil \log_2(\frac{d}{1-1/\theta}) \rceil$ bits, which is in $O(\log(d))$ provided that $\frac{1}{\theta-1} = d^{O(1)}$.*

Instead of $x\mathrm{NIR}(x) = \frac{xt'(x)}{t(x)}$ we actually approximate $\frac{xt'(y)}{t(x)}$ where $|y - x| \le \delta$ and $t'(y)$ is equal to the divided difference of (8). Thus we shall increase the above error bound $\frac{1}{8}$ by adding the upper bounds $\frac{1}{8}$ on $\alpha := |\frac{xt'(x)}{t(x)} - \frac{xt'(y)}{t(x)}|$ for $|x| = 1$ and on the rounding error β of computing $x\frac{t(x)-t(x-\delta)}{\delta t(x)}$.

In [27] we readily estimate α in terms of d, θ, and δ – simply by adjusting the first five lines of the proof of [14, Lemma 3.6]). This shows that $\alpha < \frac{1}{8}$ for δ defined with a precision $\log_2(1/|\delta|)$ of order $\log(d)$. Then we estimate β by applying straightforward error analysis and thus arrive at

Corollary 22. *Suppose that a black box oracle supplies for us approximations of a d-th degree polynomial $t(x)$ with any relative error in $1/d^{O(1)}$ for $|x| \le 1$ and let the unit disc be θ-isolated for $t(x)$ and for $\frac{1}{\theta-1} = d^{O(1)}$. Then we can perform e/i test for $t(x)$ on the unit disc by using a precision of $O(\log(d))$ bits.*

7 Conclusions

We substantially advance our progress in [25–27]. In particular we estimated and then greatly decreased the Boolean complexity of our current root-finders, where we allow to operate with the coefficients of p of (1) and depart from the case of linear divisors in [11, Thm. 3.9], implicit in [30]:

Theorem 23. *Given a positive b, the coefficients of a polynomial $\bar{t}(x) := \sum_{i=0}^{d} \bar{t}_i x^i$ such that $||\bar{t}||_1 = \sum_{i=0}^{d} |\bar{t}_i| \le 1$, and q complex points z_g, $g = 1, \ldots, q$, in the unit disc $D(0, 1)$, one can evaluate $\bar{t}(z_g)$ for $g = 1, \ldots, q$ within an error bound $1/2^b$ by using $O(\mu((d\log(d) + q)(b + q)))$ Boolean operations for $\mu(s) = O(s\log(s)\log(s)\log(\log(s)))$ (cf. (2)).*

Based on this theorem we estimate the overall Boolean cost of all e/i tests applied to $w = O(m)$ discs $D(c_\lambda, \rho_\lambda)$ at a fixed subdivision step, where $p(x)$ is evaluated at q equally spaced points at every circle $C(c_\lambda, \rho_\lambda)$, for $\rho_\lambda \geq 1/2^b$, $|c_\lambda| + \rho_\lambda \leq R$, $\lambda = 1, \dots, w$, and a fixed R. We (i) scale the variable $x \mapsto Rx$ – to map all discs $D(c_j, \rho_j)$, $j = 1, \dots, \bar{m}$, into the unit disc $D(0,1)$ and then (ii) normalize $p(Rx) \mapsto v(x) := \psi p(Rx)$ for a scalar $\psi > 0$ such that $||v(x)||_1 = 1$, $\psi = O(R^d)$, and so Theorem 23 supports the Boolean cost bound $O(\mu((d\log(d) + \bar{q})(b + \bar{q})))$ for the evaluation of $v(x)$ within error bound $1/2^b$ at all $\bar{m}$ e/i tests of that subdivision step. This is ensured if $|v(x)| \cdot |\Delta| \leq 1/2^b$ for an upper bound $\Delta = 1/d^{O(1)}$ on the relative error of the evaluation. In our e/i tests we only need to evaluate $v(x)$ at the points x of a θ_λ-isolated circles $C(c_\lambda, \rho_\lambda)$ where $1/(\theta_\lambda - 1) = \tilde{O}(m)$, and hence ([30, Eqn. (9.4)]) $\log(1/|v(x)|) = O(d\log(m) + \log(1/\rho_\lambda))$; the bound is sharp up to a constant factor; for $m = d$ and $\rho_\lambda = 1$ it is reached at $x = 1$ and the polynomial $v(x) = ((x - 1 + 1/m)/(2 - 1/m))^d$.

Substitute $v(x) = \psi p(x)$ and $\psi \leq R^d$, recall that $\min_\lambda \rho_\lambda \geq 1/2^b$, and obtain

$$\log(1/|p(x)|) = O(d\log(m) + b_\lambda\lambda \text{ for } b_\lambda = \log_2(R/\rho_\lambda) \leq b.$$

Hence we can bound the Boolean complexity of our e/i tests at any fixed subdivision step by applying Theorem 23 with q and b replaced by $\bar{q} = O(qm)$ and $\bar{b} = O(d\log(m) + b)$, respectively, and then obtain the Boolean complexity bounds of a subdivision step consisting of $O(m)$ e/i tests: $\tilde{O}((qm+d)(qm+d+b))$ where $\tilde{O}(w)$ equals $w\log^{O(1)}(w)$ and $q = m^2$ or $q = m$ for our deterministic or randomized e/i tests, respectively. Now we bound the overall Boolean cost of all $O(b)$ subdivision steps:

Corollary 24. *Our root-finders can be performed at the overall Boolean cost in* $\tilde{O}((qm + d)(qm + d + b)b)$ *where* $q = m^2$ *and* $q = m$ *in our deterministic and Las Vegas randomized e/i tests, respectively.*

So far we assumed that the input disc $D(0, R)$ is θ-isolated for $\theta - 1$ exceeding a positive constant, but if θ exceeds d^h for a reasonably large h, then $\mathrm{NIR}(x) \approx \frac{f'(x)}{f'(x)}$ where $f(x)$ is the monic factor of $p(x)$ of degree m sharing with $p(x)$ its m zeros that lie in $D(0, R)$. Hence we can decrease the above upper bound on $\log(1/|p(x)|)$ to $\log(1/|f(x)|) = O(m\log(m) + b)$. Then we can ensure relative precision in $1/d^{O(1)}$ for computing $p(x)$ if we ensure absolute precision b of order $m\log(m) + b + \log(d)$. Hence we can bound the Boolean complexity of our e/i tests at any fixed subdivision step by applying Theorem 23 with q and b replaced by $\bar{q} = O(qm)$ and $\bar{b} = O(m\log(m) + b + \log(d))$, respectively, at the price of performing root-lifting at NR cost in $O(\log(d))$.[12]

If the disc $D(0, R)$ is isolated for $p(x)$, then the unit disc $D(0, 1)$ is isolated for $v(x) = p(Rx)$, and we can ensure its θ- isolation for θ exceeding d^h. Hence we can ensure the latter bounds on $\log(1/|p(x)|$ and b if we lift the zeros of the

[12] We can also replace d by m in Theorem 23 if we only seek approximation of $\mathrm{NIR}(x)$ for $|x| = 1$ within $1/d^{O(1)}$; actually we need such approximations within relative error $1/d^{O(1)}$, which implies an absolute error bound $1/d^{O(1)}$ if $1/\mathrm{NIR}(x) = 1/d^{O(1)}$.

polynomial $v(x)$ to their hth powers for $h = O(\log(d))$. We can perform lifting at a dominated NR cost $O(\log(d))$ (e.g., apply root-squaring $O(\log(\log(d)))$ times); this increases our bound $\Phi = 1/2^b$ on $R/\min \rho_\lambda$ but at most to $\Phi^{O(\log(d)} = 1/2^{O(b\log(d))}$. We can now replace the factor $qm+d+b$ with $qm+m+b\log(d) = \tilde{O}(qm+b)$ in our bound $\tilde{O}((qm+d)(qm+d+b)b)$ of Corollary 24, to obtain

Theorem 25. *Given the coefficients of a polynomial $p(x)$ of (1) and an isolated disc $D = \{x : |x-c| \leq \rho\}$ on the complex plane for a complex c and positive ρ, let this disc contain precisely m zeros of $p(x)$. Then one can approximate all m zeros of $p(x)$ in D within $\epsilon = 1/2^b > 0$ at the overall Boolean cost in $\tilde{O}((qm+d)(qm+b)b)$. Here $q = m^2$ and $q = m$ for our deterministic and Las Vegas randomized root-finders, respectively.*

The algorithms in [25] enable further decrease of the cost bound of Theorem 25 by a factor of b – to $\tilde{O}((qm+d)(qm+b))$. For $m = o(d)$ this is *below the lower bound* of [18,21] deduced even for approximation of a single zero of p, although without assumption of any isolation of its covering disc.[13] The algorithms of [26] support the latter estimates for q decreased to a constant under Las Vegas randomization in the space of the d zeros of p.

References

1. Ahlfors, L.: Complex Analysis. McGraw-Hill Series in Mathematics. McGraw-Hill (2000). ISBN: 0-07-000657-1
2. Alman, J., Williams, V.V.: A refined laser method and faster matrix multiplication. In: ACM-SIAM SODA 2021, pp. 522–539. SIAM (2021)
3. Becker, R., Sagraloff, M., Sharma, V., Yap, C.: A near-optimal subdivision algorithm for complex root isolation based on the Pellet test and Newton iteration. J. Symbolic Comput. **86**, 51–96 (2018). https://doi.org/10.1016/j.jsc.2017.03.009
4. Pan, V.Y.: Trilinear aggregating with implicit canceling for a new acceleration of matrix multiplication, computers and mathematics (with Applications), **8**(1), 23–34 (1982)
5. Henrici, P.: Applied and Computational Complex Analysis. Volume 1: Power Series, Integration, Conformal Mapping, Location of Zeros. Wiley, New York (1974)
6. Imbach, R., Pan, V.Y.: New progress in univariate polynomial root-finding. In: Proceedings of ACM-SIGSAM ISSAC 2020, pp. 249–256, Kalamata, Greece, 20–23 July 2020. ACM Press (2020). ISBN: 978-1-4503-7100-1/20/07
7. Imbach, R., Pan, V.Y.: New practical advances in polynomial root clustering. In: Slamanig, D., Tsigaridas, E., Zafeirakopoulos, Z. (eds.) MACIS 2019. LNCS, vol. 11989, pp. 122–137. Springer, Cham (2020). https://doi.org/10.1007/978-3-030-43120-4_11
8. Imbach, R., Pan, V.Y.: Accelerated subdivision for clustering roots of polynomials given by evaluation oracles. In: Boulier, F., England, M., Sadykov, T.M., Vorozhtsov, E.V. (eds.) CASC 2022. LNCS, vol. 13366, pp. 143–164. Springer, Cham (2022). https://doi.org/10.1007/978-3-031-14788-3_9

[13] [3] claims reaching optimal estimates of [18,21] but actually requires separation of the zeros of p, both pairwise and from the origin. Neither of [18,21,25,26] imposes such restrictions.

9. Imbach, R., Pan, V.Y., Yap, C.: Implementation of a near-optimal complex root clustering algorithm. In: Davenport, J.H., Kauers, M., Labahn, G., Urban, J. (eds.) ICMS 2018. LNCS, vol. 10931, pp. 235–244. Springer, Cham (2018). https://doi.org/10.1007/978-3-319-96418-8_28
10. Karstadt, E., Schwartz, O.: Matrix multiplication, a little faster. J. ACM **67**(1), 1–31 (2020). MR 4061328, S2CID 211041916. https://doi.org/10.1145/3364504
11. Kirrinnis, P.: Polynomial factorization and partial fraction decomposition by simultaneous Newton's iteration. J. Complex. **14**, 378–444 (1998)
12. Knuth, D.E.: The Art of Computer Programming, Volume 2. Addison-Wesley (1981) (2nd edn.) (1997) (3rd edn.)
13. Luan, Q., Pan, V.Y., Kim, W., Zaderman, V.: Faster numerical univariate polynomial root-finding by means of subdivision iterations. In: Boulier, F., England, M., Sadykov, T.M., Vorozhtsov, E.V. (eds.) CASC 2020. LNCS, vol. 12291, pp. 431–446. Springer, Cham (2020). https://doi.org/10.1007/978-3-030-60026-6_25
14. Louis, A., Vempala, S.S.: Accelerated Newton iteration: roots of black box polynomials and matrix eigenvalues. In: IEEE FOCS 2016, vol. 1, pp. 732–740, January 2016. arXiv:1511.03186. https://doi.org/10.1109/FOCS.2016.83
15. McNamee, J.M.: Numerical Methods for Roots of Polynomials, Part I, XIX+354 p. Elsevier (2007). ISBN: 044452729X, ISBN-13: 9780444527295
16. McNamee, J.M., Pan, V.Y.: Numerical Methods for Roots of Polynomials, Part II, XXI+728 p. Elsevier (2013). ISBN: 9780444527301
17. Merzbach, U.C., Boyer, C.B.: A History of Mathematics, 5th edn. Wiley, New York (2011). https://doi.org/10.1177/027046769201200316
18. Pan, V.Y.: Optimal (up to polylog factors) sequential and parallel algorithms for approximating complex polynomial zeros. In: ACM STOC 1995, pp. 741–750 (1995)
19. Pan, V.Y.: Solving a polynomial equation: some history and recent progress. SIAM Rev. **39**(2), 187–220 (1997). https://doi.org/10.1137/S0036144595288554
20. Pan, V.Y.: Approximation of complex polynomial zeros: modified quadtree (Weyl's) construction and improved Newton's iteration. J. Complex. **16**(1), 213–264 (2000). https://doi.org/10.1006/jcom.1999
21. Pan, V.Y.: Univariate polynomials: nearly optimal algorithms for factorization and rootfinding. J. Symb. Comput. **33**(5), 701–733 (2002). Proceedings Version in ACM ISSAC 2001, pp. 253–267 (2001). https://doi.org/10.1006/jsco.2002.0531
22. Pan, V.Y.: Old and new nearly optimal polynomial root-finders. In: England, M., Koepf, W., Sadykov, T.M., Seiler, W.M., Vorozhtsov, E.V. (eds.) CASC 2019. LNCS, vol. 11661, pp. 393–411. Springer, Cham (2019). https://doi.org/10.1007/978-3-030-26831-2_26
23. Pan, V.Y.: Acceleration of subdivision root-finding for sparse polynomials. In: Boulier, F., England, M., Sadykov, T.M., Vorozhtsov, E.V. (eds.) CASC 2020. LNCS, vol. 12291, pp. 461–477. Springer, Cham (2020). https://doi.org/10.1007/978-3-030-60026-6_27
24. Pan, V.Y.: New progress in polynomial root-finding. arXiv 1805.12042, 30 May 2018. Accessed 31 Aug 2022
25. Pan, V.Y., Fast subdivision root-finder for a black box polynomial and matrix eigen-solvers accelerated with compression. Preprint (2023)
26. Pan, V.Y.: Fast subdivision root-finders for a black box polynomial and matrix eigen-solvers with novel exclusion tests. Preprint (2023)
27. Pan, V.Y., Go, S., Luan, Q., Zhao, L.: Fast approximation of polynomial zeros and matrix eigenvalues. arXiv 2301.11268 (2022). Accessed 2023
28. Reinke, B.: Diverging orbits for the Ehrlich-Aberth and the Weierstrass root finders. arXiv 2011.01660, 20 November 2020

29. Renegar, J.: On the worst-case arithmetic complexity of approximating zeros of polynomials. J. Complex. **3**(2), 90–113 (1987)
30. Schönhage, A.: The fundamental theorem of algebra in terms of computational complexity. Math. Dept., University of Tübingen, Tübingen, Germany (1982)
31. Storjohann, A.: The shifted number system for fast linear algebra on integer matrices. J. Complex. **21**(4), 609–650 (2005)
32. Strassen, V.: Some results in algebraic complexity theory. In: James, R.D. (ed.) Proceedings of the International Congress of Mathematicians, Vancouver, vol. 1, pp. 497–501. Canadian Mathematical Society (1974)
33. Weyl, H.: Randbemerkungen zu Hauptproblemen der Mathematik. II. Fundamentalsatz der Algebra und Grundlagen der Mathematik. Mathematische Zeitschrift **20**, 131–150 (1924). https://doi.org/10.1007/BF01188076

On the Parameterized Complexity of the Structure of Lineal Topologies (Depth-First Spanning Trees) of Finite Graphs: The Number of Leaves

Emmanuel Sam[1(✉)], Michael Fellows[1,2], Frances Rosamond[1,2], and Petr A. Golovach[1]

[1] Department of Informatics, University of Bergen, Bergen, Norway
{emmanuel.sam,michael.fellows,frances.rosamond,petr.golovach}@uib.no
[2] Western Sydney University, Locked Bag 1797, Penrith, NSW 2751, Australia

Abstract. A *lineal topology* $\mathcal{T} = (G, r, T)$ of a graph G is an r-rooted depth-first spanning (DFS) tree T of G. Equivalently, this is a spanning tree of G such that every edge uv of G is either an edge of T or is between a vertex u and an *ancestor* v on the unique path in T from u to r. We consider the parameterized complexity of finding a lineal topology that satisfies upper or lower bounds on the number of leaves of T, parameterized by the bound. This immediately yields four natural parameterized problems: (i) $\leq k$ leaves, (ii) $\geq k$ leaves, (iii) $\leq n - k$ leaves, and (iv) $\geq n - k$ leaves, where $n = |G|$. We show that all four problems are NP-hard, considered classically. We show that (i) is para-NP-hard, (ii) is hard for W[1], (iii) is FPT, and (iv) is FPT. Our work is motivated by possible applications in graph drawing and visualization.

Keywords: DFS tree · Spanning tree · Parameterized complexity

1 Introduction

For every connected undirected graph $G = (V, E)$ with vertex set $V(G)$ and edge set $E(G)$, there exists a rooted spanning tree T having the property that for every edge $xy \in E(G)$ that is not an edge of T, either x is a descendant of y with respect to T, or x is an ancestor of y. Such a tree is called a *depth first spanning tree* (or *DFS-tree* for short), as one may be computed by depth-first search (DFS), and the edges of G that are not part of T are referred to as *back edges* [21]. It has also been called *lineal spanning tree* [29], *trémaux tree* [10], and *normal spanning tree*, particularly in the case of infinite graphs [12].

The importance of the properties of such trees in the design of efficient algorithms is evident in the great variety of algorithms that employ DFS to solve graph-theoretic problems, including finding connected and biconnected components of undirected graphs [21], bipartite matching [23], planarity testing [22],

Supported by Research Council of Norway (NFR, no. 274526 and 314528).

M. Mavronicolas (Ed.): CIAC 2023, LNCS 13898, pp. 353–367, 2023.
https://doi.org/10.1007/978-3-031-30448-4_25

and checking the connectivity of a graph [15]. In the field of *parameterized complexity* [7,14], DFS has been instrumental in obtaining *fixed-parameter tractable* (*FPT*) results by way of *treedepth* [25] and bounded width *tree decompositions* of the given graph [1,16].

In the work reported herein, we refer to the triple (G, r, T), that is, a graph G together with a choice of root vertex r and a DFS tree T, as a *lineal topology* $\mathcal{T}$, or LT for short. This notion of LT corresponds to a point-set topology on the set of edges $E(G)$ (equipped with a rooted DFS tree T), where the open sets are the sets of edges of the subgraphs induced by rooted subtrees with the same root r as T. The lineal topologies of G may differ in terms of the properties of T, such as height and number of leaves. Figure 1 shows one way of representing an LT as a topological graph or drawing in the plane. Given a graph G and a DFS tree (T, r), an embedding of G in the plane so that every pair of edges that cross is a pair of back edges having at most one crossing point is an instance of an LT of G called *T-embedding* [19]. By definition, there exists a T-embedding of the graph G with no crossings points among the back edges if and only if G is a *planar graph.* This is the basis of Hopcroft and Tarjan's linear time planarity testing algorithm [22] and other algorithms for planarity testing, embedding, and Kuratowski subgraph extraction based on de Fraysseix and Rosenstiehl's Left-Right characterization of planarity [8,9].

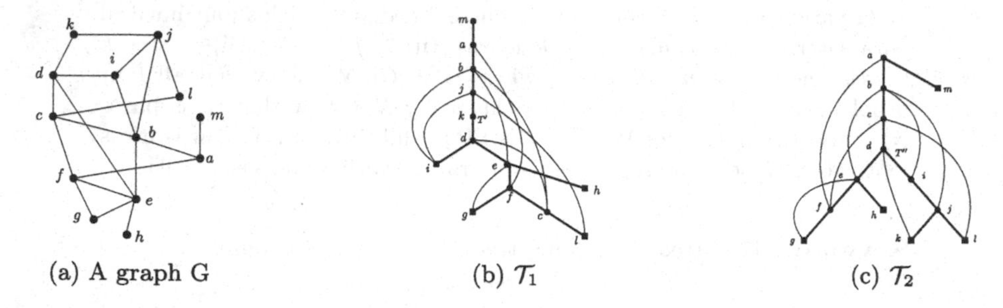

(a) A graph G (b) $\mathcal{T}_1$ (c) $\mathcal{T}_2$

Fig. 1. A given graph G, and two examples of a lineal topology of G, denoted by $\mathcal{T}_1$ and $\mathcal{T}_2$. They differ in the height and number of leaves of the DFS tree T. The leaves are shown as squares. The tree edges are shown in heavy lines, while the back edges are shown in thin curved lines.

Considering the above-mentioned applications of LT and the interesting outcomes enabled by the properties of DFS trees, it will be worthwhile to investigate how their structural properties are related to other properties of graphs, including *crossing number* [20] and *bandwidth* [3], useful in algorithms for VLSI design and graph drawing. Then, a complementary study is the complexity of finding those kinds of lineal topologies. We take the first step in this direction of research by investigating the complexity of two main classical decision problems, namely k-Minimum Leafy LT (k-Min-LLT) and k-Maximum Leafy LT (k-Max-LLT), which correspond to finding an LT with *minimum number of leaves*, and an LT with *maximum number of leaves* respectively.

Given an undirected graph G, the k-MIN-LLT and k-MAX-LLT problems ask whether G has an LT defined by a DFS tree with at most k and at least k leaves, respectively. One observation that is easy to make is that a HAMILTONIAN PATH (HP) rooted at one of the end vertices of the path defines LT with one leaf. Thus, k-MIN-LLT is clearly NP-complete because it is a generalization of the HP problem. To the best of our knowledge, the complexity of k-MAX-LLT has not been previously considered. While there are several results regarding the complexity of MAXIMUM LEAF SPANNING TREE and MINIMUM LEAF SPANNING TREE in general [2, 26–28], for DFS trees, the only available complexity results are due to Fellows et al. [17]. For a given graph G, they considered the difficulty of finding a DFS tree that satisfies upper or lower bounds on two parameters, namely $min(G, T)$ and $max(G, T)$, which stand for the minimum length of a root to leaf path of the DFS tree T, and the maximum length of such a path respectively. They showed that, for a given graph G and an integer $k \geq 0$, the following problems are NP-complete: $min(G, T) \leq k$, $min(G, T) \geq k$, $max(G, T) \leq k$, and $max(G, T) \geq k$. It was also shown that, unless $P = NP$, none of these problems admits a polynomial-time absolute approximation algorithm.

Consequently, an appropriate framework within which to study these sorts of problems is parameterized complexity (PC) [14], according to which problems can be analyzed in terms of other parameters apart from the input size. This leads to algorithms for which we pay an exponential cost in the parameter, thereby solving the problem efficiently on instances with small values of the parameter. For the basics of PC necessary to understand this paper, see Sect. 2.

We consider a parameterization of k-MIN-LLT and k-MAX-LLT, where the parameter k is the size of the solution (number of leaves), and their so-called "dual" parameterization, namely DUAL MIN-LLT (*Does G have an LT with at most $n-k$ leaves?*) and DUAL MAX-LLT (*Does G have an LT with at least $n-k$ leaves?*), where the parameter k is the number of internal vertices. These four parameterized problems are formally defined in Sect. 2. We show that while each parameterized problem and its parametric dual are trivially the same problem and NP-hard when considered classically, when analyzed in terms of PC, the tractability outcomes differ, with one being FPT and the other W[1]-hard.

1.1 Our Results

Our first result is the hardness of k-MAX-LLT. By a reduction from the MULTI-COLORED INDEPENDENT SET (MIS) problem, we show that k-MAX-LLT is hard for W[1] parameterized by k. Furthermore, we show that all four problems, considered classically, are NP-hard, and we show trivially that k-MIN-LLT parameterized k is para-NP-hard. Our main contribution is in showing the existence of an FPT algorithm for DUAL MIN-LLT and DUAL MAX-LLT, parameterized by k, via an application of Courcelle's theorem [4, 6], which relates the expressibility of a graph property using *monadic second-order logic* to its tractability in linear time on graphs with bounded *treewidth* (or *pathwidth*). For a formal definition of this logical language, see Sect. 2.4.

2 Preliminaries

Unless otherwise specified, a graph G with vertex set $V(G)$ and edge set $E(G)$ is simple, finite, undirected, and connected. For a graph G, n and m denotes the number of vertices $|V(G)|$ and the number of edges $|E(G)|$ of G, respectively. We use uv instead of $\{u, v\}$ to denote an edge in $E(G)$. For any vertex $v \in V(G)$, the set $N_G(v)$ denotes the *open neighborhood* of v, that is, the set of neighbors of v in G, and $N_G[v] = N_G(v) \cup \{v\}$ denotes its *closed neighborhood* in G. We drop the G in the subscript if the graph is clear from the context. Given any two graphs $G_1 = (V_1, E_1)$ and $G_2 = (V_2, E_2)$, if $V_1 \subseteq V_2$ and $E_1 \subseteq E_2$ then G_1 is a *subgraph* of G_2, denoted by $G_1 \subseteq G_2$. If G_1 contains all the edges $uv \in E_2$ with $u, v \in V_1$, then we say G_1 is an *induced subgraph* of G_2, or V_1 induces G_1 in G_2, denoted by $G[V_1]$. If there exists are bijective mapping $f : V_1 \to V_2$ that preserves adjacency, that is, $uv \in E_1$ if and only of $f(u)f(v) \in E_2$, then G_1 is *isomorphic* to G_2 and f is called an *isomorphism*. If G_1 is such that it contains every vertex of G_2, i.e., if $V_1 = V_2$ then G_1 is a *spanning subgraph* of G_2. Given a set of vertices $X \subseteq V$, we express the induced subgraph $G[V(G)\backslash X]$ as $G-X$. If $X = \{x\}$, we write $V(G)\backslash x$ instead of $V(G)\backslash\{x\}$ and $G - x$ instead of $G - \{x\}$. For any pair of vertices $uv \in V(G)$ in a given graph G, we denote any path from u to v by $P(u, v)$, and any path of length ℓ by P^ℓ. A vertex u is said to be *reachable* from a vertex v if there is a path $P(u, v)$. Given a graph G, a set of vertices $S \subseteq V(G)$ is a *connected vertex cover* (CVC) of G if the subgraph $G[S]$ induced by S is connected and S is a *vertex cover* of G, i.e., for every edge $uv \in E(G)$, either $u \in S$ or $v \in S$.

2.1 Lineal Topology

Here, we focus on the definitions of the substructures of lineal topologies that are relevant to our proofs. We refer the reader to [11] for details about basic tree terminologies such as root, parent, child, ancestor, etc. In all cases, a DFS tree is simply denoted by T instead of (T, r) if the root is clear from the context. For any given lineal topology $\mathcal{T} = (G, r, T)$, we denote the height of $\mathcal{T}$, that is, the maximum number of edges in any leaf-to-root path of T, by h. A *leaf* of $\mathcal{T}$ is a vertex that has no descendants but is adjacent to one or more ancestors with respect to T (see Fig. 1). We denote by Y and X the set of leaves and *internal vertices* of $\mathcal{T}$, respectively. Given a set of vertices $S \subseteq V(G)$, such that the subgraph $G[S]$ induced by S is connected, we denote the DFS tree of $G[S]$ rooted at $x \in S$ by (T_S, x). The set $E(T)$ and $E(B)$ denote the tree edges and back edges of T respectively. By definition, the set Y is an *independent set*; that is, no pair of vertices $uv \in Y$ are adjacent. This is also true for the set of vertices $U_i \subseteq V(T)$ at each level i of T. Given a vertex v, the set P_v denotes the vertices on the uniquely determined path $P(v, r)$ in T.

2.2 Parameterized Complexity

Now we review some important concepts of parameterized complexity (PC) relevant to the work reported herein. For more details about PC, we refer the reader to [7,14]. Let Σ be a fixed, finite alphabet. A *parameterized problem* is a language $P \subseteq \Sigma^* \times \mathbb{N}$. For an instance $(x, k) \in \Sigma^* \times \mathbb{N}$, $k \in \mathbb{N}$ is called the *parameter*. A parameterized problem P is classified as *fixed-parameter tractable* (FPT) if there exists an algorithm that answers the question "$(x, k) \in P$?" in time $f(k) \cdot poly(|x|)$, where f is a computable function $f : \mathbb{N} \to \mathbb{N}$. Let P and P' be two parameterized problems. A *parameterized reduction* from P to P' is an algorithm that, given an instance (x, k) of P, produces an equivalent instance (x', k') of P' such that the following conditions hold:

1. (x, k) is a YES-instance of P if and only if (x', k') is a YES-instance of P'.
2. There exist a computable function $f : \mathbb{N} \to \mathbb{N}$ such that $k' \leq f(k)$.
3. The reduction can be completed in time $f(k) \cdot poly(|x|)$ for some computable function f.

The *W-hierarchy* [13] captures the level of the intractability of hard parameterized problems. For the purpose of the discussions in this paper, it is enough to note that a problem that is hard for $W[1]$ cannot be solved in FPT running time, unless $FPT = W[1]$. A parameterized problem that is already NP-hard for some single fixed parameter value k (such as $k = 3$ for GRAPH k-COLORING) is said to be *para-NP-hard.*

2.3 Problem Definitions

We formally define the parameterized problems studied in this work as follows:

k-MIN-LLT
Input: A connected undirected graph $G = (V, E)$.
Parameter: k
Question: Does G admit an LT with $\leq k$ leaves?

k-MAX-LLT
Input: A connected undirected graph $G = (V, E)$
Parameter: k
Question: Does G have an LT with $\geq k$ leaves?

DUAL MIN-LLT
Input: A connected undirected graph $G = (V, E)$ and positive integer k
Parameter: k
Question: Does G admit an LT with $\leq n - k$ leaves?

DUAL MAX-LLT
Input: A connected undirected graph $G = (V, E)$ and positive integer k
Parameter: k
Question: Does G have an LT with $\geq n - k$ leaves?

Below, we present the definitions of the concepts used in Sect. 3, to show that DUAL MIN-LLT and DUAL MAX-LLT are FPT with respect to k.

Definition 1 (Treewidth, Pathwidth). *A tree decomposition of a given graph G is a pair (T_D, B) where T_D is a tree and B is a family of subsets $\{B_i \subseteq V(G) : i \in V(T_D)\}$, called bags, with each node in T_D associated with a bag, satisfying the following properties: (1) $\bigcup_{i\in V(T_D)} B_i = V(G)$, (2) $\forall_{uv\in E(G)}, \exists i \in V(T_D) : u \in B_i$, and $v \in B_i$, and (3) $\forall_{v\in V(G)}$, the set $T_D^i = \{i \in V(T_D) : v \in B_i\}$ gives rise to a connected subtree of T_D.*

The width of (T_D, B) is $max\{|B_i| : i \in V(T_D)\} - 1$ and the treewidth of a graph, denoted $tw(G)$, is the minimum width over all tree decompositions of G. If T_D is a path, then (T_D, B) is called the path decomposition of G and the minimum width over all path decompositions of G is its pathwidth, often denoted by $pw(G)$. For any graph G, it is a fact that $tw(G) \leq pw(G)$ [25].

2.4 Logic of Graphs

We now introduce the basic definitions and notations of the logic of graphs and MSO, the logical language with which we specify the properties associated with DUAL MIN-LLT and DUAL MAX-LLT in Sect. 3. For a thorough discussion of these topics, we refer the reader to [6,24].

Recall that *Second-Order Logic* (SO) is an extension of *First-Order Logic* (FO) that allows quantification over predicates or relations of arbitrary arity. *Monadic Second-Order Logic* (MSO) are SO formulas in which only quantification over unary relations (i.e., subsets of the domain) is allowed. To express graph properties using MSO, a graph $G = (V, E)$ can be represented either as a *logical (or relational) structure* $\lfloor G \rfloor$ whose *domain* is the vertex set V, with a binary relation *adj* on V representing the edges, or as a logical structure $\lceil G \rceil$ whose domain is formed by the disjoint union of V and E, with a binary relation *inc* representing the *incidence* between the vertices and edges of G. There are two main variants of MSO: MSO_1 and MSO_2, corresponding to $\lfloor G \rfloor$ and $\lceil G \rceil$ respectively.

Definition 2 (MSO_1 language). *The logical expressions or formulas of this language are built from the following elements:*

1. *Small variables $u, v, x_1, \ldots, x_k$ for vertices*
2. *Big variables $X, Y, U_1, \ldots, U_k$ for sets of vertices.*
3. *Predicates $adj(u, v)$ and $u \in V$ for adjacency and membership respectively, and "$=$" equality testing.*
4. *The logical connectives $\vee, \wedge, \neg, \Rightarrow$*
5. *$\forall x, \exists x$ for quantification over vertices and $\forall U$, $\exists U$ for quantification over vertex sets.*

The MSO_2 language extends MSO_1 with variables denoting edges and subset of edges, and the predicate $inc(x, e)$ for incidence, and allows quantification over edges and edge sets.

For clarity, we use $\forall_x$, $\exists_x$ instead of $\forall x$, $\exists x$, and $\forall_U$, $\exists_U$ instead of $\exists U$, and $\forall U$. Given a graph G belonging to a class of graphs $\mathcal{C}$ and a formula Φ

expressing a graph property in MSO, we denote the statement "Φ *is true of the vertices and the relation of* G" by $G \models \Phi$ (read as "G models Φ" or "Φ holds on G"). For a logical language $\mathcal{L} \in \{MSO_1, MSO_2\}$, we say that a graph property is $\mathcal{L}$*-expressible* if there exists a formula (sentence) of $\mathcal{L}$ for expressing it. The theorem below states the consequence of expressing a graph property by an MSO_2 formula.

Lemma 1 (Courcelle's theorem [4,6]). Assume that ϕ is a fixed MSO_2 formula of length ℓ expressing a graph property. Then for any graph G belonging to a graph class $\mathcal{C}$ with treewidth bounded by a fixed positive integer k, there is an algorithm that takes G and its tree decomposition as input and decides whether $G \models \phi$ in time $\mathcal{O}(f(\ell, k) \cdot n)$, for some computable function f.

By a theorem similar to Lemma 1, every graph property that is MSO_1-expressible can be decided in linear time on graphs of bounded *clique-width*, a graph complexity measure that is similar to treewidth [5]. It is worth noting that Lemma 1 also holds for MSO_1 formulas because every graph property expressible by an MSO_1 formula is also expressible by an MSO_2 formula; but the converse is not true. For example, the existence of a Hamiltonian path can only be expressed in MSO_2 [6]. Therefore, it is stronger to claim that a given property is MSO_1-expressible. In Sect. 3.2, we show that the property of having an LT with at least $n - k$ leaves or at most $n - k$ leaves is MSO_1-expressible.

3 Complexity Analysis

In this section, we consider the four natural parameterized problems defined above. We first show that k-Max-LLT is hard for W[1] parameterized by k. The reduction in the proof is polynomial and thus the problem is NP-hard. Besides, we show trivially that k–Min-LLT is para-NP-hard with respect to k. This is followed by proofs of hardness for Dual Min-LLT and Dual Max-LLT, considered classically. Moreover, we show that these two problems are FPT with respect to k. To this end, we construct MSO_1 formulas ϕ_k and φ_k to express the property of having an LT with at most $n - k$ leaves and that of having an LT with at least $n - k$ leaves. Next, we make use of the following facts on the height of a lineal topology to show the existence of an FPT algorithm for the two problems. Given a graph G and an integer $k \geq 0$, we show, for Dual Max-LLT, that if the height of the DFS tree T resulting from any DFS of G is more than $2^{k+1} - 2$, then T witnesses that the answer is NO, otherwise, G has a path decomposition of width at most $2^{k+1} - 1$. For Dual Min-LLT, we trivially show that the answer is YES if the number of internal vertices of T is at least k, otherwise, G has a path decomposition of width at most k.

3.1 Hardness Results

Theorem 1. k-Max-LLT *is W[1]-hard parameterized by* k *and NP-complete when considered classically.*

Proof. We reduce from the parameterized MULTICOLORED INDEPENDENT SET (MIS) problem parameterized by the number of colors. In the MIS problem, we are given a graph $G = (V, E)$ and a coloring of V with k colors, and the task is to determine whether G has a *k-colored independent set*, that is, a k-sized independent set containing one vertex from each color class. We trivially assume that each color class induces a clique for our argumentation. This problem is W[1]-hard with respect to k [7, chapter 13], which implies that it is unlikely that it can be solved in time $f(k) \cdot poly(n)$ for any computable function f.

Given a positive integer k, let G be an instance of the MIS problem in which $\{V_1, \ldots, V_k\}$ is a partition of the vertex set $V(G)$ such that, for each $i \in [1, k]$, V_i induces a clique and corresponds to a color class. Now we construct an instance (G', k) of the k-MAX-LLT problem from G by introducing a set of k universal vertices $U = \{u_1, u_2, \ldots, u_k\}$, i.e., every $u_i \in U$ is adjacent to every vertex in G and in $U \backslash u_i$. The completed $G' = (V', E')$ has $V' = V \cup U$, and $E' = E(G) \cup \{u_i v \mid u_i \in U, v \in V(G') \backslash u_i\}$ (see Fig. 2). The main idea of this construction, as will be argued below, is to enable a depth-first traversal of G that guarantees an LT with at least k leaves corresponding to a k-sized independent set in G, if it exists. It is not hard to see that we can construct (G', k) from (G, k) in polynomial time. Lemmas 2 and 3 below show that G' admits an LT with at least k leaves if and only if G has a k-colored independent set.

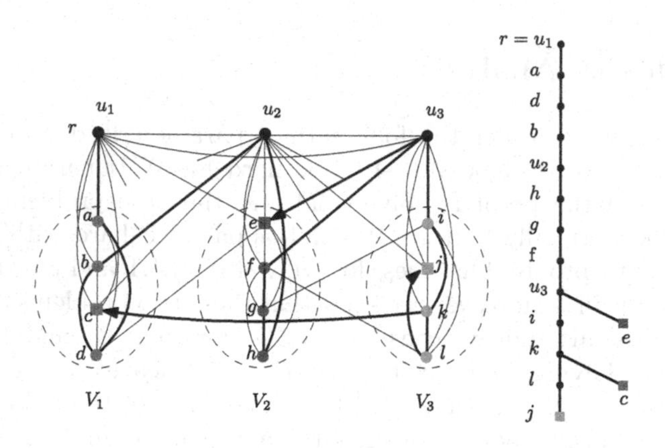

Fig. 2. An example of a reduction from an instance G of MIS, with k-colored independent set $X = \{c, e, j\}$, to an instance G' of MAX-LLT and a DFS of G' that yields a DFS tree T' with $\{c, e, j\}$ as its leaves.

Lemma 2. *If a k-colored independent set exists in G, then G' admits an LT with at least k leaves.*

Proof. Suppose that $X = \{x_1, ..., x_k\}$ is a k-colored independent set in G. Since $x_1 \in V_1, \ldots, x_k \in V_k$ with each color class V_i inducing a clique, any depth-first traversal of G' that excludes the vertices in X until all the vertices in $V(G') \backslash X$

have been visited yields an LT with the vertices in X as its leaves. One way to achieve this is to start from the vertex $u_1 \in U$ and visit every vertex in the corresponding color class V_1 except x_1. Next, choose $u_2 \in U$ and explore every vertex in V_2 except x_2. Repeat this process sequentially for each $u_i \in U$ and its corresponding color class V_i until the last vertex $u_k \in U$ is reached. Now choose the vertex x_k after exploring every vertex in the set $V_k \backslash \{x_k\}$. See Fig. 2 for an illustration of this process. At this point, every edge incident to x_k leads to a vertex already reached by the DFS because X is an independent set. Thus, x_k becomes a leaf in the resulting DFS tree T' of G'. If any vertex $x_i \in X$ is adjacent to an already visited vertex $v \in V_k$, we backtrack and choose x_i from v. Otherwise, we backtrack to $u_k \in U$, as every vertex in X is reachable from this vertex by construction. Each of the remaining vertices $\{x_1, \ldots, x_{k-1}\}$ reached by DFS becomes a leaf in T' because of the same reason as for x_k. □

Lemma 3. *If an LT with at least k leaves in G' exists, then there is a k-colored independent set in G.*

Proof. If $k = 1$ then G is obviously a YES-instance. Suppose that $k \geq 2$ and G' admits an LT in which $X = \{x_1, x_2, ..., x_k\}$ are the leaves. Observe that X is an independent set. Then, based on the following claims, we conclude that X induces a k-colored independent set in G.

Claim 3.1. Each color class V_i in G' can contain at most one vertex from X.

Proof. The set of leaves X is an independent set and, by construction, each color class V_i in G' is a clique. Therefore, there cannot be any LT of G' with two or more leaves from the same color class. ◆

Claim 3.2. None of the vertices in X is from the vertex set $U = \{u_1, \ldots, u_k\}$.

Proof. For $i \in [1, k]$, if u_i is a leaf of T', the remaining vertices in $V(G') \backslash u_i$ must necessarily be internal vertices of T' by construction. Since X contains at least 2 vertices, it follows that no vertex in U can be in X. ◆

Combining Claims 3.1 and 3.2, we conclude that X is a k-colored independent set in G.

□

Theorem 2. k-Max-LLT *is NP-complete.*

Proof. k-Max-LLT is clearly in NP. The NP-hardness of the problem follows from the proof of Theorem 1, because Multicolored Independent Set is NP-complete [7,20] and the reduction from Multicolored Independent Set to k-Max-LLT is a polynomial-time reduction.

Theorem 3. k-Min-LLT *parameterized by k is para-NP-hard.*

Proof. This follows, trivially, from the fact that k-Min-LLT is NP-hard already for $k = 1$, because for this case, k-Min-LLT is equivelent to Hamiltonian Path which is well-known to be NP-complete [20].

Theorem 4. Dual Min-LLT *and* Dual Max-LLT*, considered classically, are NP-hard.*

Proof. Dual Min-LLT can be restricted to the Hamiltonian Path problem by allowing only instances in which $k = n-1$. Similarly, $(G, n-k)$ is a YES-instance of k-Max-LLT iff (G, k) is a YES-instance of Dual Max-LLT. □

3.2 MSO Formulations

Recall that a DFS tree T is a tree formed by a set of edges $E(T)$ with a choice of a root vertex r, such that every edge not in T connects a pair of vertices that are related to each other as an ancestor and descendant in T. For any DFS tree T of a given graph G, the set of vertices U_i at each level i of T is an independent set of G. Thus, T corresponds to a partition of the vertex set $V(G)$ into a sequence of independent sets $(U_0, U_1, ..., U_h)$ with $h \in \mathbb{N}$, where the root r is the only member of U_0 and U_h contains the vertex witnessing the height h of T (see Fig. 3a). Based on this observation, we provide the following definition of a DFS tree with bounded height, which allows us to express the properties "G has an LT with at least $n-k$ leaves" and "G has an LT with at most $n-k$ leaves" in MSO_1.

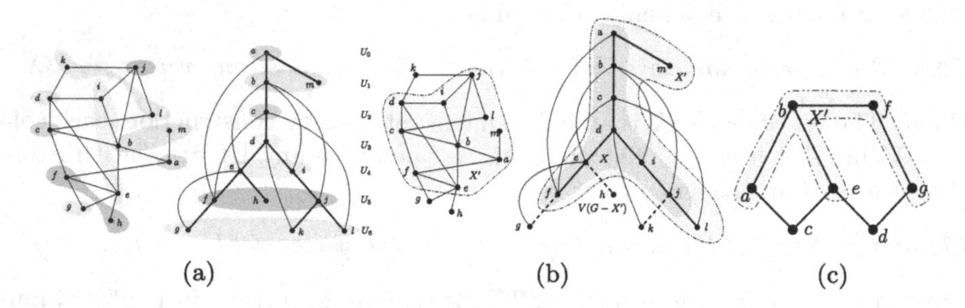

Fig. 3. (a) Color classes denoting an LT partition $(U_0, U_1, ..., U_6)$ of a graph G with $n = 13$, and a representation of the associated LT of G with height $h = 6$ and number of internal vertices $|X| = 8$. (b) A CVC X' of a graph G, and a DFS tree T of G formed by a DFS tree $(T_{X'}, a)$ for $G[X']$ and the independent set $V(G - X')$. The internal vertices X of T consist of $\{a, b, c, d, e, f, i, j\}$. Thus $|X| \leq |X'|$. (c) A CVC X' of a graph G such that $G[X']$ does not admit a DFS tree $T_{X'}$ which extends to a DFS tree T of G.

Definition 3. *Let G be a graph and h a positive integer. A tree-partition of G of height h is a sequence $(U_0, \ldots, U_h)$ with $U_0, \ldots, U_h \subseteq V(G)$ such that:*

1. *$(U_0, U_1, \ldots, U_h)$ is a partition of $V(G)$.*
2. *U_0 contains only one element r.*
3. *Every vertex $u \in U_i$ has a unique neighbor $v \in U_{i-1}$ for all $i \in [1, h]$.*

The tree associated with the tree-partition $(U_0, \ldots, U_h)$ is the rooted spanning tree T of G with root r and edge set $E(T) = \{uv : uv \in E(G), u \in U_i, v \in U_{i-1}$ for all $i \in [1, h]\}$. We say that the tree-partition $(U_0, \ldots, U_h)$ is an LT-partition of G if:

4. *For every edge uv of G, u is an ancestor of v in T or v is an ancestor of u in T.*

Lemma 4. *For every $h \in N$, a graph $G = (V, E)$ has an LT with height h if and only if G admits an LT-partition $(U_0, U_1, ..., U_h)$.*

Proof. Let T be the spanning tree of G associated with $(U_0, U_1, ..., U_h)$. Property 4 of Definition 3 ensure that every $uv \in E \setminus E(T)$ is a back edge, and, thus, for all $i \in [0, h]$, U_i is an independent set. This gives us an LT with height h.

Conversely, suppose we have an LT of G defined by a DFS tree T with height h. Then it is easy to see that the root and the vertices at each level of T constitute an LT-partition $(U_0, U_1, ..., U_h)$. □

Dual Max-LLT. Let G be a graph and $k \in \mathbb{N}$. If $k = 1$ then G is a YES-instance of Dual Max-LLT if and only if G is a *star*. Thus, in what follows, we assume that $k \geq 2$ and $|G| \geq 3$. Suppose that G admits an LT with at least $n-k$ leaves, and consider a DFS tree (T, r) witnessing that G is a YES-instance. We can readily observe that the internal vertices X of (T, r) is a CVC of G, and the subtree (T_X, r) with height $h \leq |X|$ is a DFS tree of the graph $G[X]$ induced by X. This is also true for any subtree $(T_{X'}, r)$, such that $X' \supseteq X$ and $|X'| \leq k$ (see Fig. 3b). However, given a graph $G = (V, E)$ and a CVC X of G, $G[X]$ may not have a DFS tree T_X that can be extended to a DFS tree T of G by adding the vertices in $V(G - X)$ to T_X as leaves. An example of such a CVC of a given graph is shown in Fig. 3c. Using these intuitive ideas, we characterize the graphs that admit an LT with at least $n - k$ leaves by Lemma 5. See the full version of this paper for the proof of this lemma.

Lemma 5. *A graph G has an LT with at least $n - k$ leaves if and only if it has a set of vertices X' of size at most k satisfying the following properties:*

1. *X' is a connected vertex cover of G.*
2. *$G[X']$ admits an LT partition $(U_0, ..., U_h)$ with $h \leq |X'|$ such that, for any vertex $y \in V(G) \setminus X'$, if y is adjacent to a pair of vertices $u, v \in X'$, then either u is the ancestor of v or v is the ancestor of u in the LT formed by $(U_0, ..., U_h)$.*

Theorem 5. *For all $k \in N$, there exists an MSO_1 formula ϕ_k such that for every graph G, it holds that $\lfloor G \rfloor \models \phi_k$ iff G is a YES-instance of* Dual Max-LLT.

To prove this, we construct the formula ϕ_k as a conjunction of formulas expressing the properties in Definition 3 and Lemma 5; see the full version of this paper.

Dual Min-LLT. For $k = 1$, every graph G is a YES-instance of Dual Min-LLT. Thus, henceforth, we assume that $k \geq 2$ and $|G| \geq 3$. Let G be a given graph that admits an r-rooted LT with at most $n - k$ leaves (i.e., at least k internal vertices), and consider a DFS tree T of G witnessing this fact. Then any subtree $(T_{X'}, r)$ of T, where $X' = \{x_1, \ldots, x_k\}$ is a set of k internal vertices and $r \in X'$ is the same as the root of T, is a DFS tree for the subgraph $G[X']$ induced by X'; see Fig. 4 for an example. Observe that, every leaf of $(T_{X'}, r)$ is adjacent to a vertex in the set $W = V(T) \backslash X'$, and each of the subtrees that extend $T_{X'}$ to form T is a DFS tree for some maximal connected component of $G - X'$.

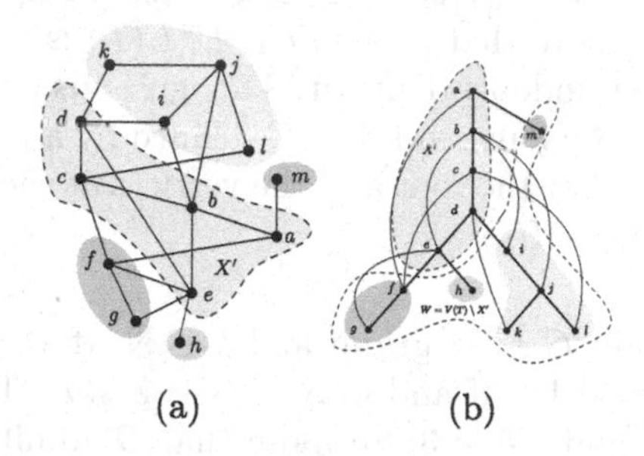

Fig. 4. (a) A graph G with a connected subset of vertices X' of size 5, and (b) a DFS tree $(T_{X'}, a)$ for $G[X']$ that extends to an LT of G with internal vertices $X' \cup \{f, i, j\}$. W consists of three maximal connected subgraphs of G shown in different colors.

As a result, we transform the problem of determining whether G admits an LT with at most $n-k$ leaves to that of deciding whether there exists a subset of k vertices X', such that, the subgraph $G[X']$ admits an r-rooted DFS tree $T_{X'}$ isomorphic to a subtree (T', r) of an LT of G witnessing that G is a YES-instance. To this end, we introduce the following definition.

Definition 4. *Let G be a graph and $X' = \{x_1, \ldots, x_k\}$, a set of k vertices that induces a connected subgraph of G. We say that a tree-partition $(U_0, \ldots, U_h)$ of $G[X']$ of height $h \leq k$ is a partial LT-partition of G or extends to an LT-partition of G if it satisfies the following property: for every $W' \subseteq V(G)$ such that $G[W']$ is a maximal connected subgraph of $G - X'$, there exists $x' \in X'$ such that any vertex in $x \in X'$ with at least one neighbor in W' is an ancestor of x'.*

In Fig. 4, it is easy to see that the subgraph induced by $\{k, j, i, l\}$ does not have a tree-partition that forms a partial LT-partition of G. For a partial LT-partition of a given subgraph $G[X']$ of size k to yield an LT of G with at most $n-k$ leaves, it is necessary that every leaf of the partial LT is adjacent to at least one vertex in $V(G) \backslash X'$. Based on these intuitive ideas, we characterize the YES-instances of DUAL MIN-LLT by Lemma 6; see the full version of this paper for the proof.

Lemma 6. *For every $k \in N$, a graph $G = (V, E)$ has an LT with the number of leaves $\leq n - k$ if and only if there exists a set of vertices X' of size at least k satisfying the following property: $G[X']$ admits a partial LT-partition $(U_0, \ldots, U_h)$ of height $h \leq k$ in which every leaf is adjacent to at least one vertex in $W = V(G) \backslash X'$.*

Now, we use Lemma 6 to obtain Theorem 6; see the full version of this paper for the proof.

Theorem 6. *For all $k \in N$, there exists an MSO_1 formula ψ_k such that for every graph G, it holds that $\lfloor G \rfloor \models \psi_k$ iff G is a YES-instance of* DUAL MIN-LLT.

3.3 FPT Algorithms for DUAL MIN-LLT and DUAL MAX-LLT

In this section, we show the existence of an FPT algorithm for DUAL MIN-LLT and DUAL MAX-LLT using the MSO formulations above, and Lemmas 7 and 8. For the proofs of these lemmas, see the full version of this paper.

Lemma 7. *Given a graph G and a positive integer t, if G admits an LT with height at most t, then G has a path decomposition of width at most t that can be computed in linear time.*

Lemma 8. *Given a graph G and a positive integer k, if G admits an LT of height at most k, then the length of any path in G is at most $2^{k+1} - 2$.*

Theorem 7. DUAL MAX-LLT *parameterized by $k \in \mathbb{N}$ is in FPT.*

Proof. Let G be a graph and k a positive integer. One observation that is easy to make is that if G is a YES-instance of DUAL MAX-LLT, then G admits an LT with height at most k. Thus, DUAL MAX-LLT can be solved as follows: (1) Construct a DFS tree T by performing a DFS of G. If the height h of T is more than $2^{k+1} - 2$, then we know, by Lemma 8 that G does not admit an LT with $h \leq k$. Therefore, return NO and stop. (2) Otherwise, use Lemma 7 to read off a path decomposition of G of width at most $2^{k+1} - 1$ from T, one bag per leaf. (3) Applying Courcelle's theorem with Theorem 5 and this path decomposition, it follows that checking whether G is a YES-instance is FPT in k. □

Theorem 8. DUAL MIN-LLT *parameterized by $k \in \mathbb{N}$ is in FPT.*

Proof. The proof follows steps analogous to that of Theorem 7. Construct any DFS tree T. If T has at most $n-k$ leaves, return YES and stop. Otherwise, we use Lemma 7, to obtain a path decomposition of G of width at most k from T. With this, Theorem 6, and Courcelle's theorem implies that we can derive an FPT algorithm that runs in time linear in n to check whether G is a YES-instance. □

4 Conclusion

In this paper, we have shown hardness results for four natural parameterized problems that have to do with finding an LT (or DFS tree) with a restricted number of leaves. Our theorem shows that k-MAX-LLT is hard for W[1]. This raises the natural question of where it belongs in the W-hierarchy with respect to membership. Is it in W[1] and thus W[1]-complete? Is it in W[P]? It seems to be

AND-compositional, like the BANDWIDTH problem, as discussed in [18] and we conjecture that k-MAX-LLT cannot be in W[P] for reasons similar to the case of BANDWIDTH. We have also shown that DUAL MIN-LLT and DUAL MAX-LLT are FPT parameterized by k. Instead of relying on Courcelle's theorem to show the existence of an FPT algorithm for these problems, we believe it should be possible to construct an algorithm that solves each problem explicitly via dynamic programming over the path decomposition returned by our algorithm. An obvious question is whether both problems admit a polynomial kernel.

On the complexity of finding an LT with restricted height h, the only known results are the NP-hardness results due to Fellows et al. [17] (see Sect. 1). A consequence of our Theorem 7 is that the problem $h \leq k$ is FPT with respect to k. The natural parameterized problem $h >= n - k$ is para-NP-complete since it is equivalent to HAMILTONIAN PATH when k=1. We plan to investigate the PC of: (i) $h \leq n - k$, and (ii) $h \geq k$, which we believe is FPT parameterized by k.

Acknowledgement. We acknowledge support from the Research Council of Norway (NFR, no. 274526 and 314528). We also thank Nello Blaser and Benjamin Bergougnoux for helping to review the initial versions of the paper.

References

1. Bodlaender, H.: On linear time minor tests with depth-first search. J. Algorithms **14**(1), 1–23 (1993)
2. Bonsma, P.S., Brueggemann, T., Woeginger, G.J.: A faster FPT algorithm for finding spanning trees with many leaves. In: Rovan, B., Vojtáš, P. (eds.) MFCS 2003. LNCS, vol. 2747, pp. 259–268. Springer, Heidelberg (2003). https://doi.org/10.1007/978-3-540-45138-9_20
3. Chinn, P.Z., Chvátalová, J., Dewdney, A.K., Gibbs, N.E.: The bandwidth problem for graphs and matrices-a survey. J. Graph Theory **6**(3), 223–254 (1982)
4. Courcelle, B.: The monadic second-order logic of graphs. I. Recognizable sets of finite graphs. Inf. Comput. **85**(1), 12–75 (1990)
5. Courcelle, B., Makowsky, J.A., Rotics, U.: Linear time solvable optimization problems on graphs of bounded clique-width. Theory Comput. Syst. **33**(2), 125–150 (2000). https://doi.org/10.1007/s002249910009
6. Courcelle, P.B., Engelfriet, D.J.: Graph Structure and Monadic Second-Order Logic: A Language-Theoretic Approach, 1st edn. Cambridge University Press, Cambridge (2012)
7. Cygan, M., et al.: Parameterized Algorithms, 1st edn. Springer, Cham (2015). https://doi.org/10.1007/978-3-319-21275-3
8. De Fraysseix, H.: Trémaux trees and planarity. Electron. Notes Discrete Math. **31**, 169–180 (2008)
9. De Fraysseix, H., De Mendez, P.O., Rosenstiehl, P.: Trémaux trees and planarity. Int. J. Found. Comput. Sci. **17**(05), 1017–1029 (2006)
10. de Fraysseix, H., de Mendez, P.O.: Trémaux trees and planarity. Eur. J. Comb. **33**(3), 279–293 (2012). Topological and Geometric Graph Theory
11. Diestel, R.: Graph Theory, 5th edn. Springer, Heidelberg (2017). https://doi.org/10.1007/978-3-662-53622-3

12. Diestel, R., Leader, I.: Normal spanning trees, Aronszajn trees and excluded minors. J. Lond. Math. Soc. **63**(1), 16–32 (2001)
13. Downey, R.G., Fellows, M.R.: Fixed-parameter tractability and completeness I: basic results. SIAM J. Comput. **24**(4), 873–921 (1995)
14. Downey, R.G., Fellows, M.R.: Fundamentals of Parameterized Complexity. Springer, London (2013). https://doi.org/10.1007/978-1-4471-5559-1
15. Even, S., Tarjan, R.E.: Network flow and testing graph connectivity. SIAM J. Comput. **4**(4), 507–518 (1975)
16. Fellows, M.R., Langston, M.A.: On search decision and the efficiency of polynomial-time algorithms. In: Proceedings of the Twenty-First Annual ACM Symposium on Theory of Computing, STOC 1989, pp. 501–512. Association for Computing Machinery, New York (1989). https://doi.org/10.1145/73007.73055
17. Fellows, M.R., Friesen, D.K., Langston, M.A.: On finding optimal and near-optimal lineal spanning trees. Algorithmica **3**(1–4), 549–560 (1988). https://doi.org/10.1007/BF01762131
18. Fellows, M.R., Rosamond, F.A.: Collaborating with Hans: some remaining wonderments. In: Fomin, F.V., Kratsch, S., van Leeuwen, E.J. (eds.) Treewidth, Kernels, and Algorithms. LNCS, vol. 12160, pp. 7–17. Springer, Cham (2020). https://doi.org/10.1007/978-3-030-42071-0_2
19. de Fraysseix, H., Rosenstiehl, P.: A characterization of planar graphs by Trémaux orders. Combinatorica **5**(2), 127–135 (1985). https://doi.org/10.1007/BF02579375
20. Garey, M.R., Johnson, D.S.: Crossing number is NP-complete. SIAM J. Algebraic Discrete Methods **4**(3), 312–316 (1983). https://doi.org/10.1137/0604033
21. Hopcroft, J., Tarjan, R.: Algorithm 447: efficient algorithms for graph manipulation. Commun. ACM **16**(6), 372–378 (1973)
22. Hopcroft, J., Tarjan, R.: Efficient planarity testing. J. ACM (JACM) **21**(4), 549–568 (1974)
23. Hopcroft, J.E., Karp, R.M.: An $n^{5/2}$ algorithm for maximum matchings in bipartite graphs. SIAM J. Comput. **2**(4), 225–231 (1973)
24. Libkin, L.: Elements of Finite Model Theory. Springer, Heidelberg (2004). https://doi.org/10.1007/978-3-662-07003-1
25. Nešetřil, J., de Mendez, P.O.: Bounded height trees and tree-depth. In: Nešetřil, J., de Mendez, P.O. (eds.) Sparsity. AC, vol. 28, pp. 115–144. Springer, Heidelberg (2012). https://doi.org/10.1007/978-3-642-27875-4_6
26. Ozeki, K., Yamashita, T.: Spanning trees: a survey. Graphs Comb. **27**, 1–26 (2011). https://doi.org/10.1007/s00373-010-0973-2
27. Prieto, E., Sloper, C.: Either/Or: using VERTEX COVER structure in designing FPT-algorithms—the case of k-INTERNAL SPANNING TREE. In: Dehne, F., Sack, J.-R., Smid, M. (eds.) WADS 2003. LNCS, vol. 2748, pp. 474–483. Springer, Heidelberg (2003). https://doi.org/10.1007/978-3-540-45078-8_41
28. Rosamond, F.: Max leaf spanning tree. In: Kao, M.Y. (ed.) Encyclopedia of Algorithms, pp. 1211–1215. Springer, New York (2016). https://doi.org/10.1007/978-1-4939-2864-4_228
29. Williamson, S.: Combinatorics for Computer Science. Dover Books on Mathematics. Dover Publications (2002). https://books.google.no/books?id=YMIoy5JwdHMC

Efficiently Enumerating All Spanning Trees of a Plane 3-Tree
(Extended Abstract)

Muhammad Nur Yanhaona[1(✉)], Asswad Sarker Nomaan[2], and Md. Saidur Rahman[2]

[1] Department of Computer Science and Engineering, BRAC University, Dhaka, Bangladesh
nur.yanhaona@bracu.ac.bd

[2] Graph Drawing and Information Visualization Laboratory, Department of Computer Science and Engineering, Bangladesh University of Engineering and Technology, Dhaka, Bangladesh
1021052002@grad.cse.buet.ac.bd, saidurrahman@cse.buet.ac.bd

Abstract. A spanning tree T of a connected, undirected graph G is an acyclic subgraph having all vertices and a minimal number of edges of G connecting those vertices. Enumeration of all possible spanning trees of undirected graphs is a well-studied problem. Solutions exist for enumeration for both weighted and unweighted graphs. However, these solutions are either space or time efficient. In this paper, we give an algorithm for enumerating all spanning trees in a plane 3-tree that is optimal in both time and space. Our algorithm exploits the structure of a plane 3-tree for a conceptually simpler alternative to existing general-purpose algorithms and takes $\mathcal{O}(n+m+\tau)$ time and $\mathcal{O}(n)$ space, where the given graph has n vertices, m edges, and τ spanning trees. This is a substantial improvement in both time and space complexity compared to the best-known algorithms for general graphs. We also propose a parallel algorithm for enumerating spanning trees of a plane 3-tree that has $\mathcal{O}(n+m+\frac{n\tau}{p})$ time and $\mathcal{O}(\frac{n\tau}{p})$ space complexities for p parallel processors. This second algorithm is useful when storing the spanning trees is important.

Keywords: Spanning Tree Enumeration · Plane 3-Tree · Time and Space Complexity

1 Introduction

A spanning tree of a connected undirected graph G is a subgraph that is a tree spanning all vertices of G. Spanning trees have always been a focus for researchers when dealing with various graph-related problems due to their wide range of applications in areas such as computer networks, telecommunications networks, transportation networks, water supply networks, and electrical grids.

The enumeration of all spanning trees of a graph G is to find all possible distinct edge combinations which minimally connect all its vertices without

M. Mavronicolas (Ed.): CIAC 2023, LNCS 13898, pp. 368–382, 2023.
https://doi.org/10.1007/978-3-031-30448-4_26

introducing any cycle. Enumerating all possible spanning trees is often crucial in electrical circuits, routing networks, and other network analysis and optimization applications [3]. For example, the current flow of an electrical network is associated with the sum of electrical admittances in its spanning trees [1,13]. The main issues with spanning tree enumeration are exponential time complexity, costly storage space requirements, and duplicate tree generation. Various algorithms with varying degrees of efficiency exist in the literature addressing these issues [2,5–7,9,10,14]. To the best of our knowledge, however, no solution exists that is optimal in all these aspects for any complex graph class.

Char's [2] 1968 algorithm is among the first works on spanning tree enumeration. For a graph G with n vertices and m edges, it considers all possible combinations of $n-1$ edges as potential spanning trees and gives unique spanning trees as output. The algorithm assigns the vertices of G increasing orders from 1 to n. It next selects the nth vertex as the root of all spanning trees. Then the algorithm repeatedly emits sequences of $n-1$ numbers as a permutation of vertex orders, where the ith number in the sequence is the endpoint of an edge originating from the vertex with order i. Although the algorithm avoids duplicate tree generation, it involves checking if a sequence forms a spanning tree or some other cyclic subgraph. If τ' and τ respectively denote the numbers of non-tree and tree sequences of G, the time complexity of the algorithm is $\mathcal{O}(m+n+\tau+\tau')$. The space required by Char's algorithm is $\mathcal{O}(nm)$. This space overhead is due to a tabular structure listing all adjacent vertices of each vertex.

The algorithm Rakshit et al. proposed in 1981 [14] is quite similar to Char's algorithm. The algorithm uses a privileged reduced incidence edge structure (PRIES). A reduced incidence edge structure (RIES) of graph G is a table with $n-1$ vertices as row headers and column entries representing the edges incident on the respective vertices. A PRIES is a table derived from a RIES. The PRIES that the algorithm proposes is effectively the exact structure of [2]. The difference between the two algorithms lies in how they detect non-tree edge sequences. While Char's algorithm repeatedly checks for cycle to prune non-tree edge sequences, Rakshit's algorithm does that by repeatedly removing two "pendant" vertices [4] from a sequence and checking the degrees of remaining vertices. The time and space complexities of this algorithm are similar to Char's algorithm. Both algorithms are superior in time complexity to their contemporary alternative enumeration algorithms [6,7,10] as they avoid duplicate edge sequences and prune non-tree sequences using the geometric properties of the input graph.

Gabow and Myers [5] and Matsui [9] improve the time complexity of spanning tree enumeration from earlier algorithms by avoiding evaluating candidate sequences of vertices that cannot form a connected subgraph. Both algorithms recursively generate spanning trees by growing them from smaller sub-trees and checking whether adding a new edge on a growing tree will produce a cycle. Gabow and Myers' algorithm traverses the input graph G in depth-first-order from a single pivot vertex that the algorithm considers the root of all spanning trees of G. As it grows a tree from that root vertex, the algorithm removes and restores edges in G in a deterministic order to avoid duplicate tree generation. The algorithm works for both directed and undirected graphs and takes $\mathcal{O}(n+m+m\tau)$ and $\mathcal{O}(n+m+n\tau)$ times for the former and the latter, respectively. The vertex and edge count multiplier over the number of spanning trees τ in this algorithm is due to checking all possible extensions to a new edge or vertex from a single vertex of the tree under construction.

Matsui's algorithm employs that one can construct a spanning tree T of a graph G by swapping an edge f of another spanning tree T' with an edge $g \notin T'$ such that $T \bigcup \{g\}$ is a cycle (known as a fundamental circuit of G). This property allows us to relate all spanning trees of G in a parent-child relationship hierarchy from a single baseline spanning tree where a child spanning tree is the result of replacing a single edge of its parent with another edge, called the pivot edge. The algorithm traverses the hierarchical domain of spanning trees by finding pivot edges using a generic enumeration algorithm proposed by Nagamochi and Ibaraki [11]. The domain traversal order can vary. However, careful edge ordering is essential to avoid returning to an ancestor-spanning tree from a descendent. Discovering pivot edges of a single spanning tree can take $\mathcal{O}(n)$ time. That gives the total time complexity $\mathcal{O}(n+m+n\tau)$. Both Gabow and Myer [5] and Matsui's algorithms only maintain a single spanning tree and a sorted sequence of edges in memory. Therefore, their space complexity is $\mathcal{O}(n+m)$. One can view these algorithms as optimizations to Read and Tarjan's generic backtracking-based algorithm for listing specific types of subgraphs of a graph [15], which provides $\mathcal{O}(n+m+m\tau)$ time and $\mathcal{O}(n+m)$ space bounds for listing spanning trees.

Onete et al.'s [13] more recent algorithm of spanning tree enumeration achieves the same time and space bounds as Matsui's algorithm. However, the algorithm involves no backtracking. Onete et al. found that all and only non-singular submatrices with a specific structure of a reduced incidence matrix of a graph represent spanning trees. Their algorithm utilizes this finding by generating unique submatrices of that kind and spewing corresponding spanning trees in the incidence matrix format. Since checking for non-singularity is a linear process in the number of vertices, the time complexity of their algorithm is also $\mathcal{O}(n+m+n\tau)$.

Kapoor and Ramesh [8] and Shioura and Tamura [16] provide two alternative algorithms to improve the time complexity from $\mathcal{O}(n+m+n\tau)$ to $\mathcal{O}(n+m+\tau)$ when the individual spanning trees need not be output. Instead, their difference from a baseline tree as a list of edge exchanges is sufficient. Both algorithms implicitly traverse the network $S(G)$ formed by all spanning trees of G where the nodes of $S(G)$ are the spanning trees of G and there is an edge between two nodes of $S(G)$ if the corresponding trees differ by an edge exchange. Both algorithms maintain a dynamic, ordered list of edges associated with fundamental cycles of G to optimize $S(G)$ traversal and avoid duplicate tree generation. Although the underlying list data structures are different, their maintenance increases both algorithms' space complexity to $\mathcal{O}(nm)$. Furthermore, recreating the individual spanning trees from the description of differences sacrifices time optimization.

All the above algorithms involve checking for cycle formation or tracking cyclic relations among edges of the graph, which is the chief contributor to their time or space complexity. Avoiding cycle tracking during spanning tree enumeration of a general graph is challenging. However, alternative approaches that utilize a graph's structure to enumerate spanning trees without cycle tracking can lead to better algorithms for specific graph classes.

In this paper, we consider the problem of enumerating spanning trees of a plane 3-tree. Plane 3-trees are a class of planar graphs formed from repeated triangulation of phases of an initial triangle. Formally, a *plane 3-tree* G_n is a triangulated plane graph G with $n \geq 3$ vertices such that if $n > 3$ then G has a vertex x whose deletion gives a plane 3-tree G' of $n-1$ vertices. Figure 2(a) shows an example of plane 3-tree G_{10} containing 10 vertices.

Our approach uses the face decomposition structure of a plane 3-tree to order its vertices, then inductively generates all its spanning trees without duplication from that vertex order. The algorithm using breadth-first propagation of inductive expansion has $\mathcal{O}(n+m+\tau)$ time and $\mathcal{O}(n\tau)$ space complexities. When space is not an issue, this algorithm has the advantage of being easily parallelizable. Our second algorithm captures the logic of inductive expansion into a tree alteration scheme like Matsui's to output all spanning trees using dynamic programming (DP) in $\mathcal{O}(n)$ space and $\mathcal{O}(n+m+\tau)$ time.

The structure of the rest of the paper is as follows. Section 2 presents preliminaries on various terms and concepts related to graphs, plane 3-trees, and our algorithms. Section 3 introduces some lemmas and theorems related to properties of spanning trees of a plane 3-tree. Section 4 presents our first and inductive algorithm for spanning tree enumeration. Section 5 describes and analyzes the DP algorithm. The paper concludes with a discussion of future research directions.

2 Preliminaries

In this section, we define some terms we will use for the rest of the paper and present some preliminary findings.

Let $G = (V, E)$ is a connected simple graph with vertex set V and edge set E. A *subgraph* of G is a graph $G' = (V', E')$ such that $V' \subseteq V$ and $E' \subseteq E$. G' is a *spanning subgraph* of G when $V' = V$. G' is an *induced subgraph* of G when G' contains all edges of E having both endpoints in V'. When each edge $(u, v) \in E$ is directional, i.e., $(u, v) \neq (v, u)$ then G is a *directed graph*, otherwise, G is *undirected*. In this paper, we only consider undirected graphs. The *degree* of a vertex $v \in G$, denoted by $degree(v)$, is the number of edges incident to it. The *neighbors* of v are the vertices that share an edge with v. That is, if an edge $(u, v) \in E$ then u and v are neighbors. We also say u and v are *adjacent* in G.

A *tree* $T = (V, E)$ is a connected graph with no cycles. T is *rooted* when one vertex $r \in V$, called the *root*, is distinguished from others. A *spanning tree* of a graph G is a spanning subgraph that is a tree. We use the term *node* and vertex interchangeably to refer to a vertex of a spanning tree. The unique *path* between two nodes u, w in T is the sequence of edges of the form

$(u, v_1), (v_1, v_2), \cdots, (v_i, w)$ that connects u and w in the tree. We use the notations $\rho_{u,v_1,\cdots,w}$ or $\rho_{u,w}$ interchangeably to denote the path between u and w. The *length* of a path is the number of edges in it. Let T be a rooted tree with root r. The *parent* u of any node v other than r in T is the immediate predecessor of v on the path from r to v. Conversely, v is a *child* of u. A *leaf* of T is a node with no children; otherwise, it is an *internal node*. A *sub-tree* T' of T is an induced proper subgraph of T that is also a tree. A node w is an *ancestor* of another node v in T when the unique path from v to the root r contains w; conversely, v is a *descendent* of w. A *fundamental cycle* of a graph G is a cycle formed by adding a non-tree edge $(u, v) \in G$ on its spanning tree.

A graph G is *planar* if one can embed it on a plane without any edge crossing. We call such an embedding a plane embedding of G. A *plane* graph is a planar graph with a fixed embedding, where the connected regions of a plane embedding are its *faces*. The unbounded region is the *outer face*, and all others are the *inner faces*. The vertices of the outer face are the *outer vertices*, while the rest are *inner vertices*. A plane graph G is a *triangulated plane graph* when every face has precisely three vertices.

A *plane 3-tree* is a triangulated plane graph whose plane embedding is a repeated triangulation of the faces starting from a single outer triangle. Therefore, a *plane 3-tree* G_n is a triangulated plane graph G with $n \geq 3$ vertices such that if $n > 3$ then G has a vertex x whose deletion gives a plane 3-tree G_{n-1} of $n-1$ vertices. We call x the last, or $(n-3)^{th}$, *dividing vertex* of G_n. Note that $degree(x) = 3$. We call the three neighbors u, v, w of x, the *face vertices* of x in G_n and denote their set by ζ_x. If $n > 4$ then the last dividing vertex y of G_{n-1} is the $(n-4)^{th}$ dividing vertex of G_n. Here $n-3$ and $n-4$ are the *dividing orders* of x and y respectively. Our notion of a dividing vertex is similar to that of a *representative vertex* described in [12].

Fact 1. *One can describe the repeated triangulation-based embedding of a plane 3-tree G_n of $n > 3$ vertices unambiguously by an ascending sequence of dividing vertices $v_1, v_2, \cdots, v_{n-3}$ and their face vertex sets $\zeta_{v_1}, \zeta_{v_2}, \cdots, \zeta_{v_{n-3}}$.* □

Note that multiple sequences of dividing vertices exist for a plane 3-tree G_n when it has multiple degree-3 vertices. However, all such sequences describe the same plane embedding. For a dividing vertex sequence $d_s = v_1, v_2, \cdots, v_{n-3}$ of a plane 3-tree G_n with $n > 3$, we use $G_1^{im}, G_2^{im}, \cdots, G_{n-3}^{im}$ to represent the sequence of intermediate plane-3 trees we get by introducing the dividing vertices in corresponding order starting from G_3 (note that $G_{n-3}^{im} = G_n$). We use the notation $\delta(v, d_s)$ to represent the dividing order of vertex v in d_s.

Fact 2. *If w is the last dividing vertex among three vertices u, v, w appearing in any dividing vertex sequence of a plane 3-tree G_n of $n \geq 3$ and u and v are adjacent to w then u and v are also mutually adjacent.* □

We next define some operations on trees that are essential for describing our algorithms. *Edge extension* is the operation of adding a new leaf node to a tree. The reverse operation of edge extension is *edge removal* which delete a leaf and its incident edge. The operation of adding a new vertex of degree 2 in the edge between two tree nodes is *edge subdivision*, and the reverse operation is *path contraction* which removes a node of degree two and connects its two neighbors by an edge. Finally, replacing a path of length two in a tree with a new node connected to all three nodes of the path is *root extension*, and the reverse operation is *root flattening*. Figure 1 illustrates these operations.

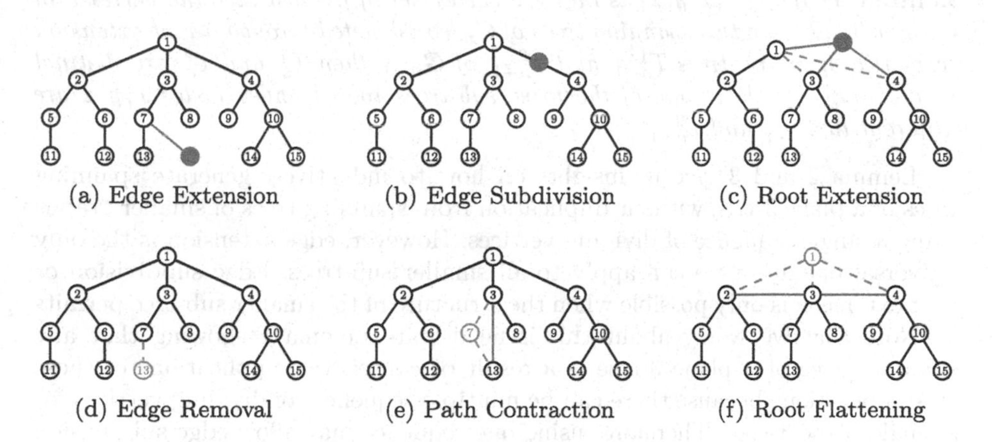

Fig. 1. Tree operations

3 On Spanning Trees of a Plane 3-Tree

In this section, we present some properties of the spanning trees of a plane 3-tree in various lemmas. We omit their proofs due to space shortage. We use these lemmas in subsequent sections to prove the correctness of our tree enumeration algorithms.

Lemma 1. *Any spanning tree T_n of a plane 3-tree G_n with $n > 3$ results from consecutive applications of edge extension, edge subdivision, and root extension starting from a path of length two connecting the outer vertices of G_n.* □

A corollary of Lemma 1 is as follows.

Corollary 1. *Any spanning tree T_n of a plane 3-tree G_n with $n > 3$ is an edge extension, edge subdivision, or root extension of a spanning tree of another plane 3-tree G_{n-1}.*

Lemma 1 and its corollary provide a convenient way to generate spanning trees of larger plane 3-trees from that of smaller plane 3-trees. However, we need to know when edge extension, edge subdivision, and root extension create duplicate trees and when they do not. The following lemmas guide us in that regard. For the sake of notational brevity, we now use G_{n-1} to represent G^{im}_{v-4} of the larger plane 3-tree G_n. That is, G_{n-1} is the plane 3-tree that results from removing the last dividing vertex of G_n and its incident edges.

Lemma 2. *If two spanning trees T^1_{n-1} and T^2_{n-1} of G_{n-1} are distinct, then an edge extension or edge subdivision to them to add a node for the last dividing vertex v of G_n will always produce two distinct spanning trees T^1_n and T^2_n of G_n.*

Lemma 2 is trivially true due to the precondition of distinctness of the spanning trees of G_{n-1}. We now have Lemma 3 (proof omitted).

Lemma 3. *If $\zeta_v = \{x, y, z\}$ is the face vertex set of the last dividing vertex v of G_n and T^1_n, T^2_n are two spanning trees of G_n constructed through a root extension from two spanning trees T^1_{n-1} and T^2_{n-1} of G_{n-1} then T^1_n and T^2_n are distinct if and only if at least one of the three sub-trees rooted under nodes x, y, z are different in T^1_{n-1} and T^2_{n-1}.* □

Lemma 2 and 3 give us insights on how to inductively generate spanning trees of a plane 3-tree without duplication from spanning trees of smaller graphs using a single sequence of dividing vertices. However, edge extension is the only universal operation we can apply to all smaller sub-trees. Edge subdivision or root extension is only possible when the structure of the smaller sub-tree permits it. Note that we worry about this issue despite Lemma 1 showing that any spanning tree of a plane 3-tree is a result of consecutive applications of these three operations because there can be multiple sequences of dividing vertices for a single plane 3-tree. Therefore, using one sequence may allow edge subdivision or root extension at a specific case, while another does not.

Assume $\tau_{n-1} = \{T^1_{n-1}, T^2_{n-1}, \cdots, T^N_{n-1}\}$ is the set of all spanning trees of G_{n-1}. Consider any arbitrary T^i_{n-1} that has no pair of nodes from the face vertex set $\zeta_v = \{x, y, z\}$ of the last dividing vertex v of G_n connected by an edge in the tree. We can construct a spanning tree T_n for G_n from T^i_{n-1} where node v is adjacent to two nodes corresponding to its face vertices as follows:

1. First, add v by an edge extension.
2. Then connect leaf v with another face vertex node by an edge to form a fundamental cycle.
3. Then remove any one edge from that cycle other than those incident to v.

To produce a T_n where v shares an edge with all of x, y, z, we can repeat the above process for the remaining face vertex. One can convince thyself that this alternative process generates all spanning trees [9,16]. The vital concern here is that we can generate spanning trees describable using edge subdivisions and root extensions that we did not generate using those operations in the first place. Let us call this alternative approach of spanning tree generation *cycle breaking*. The following two lemmas (proofs omitted) prove that cycle breaking generates trees that are replicas of other trees inductively generated using edge subdivision or root extension following a single dividing vertex sequence.

Lemma 4. *Assume $\tau_{n-1} = \{T^1_{n-1}, T^2_{n-1}, \cdots, T^N_{n-1}\}$ is the set of all spanning trees of G_{n-1} and $T^i_{n-1} \in \tau_{n-1}$ has no edge (u, v) such that u, v are two face vertices of the last dividing vertex x of G_n. If T_n having x adjacent to both u and v is a spanning tree generated from T^i_{n-1} using cycle breaking then there is another tree $T^j_{n-1} \in \tau_{n-1}$ that we can convert to T_n applying an edge subdivision.* □

Lemma 5. *Assume* $\tau_{n-1} = \{T^1_{n-1}, T^2_{n-1}, \cdots, T^N_{n-1}\}$ *is the set of spanning trees of* G_{n-1} *and* $T^i_{n-1} \in \tau_{n-1}$ *does not have a pair of edges* $(x, y), (y, z)$ *where* x, y, z *are the face vertices of the last dividing vertex* v *of* G_n. *If* T_n *having* (v, x), (v, y) *and* (v, z) *edges is a spanning tree generated from* T^i_{n-1} *using cycle breaking then there is another tree* $T^j_{n-1} \in \tau_{n-1}$ *that we can convert to* T_n *applying a root extension.* □

Lemma 4 and 5 imply that one can add the last vertex of a dividing vertex sequence $d_s = v_1, v_2, \cdots, v_{n-3}$ of G_n to inductively generate all spanning trees of G_n from the spanning trees of G_{n-1} using only edge extension, edge subdivision, and root extension regardless of the dividing order of earlier vertices. However, if G_n has more than one degree 3 vertices, then the last dividing vertex itself can be different between two $d^1_s = v^1_1, v^1_2, \cdots, v^1_{n-3}$ and $d^2_s = v^2_1, v^2_2, \cdots, v^2_{n-3}$ describing G_n. Then $G^{im}_{n-4} = G_{n-1}$ for the two sequences differ by at least one pair of vertices. Consequently, their set of spanning trees τ^1_{n-1} and τ^2_{n-1} has no common tree, which may cause us to miss some spanning trees if we only use one of d^1_s and d^2_s for tree enumeration. The following lemma (proof omitted) eliminates this concern by showing that the ordering differences of vertices among two dividing vertex sequences of G_n do not impact spanning tree generation.

Lemma 6. *Assume that two dividing vertex sequences* $d^1_s = v^1_1, v^1_2, \cdots, v^1_{n-3}$ *and* $d^2_s = v^2_1, v^2_2, \cdots, v^2_{n-3}$ *of a plane 3-tree* G_n *start diverting at index* k. *That is, the sub-sequences* $d^1_{sub} = v^1_k, v^1_{k+1}, \cdots, v^1_{n-3}$ *and* $d^2_{sub} = v^2_k, v^2_{k+1}, \cdots, v^2_{n-3}$ *are different from their beginning. Assume* τ_{k-1} *is the set of all spanning trees of* G^{im}_{k-1}. *Then for each tree* $T_1 \in \tau_{k-1}$ *and a series of edge extension, edge subdivision, and root extension to* T_1 *in the order of* d^1_{sub} *forming a spanning tree* T_n*; there is another tree* $T_2 \in \tau_{k-1}$ *such that an alternative sequence of those operations in the order of* d^2_{sub} *generates the same spanning tree of* G_n. □

4 Inductive Algorithm for Spanning Tree Enumeration

Lemma 1, 2, 3, 4, 5 and 6 give us all the insights we need to generate all spanning trees of a plane 3-tree inductively from spanning trees of smaller constituent plane 3-trees. This section presents an inductive algorithm for spanning tree enumeration that we will use as the blueprint for a better DP algorithm in the next section.

A General Outline of the Algorithm: We will take a sequence of dividing vertices, d_s, and corresponding face vertex set sequence, f_s, of the plane 3-tree G_n as inputs. We initiate a list with the spanning trees for G_3. Then we iterate over the dividing vertex sequence. In each iteration, we take one dividing vertex, v, and apply edge extension to all trees in the list, edge subdivision to those trees where the operation is permitted, and root extension to those trees where the

Algorithm 1: Inductive Algorithm for Spanning Tree Enumeration

Input: d_s, f_s - a dividing vertex and face vertex set sequence pair of G_n
Output: τ - the set of all spanning trees of plane 3-tree G_n

```
{x, y, z} ← f_s[0]
τ ← {ρ_{x,y,z}, ρ_{y,x,z}, ρ_{x,z,y}}
for ( i = 0; i < length(d_s); i = i + 1 ) do
    ν ← d_s[i], ζ ← f_s[i]
    [α, β, γ] ← sortByDividingOrder(ζ)
    τ' ← ∅
    foreach T ∈ τ do
        foreach μ ∈ {α, β, γ} do
            τ' ← τ' ⋃{T ∪ {(μ, ν)}}
        end
        foreach (μ, ω) ∈ T & {μ, ω} ⊂ ζ do
            τ' ← τ' ⋃{(T − {(μ, ω)}) ∪ {(μ, ν), (ν, ω)}}
        end
        if ρ_{α,β,γ} ∈ T then
            τ' ← τ' ⋃(T − ρ_{α,β,γ}) ∪ {(ν, α), (ν, β), (ν, γ)}
        end
    end
    τ ← τ'
end
```

face vertices of v form a specific path of length two. We store the result of these operations in a new list that replaces the old list at the end of the iteration. When the iterative process completes, the list contains all spanning trees of G_n without duplication. We use root extension with caution due to Lemma 3 showing that some sub-trees under the face vertices of a dividing vertex need to be different to avoid duplicate spanning tree generation. In that regard, if v is the dividing vertex under consideration at a particular iteration and x, y, z are v's face vertices with $\delta(x, d_s) < \delta(y, d_s) < \delta(z, d_s)$, we apply root extension for v on a tree only if it has the path $\rho_{x,y,z}$. To use this logic for the three outer face vertices, we assign them arbitrary but fixed dividing orders $-2, -1$, and 0.

Algorithm 1 presents a pseudo-code of the spanning tree enumeration process. The correctness of Algorithm 1 is evident from the various lemmas of Sect. 3. Hence, we focus on its time and space complexities. For that, we need to understand the cost of generating a dividing vertex sequence and modifying a spanning tree of the smaller subgraph to produce one for a larger graph/subgraph.

Dividing Vertex Sequence Generation: We can generate a d_s, f_s pair for a plane 3-tree G_n by using a graph traversal of G_n starting from any vertex and populating a pair of stacks. Anytime the graph traversal reaches a vertex v of degree three, we push v in one stack and its neighbor set to the other. Then we remove v and its incident edges from the graph and continue the traversal. The traversal ends when we find a vertex with degree two. Then we generate the d_s, f_s pair by popping one element at a time from both stacks and appending the elements in the growing list of dividing vertex and face vertex set sequences. The whole process requires a single traversal of G_n that would take $\mathcal{O}(n+m)$ time where m is the number of edges in G_n and $\mathcal{O}(n+3n) = \mathcal{O}(n)$ space beyond the initial storage for the input graph G_n.

Data Structure for Spanning Trees: We can represent the spanning trees using a simple array of $n-1$ vertices where the entry at i^{th} index is the other endpoint of an edge incident to the vertex with dividing order $i-1$. We adopt the following approach of modifying a spanning tree for G_{j-1}^{im} to form a spanning tree for G_j^{im} to ensure that each entry in the array refers to a distinct edge.

1. If we add the new vertex v with $\delta(v, d_s) = j$ to the spanning tree using an edge extension from existing vertex x then we enter $\delta(x, d_s)$ at index $(j+1)$.
2. If we add v by subdividing the edge (x, y) then we investigate index $\delta(x, d_s)+1$ and $\delta(y, d_s)+1$ to determine which entry refers to the edge (x, y), replace the current value with j in that entry, and then write the dividing order of the other entry at index $(j+1)$.
3. Finally, if we use root extension to x, y, z to add v in the spanning tree, then we modify two existing entries using the approach of (2) and add a new entry for the remaining vertex at index $(j+1)$.

Note that the above approach works even when one of the face vertices of v is the first outer face vertex, which has no entry in the array. We now prove the following theorem of Algorithm 1 and time complexities.

Theorem 1. *When replicating a fixed-size array in memory is a constant time operation, Algorithm 1 enumerates all spanning trees of a plane 3-tree G_n of n vertices, m edges, and τ spanning trees in $\mathcal{O}(n+m+\tau)$ time and $\mathcal{O}(n\tau)$ space.*

Proof. Assuming memory replication of a fixed-length array is a constant time operation, and the storage structure of a spanning tree is as we described above; generating a new spanning tree T' for G_{i+1}^{im} from a spanning tree T of G_i^{im} is a constant time operation. Furthermore, evaluating whether an edge subdivision or root extension is admissible in T also takes a constant time (we can investigate the three indexes for the face vertices of the current vertex in T to determine their adjacency). Moreover, the algorithm considers each T exactly once and generates at least three new trees. Hence, the algorithm's running time is proportional to the number of spanning trees generated through the iterative process.

If τ_{i+1} represents the number of spanning trees of G_{i+1}^{im} then $\tau_i \leq \frac{\tau_{i+1}}{3}$ for all $i < n$. Thus, the total number of trees the algorithm generates is $\leq \tau + \frac{\tau}{3} + \frac{\tau}{3^2} + \cdots + \frac{\tau}{3^n} = \frac{\tau(3^{n+1}-1)}{2\times 3^n} < 2\tau$, where τ is the number of spanning trees of G_n. Thus, the algorithm's running time is $\mathcal{O}(\tau)$. After adding the computation time for generating a dividing vertex and its face vertex set sequences, the overall time complexity of enumerating the spanning trees of an input plane 3-tree G_n becomes $\mathcal{O}(n+m+\tau)$, where m is the number of edges in G_n. The algorithm only retains the two input sequences and the set of spanning trees of the current plane 3-tree. Therefore, the space complexity is $\mathcal{O}(n+n\tau) = \mathcal{O}(n\tau)$. □

Suppose we relax the assumption that memory replication of a fixed-length array is a constant time operation. If an array copy takes time linear to the number of elements in the array, then constructing each spanning tree takes $\mathcal{O}(n)$ time. Then the running time of Algorithm 1 becomes $\mathcal{O}(n + m + n\tau)$, which is equivalent to the best-known algorithm for spanning tree enumeration in a general graph. In the next section, we will use dynamic programming to convert the inductive tree generation process into a tree alternation process that avoids array copying and reduces the space required to $\mathcal{O}(n)$. However, one should notice that Algorithm 1 involves no coordination among the elements of the spanning tree set in each iteration. That makes it perfectly parallelizable. After the generation of the two input sequences and a few iterations of smaller spanning tree generation; different parallel processors can independently continue the inductive process from the initial spanning tree set. That gives the time complexity of $\mathcal{O}(n + m + \frac{n\tau}{p})$ for p-processor parallel version of the algorithm. The space requirement in individual processors is $\frac{n\tau}{p}$. Thus the following theorem holds.

Theorem 2. *Assuming copying a fixed-size array is a linear time operation to the size of an array, a p-processor parallel version of Algorithm 1 enumerates all spanning trees of a plane 3-tree G_n of n vertices, m edges, and τ spanning trees in $\mathcal{O}(n + m + \frac{n\tau}{p})$ time and $\mathcal{O}(\frac{n\tau}{p})$ space.* □

5 DP Algorithm for Spanning Tree Enumeration

Algorithm 1 of the previous section is close to optimal because it involves no duplicate spanning tree generation and churns out new spanning trees in constant time from other intermediate spanning trees. However, it must keep all the intermediate spanning trees of the immediately previous step in memory and involves a tree copying overhead during a new spanning tree generation.

We can eliminate both limitations by converting the inductive tree-generation process into a recursive tree mutation process. Note that if τ is the set of all spanning trees of the plane 3-tree G_n, one can view each spanning tree $T \in \tau$ as a unique configuration of edge extension, edge subdivision and root extension operations. Furthermore, for all $T \in \tau$, these operations are applied on a path of length two connecting the outer face vertices in the same order vertices appear in a dividing vertex sequence. Consequently, we can consider successive trees added in τ as migration from one configuration to the next. A scheme that can alter between configurations in constant time, traverse all configurations, and only emit unique configurations should emulate Algorithm 1. We only need to maintain a single spanning tree corresponding to the current configuration when traversing the domain of all spanning tree configurations. We can apply a clever trick of using the list of edges of the current spanning tree as the representation of the underlying configuration. In that case, there is no additional space overhead beyond that of a single spanning tree and two lists for the dividing vertex and face vertex set sequences. That is the logic of our DP algorithm in this section.

First, we need an initial configuration for the described scheme to work. Assume a, b, and c are the outer face vertices with our assigned respective dividing orders $-2, -1$, and 0. Then the initial configuration has the path $\rho_{a,b,c}$ and for each dividing vertex x in the sequence d_s the edge (w, x) such that $w \in \zeta_x$ and $\delta(w, d_s)$ is the largest for all of x's face vertices. Figure 2(a) and (b) illustrate a plane 3-tree and its spanning tree for the initial configuration.

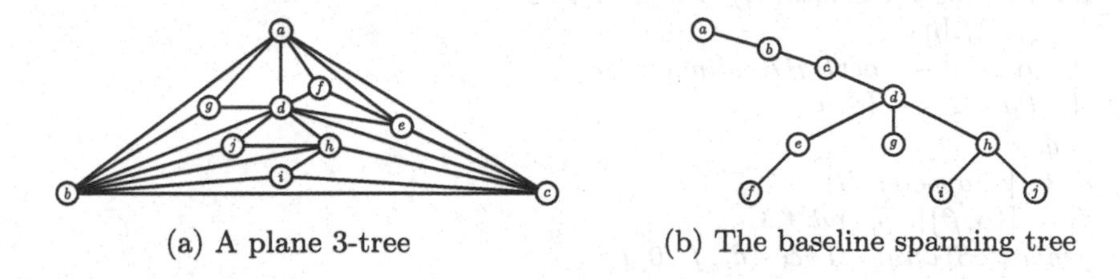

(a) A plane 3-tree

(b) The baseline spanning tree

Fig. 2. An example plane 3-tree and its baseline spanning tree

If $\zeta_x = \{u, v, w\}$, then alternative configurations for x are to have edge extension from u or v, subdividing edges $(u, v), (v, w)$ or (u, w), and root extending the path $\rho_{u,v,w}$. Among them, the admissibility of edge subdivision or root extension depends on the current configuration for u, v, and w. Let us call moving to any alternative configuration for x from its default configuration in any tree T a *mutation* and number the mutations in the order we specified from zero to five. Then we can define the following functions.

$$mutateTree(T, v, \zeta_v, i) = \begin{cases} T' & \text{mutation } i \text{ can be applied to } T \text{ for } v \text{ to create } T' \\ \emptyset & \text{otherwise} \end{cases}$$

$$reverseMutation(T', v, \zeta_v, i) = \begin{cases} T & T' \text{results applying mutation } i \text{ for } v \text{ to } T \\ T' & \text{otherwise} \end{cases}$$

If we use the data structure described in Sect. 4 for the spanning trees, then both *mutateTree* and *reverseMutation* functions should take constant time. Both require a constant number of index checking and entry updates in the list of edges sorted by the dividing order of vertices.

To permit configuration changes for the initial path connecting the outer face vertices, we only need to support an alternative edge extension and an edge subdivision for the outer vertex with order zero as an exceptional case. Finally, note that the *mutateTree* function updates the argument tree T when the mutation is admissible; otherwise, it leaves T unchanged. Now we can define our DP algorithm for spanning tree enumeration, as illustrated in Algorithm 2.

Now we prove the following theorem on the correctness and time and space complexities of Algorithm 2.

Algorithm 2: DP Algorithm for Spanning Tree Enumeration

```
Input: d_s, f_s - a dividing vertex and face vertex set sequence pair of G_n
Output: τ - the set of all spanning trees of plane 3-tree G_n
ζ0 ← f_s[0]
[α0, β0, γ0] ← sortByDividingOrder(ζ0)
T ← []
T[0] ← δ(α0, d_s), T[1] ← δ(β0, d_s)
for ( i = 0; i < length(d_s); i = i + 1 ) do
    ζ ← f_s[i]
    [α, β, γ] ← sortByDividingOrder(ζ)
    T[i + 2] ← δ(γ, d_s)
end
d_s ← [γ].append(d_s)
f_s ← [{α, β}].append(f_s)
generateSpanningTrees(d_s, f_s, 0, T)
Function generateSpanningTrees(d_s, f_s: List, i: Integer, T: Array) is
    if i == length(d_s) then
        output(T)
    else
        ν ← d_s[i], ζ ← f_s[i]
        for ( m_i = 0; m_i ≤ 5; m_i = m_i + 1 ) do
            T' ← mutateTree(T, ν, ζ, m_i)
            if T' ≠ ∅ then
                generateSpanningTrees(d_s, f_s, i + 1, T')
                T ← reverseMutation(T', ν, ζ, m_i)
        end
        generateSpanningTrees(d_s, f_s, i + 1, T)
    end
end
```

Theorem 3. *Algorithm 2 enumerates all spanning trees of a plane 3-tree G_n with n vertices, m edges, and τ spanning trees without duplication in $\mathcal{O}(n+m+\tau)$ time and $\mathcal{O}(n)$ space.*

Proof. Algorithm 2 first generates a spanning tree of G_n for the initial configuration, updates the dividing vertex and face vertex set sequences to include one outer face vertex, and then invokes the function *generateSpanningTrees* that implements the DP algorithm. Therefore, it suffices that we only analyze function *generateSpanningTrees* for DP scheme's accuracy and complexity.

Each time *generateSpanningTrees* output a spanning tree, it is for a different configuration of mutations of vertices from d_s. Hence, the algorithm only outputs distinct spanning trees of G_n. The function removes mutation in the reverse order it applies them. Thus, a mutation of any vertex is reversible during a backtracking step if it was admissible during the recursive unfolding. We only wonder whether or not *generateSpanningTrees* includes all possible mutations of each vertex. The answer lies in the default configuration for dividing vertices.

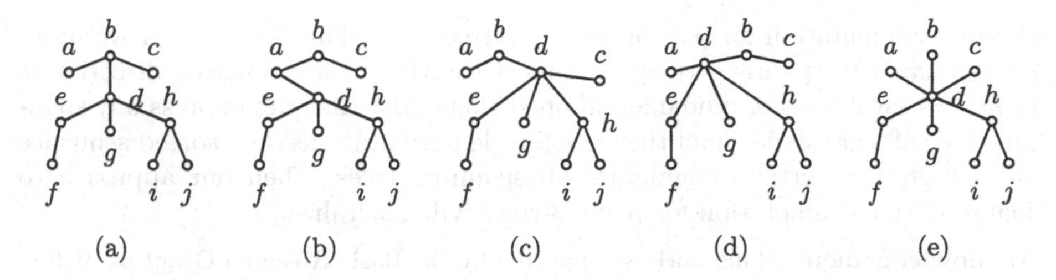

Fig. 3. Possible mutations of vertex d from the baseline tree of Fig. 2(b)

Let u be the vertex with dividing order i. Anytime *generateSpanningTrees* tries to apply a mutation for u from its default configuration, all vertices v with $\delta(v, d_s) > \delta(u, d_s)$ are attached via an edge extension to their respective face vertex with the highest dividing order. The critical characteristic of edge extension from a vertex x to y where $\delta(x, d_s) < \delta(y, d_s)$ is that it does not preclude any mutation of x involving vertices with lower dividing orders. Figure 3 illustrates this fact for vertex d of the plane 3-tree of Fig. 2(a) (Notice that one mutation for d, i.e., subdividing the edge (a, c) is not admissible from the initial configuration as that requires a mutation of c.). Consequently, admissible mutations for u depend only on the current configuration of vertices that appeared earlier in d_s. Therefore, applying a mutation or skipping all mutations for u and then progressing to the next vertex in Algorithm 2 is equivalent to generating all possible spanning trees of G_i^{im} from a fixed spanning tree of G_{i-1}^{im} in Algorithm 1. Thus, Algorithm 2 is behaviorally equivalent to Algorithm 1 and correctly enumerates all spanning trees of G_n without duplication.

Algorithm 2 only maintains a single spanning tree of G_n and two sequences for dividing vertices and their face vertex sets. Hence, we can derive from the analysis of Algorithm 1 that Algorithm 2's time complexity is $\mathcal{O}(\tau)$ and space complexity is $\mathcal{O}(n)$. Combining that with the initial cost of generating the dividing vertex and face vertex set sequences, the overall cost of spanning tree enumeration of a plane 3-tree G_n using DP is $\mathcal{O}(n + m + \tau)$ in time and $\mathcal{O}(n)$ in space, where m is the number of edges and τ is the number of spanning trees. □

6 Conclusions

In this paper, we propose an algorithm for enumerating all spanning trees of a plane 3-tree G_n of n vertices, m edges, and τ spanning trees in $\mathcal{O}(n + m + \tau)$ time and $\mathcal{O}(n)$ space. Our algorithm substantially improves both time and space bounds of the best general-purpose spanning tree enumeration algorithm that takes $\mathcal{O}(n + m + n\tau)$ time and $\mathcal{O}(nm)$ space for this specific graph class. The central idea underlying this improvement is identifying a few low-cost extension/mutation operations that we can inductively use to extend spanning trees of smaller induced subgraphs of a plane 3-three to that of larger subgraphs.

There are several avenues for future research from our current findings. First, one can investigate the possibility of a similar spanning tree enumeration technique for other graph classes, such as regular graphs with small vertex degrees. Second, one can try to relate our tree extension/mutation operations with edit distance computation for pair of spanning trees of a plane 3-tree or prove properties related to spanning trees. The most exciting future research direction is to prove that if a set of fundamental operations exist that can express any spanning tree of a graph, then all their applicable permutations on a sorted sequence of that graph's vertices enumerate all spanning trees. Then our approach to spanning tree enumeration for plane 3-trees will generalize.

Acknowledgement. This work is supported by the Basic Research Grant of BUET.

References

1. Brooks, R., Smith, C., Stone, A., Tutte, W.: Determinants and current flows in electric networks. Discret. Math. **100**(1), 291–301 (1992). https://doi.org/10.1016/0012-365X(92)90648-Y
2. Char, J.: Generation of trees, two-trees, and storage of master forests. IEEE Trans. Circ. Theory **15**(3), 228–238 (1968). https://doi.org/10.1109/TCT.1968.1082817
3. Chen, W.K.: Topological analysis for active networks. IEEE Trans. Circ. Theory **12**(1), 85–91 (1965). https://doi.org/10.1109/TCT.1965.1082396
4. Deo, N.: Graph Theory with Applications to Engineering and Computer Science. Prentice Hall Series in Automatic Computation, Prentice-Hall Inc., USA (1974)
5. Gabow, H.N., Myers, E.W.: Finding all spanning trees of directed and undirected graphs. SIAM J. Comput. **7**(3), 280–287 (1978). https://doi.org/10.1137/0207024
6. Hakimi, S., Green, D.: Generation and realization of trees and k-trees. IEEE Trans. Circ. Theory **11**(2), 247–255 (1964). https://doi.org/10.1109/TCT.1964.1082276
7. Hakimi, S.: On trees of a graph and their generation. J. Franklin Inst. **272**(5), 347–359 (1961). https://doi.org/10.1016/0016-0032(61)90036-9
8. Kapoor, S., Ramesh, H.: Algorithms for enumerating all spanning trees of undirected and weighted graphs. SIAM J. Comput. **24**(2), 247–265 (1995). https://doi.org/10.1137/S009753979225030X
9. Matsui, T.: An algorithm for finding all the spanning trees in undirected graphs (1998)
10. Mayeda, W., Seshu, S.: Generation of trees without duplications. IEEE Trans. Circ. Theory **12**(2), 181–185 (1965). https://doi.org/10.1109/TCT.1965.1082432
11. Nagamochi, H., Ibaraki, T.: A linear-time algorithm for finding a sparse k-connected spanning subgraph of a k-connected graph. Algorithmica **7**(5&6), 583–596 (1992). https://doi.org/10.1007/BF01758778
12. Nishat, R.I., Mondal, D., Rahman, M.S.: Point-set embeddings of plane 3-trees. In: Brandes, U., Cornelsen, S. (eds.) GD 2010. LNCS, vol. 6502, pp. 317–328. Springer, Heidelberg (2011). https://doi.org/10.1007/978-3-642-18469-7_29
13. Onete, C.E., Onete, M.C.C.: Enumerating all the spanning trees in an un-oriented graph - a novel approach. In: 2010 XIth International Workshop on Symbolic and Numerical Methods, Modeling and Applications to Circuit Design (SM2ACD), pp. 1–5 (2010). https://doi.org/10.1109/SM2ACD.2010.5672365
14. Rakshit, A., Sarma, S.S., Sen, R.K., Choudhury, A.: An efficient tree-generation algorithm. IETE J. Res. **27**(3), 105–109 (1981). https://doi.org/10.1080/03772063.1981.11452333
15. Read, R., Tarjan, R.: Bounds on backtrack algorithms for listing cycles, paths, and spanning trees. Networks **5**(3), 237–252 (1975). https://doi.org/10.1002/net.1975.5.3.237
16. Shioura, A., Tamura, A.: Efficiently scanning all spanning trees of an undirected graph. J. Oper. Res. Soc. Jpn. **38**(3), 331–344 (1995). https://doi.org/10.15807/jorsj.38.331

Communication-Efficient Distributed Graph Clustering and Sparsification Under Duplication Models

Chun Jiang Zhu(✉)

University of North Carolina at Greensboro, Greensboro, USA
chunjiang.zhu@uncg.edu

Abstract. In this paper, we consider the problem of clustering graph nodes and sparsifying graph edges over distributed graphs, when graph edges with possibly edge duplicates are observed at physically remote sites. Although edge duplicates across different sites appear to be beneficial at the first glance, in fact they could make the clustering and sparsification more complicated since potentially their processing would need extra computations and communications. We propose the first communication-optimal algorithms for two well-established communication models namely the message passing and the blackboard models. Specifically, given a graph on n nodes with edges observed at s sites, our algorithms achieve communication costs $\tilde{O}(ns)$ and $\tilde{O}(n+s)$ ($\tilde{O}$ hides a polylogarithmic factor), which almost match their lower bounds, $\Omega(ns)$ and $\Omega(n+s)$, in the message passing and the blackboard models respectively. The communication costs are asymptotically the same as those under non-duplication models, under an assumption on edge distribution. Our algorithms can also guarantee clustering quality nearly as good as that of centralizing all edges and then applying any standard clustering algorithm. Moreover, we perform the first investigation of distributed constructions of graph spanners in the blackboard model. We provide almost matching communication lower and upper bounds for both multiplicative and additive spanners. For example, the communication lower bounds of constructing a $(2k-1)$-spanner in the blackboard with and without duplication models are $\Omega(s+n^{1+1/k}\log s)$ and $\Omega(s+n^{1+1/k}\max\{1,s^{-1/2-1/(2k)}\log s\})$ respectively, which almost match the upper bound $\tilde{O}(s+n^{1+1/k})$ for both models.

Keywords: Distributed Graph Clustering · Graph Sparsification · Spectral Sparsifiers · Graph Spanners

1 Introduction

Graph clustering is one of the most fundamental tasks in machine learning. Given a graph consisting of a node set and an edge set, graph clustering

This work was supported by UNC-Greensboro start-up funds.

M. Mavronicolas (Ed.): CIAC 2023, LNCS 13898, pp. 383–398, 2023.
https://doi.org/10.1007/978-3-031-30448-4_27

asks to partition graph nodes into clusters such that nodes within the same cluster are "densely-connected" by graph edges, while nodes in different clusters are "loosely-connected". Graph clustering on modern large-scale graphs imposes high computational and storage requirements, which are too expensive to obtain from a single machine. In contrast, distributed computing clusters and server storage are a popular and cheap way to meet the requirements. Distributed graph clustering has received considerable research interests, *e.g.*, [CSWZ16, SZ19, ZZL+19]. Interestingly, these works show their close relationships with (distributed) graph sparsification.

Graph sparsification is the task of approximating an arbitrary graph by a sparse graph that has a reduced number of edges while approximately preserving certain property. It is often useful in the design of efficient approximation algorithms, since most algorithms run faster on sparse graphs than the original graphs. Several notions of graph sparsification have been proposed. Spectral sparsifiers [ST11] well approximate the spectral property of the original graphs and can be used to approximately solve linear systems over graph Laplacian, and to approximate effective resistances, spectral clustering, and random walk properties [SS11, CSWZ16]. On the other hand, graph spanners are a type of graph sparsifiers that well approximate shortest-path distances in the original graph. A subgraph H of an undirected graph G is called a k-spanner of G if the distance between any pair of vertices in H is no larger than k times of that in G, and k is called the *stretch* factor. It is well known that for any n-vertex graph, there exists a spanner of stretch $2k-1$ and size (the number of edges) $O(n^{1+1/k})$ [TZ05]. This is optimal if we believe the Erdos's girth conjecture [Erd64]. Many research efforts were then devoted to *additive spanners*, where the distance between any vertex pair is no larger by an additive term β instead of a multiplicative factor. Here the spanner is called a $+\beta$-spanner. There have been different constructions of +2-, +4-, +6-spanners of size $O(n^{3/2})$, $O(n^{7/5})$, and $O(n^{4/3})$, respectively [BKMP10, Che13]. Spanners have found a wide range of applications in network routing, synchronizers and broadcasting, distance oracles, and preconditioning of linear systems [TZ05, ABS+20].

In an n-vertex distributed graph $G(V, E)$, each of s sites, S_i, holds a subset of edges $E_i \subseteq E$ on a common vertex set V and their union is $E = \cup_{i=1}^{s} E_i$. We consider two well-established models of communication, the *message passing* model and *blackboard* model, following the above work. In the former, there is a communication channel between every site and a distinguished coordinator. Each site can send a message to another site by first sending to the coordinator, who then forwards the message to the destination. In the latter, sites communicate with each other through a shared blackboard such as a broadcast channel. The models can be further considered in two settings: edge sets of different sites are disjoint (*non-duplication* models) and they can have non-empty intersection (*duplication* models). Here the major objective is to minimize the communication cost that is usually measured by the total number of bits communicated.

A typical framework of distributed graph clustering is to employ graph sparsification tools to significantly reduce the size of edge sets of different sites while

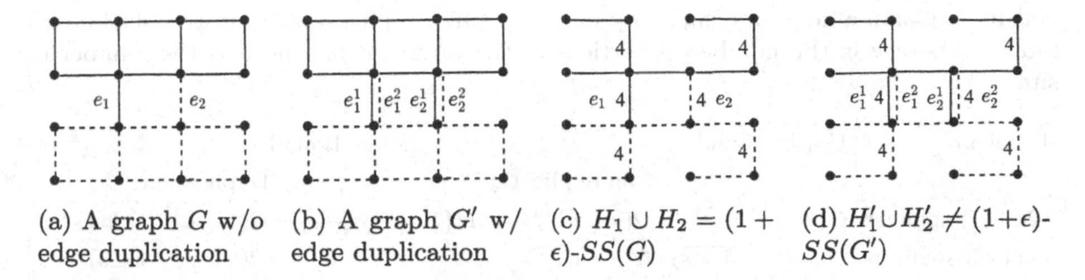

(a) A graph G w/o edge duplication (b) A graph G' w/ edge duplication (c) $H_1 \cup H_2 = (1+\epsilon)$-$SS(G)$ (d) $H_1' \cup H_2' \neq (1+\epsilon)$-$SS(G')$

Fig. 1. An illustrating example for challenges in processing edge duplicates across sites. For all subfigures, edge weights are one unless stated explicitly and edges are distributed at two sites: solid edges are in site S_1 and dash edges are in S_2. (a): a graph G without edge duplication. The graph G' in (b) is similar to G but edge e_1 (and e_2) appears in both sites S_1 and S_2 as e_1^1 and e_1^2 (e_2^1 and e_2^2), respectively. (c) shows the decomposability. Each site S_i constructs a spectral sparsifier H_i of its local graph and their union is a spectral sparsifier of G. However, the decomposability does not work for G' as in (d). It is unknown how to process the two "duplicates" of e_1 and e_2, *e.g.*, e_1^1 and e_1^2 with different weights 4 and 1.

keeping structural properties. [CSWZ16] proposed to compute spectral sparsifiers for the graphs at different sites and transmit them to the coordinator. Upon receiving all sparsifiers, the coordinator takes their union and applies a standard clustering algorithm, *e.g.*, [NJW01]. However, all the existing methods that follow this framework such as [CSWZ16, ZZL+19] only work in non-duplication models. The assumption that edge sets of different sites are disjoint is crucial to get the *decomposability* of spectral sparsifiers: the union of spectral sparsifiers of subgraphs at different sites is a spectral sparsifier of the distributed graph. Unfortunately, the decomposability does not work in duplication models. When edge sets of different sites have non-empty intersection, it is unclear how to process edge "duplicates" that are possible to have different edge weights after sparsification. See Fig. 1 for a concrete example. To the best of our knowledge, none of the existing algorithms can perform distributed graph clustering in the more general duplication models with reasonable theoretical guarantees on both communication cost and clustering quality. Instead of restoring the decomposability and turning to the framework, our algorithms are built based on the construction of spectral sparsifiers by graph spanners [KX16]. The adaptation of the algorithm to the duplication models need new algorithmic procedures such as weighted graph spanners and uniform sampling.

Although distributed constructions of graph spanners have been studied in message passing and CONGEST models [CHKPY18, FWY20, ZLB21], unfortunately they have not been systematically studied in the blackboard model. The blackboard model represents distributed systems with a broadcast channel. It can be viewed as a model for single-hop wireless networks and has received increasingly growing research [CSWZ16, DOR21, VWW20]. In the second part of this paper, we also investigate the problem of constructing graph spanners under the blackboard with both duplication and non-duplication models and obtain several almost matching communication lower and upper bounds.

Table 1. Communication complexity of computing graph spanners in the blackboard model, where n is the number of vertices in the input graph and s is the number of sites.

Problem	Upper Bound	Lower Bound	
		Non-duplication	Duplication
$(2k-1)$-spanner	$\tilde{O}(s+n^{1+1/k})$	$\Omega(s+n^{1+1/k}\max\{1, \frac{\log s}{s^{(1+1/k)/2}}\})$	$\Omega(s+n^{1+1/k}\log s)$
+2 or 3-spanner	$\tilde{O}(s+n\sqrt{n+s})$	$\Omega(s+n^{3/2})$	$\Omega(s+n^{3/2}\log s)$
$+k$-spanner	$\tilde{O}(s+n\sqrt{n+s})$	$\Omega(s+n^{4/3-o(1)})$	$\Omega(s+n^{4/3-o(1)}\log s)$

Table 2. Communication complexity of computing graph spanners in the message passing model [FWY20].

Problem	Upper Bound		Lower Bound	
	Non-duplication	Duplication	Non-duplication	Duplication
$(2k-1)$-spanner	$\tilde{O}(ks^{1-2/k}n^{1+1/k}+snk)$	$\tilde{O}(sn^{1+1/k})$	$\Omega(ks^{1/2-1/(2k)}n^{1+1/k}+sn)$	$\Omega(sn^{1+1/k})$
+2 or 3-spanner	$\tilde{O}(\sqrt{s}n^{3/2}+sn)$	$\tilde{O}(sn^{3/2})$	$\Omega(\sqrt{s}n^{3/2}+sn)$	$\Omega(sn^{3/2})$
$+k$-spanner	$\tilde{O}(\sqrt{s/k}n^{3/2}+snk)$	$\tilde{O}(sn^{3/2})$	$\Omega(n^{4/3-o(1)}+sn)$	$\Omega(sn^{4/3-o(1)})$

Our Contributions. We perform the first investigation of distributed graph clustering and spectral sparsification under duplication models. We propose communication-optimal (up to polylogarithmic factor) algorithms with communication cost $\tilde{O}(ns)$ and $\tilde{O}(n+s)$ in the message passing and blackboard with duplication models, respectively. Interestingly, the communication costs are asymptotically the same as the those in the non-duplication models under an assumption on edge distribution: the probability of an edge residing at each of the sites is a known value. This is practical when the popularity or degree of duplication of edges is obtainable. It is guaranteed that the quality of our clustering results is nearly as good as the simple method of centralizing all edge sets at different sites and then applying a standard clustering algorithm, *e.g.*, [NJW01].

Furthermore, we study distributed constructions of graph spanners in the blackboard models with and without edge duplication in order to improve our poor understanding on the communication complexity. Table 1 summarizes our main findings and Table 2 provides the communication complexity in the message passing model [FWY20]. We confirm that the blackboard model is able to significantly reduce the communication complexity compared to the message passing model. Unlike the problem of distributed clustering and spectral sparsification, edge duplication potentially brings more communications for distributed spanner construction problem. See detailed discussions in Sect. 4.

Related Work. There have been extensive research on graph clustering in the distributed setting, *e.g.*, [YX15, CSWZ16, SZ19, ZZL+19]. [YX15] proposed a divide and conquer method for distributed graph clustering. [CSWZ16] used spectral sparsifiers in graph clustering for two distributed communication models to reduce communication cost. [SZ19] presented a computationally and communication efficient node degree based sampling scheme for distributed graph clustering. [ZZL+19] studied distributed dynamic graph clustering based on the monotonicity property of graph sparsification. However, all these methods assume that there are no edge duplicates across different sites and do not work in the more general duplication setting. Graph spanners have been studied in the non-distributed model [TZ05, AB16] and a few distributed models [CHKPY18, FWY20]. [CHKPY18] studied distributed constructions of pairwise spanners that approximate distances only for some pairs of vertices in the CONGEST model. [FWY20] studied distributed construction of a serials of graph spanners in the message passing with and without duplication models. But, there exists no prior work considering such construction in the blackboard model, which has been a widely adopted communication model [BO15, VWW20, DOR21].

2 Definitions and Notations

A weighted undirected graph $G(V, E, W)$ consists of a vertex set V, an edge set E and a weight function W which assigns a weight $W(e)$ to each edge $e \in E$. W can be omitted from the presentation if it is clear from the context. Throughout the paper let $n = |V|$ and $m = |E|$ denote the number of vertices and the number of edges in G respectively, and s be the number of remote sites G is observed. Let w be the maximum edge weight in G, *i.e.*, $w = \max_e W(e)$. We denote by $d_G(u, v)$ the *shortest-path distance* from u to v in G. A α-spanner and $+\beta$-spanner for G are a subgraph $H(V, E' \subseteq E)$ of G such that for every $u, v \in V$, $d_H(u, v) \leq \alpha * d_G(u, v)$ and $d_H(u, v) \leq d_G(u, v) + \beta$, respectively.

3 Distributed Graph Clustering

In this section, we state our distributed graph clustering algorithms in the message passing and blackboard with duplication models. We first discuss challenges introduced by edge duplicates presenting at different sites and then show how we overcome the challenges.

Definitions. Define the graph *Laplacian* of a graph G as $L = D - A$ where A is the adjacency matrix of G and D is the degree matrix, *i.e.*, a diagonal matrix with the i-th diagonal entry equal to the sum over the i-th row of A. A $(1+\epsilon)$-*spectral sparsifier* of G, denoted as $(1+\epsilon)$-$SS(G)$, is a (possibly re-weighted) subgraph H of G such that for every $x \in R^n$, the inequality

$$(1-\epsilon)x^T L_G x \leq x^T L_H x \leq (1+\epsilon)x^T L_G x$$

holds. Each edge e in G has *resistance* $R(e) = 1/W(e)$, and the *effective resistance* between any two vertices u and v in G, denoted as $R_G(u, v)$, is defined as the potential difference that has to be applied between them in order to drive one unit of current through the network G.

Challenges. Distributed graph clustering algorithms designed for non-duplication models cannot be easily extended to duplication models. We explain the fact using [CSWZ16] in the message passing model as an example: every site S_i constructs a spectral sparsifier of its local graph $G_i(V, E_i)$ as a synopsis H_i and then transmits H_i, instead of G_i, to the coordinator. Upon receiving H_i from all sites, the coordinator takes their union, $H = \cup_{i=1}^{s} H_i$ as the constructed structure. The algorithm is based on the decomposability property of spectral sparsifiers. To see this, for every $i \in [1, s]$, by definition of spectral sparsifiers, we have for every vector $x \in R^n$, $(1-\epsilon)x^T L_{G_i} x \leq x^T L_{H_i} x \leq (1+\epsilon)x^T L_{G_i} x$. Summing all inequalities for $i \in [1, s]$, we get that

$$(1-\epsilon) \sum_{i\in[1,s]} x^T L_{G_i} x \leq \sum_{i\in[1,s]} x^T L_{H_i} x \leq (1+\epsilon) \sum_{i\in[1,s]} x^T L_{G_i} x.$$

In the non-duplication model, it is easy to check that $\sum_{i=1}^{s} L_{G_i} = L_G$ by the definition of Laplacian matrix. Then the above inequality is equivalent to

$$(1-\epsilon)x^T L_G x \leq x^T L_H x \leq (1+\epsilon)x^T L_G x, \tag{1}$$

which concludes that H is a $(1+\epsilon)$-spectral sparsifier of G. Under the duplication model, however, it is clear that $\sum_{i=1}^{s} L_{G_i} \neq L_G$ and thus Inequality (1) does not hold any longer. In other words, the structure H constructed using the same principle is not a spectral sparsifier of G. See Fig. 1 for an illustrating example.

Proposed Method. Restoring the decomposability of spectral sparsifiers in the duplication models appears to be quite challenging. We avoid it by asking every site cooperates to construct a spectral sparsifier of the distributed graph in the coordinator, who can then get clustering results by any standard clustering algorithm. A standard method of computing spectral sparsifiers [SS11] is to sample each edge in the input graph with a probability proportional to its effective resistance and then include the sampled edges (after appropriate weight rescaling) into the sparsifier. But, when there are duplicated edges across different sites, an edge (u, v) may get sampled more than once at different sites, thereby resulting in multiple edges of possibly different weights between u and v, *e.g.*, edges e_1^1 and e_1^2 in Fig. 1. It is unclear how to process these edges to guarantee the resulting structure is always a spectral sparsifier. As in Fig. 1, simply taking union by summing edge weights does not produce a valid spectral sparsifier.

Instead of using the classic sampling method, we propose to make use of the fact that spectral sparsifiers can be constructed by graph spanners [KX16] to compute spectral sparsifiers in the coordinator. The connection between spectral sparsifiers and graph spanners allows us to convert spectral sparsification to graph spanner construction and uniform sampling under duplication models. In the followings, we first introduce the algorithm of [KX16] and then discuss how to adapt the algorithm in the message-passing and blackboard under duplication models.

The Algorithm of [KX16]**.** Given a weighted graph, their algorithm first determines a set of edges that has small effective resistance through graph spanners. Specifically, it constructs a t-bundle $\log n$-spanner $J = J_1 \cup J_2 \cup \cdots \cup J_t$, that is, a sequence of $\log n$-spanners J_i for each graph $G_i = G - \cup_{j=1}^{i-1} J_j$ with $1 \leq i \leq t = O(\epsilon^{-2} \log n)$. Intuitively, it peels off a spanner J_i from the graph G_i to get G_{i+1} before computing the next spanner J_{i+1}, *i.e.*, J_1 is a spanner of G, J_2 is a spanner of $G - J_1$, *etc.* The t-bundle spanner guarantees that each *non-spanner* edge (edge not in the spanner) has t edge-disjoint paths between its endpoints in the spanner (and thus in G), serving as a certificate for its small effective resistance. The algorithm then uniformly samples each non-spanner edge with a fixed constant probability, *e.g.*, 0.25 and scales the weight of each sampled edge proportionally, *e.g.*, by 4 to preserve the edge's expectation. By the matrix concentration bounds, it is guaranteed that the spanner together with the sampled non-spanner edges are a moderately sparse spectral sparsifier, in which the number of edges has been reduced by a constant factor. The desirable spectral sparsifier can be obtained by repeating the process until we get a sufficient sparsity, which happens after logarithmic iterations.

Weighted Graph Spanners. An important building block in [KX16] is the construction of graph spanners of stretch factor $\log n$, which can be used to construct the t-bundle $\log n$-spanner. Unfortunately, there is no algorithm that can generate such a spanner under the duplication models. [FWY20] developed an algorithm for constructing $(2k-1)$-spanners in unweighted graphs under the message passing with duplication model through the implementation of the greedy algorithm [ADD+93]. But the algorithm does not work in weighted graphs, where the greedy algorithm would need to process the edges in nondecreasing order of their weights. This seems to be a notable obstacle in both the message passing model and the blackboard model.

In this paper, we first propose an algorithm for constructing $(4k-2)$-spanners in weighted graphs under the message passing with duplication model. We are able to overcome the challenge in weighted graphs at the expense of a larger stretch factor $4k - 2$. However, this is sufficient for the construction of $\log n$-spanners in weighted graphs by setting the parameter $k = O(\log n)$.

Specifically, we divide the range of edge weights $[1, w]$ into logarithmic intervals, where the maximum edge weight w is assumed to be polynomial in n[1]. Then we process edges in each logarithmic scale $[2^{i-1}, 2^i)$, where $1 \leq i \leq log_2(nw)$, as follows. Each site S_j in order decides which of its edge $e \in E_j$ of weight in $[2^{i-1}, 2^i)$ to include into the current spanner H. If including the edge e results in a cycle of at most $2k - 1$ edges, then the shortest distance between e's endpoints in the current spanner is guaranteed to be less than $(4k-2)W(e)$ (see our proof below). Thus the edge can be discarded. Otherwise, we update the current spanner H by including e. After completing processing of E_j, S_j forwards the possibly updated spanner H to the next site. The algorithm is summarized in Algorithm (Alg.) 1.

Theorem 1. *Given a weighted graph and a parameter $k > 1$, Algorithm 1 constructs a $(4k-2)$-spanner using communication cost $\tilde{O}(sn^{1+1/k})$ in the message passing with or without duplication model.*

[1] This is a common and practical assumption for modern graphs.

Algorithm 1. *Spanner*(G, k): $(4k-2)$-spanners under duplication models

Input: Graph $G(V, E, W)$ and a parameter $k > 1$
Output: Spanner H

```
1: H ← ∅
2: for i ∈ [1, log2(nw)] do
3:   for each site S_j do
4:     Wait for H from site S_{j-1}
5:     for each edge e ∈ E_j of weight in [2^{i-1}, 2^i) do
6:       if (V, H ∪ {e}) does not contain a cycle of ≤ 2k edges then
7:         H ← H ∪ {e}
8:       end if
9:     end for
10:    Transmit H to the next site S_{j+1}
11:  end for
12: end for
13: return H;
```

Proof. We first prove that the stretch factor is $4k-2$. For each edge $(u, v) \in E$, if $(u, v) \notin H$, it must be that including the edge (u, v) would close a cycle of length $\leq 2k$. That is, there exists a path P of $\leq 2k-1$ edges between u and v in H. Since we process edges in logarithmic scale, the edge weights in P cannot be larger than $2W(u, v)$. Thus the path length of P is at most $(4k-2)W(e)$. Therefore, the output H is a $(4k-2)$-spanner.

We then prove the communication cost. By construction, the output graph H has girth (the minimum number of edges in a cycle contained in the graph) larger than $2k$. It is well known that a graph with girth larger than $2k$ have $O(n^{1+1/k})$ edges [ADD+93]. Then H always has $O(n^{1+1/k})$ edges throughout the processing of each logarithmic interval. Thus the total communication cost is $\tilde{O}(sn^{1+1/k})$. The algorithm works for both with and without duplication settings, which do not affect the communication complexity. □

Algorithm 1 can be extended to the blackboard model with the following modification: In Line 10, if site S_j does change H by adding some edge(s), it transmits the updated spanner H to the blackboard, instead of the next site; otherwise, it sends a special marker of one bit to the blackboard to indicate that it has completed the processing. The results are summarized in Theorem 2. In Sect. 4, we will show that the communication cost can be reduced to $2k-1$ in unweighted graphs.

Theorem 2. *The communication complexity of constructing a* $(4k-2)$*-spanner in weighted graphs under the blackboard with or without duplication model is* $\tilde{O}(s + n^{1+1/k})$. *In unweighted graph, the stretch factor can be reduced to* $2k-1$.

Constructing t-bundle $\log n$-spanner. Recall that a t-bundle $\log n$-spanner $J = J_1 \cup J_2 \cup \cdots \cup J_t$, where J_i is a $\log n$-spanner for graph $G_i = G - \cup_{j=1}^{i-1} J_j$, for $1 \leq i \leq t$. When $i = 1$, $G_1 = G$ is a distributed graph with each site S_j having edge set E_j. We can use Algorithm 1 with $k = (2 + \log n)/4$ to compute a $\log n$-spanner J_1 of G_1. For $2 \le i \le t$, $G_j = G_{j-1} - J_j$ is again a distributed graph: each site S_j knows which of its edges E_j was included in $J_1, J_2, \cdots, J_{i-1}$ and those edges are excluded from its edge set $E_j - J_1 - J_2 - \cdots - J_{i-1}$. Therefore, the construction of a t-bundle $\log n$-spanner invokes Algorithm 1 for t times. Because of $t = O(\epsilon^{-2} \log n)$ and Theorems 1 and 2, the total communication costs in the message passing and blackboard with duplication models are $\tilde{O}(sn)$ and $\tilde{O}(s + n)$, respectively.

Uniform Sampling. After the spanner construction, the algorithm of [KX16] then uniformly samples each non-spanner edge with a fixed probability, *e.g.*, 0.25 and scales the weight of each sampled edge proportionally, *e.g.*, by 4. We observe that sampling with a fixed probability is much more friendly to edge duplicates as compared to sampling with a varied probability used in traditional methods such as [FHHP11]. For example in Fig. 1, if the duplicates e_1^1 and e_1^2 of e_1 are both sampled (under a fixed probability 0.25), they still have the same weight $4W(e_1)$ and are edge duplicates again in the next iteration. If one of them, say e_1^1, is not sampled, it is removed from the (local) graph at site S_1 and will not formulate duplicates with e_1^2 at site S_2. In contrast, non-uniform sampling could result in sampled edges of rather different weights, which may not be even considered as duplicates. However, uniform sampling under duplication models is still very challenging: if a fixed probability is used for every edge, an edge with d duplicates across different sites is processed/sampled for d times, each at one of the d sites, and thus has a higher probability being sampled than another edge with smaller duplicates. This results in a non-uniform sampling.

To achieve the uniform sampling, we suppose that the probability of an edge e residing at each of the sites is a known value r_e. If we set the probability of random sampling at each site as p_e, then the probability that the edge is not sampled at each site is $1 - p_e * r_e$. It can be derived that the probability that e is sampled by *at least* one site is $p = 1 - (1 - p_e * r_e)^s$. Since the values of r_e and s are known, we can tune the value of p_e to get the expected sampling probability $p = 0.25$. At some site, if e is sampled and added to H, we update its presenting probability as $p_e * r_e$, which will be used in the next iteration. Otherwise (if e is not sampled), it is discarded and will not participate in the next iteration. See the details in Algorithms 2 and 3.

The main algorithm, Algorithm 3 computes $(1 + \epsilon)$-spectral sparsifier in $\lceil \log \rho \rceil$ iterations of *Light-SS*, where ρ is a sparsification parameter. The communication cost of *Light-SS* is composed of the cost for the bundle spanner construction and the cost for non-spanner edge sampling. If the sampled edges are transmitted to the coordinator, the communication cost $\tilde{O}(m)$ could be prohibitively large. To see this, the number of edges in the output G_i after each iteration is only reduced by a constant factor because the uniform sampling removes 3/4 of the non-spanner edges in expectation. To improve the communication cost, we keep sampled edges in each iteration at local sites and do not transmit them to the coordinator except for the very last iteration. Then similar to the input graph G, the output G_i for each iteration $i \in [1, \lceil \log \rho \rceil - 1]$ are also

Algorithm 2. *Light-SS* under duplication models

Input: $G(V, E), \epsilon \in (0, 1)$, and probability r_e for each edge e
Output: H with updated r'_e for each edge $e \in H$

1: $G_1 \leftarrow G$; $J \leftarrow \emptyset$
2: **for** $i \in [1, 24 \log^2 n/\epsilon^2]$ **do**
3: $J_i \leftarrow Spanner(G_i, (2 + \log n)/4)$
4: $G_{i+1} \leftarrow G_i - J_i$
5: **end for**
6: $H \leftarrow J$; $r'_e \leftarrow r_e$
7: **for** each site S_i **do**
8: **for** each edge $e \in E_i - J$ **do**
9: Sample the edge e with probability p_e such that $1 - (1 - p_e * r_e)^s = 0.25$; if e is sampled, adds e to H with a new weight $4W(e)$ and set r'_e to $p_e * r_e$
10: **end for**
11: **if** it is the last iteration of the for-loop in Line 2 of Alg. 3 **then**
12: Transmit the sampled edges to the coordinator
13: **end if**
14: **end for**
15: **return** H;

Algorithm 3. $(1 + \epsilon)$-*SS* under duplication models

Input: $G(V, E)$, probability r_e for each edge e, and parameters $\epsilon \in (0, 1)$ and $\rho > 1$
Output: H

1: $G_0 \leftarrow G$
2: **for** $i \in [1, \lceil \log \rho \rceil]$ **do**
3: $G_i \leftarrow Light\text{-}SS(G_{i-1}, \epsilon/\lceil \log \rho \rceil, r_e)$
4: **end for**
5: $H \leftarrow G_{\lceil \log \rho \rceil}$ {H is already transmitted to and known by the coordinator}
6: **return** H;

a distributed graph with possible edge duplication. Edge duplicates come from two sources: either the edge is included into the bundle spanner, or the edge is sampled by more than one site. In this way, the communication cost of *Light-SS* (except for the last iteration) contains only the cost of constructing the bundle spanner. In the last iteration, the number of sampled edges must be small $\tilde{O}(n)$, which is also the communication cost of their transmission. Therefore, the communication costs of Algorithm 3 in the message passing and blackboard under duplication models are $\tilde{O}(ns)$ and $\tilde{O}(n + s)$, respectively. Putting all together, our results for distributed spectral sparsification under duplication models are summarized in Theorem 3 with its formal proof deferred to Appendix.

Theorem 3 (Spectral Sparsification under Duplication Models). *For a distributed graph G and parameters $\epsilon \in (0, 1)$ and $\rho = O(\log n)$, Algorithm 3 can construct a $(1+\epsilon)$-spectral sparsifier for G of expected size $\tilde{O}(n)$ using communication cost $\tilde{O}(ns)$ and $\tilde{O}(n+s)$ in the message passing and blackboard with duplication models respectively, with probability at least $1 - n^{-c}$ for constant c.*

Clustering in the Sparsifier. After obtaining the spectral sparsifier of the distributed graph, the coordinator applies a standard clustering algorithm such as [NJW01] in the sparsifier to get the clustering results. We can guarantee a clustering quality nearly as good as the simple method of centralizing all graph edges and then performing a clustering algorithm. Before formally stating the results, we define a few notations.

For every node set S in a graph G, let its *volume* and *conductance* be $vol_G(S) = \sum_{u \in S, v \in V} W(u,v)$ and $\phi_G(S) = (\sum_{u \in S, v \in V-S} W(u,v))/vol_G(S)$, respectively. Intuitively, a small value of conductance $\phi(S)$ implies that nodes in S are likely to form a cluster. A collection of subsets $A_1, \cdots, A_k$ of nodes is called a *(k-way) partition* of G if (1) $A_i \cap A_j = \emptyset$ for $1 \leq i \neq j \leq k$; and (2) $\cup_{i=1}^{k} A_i = V$. The *k-way expansion constant* is defined as $\rho(k) = \min_{partition A_1, \cdots, A_k} \max_{i \in [1,k]} \phi(A_i)$. A lower bound on $\Upsilon_G(k) = \lambda_{k+1}/\rho(k)$ implies that G has exactly k well-defined clusters [PSZ15], where λ_{k+1} is the $k+1$ smallest eigenvalue of the normalized Laplacian matrix. For any two sets X and Y, their symmetric difference is defined as $X \Delta Y = (X - Y) \cup (Y - X)$.

Theorem 4. *For a distributed graph G with $\Upsilon_G(k) = \Omega(k^3)$ and an optimal partition $P_1, \cdots, P_k$ achieving $\rho(k)$ for some positive integer k, there exists an algorithm that can output partition $A_1, \cdots, A_k$ at the coordinator such that for every $i \in [1,k]$, $vol(A_i \Delta P_i) = O(k^3 \Upsilon^{-1} vol(P_i))$ holds with probability at least $1 - n^{-c}$ for constant c. The communication costs in the message passing and blackboard with duplication models are $\tilde{O}(ns)$ and $\tilde{O}(n+s)$, respectively.*

To the best of our knowledge, this is the first algorithm for performing distributed graph clustering in the message passing and blackboard with edge duplication models. Remarkably, we can show that the communication costs are *optimal*, almost matching the communication lower bounds $\Omega(ns)$ and $\Omega(n+s)$, respectively. It is interesting to see that the communication costs incurred under duplication models are asymptotically the same as those under non-duplication models. In other words, edge duplication does not incur more communications in the graph clustering task, unlike other problems such as graph spanner construction as we will show in Sect. 4. Although we make an assumption on the edge distribution probability, we conjecture that when the assumption is relaxed, *i.e.*, graph edges are presenting at different sites arbitrarily, the communication upper bounds remain the same in duplication models. We leave the study as an important future work.

4 Spanner Constructions in the Blackboard Model

In this section, we study distributed constructions of graph spanners in the blackboard models with and without edge duplication. This, unfortunately, has not been investigated by prior work yet. We prove several interesting communication upper and lower bounds for typical graph spanners as summarized in Table 1. Due to limit of space, we cannot enumerate every result in Table 1. Hence, here we only describe the general $(2k-1)$-spanners and move the additive spanners to the Appendix. We start with the duplication model, followed by the non-duplication model. The lower bounds obtained in Theorems 5 and 6 hold in both weighted and unweighted graphs and the rest results are on unweighted graphs.

Duplication Model. In Sect. 3, we have provided the communication upper bound, $\tilde{O}(s + n^{1+1/k})$, of constructing $(2k-1)$-spanners in unweighted graphs in Theorem 2. We now show that the communication lower bound is $\Omega(s + n^{1+1/k} \log s)$.

Theorem 5. *The communication lower bound of constructing a* $(2k-1)$*-spanner in the blackboard with duplication model is* $\Omega(s + n^{1+1/k} \log s)$.

Proof. To prove this, we target a more general statement that works for every spanner.

Lemma 1. *Suppose there exists an* n*-vertex graph* F *of size* $f(n)$ *such that* F *is the only spanner of itself or no proper subgraph* F' *of* F *is a spanner. Then the communication complexity of computing a spanner in the blackboard with duplication model is* $\Omega(s + f(n) \log s)$ *bits.*

Proof. Our proof is based on the reduction from the Multiparty Set-Disjointness problem ($DISJ_{m,s}$) to graph spanner computation. In $DISJ_{m,s}$, s players receive inputs $X_1, X_2, \cdots, X_s \subseteq \{1, \cdots, m\}$ and their goal is to determine whether or not $\cap_{i=1}^{s} X_i = \emptyset$. Now we construct a distributed graph G from the graph F and an instance of $DISJ_{f(n),s}$ as follows. We add edge e_j in F to site i if $j \notin X_i$ for $1 \le j \le f(n)$. If the coordinator outputs F as the spanner, we report $\cap_{i=1}^{s} X_i = \emptyset$; otherwise we report $\cap_{i=1}^{s} X_i \neq \emptyset$. It can be seen that the coordinator outputs F iff all its edges appear at some site, which is the case $\cap_{i=1}^{s} X_i = \emptyset$. Finally, according to the communication lower bound of $DISJ_{m,s}$ in the blackboard model [BO15], $\Omega(s + m \log s)$, the communication complexity of computing a spanner is $\Omega(s + f(n) \log s)$. □

For the lower bound of $(2k-1)$-spanners, the Erdos's girth conjecture states that there exists a family of graphs F of girth $2k+1$ and size $\Omega(n^{1+1/k})$ [Erd64]. This implies that there exists only one $(2k-1)$-spanner of F, that is F itself. It is because the deletion of any edge in F would result in that the distance between the endpoints of the edge becomes at least $2k$. Then by Lemma 1, we get the lower bound $\Omega(s + n^{1+1/k} \log s)$. □

Non-Duplication Model. In the non-duplication model, we prove a lower bound via a reduction from the lower bound for the duplication model.

Theorem 6. *The communication complexity of constructing a* $(2k-1)$*-spanner in the blackboard without duplication model is* $\Omega(s + n^{1+1/k} \max\{1, s^{-1/2-1/(2k)} \log s\})$.

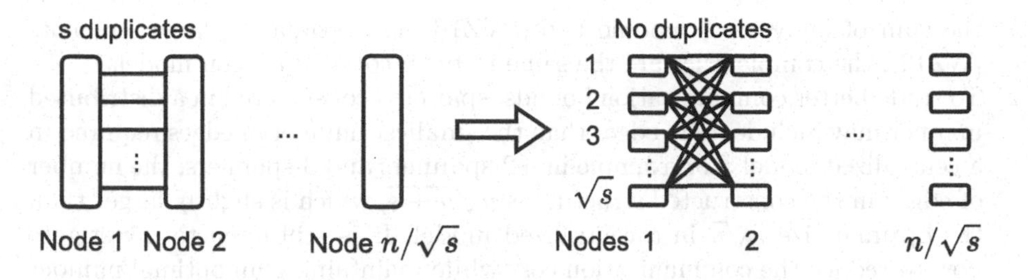

Fig. 2. Converting a graph with duplication on s sites and $n/\sqrt{s}$ vertices into a graph without duplication on s sites and n vertices

Proof. We can construct an instance of the $(2k-1)$-spanner problem without duplication on s sites and n vertices from an instance of the $(2k-1)$-spanner problem with duplication on s sites and $n/\sqrt{s}$ vertices. Specifically, we construct a graph G' with no duplication by replacing each vertex v by a set of vertices S_v of size $\sqrt{s}$. Since there are at most s copies of an edge (u, v) in the original graph G across the s sites, we can assign each server's copy to a distinct edge $(u', v') \in S_u \times S_v$ in G'. See Fig. 2 for an illustrating example of the construction. Then we apply an algorithm for the without duplication model, *e.g.*, the algorithm in Theorem 2, to get a $(2k-1)$-spanner H' of G'. Finally, the coordinator computes a $(2k-1)$-spanner H of G by including an edge (u, v) in H if there is at least one edge between S_u and S_v in H'.

To show the constructed H is a $(2k-1)$-spanner of G, let us consider an edge $(u, v) \in G$. By construction, there must be an edge $(u', v') \in S_u \times S_v$ in G'. Because H' is a $(2k-1)$-spanner of G', it contains a path P' of length at most $(2k-1) \cdot W(u, v)$ between u' and v'. For every edge (x', y') in P' where $x' \in S_x, y' \in S_y$, we have included an edge (x, y) in H. Therefore, there exists a path P of length at most $(2k-1) \cdot W(u, v)$ between u and v in H and thus H is a $(2k-1)$-spanner of G. Since the lower bound in the duplication model is $\Omega(s + n^{1+1/k} \log s)$ (Theorem 5), we have that the lower bound for the non-duplication model is $\Omega(s + (n/\sqrt{s})^{1+1/k} \log s) = \Omega(s + n^{1+1/k} s^{-1/2-1/(2k)} \log s)$.

Since representing the result itself needs $\Omega(n^{1+1/k})$, combining this with the above result get the final lower bound, $\Omega(s + n^{1+1/k} \max\{1, s^{-1/2-1/(2k)} \log s\})$. □

Discussions. We highlight several interesting observations from our results in Table 1 and prior results in Table 2.

1. We demonstrate that for graph spanner constructions, the blackboard model is powerful to significantly reduce the communication complexity compared to the message passing model. For instance in duplication models, computing the $(2k-1)$-spanners incurs communication cost $\tilde{O}(sn^{1+1/k})$ in the message passing model but only $\tilde{O}(s + n^{1+1/k})$ in the blackboard model. This is not necessarily the case for all computing problems. For example, for computing

the sum of bit vectors modulo two [PVZ16] and estimating large moments [WZ12], the complexities are the same in both communication models.
2. To trade better communication bounds, spanners constructed in a distributed manner may include more edges than the smallest number of edges required in a centralized model. For example in +2-spanners and 3-spanners, the number of edges in the constructed structure is $n\sqrt{n+s}$, which is slightly larger than the optimal size $n\sqrt{n}$ in a centralized model. It is still open to investigate how to reduce the communication cost while maintaining an optimal number of edges in the spanner.
3. For constructing $(2k-1)$-spanners, the upper bound $\tilde{O}(s+n^{1+1/k})$ with a logarithmic factor hidden is very close to the lower bound $\Omega(s+n^{1+1/k}\log s)$. There is a small gap between the upper bound $\tilde{O}(s+n\sqrt{n+s})$ and lower bound $\Omega(s+n^{3/2}\log s)$ for +2 or 3-spanners. The gap is larger in $+k$-spanners (for $k>2$) where the lower bound becomes $\Omega(s+n^{4/3-o(1)}\log s)$. But this problem also happens in the message passing model. The construction of $+k$-spanners often involves more complex operations and might not be easy to adapt to distributed models.

5 Conclusions and Future Work

In this paper, we propose the first set of algorithms that can perform distributed graph clustering and spectral sparsification under edge duplication in the two well-established communication models, the message passing and the blackboard models. We show the optimality of the achieved communication costs while maintaining a clustering quality nearly as good as a naive centralized method. We also perform the first investigation of distributed algorithms for constructing graph spanners in the blackboard under both duplication and non-duplication models.

As the future work, we will study how to achieve the optimal communication complexity for distributed graph clustering while relaxing the assumption made. Furthermore, most of the existing work concentrate on global clustering but ignore local clustering which only returns the cluster of a given seed vertex. We will devise a local clustering method that hopefully enjoys communication cost not dependent on the size of the input graph and is more communication-efficient than traditional global graph clustering methods.

Cut sparsifiers are another type of graph sparsifiers and they can approximately preserve all the graph cut values in the original graph. Although spectral sparsifiers are also cut sparsifiers, the latter might have smaller number of edges. Because the algorithm of [KX16] can be generalized to cut sparsifiers, it is promising to adapt the techniques in this work to the new problem. Finally, it is an intriguing open problem to improve the upper bounds or lower bounds and close their gap in both duplication and non-duplication models.

References

[AB16] Abboud, A., Bodwin, G.: The 4/3 additive spanner exponent is tight. In: Proceedings of the ACM STOC Conference, pp. 351–361 (2016)

[ABS+20] Ahmed, R., et al.: Graph spanners: a tutorial review. Comput. Sci. Rev. **37**, 100253 (2020)

[ADD+93] Althöfer, I., Das, G., Dobkin, D., Joseph, D., Soares, J.: On sparse spanners of weighted graphs. Discret. Comput. Geom. **9**(1), 81–100 (1993). https://doi.org/10.1007/BF02189308

[BKMP10] Baswana, S., Kavitha, T., Mehlhorn, K., Pettie, S.: Additive spanners and (α, β)-spanners. ACM Trans. Algorithms **7**(1) (2010)

[BO15] Braverman, M., Oshman, R.: The communication complexity of number-in-hand set disjointness with no promise. Electron. Colloq. Comput. Complex **22**, 2 (2015)

[Che13] Chechik, S.: New additive spanners. In: Proceedings of the SIAM SODA Conference, pp. 498–512 (2013)

[CHKPY18] Censor-Hillel, K., Kavitha, T., Paz, A., Yehudayoff, A.: Distributed construction of purely additive spanners. Distrib. Comput. **31**(3), 223–240 (2018)

[CSWZ16] Chen, J., Sun, H., Woodruff, D.P., Zhang, Q.: Communication-optimal distributed clustering. In: Proceedings of the NIPS Conference, pp. 3720–3728 (2016)

[DOR21] Dershowitz, N., Oshman, R., Roth, T.: The communication complexity of multiparty set disjointness under product distributions. In: Proceedings of the ACM STOC Conference, pp. 1194–1207 (2021)

[Erd64] Erdos, P.: Extremal problems in graph theory. Theory of Graphs and Its Applications, pp. 29–36 (1964)

[FHHP11] Fung, W.S., Hariharan, R., Harvey, N.J.A., Panigrahi, D.: A general framework for graph sparsification. In: Proceedings of the ACM STOC Conference, pp. 71–80 (2011)

[FWY20] Fernandez, M.V., Woodruff, D.P., Yasuda, T.: Graph spanners in the message-passing model. In: Proceedings of the ITCS Conference (2020)

[KX16] Koutis, I., Xu, S.C.: Simple parallel and distributed algorithms for spectral graph sparsification. ACM Trans. Parallel Comput. **3**(2), 14 (2016)

[NJW01] Ng, A.Y., Jordan, M.I., Weiss, Y.: On spectral clustering: analysis and an algorithm. In Proceedings of the NIPS Conference, pp. 849–856 (2001)

[PSZ15] Peng, R., Sun, H., Zanetti, L.: Partitioning well-clustered graphs: spectral clustering works! In: Proceedings of the COLT Conference, pp. 1423–1455 (2015)

[PVZ16] Phillips, J.M., Verbin, E., Zhang, Q.: Lower bounds for number-in-hand multiparty communication complexity, made easy. SIAM J. Comput. **45**(1), 174–196 (2016)

[SS11] Spielman, D.A., Srivastava, N.: Graph sparsification by effective resistances. SIAM J. Comput. **40**(6), 1913–1926 (2011)

[ST11] Spielman, D.A., Teng, S.-H.: Spectral sparsification of graphs. SIAM J. Comput. **40**(4), 981–1025 (2011)

[SZ19] Sun, H., Zanetti, L.: Distributed graph clustering and sparsification. ACM Trans. Parallel Comput. **6**(3), 17 (2019)

[TZ05] Thorup, M., Zwick, U.: Approximate distance oracles. J. ACM **52**(1), 1–24 (2005)

[VWW20] Vempala, S.S., Wang, R., Woodruff, D.P.: The communication complexity of optimization. In: Proceedings of the SIAM SODA Conference, pp. 1733–1752. SIAM (2020)

[WZ12] Woodruff, D.P., Zhang, Q.: Tight bounds for distributed functional monitoring. In: Proceedings of the ACM STOC Conference, pp. 941–960 (2012)

[YX15] Yang, W., Xu, H.: A divide and conquer framework for distributed graph clustering. In: Proceedings of the ICML Conference, pp. 504–513 (2015)

[ZLB21] Zhu, C., Liu, Q., Bi, J.: Spectral vertex sparsifiers and pair-wise spanners over distributed graphs. In: Proceedings of the ICML Conference, pp. 12890–12900 (2021)

[ZZL+19] Zhu, C.J., Zhu, T., Lam, K.Y., Han, S., Bi, J.: Communication-optimal distributed dynamic graph clustering. In: Proceedings of the AAAI Conference, pp. 5957–5964 (2019)

Author Index

M. Mavronicolas (Ed.): CIAC 2023, LNCS 13898, pp. 399–400, 2023.
https://doi.org/10.1007/978-3-031-30448-4

Printed in the United States
by Baker & Taylor Publisher Services

Printed in the United States
by Baker & Taylor Publisher Services